PRACTICAL ELECTRICAL WIRING

Residential, Farm, Commercial, and Industrial

19th edition

BASED ON THE 2005 NATIONAL ELECTRICAL CODE®

Herbert P. Richter

Frederic P. Hartwell

With a foreword by Wilford "Bill" Summers

PARK PUBLISHING, INC.
Minneapolis, Minnesota
New Richmond, Wisconsin

PARK PUBLISHING, INC.
511 Wisconsin Drive
New Richmond, WI 54017

Printed in the United States of America

ISBN 0-97197-791-7

Important notice
Information in this book has been obtained from sources believed to be reliable, but its accuracy and completeness are not guaranteed. The publisher and authors disclaim liability, whether for personal injury or property damage, arising out of typographical error or the use or misuse of this publication regardless of the fault or negligence of the publisher or authors. The instructions in this book are based on the 2005 *National Electrical Code®*. This book provides information, but Park Publishing and its authors are not supplying engineering or other professional services. If needed, these services should be obtained from a professional. For your safety, and the safety of others, you must use caution, care, and good judgment when following the procedures described in this book. You must adapt the procedures to your particular situation.

Executive editor: Margaret R. Wolfe
Editor: Carol J. Frick
Indexer: Robert J. Richardson

Cover design: Silverchair Science+Communications
Cover photo: InsideOut Studios
Illustrations: Farkas Graphic Resources
Layout: Bay Graphics, Inc.

For his review of the manuscript and helpful suggestions for updating Chapters 14 and 29, thanks to Joseph R. Knisley, a former colleague of Fred Hartwell at EC & M Magazine.

National Electrical Code® and *NEC®* are registered trademarks of the National Fire Protection Association.

Publisher's Cataloging-in-Publication Data
Richter, H. P. (Herbert P.), 1900–1978.
     Practical electrical wiring : residential, farm, commercial / Herbert
P. Richter, Frederic P. Hartwell.-- 19th ed.
        p. cm.
"Based on the 2005 *National Electrical Code*."
Includes bibliographical references and index.
    ISBN 0-97197-791-7
    1. Electric wiring, Interior. I. Hartwell, Frederic P. II. Title.
    TK3271 .R48 2005
    621.319'24--dc21

Brief contents

Detailed contents

Tables

Chapter tables

Appendix tables

Foreword to the eighteenth edition

I am pleased to have the opportunity to comment on this excellent expansion of H. P. Richter's original book. Frederic P. Hartwell is eminently qualified to undertake a project of this magnitude. He is renowned among his peers as an electrical code expert with vast experience as an electrician, electrical inspector, *NEC* and state electrical code panel member, and editor of EC&M Magazine.

This tome is a solid winner.

The art of electrical installations has become immensely more complicated in recent years due to the electronic revolution. The incorporation of electronic components into electrical devices has opened up entirely new concepts of control and operation of electrical circuits. Current practices are well covered in this new edition of *Practical Electrical Wiring*.

Earliest and perhaps most important of the electronic advances in electrical safety was the ground-fault circuit interrupter. Present GFCI standards and requirements are covered in great detail in Chapter 9 of *Practical Electrical Wiring*, and the developments leading to today's regulations provide useful background. Initially, such devices were installed in selected locations throughout the United States and were monitored for efficiency. The early devices were susceptible to lightning storms and were subsequently modified to eliminate nuisance tripping from such sources.

Construction sites were the first mandatory application for GFCIs. This caused consternation among electrical contractors, who generally were held responsible by general contractors for work stoppages due to tripped GFCIs. The devices were roundly criticized even though they were performing their intended function, that is, when sufficient leakage currents occurred the devices would trip, requiring action to eliminate the cause. The devices were set to trip at 4 to 6 milliamperes, permitting current sufficient to be felt but not strong enough to cause heart fibrillation. The solution to reducing job site outages due to widespread "nuisance tripping" was to clean up faulty drills and other electrical equipment and to locate the GFCIs closer to end use—the longer the portable cord, the greater the likelihood of it being subjected to wet conditions sufficient to cause a leakage current of 6 milliamperes.

OSHA held hearings in Washington, D.C. As Secretary of the *National Electrical Code* Committee for NFPA at the time, I testified to the safety advantages of the device. The outcome was that OSHA upheld the use of the devices and job site conditions improved until OSHA concluded that alternative safety programs would be permitted under strict conditions of compliance. As OSHA records reflect, many lives have been saved over the years. With succeeding Code editions, the mandated use of GFCIs has expanded to include residential properties, swimming pools, other occupancies, remodeling, maintenance, repair and demolition activities and other locations.

Extensive coverage of another advance in the application of electronic components, the development of Ground Fault Protection Equipment Systems, is presented in Chapter 28 of *Practical Electrical Wiring*. Fire records showed that 480Y/277-volt systems were vulnerable to catastrophic burndowns from arcing faults that did not generate enough amperage to clear the fault but were sufficient to cause great damage to equipment and buildings. The ground fault protection system is set to detect arcing currents below the level of traditional fuses and circuit breakers thereby shutting off the circuit before extreme damage could occur.

Chapters 14 and 29 of this book include in-depth coverage of fluorescent lamps. Federal legislation has required that fluorescent lighting be more energy saving. The result has been the elimination of many of the old standby lamps and the manufacture of many new energy-saving types. Electronic ballasts have proven to be much more efficient than the old electromagnetic type using the ballast. In addition, they have virtually eliminated the undesirable hum from the windings of the electromagnetic types. Other advances in fluorescent lamps relate to their color temperature and color rendition index, with colors closely matching outdoor sunshine showing a positive relationship to production levels of office workers and even the ability of school children to learn.

Wilford "Bill" Summers

Preface to the ninth edition

In preparing this book it has been the author's aim to make it simple enough for the beginner, yet complete enough so that it will be of value also to those already engaged in electrical work. It is intended to be not a manual that merely recites the methods used in wiring buildings for the use of electricity, but rather a book that explains the subject in such fashion that the reader will learn both the *way* things are done and *why* they are done in that particular way. Only in this manner can the student master the subject in order to solve unique problems that will be encountered in actual practice, for no book can possibly cover all the different problems that are likely to arise.

Because this book is not intended to include the subject of electrical engineering, only so many basic engineering data as are essential have been included, and these so far as possible have been boiled down to ABC proportions.

All methods shown are in strict accordance with the *National Electrical Code*, but no attempt has been made to include a detailed explanation of all subjects covered by the Code. The Code is written to include any and all cases that might arise in wiring every type of structure from the smallest cottage to the largest skyscraper; it covers ordinary wiring as well as those problems that come up only very rarely. The scope of this book has been limited to the wiring of structures of limited size and at ordinary voltages, at 600 volts or less. Skyscrapers and steel mills and projects of similar size involve problems that the student will not meet until long after mastering the contents of this book.

H. P. Richter

Preface to the nineteenth edition

Practical Electrical Wiring has achieved the status of a classic over a history that reaches back more than 65 years. First written and continued for ten editions by H. P. Richter, the book has an unparalleled ability to do two things at the same time. First, using everyday language and by avoiding any higher mathematics, it makes the majority of electrical installation practices accessible to the beginner. Second, it includes enough detail to maintain its usefulness as an essential reference work on the shelf of any who continue in the trade.

This edition and the one before it follow seven editions prepared by W. Creighton Schwan, who ably continued H. P. Richter's original work. Changes in the electrical trade prompted a decision to produce an extensively rewritten and reorganized work to reflect present conditions and emerging challenges. For example, we may have already passed the point at which the majority of electric energy consumed in the United States is processed electronically, and therefore involves substantial harmonic components. Today's apprentices need some understanding of how these loads affect wiring decisions, or risk being left behind.

Nevertheless, I have tried to keep the presentation accessible to all, for a personal reason. In 1957, at the age of ten, I helped my father extend a branch circuit in our home. My father, although handy, was never in the trade. Yet he correctly understood my intuitive fascination with what we were doing, and the next day I came home from school to a brand new copy of the fifth edition, just out, of *Practical Electrical Wiring*. I was riveted, and read the entire book cover to cover. It seemed as if I had been given the key to the inside door to a training room of a wonderful profession. Not only did I know why wiring had to be installed in a certain way, I was no stranger to the tools and equipment used to make the installations.

I didn't actually enter the trade until many years later, but I never got the book out of my mind. After college, I returned to my true calling, and I returned to this book as well, always an invaluable reference. It is now my great honor to carry on a wonderful tradition. This edition is based on the 2005 *National Electrical Code*, and will continue to be revised with each succeeding edition of that document.

Frederic P. Hartwell

Introduction and guide

Practical Electrical Wiring is back in a new version based on the 2005 *National Electrical Code® (NEC®)**. The advanced level of discussion reflects the greater technical complexities of today, yet the style remains highly accessible to a wide range of readers. Drawings and photos continue to be revised and updated, making instructions easy to understand and follow. In addition to updating this edition to conform to the 2005 *NEC*, author and code expert Frederic P. Hartwell worked with lighting expert Joseph R. Knisley in extensively revising the two chapters on residential and nonresidential lighting. Hartwell has been responsible for nearly one thousand successful proposals and public comments regarding changes in the *National Electrical Code* over the past seven Code-making cycles and continues as a principal member of the *National Electrical Code* Committee. He brings years of experience and practical hints gained as a licensed master electrician and former head electrician at a college in Massachusetts.

Intended audience

Practical Electrical Wiring is aimed at the person who wants to make a serious study of the field of electrical wiring—

- Practitioners—electricians, contractors, designers, inspectors working in residential, farm, commercial, or industrial settings—will find what they need to know in one handy, compact volume that can be easily carried to the job site.
- Students and apprentices will appreciate the way the book captures all the basics in an accessible way, useful while learning the trade and later as a reference.
- Seekers of information about careers in the electrical industry will learn what is involved in different segments of the industry.
- General readers can learn the basics of the trade in order to do things right and become aware of what they don't know so they can make a considered decision about how to proceed.

*National Electrical Code® and *NEC®* are registered trademarks of the National Fire Protection Association, Inc., Quincy, MA 02269.

Guide to using this book

Practical Electrical Wiring covers all phases of a wiring installation—beginning with the necessary background theory and principles and on to the planning, installing, and finishing of residential, farm, commercial, and industrial jobs. Readers of previous editions will note a more advanced level of discussion reflecting the greater technical complexities of today. All new drawings and illustrations plus a new page design make instructions easier to follow and understand.

To help you progress in your knowledge and skills, the book is divided into three parts—Wiring Principles; Residential and Farm Wiring; Commercial and Industrial Wiring. All three parts feature step-by-step instructions for quality work, with 450 drawings and photographs to enhance understanding.

Part 1, WIRING PRINCIPLES, provides the basic foundation for the parts that follow—the safety, practical, and legal considerations involved in wiring decisions and practices. This is essential reading before undertaking any electrical work. The standards and codes that help ensure safety are explained. Selecting proper conductors and how to connect them is covered in detail, followed by how to safely ground the installation. Specific wiring methods and their latest *NEC* rules are detailed. The explanation of how electrical power is measured and delivered to users connects your installation to the broader picture and helps in understanding regional variations in electric rates. Chapters on planning residential installations, principles of good lighting, and an overview of residential and farm motors conclude Part 1. The principles and theory presented in Part 1 provide the background that can be applied to solving problems that may be encountered in unique situations.

Students and novices should read and thoroughly understand the background information in Part 1 before tackling wiring installations in Part 2. Experienced practitioners who want to review specific basics, check on 2005 *NEC* regulations, or benefit from the practical hints and shortcuts offered by author and NEC expert Frederic P. Hartwell, will find it easy to go directly to chapters and topics of particular interest.

Part 2, RESIDENTIAL AND FARM WIRING, puts the principles established in Part 1 to work in the kind of wiring installations most commonly encountered in homes and on farms. Materials and step-by-step techniques are covered for every phase of the installation from installing and grounding service equipment, to roughing in a variety of outlets, and finishing with outlets, switches, and receptacles. Separate chapters cover specialized topics including limited-energy wiring, multi-circuits, old work, farms, on-site power generation, RVs and parks, and apartment buildings.

Part 3, COMMERCIAL AND INDUSTRIAL WIRING, opens with Hartwell's unique approach to understanding one of the primary responsibilities in the electrical trade—selecting the minimum allowable size of wire for a given load. This approach (in the 18th edition) served as the basis for the new wire selection example in the 2005 *NEC* [Annex D, Example 3(a)], and this edition has been updated to track the actual *NEC* calculation. This chapter is followed by discussions of how wiring methods, planning, and lighting differ for non-residential installations compared with the residential and farm installations covered earlier in the book. The lighting chapters, both in this part and Part 1, have been completely updated to reflect current technology and engineering practice. The chapter on the wiring of motors in commercial and industrial applications (at not over 600 volts) builds on the motor basics provided in Part 1. The final chapter gives a preview of more advanced work in wiring specific installations that should be undertaken only by those who have had much experience with more ordinary installations detailed in Parts 1 and 2, or who have received expert training on installations similar to those in this chapter.

Your keys to locating topics quickly and easily are the table of contents and the index. The detailed contents at the front of the book provides an overview of the wide range of material covered, and the remarkably comprehensive index at the back leads you to discussions of specific topics. With its many *see also* cross references to related topics, the index also serves a tutorial function in helping you see how a specific topic fits into the broader picture.

CHAPTER 1

NEC®, Product Standards, and Inspection

THE FIRST PART of *Practical Electrical Wiring* provides the framework for understanding the theory and purpose behind the wiring rules that assure safety. Then in Parts 2 and 3 the book moves on to the actual practice of wiring—how to select conductors and the correct wiring methods for installing them and other electrical equipment in accordance with those rules. Practical design guidance is also provided so that installed equipment will provide the benefits we have come to expect from electrical systems in residential, farm, commercial, and industrial settings.

Electricity is powerful. Under control, electricity performs an endless variety of work in a safe manner. But uncontrolled, it can be destructive. Electricity can be controlled if you use the right kinds of wiring and equipment and install them the right way. It can be dangerous if you use the wrong kinds of wiring or equipment, or if the right kinds are installed improperly. Even properly installed wiring and equipment can become dangerous if you don't properly maintain them. Improper electrical installations can be uneconomical and inconvenient, and they can result in higher insurance rates. Worst of all, unsafe installations can burn, maim, and kill.

Safe wiring practices are the focus of the *National Electrical Code*® (*NEC*), but the *NEC* alone can't provide effective electrical safety. It is only one side of a triangle that comprises the essential elements of electrical safety. Product standards form the second side of this triangle. When you make an electrical installation, you use various electrical products. The most careful electrical installation soon fails if the products used to complete it haven't been adequately designed for their purpose. Inspection provides the third side of the triangle. Because it's human nature to make errors, you can't be sure your electrical installation is really safe until a disinterested party reviews your work. Take away any one of the sides of the triangle, and the electrical safety system is dangerously incomplete. Although the balance of this book focuses on *NEC* rules, an appreciation of the larger picture helps in understanding the reasons for the rules.

NATIONAL ELECTRICAL CODE (NEC)

The *National Electrical Code* is a set of rules detailing the proper methods for installing electrical materials so that the finished job will be safe. It is published by the National Fire Protection Association, Inc. (NFPA), 1 Batterymarch Park, Quincy, Massachusetts 02269. The *National Electrical Code* is referred to in this book as the *NEC*. As you read this book, you will need to refer frequently to your copy of the 2005 *NEC* in order to follow the actual electrical requirement under discussion.

Formulation of the *NEC* by the *NEC* Committee   The *National Electrical Code* Committee of NFPA formulates the *NEC*. The *NEC* Committee has members representing all facets of the electrical industry. The *NEC* Committee consists of the Technical Correlating Committee and 19 sectional committees called Code-making panels. The members of each panel are nationally recognized experts in one or more fields. These experts revise and update the *NEC* at three-year intervals, covering new materials and methods and deleting any obsolete rules. In the process, some section numbers of the *NEC* may change even where the text itself remains unchanged. For example, the *NEC* article covering grounding— one of the most important in the entire document—was completely reorganized for the 1999 *NEC* resulting in every section number being changed. Therefore, a reference to an *NEC* rule in this book may have a different section number from that used in a former edition of this book in referring to the same *NEC* rule.

By NFPA rules, no more than one-third of a Code-making panel can represent a single interest, and it takes the agreement of two-thirds to accept a change in the *NEC*. That forces compromises that work toward consensus in developing a new *NEC*. After the panels meet, the Technical Correlating Committee reviews their work to avoid inadvertent conflicts among *NEC* articles. Anyone can submit a proposal, and after the initial proposal review NFPA publishes the proposed changes in its official *Report on Proposals* (available without charge as long as it's in print). Anyone can comment on the changes as well. The public comments go back to the Code-making panels for their review prior to a vote at the NFPA annual meeting.

This extraordinary degree of open participation by all interested parties makes the *NEC* an authoritative code, and compliance with it will result in a safe instal-lation if the installation is also maintained to remain in compliance. The word "authoritative" as used here means "an accepted source of expert information" rather than "the power to enforce," since the *NEC* does not become law unless legally adopted as a law or code in a particular locality.

How local codes relate to the *NEC*   In the United States, it is the states rather than the federal government that regulate electrical work. Some states delegate that authority to municipalities; others retain it on the state level. However, almost every locality has an electrical code or ordinance. Any of these local codes will have the *National Electrical Code* as its basis. It isn't practical for a local or state government to recreate from scratch an entire electrical code. Electrical installations in any locality must meet the requirements of the electrical code in that locality. Many other countries use the *NEC* as well. For them, the extent to which there are local *NEC* amendments depends on local needs, culture, and political history.

Some local code rules go beyond *NEC* requirements; some relax them; some address local conditions that are beyond the usual scope of the *NEC*—for example, in the earthquake-prone areas of the west coast you'll often find modifications to address seismic problems. Other localities adopt the *National Electrical Code* without any changes.

Enforcement of the *NEC*   When the *NEC* is legally adopted by a city, town, state, or other governmental body as an official code of that governmental body, it becomes law. Compliance with the *NEC* in that locality then becomes mandatory on the effective date determined at the time of official adoption.

The *NEC* concerns itself with minimum safety, not good design   The *NEC* is concerned with safety only, as made clear by the following statements excerpted from *NEC* 90.1:

> This Code contains provisions that are considered necessary for safety. Compliance therewith and proper maintenance will result in an installation that is essentially free from hazard but not necessarily efficient, convenient, or adequate for good service or future expansion of electrical use.

> This Code is not intended as a design specification nor an instruction manual for untrained persons.

The safety measures advocated in this book are based on *NEC* requirements as the writer interprets the 2005 *NEC*. Because the *NEC* is not intended as a design or instruction manual, this book also covers practical design and installation methods, and the economic and convenience aspects of the art as well as safety.

Think of the *NEC* as an essential tool on your belt, just as essential as a screwdriver. Electrical work constantly changes. Every day you're likely to find a field application that wasn't quite the same as your instructor or this book might have anticipated. The *NEC*, if you know how to use it, gives you access to more than one hundred years of experience in terms of making safe electrical installations. That's why every person who installs electrical wiring and equipment absolutely needs to have and use a copy of the *NEC*. One of the objectives of this book is to make it easier for you to understand and use the *NEC* so you can properly apply its rules to that unanticipated situation.

Format of the *NEC*   The *NEC* consists of nine chapters. Each chapter contains several articles, except Chapter 9, which contains general tables. The first four chapters apply generally to all electrical work. The next three chapters cover special occupancies and special equipment and conditions, and supplement or modify the general rules. The eighth chapter covers communications circuits, and generally stands on its own. Some articles are divided into parts designated by Roman numerals, such as Part I, Part II, etc. All articles are divided into sections, but effective with the 2002 *NEC* the word "section" is no longer used to refer to specific provisions of the *NEC*. Internal references simply point to the appropriate numerical reference, for example "as covered in 90.1." Many sections have subsections, and some have sub-subsections. Some sections or their subdivisions have numbered paragraphs. Some sections, subsections, etc., are followed by fine-print notes. All mandatory rules of the *NEC* are in full-size print. The fine-print notes contain explanatory or informational material and are not mandatory rules. If you think a fine-print note you're reading changes the rule it follows, go back and reread the rule to find your error. Fine-print notes only explain and inform; they never change a rule.

History of the *NEC*   The first electrical code was the "Standard for Electric Light Wires, Lamps, Etc." adopted by the New York Board of Fire Underwriters on October 19, 1881. It contained seven rules plus instructions for applying for permission to use electric lights. That was the ancestor of the *NEC*.

In May 1882, the National Board of Fire Underwriters adopted the rules of the New York Board of Fire Underwriters. Later, similar rules were adopted by various other groups. Finally, a committee consisting of representatives from the fire-insurance groups, the electric utility company association, and a number of other national organizations produced the first *National Electrical Code* in 1897. That was the first electrical code that represented a nationwide consensus of the entire electrical industry without any single group having a dominant position.

NFPA—a worldwide fire safety organization   The National Fire Protection Association is a nonprofit voluntary membership organization that was formed in 1896 with the primary objective of achieving a fire-safe environment. It has members not only in the United States but all over the world. No single group has a dominant role. In addition to the fire services, the NFPA membership is representative also of architects, engineers, and other professions; of industry and commerce; of government agencies at local, state, regional, and national levels, including the military branches; of hospital and school administrators; and of others that have vocational or even avocational interests in achieving a fire-safe environment. (As used here, the term "fire safe" is used in its broadest sense and includes all facets of fire safety, including but not limited to electrical safety.)

The many codes, standards, recommendations, and manuals that have been developed by NFPA technical committees, which are composed of nationally recognized experts in their respective fields, have been widely adopted by national and local governments and by private industry. One of these, the *NEC*, is the most widely adopted and used construction code in the world.

Interpretations of the *NEC*   Unfortunately, many parts of the *NEC* are interpreted differently by different people. The interpretations in this book are the opinions of the writer, based on experience and the opinions of others. There is no "official interpretation" of the *NEC* as a whole, and very few interpretations of specific parts of the *NEC* are handed down. (NFPA, in its *Regulations Governing Committee Projects*, outlines how such interpretations may be obtained; it is not a simple procedure.) *NEC* 90.4 makes it entirely clear that the local electrical inspector has the final word in any situation. Most jurisdictions have administrative appeal mechanisms in place, however, to prevent an inspector from acting purely on whim.

PRODUCT STANDARDS AND CERTIFICATION

The National Electrical Manufacturers Association (NEMA) has representatives on all of the nineteen *NEC* Code-making panels for a simple reason: It does no good for the *NEC* to consider a new rule if products, for various reasons, could not economically be made to conform to it. Conversely, when a rule does change, the manufacturers need to know why so they can respond: demands of the market change with each new *NEC* edition. Just as the *NEC* is the level playing field for electrical designers and installers, the product standards constitute the level playing field for manufacturers.

Operating within the policy framework established by the *NEC*, industry product standards set the minimum standard for electrical product design. Although most widely known as a testing laboratory, Underwriters Laboratories (UL) generates the majority of electrical product standards used in the United States. For this reason, UL also has representatives on all of the nineteen Code-making panels.

Even if its present size were tripled, the *NEC* could not possibly encompass the level of detail in a product standard. For example, various *NEC* rules require luminaires (lighting fixtures) to be marked with certain information. Only the product standards, however, translate that policy requirement into actual markings. The product standards include such details as size of lettering, durability of label adhesive under specific temperature exposure, etc.

Product standards are meaningful   Product standards generally establish the testing procedures a product must endure. In order for a test to be meaningful, the product must be tested under the most severe conditions that would be encountered when the product is being used as intended. There must also be a safety-factor allowance. A fuse, as one example, must be able to operate (blow) and open the circuit (turn the circuit off as a switch does) within a predetermined time when, for instance, a wire is overloaded beyond its safe capacity; it must also blow if there is a short circuit, in which case the current might be a hundred or more times greater than in an overloaded wire. In other words it must be able to blow when the current is the maximum amount available, and it must do that without exploding or disintegrating, which could start a fire.

Fig. 1–1 UL tested both fuses. It listed the one at the bottom. The fuse at the top had no chance of becoming a listed item. *(UL Inc.)*

As an example, the old plug fuses, now largely obsolete but still manufactured, are made only in ratings up to (but not exceeding) 30 amps. However, they must function properly in a circuit that can deliver 10 000 amps, which is the amount of current used by UL to test the two fuses shown in Fig. 1–1. Each fuse had a handful of cotton placed on top of it when the fuse was tested inside a fireproof vault. The cotton on the lower fuse did not ignite when the fuse operated (blew). The fuse shown at the top not only burned up the cotton, but it severely damaged the fuseholder as well as itself. Some fuses explode while undergoing this test, sending molten particles flying against the walls and ceiling of the test vault. The fuse at the bottom in Fig. 1–1 was one of a group that had been submitted for testing by a reliable manufacturer. Only that fuse complied with the product standard and was therefore eligible to receive a listing. Similar meaningful tests are built into all product standards governing potential submittals for listing, such as wires, cables, and appliances.

Testing laboratories, the listing process, and approval   Just as the *NEC* requires electrical inspection to come alive, the product standards require enforcement through the listing process to have meaning. The public also has a right to expect disinterested evaluations; that is, just as we expect an inspection agency to have no financial interest in the outcome of an inspection, we also expect that the agency administering product testing must have no financial interest in that outcome either. That's why testing laboratories are independent entities. Although approval of all installations is the responsibility of the authority having jurisdiction (the inspector or the inspector's boss), that person may (and normally does) base approval of wiring materials and equipment on prior listing of the materials and equipment by a nationally recognized testing laboratory, as provided in *NEC* 90.7, since the inspector has no means of making suitable tests of most types of material and equipment. After the wiring and materials have been installed, it is the inspector who must determine whether acceptable materials have been used and properly installed. The inspector then approves the completed installation, or turns it down until any deficiencies have been corrected.

There are many recognized testing laboratories; the testing laboratory most widely used in the United States is the Underwriters Laboratories Inc. (UL). Note that testing laboratories do not "approve" anything. If, for example, UL tests an item and it is found to meet the product standard, UL lists it in a book (discussed in more detail later in this chapter), and the item is then described as "listed by UL" or "UL listed." It is therefore totally incorrect to say "approved by UL." The *NEC* in Article 100 defines "approved" as "acceptable to the authority having jurisdiction." For brevity, in this book we substitute "inspector" for "authority having jurisdiction." The inspector approves an item after UL (or any other testing laboratory recognized by that jurisdiction) has listed it.

Not every electrical product requires listing. Code-making panels have robust discussions every cycle about what standard of approval should apply to given items. Many products require careful evaluation in a testing laboratory to determine suitability, hence the *NEC* stipulates listing in those cases. In other cases that is judged excessive. Although most electrical boxes installed under the *NEC* are in fact listed, the *NEC* only requires listing in the case where special alloys or constructions result in variation from the basic stated rules on material thickness. The same panel rejected a proposal to require listing on large pull boxes because they are frequently made up in local sheet metal shops to particular dimensions suited to a particular job, and the expense of a listing would be excessive.

Identified   The *NEC* also uses the term "identified," defined in Article 100 as "recognizable as suitable for the specific purpose, function, use, environment, application, etc., where described in a particular Code requirement." It does not mean the same thing as "marked." In a reputable manufacturer's catalog, you might see a new product specifically designed to meet a particular electrical need. Although frequently the product is listed in addition, the design alone would probably qualify it as being "identified" within the meaning of this definition.

Labeled   The term "listed" does not literally mean labeled, nevertheless we all look for the label to determine whether a product has been listed. The qualified testing laboratories tightly control the use of their marks. Two of their policies, uniform across all the testing laboratories, bear repeating here. All listed equipment is labeled, and the marks only issue as part of the manufacturing process or under the direct supervision of a test lab employee. Thus if a label falls off a piece of equipment, that equipment is no longer listed. Furthermore, a new mark, if it is to be applied to reestablish a "listed" status, must be applied in the presence of a test lab employee. The label cannot be simply sent through the mail. All the test labs have field personnel who will witness field marking as required.

Product directories   All qualified testing laboratories maintain product directories of their listed products. Due in part to the UL's role as developer of the majority of product standards, the UL directories are the most extensive. Tested products that meet UL standards are listed under various categories. The listed products that are of interest to readers of this book are contained in UL's *Electrical Construction and Materials Directory, Electrical Appliance and Utilization Equipment Directory,* and *Hazardous Location Equipment Directory.* These directories, historically known in the field by their colors (respectively green, orange, and red), have been a vital part of an electrical inspector's library. However, there is a substantial expense involved in the publication of these directories, and UL is in the process of discontinuing them in favor of simply maintaining the information on its Internet site. In this way the information can be continually updated without waiting for an annual publication.

Guide card information   The UL product directories have another even more important function. They tell you about the assumptions, regarding uses and conditions, that the testing lab made when it evaluated the equipment; these assumptions are listed on "guide cards." That means if you use a product in a way the testing laboratory didn't anticipate during its testing, that product may fail even though it is a listed product. There is a provision in the *NEC* that requires all products to be installed and used in accordance with any directions that come out of the manufacturing and listing process.

That means simply follow directions. It also means that every item on a guide card generated pursuant to a product standards is, in effect, incorporated into the *NEC* by reference. For example, suppose you supply a cord in the field to connect a swimming pool pump motor for an outdoor installation. If that cord isn't marked in a certain way, the installation will be rejected. That's true even if the cord complies with the size and length restrictions set in the *NEC* for that application. Why? Because when you look at the guide card information, you'll find that only cords with particular markings have been tested for outdoor exposures.

This guide card information is so critical that, as a public service, UL publishes another directory—called the White Book—with just the guide card information extracted from its green and red books and selected information from the orange book. If you're planning to become an electrical contractor, you must obtain at least this directory. Much of its information is also available on the Internet. Note that UL plans to continue publication of this critical resource.

Identifying UL-listed products   The various product lists, directories, quarterly supplements, etc., should be used to find out which manufacturers have listed products of the kind in question. Do not depend on such lists entirely, however, because a manufacturer might produce both listed and unlisted products under different model numbers. Therefore, it is essential that you look for the listing mark (or label) itself, which indicates that a product is produced under the UL's Follow-up Service (explained on the next page).

For some small items such as snap switches, receptacles, lampholders (sockets), and outlet boxes, the listing mark or symbol as shown in Fig. 1–2 is stamped or molded on the surface of each item. For larger items such as raceways, lighting fixtures, panelboards, and motor controllers, a label of one of the types shown in Fig. 1–3 is attached to the surface of each item. The stamp or label may be on an interior or exterior surface, such as the interior surface of a panelboard enclosure and the exterior surface of a length of conduit or of a fuse. Some products have a label on the carton or reel.

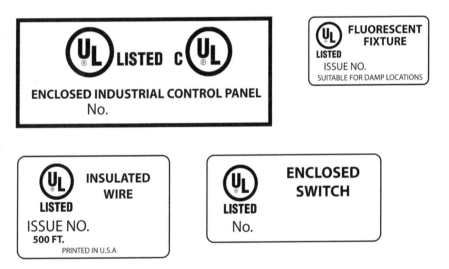

Fig. 1–2 These labels are examples of many authorized to be applied to listed equipment. All qualified test labs have comparable labels.

◀ **Fig. 1–3** If because of size, shape, material, or surface texture the product bears only the round symbol at left, a complete label with the four elements shown at right will be found on the smallest unit container in which the product is packaged.

Flexible cords sold by the foot have a label on the spool or on precut lengths varying from 6 to 30 ft. A "replacement power supply cord" is a length of flexible cord having an attachment plug on one end and having the other end prepared for permanent attachment to an appliance, such as a vacuum cleaner whose original cord has deteriorated or been damaged. Listed cords of this type have a tan flag-type label as shown in Fig. 1–4.

A general-use "cord set" is a length of cord at least 6 ft long with an attachment plug on one end and the female member of a separable connector on the other end. This type of cord, commonly called an "extension cord," is for temporary use only and should be disconnected when the tool or appliance with which it is used has been turned off. An extension cord should never be used as a permanent extension of a branch circuit. Listed cords of this type have either a blue flag label or a doughnut-shaped label. The flag label is shown in Fig. 1–4.

Fig. 1–4 Cord assemblies, if listed, bear a flag label similar to this.

Listed appliances have the mark shown in Fig. 1–2 combined with the word LISTED on the nameplate of the appliance. UL also requires a control number, and the manufacturer's name, trademark, or similar identifying symbol on listed products. Sometimes the equipment name is also included to indicate under what category the manufacturer submitted the product. Most household appliances do not need this because their identity is obvious (a toaster, for example).

UL Follow-up Service   Each manufacturer that obtains UL listing of its products agrees to the UL Follow-up Service program, enabling a UL inspector to visit the factory at random or, in some large plants, to conduct daily inspections at the factory to see that the products are manufactured and tested in accordance with the UL standard and test procedures by which the original listing was obtained. The UL inspectors make or witness follow-up tests at the plant, and sometimes send randomly selected samples to a UL testing laboratory for follow-up testing. Occasionally, UL inspectors purchase listed products on the open market for follow-up testing. If the safety level of the product is not maintained, the UL listing is withdrawn and the manufacturer is prohibited further use of the UL listing mark until the safety requirements are again met and maintained. Other testing laboratories maintain similar arrangements.

What a listing means   Like the *NEC*, product standards address safety only, and not efficiency or convenience. Hence, a listed product of one manufacturer may be superior to the same type of product of another manufacturer that is also listed. Yet both products meet the minimum standards of safety. Use good judgment when buying merchandise. A listing is not the only factor to be considered, even though it is an essential factor.

Finally, it is important to remember that a listed cord attached to an appliance does not mean that the appliance itself is listed. Product tests involve other features besides electrical insulation, clearances, etc. Hence, if an appliance is unsafe to use for its intended purpose in the intended manner, it will not receive a listing even if the electrical parts are safe, because product testing covers total safety, which includes (but is not limited to) electrical safety. Be sure that an entire assembly is listed.

Recognized component   Never confuse a component recognition mark with a listing. UL uses its component recognition program to simplify its listing evaluations. Consider swimming pool filtering equipment, for example. Typically a manufacturer supplies this equipment as an assembly with a circulating motor, filter, plumbing hook-ups, etc. However, usually the manufacturer buys the motor from another manufacturer who specializes in motors. In the past, UL has been reluctant to list motors because it can't judge the shaft load on unknown applications (the UL exception for explosionproof motors isn't relevant here). That might seem to put UL in the position of having to constantly reevaluate internal characteristics of every motor used by this filter manufacturer, even though this manufacturer never did anything to the motor other than take it out of its shipping carton.

To address this, although UL may not list the motor, it can and frequently does extend component recognition to the motor. The motor manufacturer, subject to UL review, builds the motor to recognized standards, and puts the RECOGNIZED COMPONENT mark on it, as shown in Fig. 1–5. If the UL inspector in the pool equipment plant sees a motor with component recognition, he knows that he doesn't have to tear the motor apart; he only has to look at the way his listee completed the assembly. UL deliberately drew the component recognition mark to avoid confusion with its listing marks; nevertheless, it constantly happens that the component recognition mark is mistakenly considered synonymous with a listing mark. Recognized components are only components. They lack essential features required to complete a listing evaluation.

Although UL is now expanding its listing activities on motors, the principle covered here on the uses and limitations of component recognition still applies.

Classified products   In some cases electrical equipment will be evaluated with respect to a specific application or certain performance criteria, beyond or instead of a general listing. For example, some circuit breakers have been

Fig. 1–5 Don't confuse this component recognition mark with a listing mark. *(UL Inc.)*

evaluated as to performance in panelboards of competing manufacturers. These breakers carry a classification designation indicating the specific panels you can install them in safely. As long as you follow the installation instructions on this equipment as covered in its classification, you may install those circuit breakers in a competing panelboard. You have to observe all restrictions that come with the classification. For example, in the case of the classified circuit breakers, you usually can't use them in circuits that could be subjected to very high fault currents, even though a native circuit breaker of the same size could be used in the same panelboard under equivalent conditions. There are very good reasons for these distinctions.

INSPECTION

The International Association of Electrical Inspectors (IAEI) has as its logo the keystone of a bridge, underscoring the essential nature of disinterested, qualified, third-party electrical inspection to safe electrical installations. In most every jurisdiction, electrical work must be inspected in progress before it gets covered, and again upon completion. Inspection provides two principal benefits. The first and most obvious benefit is active, consistent enforcement of the minimum safety requirements in the *NEC* to assure an installation essentially free from hazard. The public generally understands and supports this function. The second benefit serves reputable manufacturers and electrical contractors. The electrical inspector, by enforcing the rules, establishes a level playing field. Consistent enforcement is the only effective way that the electrical contractor who performs shoddy work and the manufacturer of substandard products will be forced out of the market, because the general public lacks the specialized training to detect such problems before they lead to injury or damage.

This book usually uses the term "electrical inspector" or "inspector" to describe the person performing this function, because that is the usual case. However, the *NEC* uses the term "authority having jurisdiction" or "AHJ" because in some instances the responsible authority may be a fire department chief or some other person, perhaps representing an insurance firm. The important point is the requirement for disinterested third party review of electrical work.

International Association of Electrical Inspectors (IAEI)   Membership in the IAEI is not limited to inspectors. In fact, the great majority of IAEI members aren't inspectors. If you plan to make a living in the electrical field, you should join your local chapter as an associate member. You'll have a vote in most chapter actions, and you'll get the bimonthly *IAEI News*, which is packed with discussions about the *NEC*. In addition, you'll be in the same room with other electricians and inspectors at every chapter meeting, discussing *NEC* topics of local interest.

Permits   In many places it is necessary to get a permit from city, county, or state authorities before a wiring job can be started. The fees charged for permits generally are used to pay the expenses of electrical inspectors, whose work leads to safe, properly installed jobs. Power suppliers usually will not furnish power until an inspection certificate has been turned in. The permits also provide the means for notification to the local inspection authority that a property owner plans to make an electrical installation.

Licenses   In some localities, electrical installations can legally be made only by licensed electricians or by people supervised by a licensed electrician. Many localities in which licensing is required make an exception to the licensing law, allowing you to install wiring for your own home or farm. Other localities do not have such an exception. Ask the inspector in your locality about local rules, and get a copy of the local code.

WHERE DO YOU FIT IN?

When your state or local government adopted some version of the *NEC* as its electrical code and appointed an electrical inspector, it did so as an exercise in self-government anticipated in the constitutional framework of our nation. Over the last hundred years, these voluntary arrangements have coalesced into a system that works well as long as its participants remain mindful of their roles and engaged in its outcome. Now that you have a better idea of where you fit in the larger picture, and with all sides of our safety triangle in place, we can move on to some basic electrical concepts, and then to the art and science of the trade.

CHAPTER 2
Numbers, Measurements, and Electricity

THE SPECIFICS OF practical electrical wiring begin in Chapters 3 and 4. But just as Chapter 1 contains basic information on wiring practice and regulation that you will need before starting to plan and install electrical wiring and equipment, this chapter contains much background information that you will need about numbers and basic principles. The information about numbers (including the metric system) will help you understand what you read. After studying this chapter, if you see a prefix such as "milli-" or "kilo-" as part of a word, you will know what it means. If you forget, you can refer to Table 2–1. In addition, the basic principles of electricity and magnetism are very briefly reviewed.

INTERNATIONAL NUMBER STYLE

In this book the style of writing numbers conforms to the international standard used today in all branches of science and technology. This practice eliminates confusion resulting from the various traditional number styles used in different countries. The traditional styles continue to be used in many nonscientific situations and, as many international travelers have discovered while studying a restaurant or hotel bill, these style differences continue to cause confusion across cultures.

For instance, in the United States and some other countries, large numbers traditionally have been separated by commas at intervals of three digits; in some other countries periods are used instead of commas. In the United States and some other countries, a period serves as a decimal point while other countries use a comma for that purpose. In some countries, including the United States, a billion is a thousand millions; in other countries, a billion is a million millions.

Such variations are tolerated for some things, but with international trade growing in a world that's ever more closely interconnected, it is desirable—essential in science and technology—to have international standardization. Like other branches of science and technology, electricity knows no political boundaries. Its laws are the laws of physics and chemistry, which are the same everywhere.

Table 2–1 ORDERS OF MAGNITUDE: NOTATION FOR LARGE AND SMALL NUMBERS

Unit	Multiple	Prefix	Symbol
1	(one)	None	None
$10 = 10^1$	(ten)	deka	da
$0.1 = 10^{-1}$	(one-tenth)	deci	d
$100 = 10^2$	(one hundred)	hecto	h
$0.01 = 10^{-2}$	(one-hundredth)	centi	c
$1000 = 10^3$	(one thousand)	kilo	k
$0.001 = 10^{-3}$	(one-thousandth)	milli	m
$1\ 000\ 000 = 10^6$	(one million)	mega	M
$0.000\ 001 = 10^{-6}$	(one-millionth)	micro	μ*
$1\ 000\ 000\ 000 = 10^9$	(one billion)	giga	G
$0.000\ 000\ 001 = 10^{-9}$	(one-billionth)	nano	n
$1\ 000\ 000\ 000\ 000 = 10^{12}$	(one trillion)	tera	T
$0.000\ 000\ 000\ 001 = 10^{-12}$	(one-trillionth)	pico	p

*This is the lowercase Greek letter mu.

By international agreement, a space is now used instead of a comma, period, or any other mark between groups of three digits to both the right and left of the decimal (with no spaces directly on either side of the decimal). An exception is numbers consisting of four or fewer digits, in which case the digits are grouped together without a space. A period is used only for a decimal point. Examples: 7; 21; 356; 1245; 49 681; 0.496 81; 6 348 907.301 25; 0.634 890 7; 75 372.860 4. Always use a zero before a decimal point if there is no other number preceding it, thus: 0.34. Never this way: .34.

Numbers are based on the unit 1 and multiples of 10. Each multiple of 10 has a prefix, as shown in Table 2–1. Terms such as "ampere," "watt," and similar terms used here to explain numbers or abbreviations are defined and explained later in this chapter, but you can readily see from Table 2–1 that if the unit is 1 ampere, then $\frac{1}{1000}$ (0.001 or one-thousandth) of an ampere is 1 milliampere. Similarly, if the unit is 1 watt, then 1000 watts is 1 kilowatt, etc.[1]

ABBREVIATIONS

The international agreement also requires that most abbreviations be in capital letters for terms bearing or derived from the name of a person, and in lowercase letters for other terms. (The terms used here in explaining abbreviations will

---

1. For a complete treatise on the material shown in brief in Table 2–1, see *Letter Symbols for Units Used in Science and Technology* (IEEE No. 260; ANSI Y1O.19—1978). IEEE is the Institute of Electrical and Electronics Engineers. ANSI stands for American National Standards Institute. The above standard is sponsored by the IEEE and has been adopted as an ANSI standard.

themselves be explained later as they are used.) For example, the "watt" is named for James Watt, so it is abbreviated with a capital W. Similarly, such terms as "ampere," "volt," and "ohm" (each derived from surnames of people) are abbreviated with a capital A, capital V, and Ω (the Greek capital letter omega). But such terms as "kilo" and "hour" are not derived from people's names and are abbreviated with lowercase letters (k and h for the examples given). So "kilowatthour" is abbreviated kWh, and "kilovoltampere" is abbreviated kVA.

Some exceptions to this capitalization practice are made in order to avoid confusion. The abbreviation for "mega" is a capital M, even though mega is not derived from a person's name. See Table 2–1. Capital M is used for mega because lowercase m represents the more commonly used term "milli." Another example is the term "kilovar," which is abbreviated kvar even though the term (which stands for kilovoltampere-reactive) contains the words "volt" and "ampere," both derived from surnames. Again, this and other seeming contradictions are necessary to avoid confusion where infrequently used terms have the same letter abbreviations as those used for more frequently used terms. When terms are spelled out, they are never capitalized (unless the term is the first word of a sentence) even if derived from names of people. In addition to the abbreviations shown in Table 2–1, some of the more common terms and their abbreviations, together with brief explanations of their meanings, are listed in Table A–6 in the Appendix. A complete list is available in *Letter Symbols for Units Used in Science and Technology* as cited in the first footnote in this chapter.

METRIC SYSTEM

You already have some familiarity with using the metric system—American money (among other things) is based on the decimal principles underlying it. The metric system, which originated in France, is usually referred to as "SI" for Système International d'Unités, or International System of Units. As the world shrinks economically, the metric system of measurement increasingly becomes the universal system, and the *NEC* now includes SI units with most customary English units. Remember that the *NEC* is used in many countries that use only the metric system. Table 2–2 shows some commonly used English-metric equivalents.

In a continuing push to expand its international usability, the 2005 *NEC* uses the metric dimension first in its requirements, followed by the customary English units in parentheses. In addition, in most rules the conversion is a *hard,* not *soft* conversion. That means (and this will be counterintuitive for most people) the *NEC* uses *inexact,* not *exact* conversions. Editions prior to 2002 placed the metric unit in parentheses, after the English system unit, and in all cases used a *soft (exact)* conversion. In some cases the soft conversion has been retained, as in cases where the hard conversion would change a dimension in a way that would compromise safety, but those occasions are comparatively unusual. The topic is covered in *NEC* 90.9. You may well be having trouble thinking of something hard being inexact and something soft being exact. One way to think about the terminology is to think of hard metric conversions as something you would expect to see from a hard-core metric believer. For example, the basic minimum rule for spacing receptacles on dwelling unit wall areas is that no point can be more than 6 ft from a receptacle. The soft conversion for 6 ft is 1.83 m. If you were hard-core metric and wrote this rule, surely you would round the metric dimension and then tack on the English equivalent. The 2005 *NEC* does exactly that, reading "1.8 m (6 ft)."

Table 2–2 COMMONLY USED ENGLISH–METRIC EQUIVALENTS

English to Metric	Metric to English
Length	
1 in = 25.4 mm (millimeters)	1 mm = 0.0394 in
1 in = 2.54 cm (centimeters)	1 cm = 0.394 in
1 ft = 304.8 mm	1 cm = 0.033 ft
1 ft = 30.48 cm	1 m = 39.37 in
1 ft = 0.305 m (meters)	1 m = 3.28 ft
1 yd = 0.915 m	1 km = 3280.83 ft
1 mi = 1609.34 m	1 km = 0.621 mi (miles)
1 mi = 1.609 km (kilometers)	
Area	
1 sq in = 645.16 sq mm (mm²)	1 sq cm = 0.155 sq in
1 sq in = 6.45 sq cm (cm²)	1 sq cm = 0.0011 sq ft
1 sq ft = 929.03 sq cm (cm²)	1 sq m = 10.764 sq ft
1 sq ft = 0.093 sq m (m²)	1 sq m = 1.2 sq yd (yards)
Volume	
1 cu in = 16.38 cc (cm³)	1 cc = 0.061 cu in
1 cu in = 0.016 liters	1 liter = 61.02 cu in
1 cu ft = 28.32 liters	1 liter = 0.035 cu ft
1 liquid qt = 0.9475 liters	1 liter = 1.056 liquid qt
1 liquid gal = 3.79 liters	1 liter = 0.264 liquid gal
Miscellaneous conversions	
946 cubic centimeters (cm³) = 1 liquid qt (quart)	
1000 cubic centimeters (1000 cm³) = 1 cubic decimeter = 1 liter	
1 lb = 0.454 kg (kilogram) 1 kg = 2.2 lb (pounds)	
1 oz (ounce avoirdupois) = 28.349 g (grams)	

TEMPERATURE CONVERSION

It will be helpful for you to know how to convert a given temperature from the Fahrenheit (°F) to the Celsius (°C) scale, and vice versa. The temperature scale that was formerly (and sometimes still is) called the "centigrade" scale is now correctly called the "Celsius" scale in honor of the scientist who originated it. The word "centigrade" is a Latin word meaning "hundred grades" or "hundred marks"; in that scale water freezes at 0° and boils at 100°. In the Fahrenheit scale water freezes at 32° and boils at 212°.

The formulas for converting temperatures from one scale to the other are as follows:

(degrees F – 32)(⁵⁄₉) = degrees Celsius

(degrees C)(⁹⁄₅) + 32 = degrees Fahrenheit

In other words, to convert degrees F to degrees C, first subtract 32 from degrees F, then multiply the remainder by ⁵⁄₉. The answer is the same temperature on the Celsius scale. Example: Convert 86°F to degrees C: 86 – 32 = 54; then

$$(54)(\text{⁵⁄₉}) = \frac{(54)(5)}{9} = \frac{270}{9} = 30°C$$

To convert degrees C to degrees F, first multiply the degrees C by ⁹⁄₅, then add 32. The answer is the same temperature on the Fahrenheit scale. Example: Convert 30°C to degrees F:

$$(30°C)(\text{⁹⁄₅}) = \frac{(30)(9)}{5} = \frac{270}{5} = 54$$

Then add 32; 54 + 32 = 86°F.

But be careful if you are trying to convert *the change in number of degrees* on a Celsius thermometer into the change in number of degrees on a Fahrenheit thermometer (or vice versa), rather than the actual reading in degrees on the two different thermometers. This is confusing to many people, but note:

	Celsius	Fahrenheit
Water boils at	100°	212°
Water freezes at	0°	32°
Difference	100°	180°

From this you can see that each degree of change on the Celsius scale is equivalent to ¹⁸⁰⁄₁₀₀ or ⁹⁄₅ or 1.8° change on the Fahrenheit scale; a degree of change on the Fahrenheit scale is equivalent to ¹⁰⁰⁄₁₈₀ or ⁵⁄₉° of change on the Celsius scale.

ARITHMETIC REFRESHER

These are the basic math skills you will need in order to do the calculations in this book, which are those commonly encountered in electrical work.

Multiplication   When two or more numbers, each in parentheses, follow each other, it means they are to be multiplied. Thus (5)(11)(13) means 5 × 11 × 13, or 715; and 2X means (2)(X), which means 2 × X.

But when letters that are abbreviations of terms (such as V for volts and A for amperes) are to be multiplied, they need not be (but may be) placed in parentheses. Thus VA means (V)(A) or V × A, all indicating that the volts must be multiplied by the amperes.

Order of operations   Where multiplication, division, addition, and subtraction are shown in a formula, you must perform all multiplications and divisions first, and then perform the remaining operations. This is true except where additions or subtractions are enclosed in parentheses ( ) or brackets [ ], in which case you must perform the enclosed operations first. Examples:

$$(4)(6) - 2 = 24 - 2 = 22$$
$$(4)(6 - 2) = (4)(4) = 16$$

$$(4/5)(5/8)(3) + 3 - 1 = \frac{(4)(5)}{(5)(8)}(3) + 3 - 1$$

$$= \frac{20}{40}(3) + 3 - 1$$

$$= \frac{60}{40} + 3 - 1 = 1.5 + 3 - 1$$

$$= 3.5$$

$$(4/5)(5/8)(3 + 3 - 1) = \frac{(4)(5)}{(5)(8)}(3 + 3 - 1)$$

$$= \frac{20}{40}(5) = \frac{100}{40} = \frac{5}{2}$$

$$= 2.5$$

$$[(4) + (6 - 2)](3) = (4 + 4)(3) + (8)(3) = 24$$

Square roots   When a square root is to be extracted, as indicated by the radical sign $\sqrt{}$, everything under the sign must be done first, then the square root is extracted, and finally any other indicated operations are performed. Examples: $\sqrt{4} = 2$ (because 2 multiplied by itself equals 4).

$$\sqrt{(2)(8)} = \sqrt{16} = 4$$
$$2\sqrt{(4)(4)} = 2\sqrt{16} = (2)(4) = 8$$

Exponents   Where a number is followed by a second number that is placed just above and to the right of the first number, the second number is called an exponent. It means that the first number is to be used as a factor in multiplication as often as indicated by the exponent. Thus 2^2 means 2 "to the second power" or 2 "squared" which means $2 \times 2 = 4$; 3^4 means 3 raised to the fourth power, or $3 \times 3 \times 3 \times 3 = 81$; 10^5 means 10 raised to the fifth power, or $10 \times 10 \times 10 \times 10 \times 10 = 100\ 000$. It is common practice in speech to omit the word "power" by saying, for example, "10 to the fifth," meaning 10 raised to the fifth power, or 10^5.

Working with units   Be sure to carry the units of measurement through your calculations. For example, suppose you're driving a car 100 km/h (about 60 mph) and you want to know how far you'll get after 3 hours. Should you multiply by 3 or divide by 3? It's easy to make a careless mistake, but if you discipline yourself always to carry the units throughout, you'll catch your errors. In this case, multiplying 100 km/h times 3 hours equals (300 km/h)h. Note that both the numerator and the denominator of the resulting fraction contain the unit "h" which thus cancels itself, leaving only the unit of distance. A distance measurement is what we were looking for, so multiplication was the correct operation. Division would have given km/h/h, or km/h^2, which is a measure of acceleration. Although you might think it a waste of time to carry the units of measurement through every calculation, it prevents countless errors. Remember, when you work with electrical quantities, you're working with values that are far more abstract and less intuitive than simple time/speed/distance driving problems.

UNITS OF ELECTRICAL MEASUREMENT

In the study of electricity you will meet many terms that have to do with measurement: "volts," "amperes," "watts," and others. It is much easier to learn how one is related to another than to get an idea of the absolute value of each. Electric power is measured in units that cannot be compared directly with meters, feet, grams, pounds, liters, quarts, or any other measure familiar to you. If you consult your dictionary, you will find definitions like these:

volt—the pressure required to force one ampere through a resistance of one ohm

ampere—the electric current that will flow through one ohm under a pressure of one volt

ohm—the resistance through which one volt will force one ampere

These definitions show a clear interrelationship among the three items, but unfortunately define each in terms of the other two. How "big" is each unit? It is as confusing as the beginner's first encounter with the metric system: 10 millimeters make a centimeter, 400 centimeters make a meter, 1000 meters make a kilometer; further, 1000 cubic centimeters make a liter; a liter of water weighs a kilogram. Now pretend you are not familiar with the information in Table 2–2 in this chapter. These metric terms would be meaningless to you unless you could translate them into more familiar terms such as inches, miles, pounds, quarts.

After using these metric terms for a while, you indeed begin to see some relationship between them and the more familiar terms you are accustomed to using. Similarly, in electrical work you will begin to see some meaningful relationships among the various terms such as volts, amperes, ohms, and others.

Unfortunately, in electrical work it is difficult to translate an ampere, a volt, or an ohm into something that is familiar to you. But we can compare these terms with other measures that behave in a similar fashion. Though such comparisons are rarely entirely adequate, they may be of some help in conveying abstract concepts.

Electric fields and the nature of electricity   You may be aware that electricity moves through a conductor at essentially the speed of light, or 186 000 mi/sec (300 000 km/sec). Thus, forgetting about losses in the wire for a moment, if you connected a load to the end of a wire that went around the earth seven times, and turned the switch to ON (in the trade this is called "closing" the switch), the load would receive the electric energy only a second later. This degree of speed occurs because closing the switch creates an electric field through the conductor, and that field is part of the same electromagnetic spectrum as radio, radar, infrared, visible light, ultraviolet, x-rays, and gamma radiation. Even though the details on how these fields work are well beyond the scope of this book, it might be interesting to note that this is a speed of just under 1 billion feet per second, which would seem fast enough for any purpose. But modern computers perform operations in billionths of a second. If, in the circuits within a computer, the current has to travel over long wires, every foot of extra length will slow down an operation by a billionth of a second. Today's computer and chip designs take this into account.

Effects of electricity   The endless assortment of things that electricity does can, in large part, be categorized into forms or combinations of three basic effects: thermal, magnetic, and chemical.

Thermal effect   The thermal effect of electricity is heat. A current cannot flow without causing some heat. Sometimes heat is not desired, as for example in the case of the unavoidable power loss referred to in the examples in this chapter. In an ordinary incandescent lamp, over 90% of the current is wasted as heat, and less than 10% is converted into light; but the light is not possible without the heat. In a toaster or clothes iron, only the heat is desired.

Magnetic effect   The magnetic effect can be stated very simply. When a current flows through a wire, the wire is surrounded by a magnetic field—the area immediately around the wire becomes magnetized. Bring a small compass near a wire that is carrying current and the needle will move just as it will when you bring it near an ordinary magnet. Wrap a wire a number of times around a piece of soft iron that is not magnetic; during the time that a current flows through the wire, the soft iron becomes a magnet, weak or powerful depending on such factors as the number of turns of wire and the number of amperes flowing. The moment the current stops flowing, the iron ceases to be magnetic. It is this magnetic effect that causes doorbells to ring, motors to run, telephones and radio speakers to operate, and magnetic computer drives to operate.

Chemical effect   The chemical effects are of great variety, including the electroplating of metals, the charging of storage batteries, and the electrolytic refinement of metals. In a dry-cell battery there is the reverse effect. A chemical action produces an electric current.

Coulomb—basic unit of electricity   The analogies to water and plumbing that follow are intended to help you understand the basic concepts, but they are only analogies. Electrical conductors aren't hoses containing uncountable electrons moving at the speed of light. There are free electrons in conductors, and they can migrate over time, but what actually does the work is changes in an electric field that move through the conductor almost instantaneously. With that understood, the plumbing analogy may be useful as you develop your understanding of the concepts.

We can measure water in pounds or liters or cubic feet or acre-feet or in many other ways. The measure commonly used in the United States is the gallon. A gallon of water is a specific quantity of water. In measuring electric current, "coulomb" is the term that corresponds to the gallon. Ask all the people you know in the electrical business, "How much is a coulomb of electric power?" Ninety-nine percent or more are likely to answer, "Coulomb? I vaguely remember the term from way back when, but I don't know what a coulomb is." While very few people remember the definition of that term, it is nevertheless helpful in beginning to understand other electrical terms. Accept it as a fact that a coulomb is a definite quantity of electric current (actually, the charge represented by 6 280 000 000 000 000 000, or 6.28×10^{18} electrons); don't try to understand how big that quantity is.

Ampere—rate of flow   Gallons or liters of water standing in a tank are just quantities of water. But if there is a small hose connected to the tank, water will flow out of it. If there is a big hose water will flow out of it faster than out of the small hose. If you want to talk about how much faster it flows out of the big hose than out of the small one, you must use some measure to denote the rate of flow, so usually you talk about gallons or liters per minute. This indicates the quantity (gallons or liters) and the time (per minute). Together, quantity and time indicate the rate of flow.

Instead of a tank of water, consider a battery, a generator, or other source of electric power. Instead of a hose connected to the tank, think of a wire through which electric current will flow. Coulombs of electric power will flow through that wire, just as gallons of water flowed through the hose. Instead of gallons of water per minute, we have coulombs of electric current per second. Rather than saying "10 coulombs per second" we say "10 amperes." An ampere (named for an early 19th century French scientist who did much of the original research on electric current) is defined as a flow of one coulomb per second.

Don't say that current is flowing at "10 amperes per second"—that would be the same as saying it's flowing at the rate of 10 coulombs per second per second. Just remember that a coulomb is a quantity; an ampere is a rate of flow of that quantity. Once you clearly understand that, you can forget the coulomb and think in terms of amperes. The term "amperage" for the amount of current in amperes is not a formal term, but is generally considered acceptable for informal usage. Similarly, the abbreviations "amp" and "amps" are not formal abbreviations, but are regularly used and will be used in the text of this book (example: a 30-amp fuse blows on a current of 30 amps); in formulas the letter "A" will be used to signify amperes.

Volt—measure of pressure   A gallon of water standing in a tank is an inert, static quantity. Water dribbling out of your not-quite-shut-off garden hose at a gallon per minute is just a nuisance. Water coming out of your sprinkler on the lawn at a few gallons per minute waters your lawn and can be a delightful shower if you stroll through it on a hot day. The same gallons-per-minute flow coming out of a hose in the form of one tiny stream but at a higher pressure would be painful instead of delightful if directed at you.

Suppose there are three water tanks located 10, 20, and 100 ft above ground level. A pipe runs from each tank to the ground level, and a pressure gauge is connected at the ground level. Disregarding friction losses, the gauges on the three tanks will show pressures of 4.3, 8.6, and 43 lb per sq in. If a tank were located 1000 ft above ground level, the pressure would be 430 lb per sq in.

One gallon per minute running out of the first tank would be a nuisance; a gallon per minute out of the last tank at 430 lb per sq in. could do a lot of damage. The difference is in the amount of pressure, which in the case of water is measured in pounds per square inch.

Electric power is also under pressure, but instead of being measured in pounds per square inch, it is measured in volts. One volt (named for an Italian scientist who at the turn of the 19th century discovered the principle of batteries) is a very low pressure. An ordinary flashlight dry-cell battery (regardless of size) will, if fresh, develop approximately 1½ volts. A single cell of a storage battery develops about 2 volts when fully charged; the six cells of an automobile battery develop 6 × 2, or 12 volts. An ordinary house-lighting circuit operates at about 120 volts. The high-voltage lines feeding the transformers in alleys or city streets usually operate at 4160 to 13 800 volts.

The amount of electric current in amperes depends on the number of electrons flowing past a given point in one second. The pressure forcing the electrons or electric current to flow is called the voltage. Compare the electrons with bullets. If you toss ten bullets at a plywood target every second, they will hit at a low pressure and cause no damage. If you fire the same ten bullets from a machine gun firing ten rounds per second, they'll hit the target at a high pressure and will likely demolish it. The difference lies in the amount of power associated with the moving objects.

However, remember the earlier discussion of electric fields. Voltage does not affect the speed at which electric power flows, which is about the same as the speed of light.

Watt—measure of rate of power   To explain watts using our imperfect plumbing analogy, suppose that each of the water tanks we have talked about has a capacity of 100 gal and that each tank is empty. You must fill each tank using a hand pump (pumping out of a source of water at pump level, to simplify the problem). Your ability to fill each tank in an allotted time relates not just to the speed of water delivery (100 gal in 5 min), but also to the amount of force you need to apply to overcome backpressure from the vertical column of water. As the tanks get higher, you need ever-increasing effort to pump water at the same rate.

Suppose it takes you 5 min to fill the 100-gal tank located 10 ft above ground level. Then to fill a tank ten times as large (1000 gal) located at the same level should take you ten times as long, or 50 min. To fill the original 100-gal tank but now located ten times as high (100 ft) should take you ten times as long as at the 10-ft level, again 50 min. In other words, it would take you 50 min to fill a 1000-gal tank located 10 ft above ground level or a 100-gal tank located 100 ft above ground level. This is probably easier to grasp in tabular form.

Capacity of tank, gal	Height above ground, ft	Approx pressure, lb per sq in.	Time to fill, min
100	10	4.3	5
1000	10	4.3	50
100	100	43.0	50

This discussion demonstrates that gallons per minute do not determine the amount of power involved in filling the tank; and pressure alone does not determine it; but the two in combination do. In other words, gallons per minute times pressure per square inch equals the amount of power being consumed or used at a specific moment in pumping.

In measuring electric power, neither the current in amperes nor the pressure in volts tells us the amount of power in a circuit at any moment. A combination of the two does tell us the answer very simply, because volts × amperes = watts. Watts equal the total power of a circuit at any given moment just as horsepower equals the power developed by an engine at any given moment. Indeed, horsepower and watts are merely two different ways of measuring or expressing the rate of work or the power; 746 watts equal 1 horsepower.

If a lamp consumes 746 watts, you would be technically correct if you called it a 1-hp lamp, although that method of designating lamps is never used. Similarly you would be correct in saying that a 1-hp motor is a 746-watt motor, although that method of designating power of motors is not used (except in the case of "flea power" motors, which are sometimes rated, for example, "approximately 1½ watts output" instead of being rated as "¹/₅₀₀ hp").

Note, however, that a motor that delivers 1 hp or 746 watts of power actually consumes closer to 1100 watts from the power line. The difference between the 1100 watts consumed and the 746 watts delivered as useful power is consumed through such factors as heat in the motor, overcoming bearing friction, and air resistance of the moving parts. In addition, some of the 1100 watts is "apparent power" that does not involve overcoming whatever mechanical resistance exists to shaft rotation. This portion of the current is real in that the electrical system must allow for it, but it does no useful work, existing only to magnetize the internal parts of the motor so it will turn as expected. Since wiring must accommodate actual current, the *NEC* and this book refer when appropriate to a parallel term "volt-amperes" (VA). This is the product of actual current and voltage, and while accurate for predicting current flow, it overstates actual power (and energy) consumed. The 1-hp motor just mentioned, if fully loaded for an hour, will add about 1100 VA to the electrical system, but a conventional electric meter would record only about 800 watts of energy consumed in the same period. The magnetizing current is returned to the line instead of being converted into mechanical energy. Refer to the discussion of "power factor" in the next chapter.

Again note that the electrical term is simply "watts" (or volt-amperes), not "watts per hour." You would not say the engine in your automobile delivers 300 hp per h; rather at any given moment it delivers 300 hp. Both watts and horsepower denote a rate at which work is being done at a particular moment, not a quantity of work being done during a given time. Many people use the term "wattage" for the amount of power in watts; this is not a formal term but is generally considered acceptable for informal usage. From the preceding you can see that a given amount of power in watts may be obtained from a combination of various voltage and current values. For example:

> 3 volts × 120 amps = 360 watts
>
> 6 volts × 60 amps = 360 watts
>
> 12 volts × 30 amps = 360 watts
>
> 60 volts × 6 amps = 360 watts
>
> 120 volts × 3 amps = 360 watts
>
> 360 volts × 1 amp = 360 watts

Carrying the illustration further, a lamp in an automobile headlight consuming 5 amps from a 12-volt battery consumes a total of 5 × 12, or 60 watts; a lamp consuming ½ amp from a 120-volt lighting circuit in a home consumes a total of ½ × 120—also 60 watts. The voltage and the current values differ widely, but the power in watts of the two lamps is the same.

This simple formula is not correct under all circumstances; the exceptions are covered in a later chapter.

Kilo-, mega-, milli-, micro-   A watt is a very small amount of power; it is only ¹⁄746 hp. As given in Table 2–1, the term "kilo-" means thousand, so when we say "1 kilowatt," it is just another way of saying "1000 watts." The abbreviation is kW, and 25 kW then means 25 000 watts. And 20 kilovolts (kV) means 20 000 volts, and so on.

Practical examples of the use of other prefixes defined in Table 2–1 are:

- *Mega-:* a million. Examples: megawatts, megacycles. Thus 400 mega-watts means 400 000 kW or 400 000 000 watts. The abbreviation for mega is the capital letter M.

- *Milli-:* one-thousandth. Examples: milliamperes, milliwatts. Thus 25 milliamperes means ²⁵⁄1000 amp. The abbreviation for milli is the lowercase letter m.

- *Micro-:* one-millionth. Examples: microvolts, microamperes, micro-seconds. Thus 25 microamperes means ²⁵⁄1 000 000 amp. The abbreviation for micro is the lowercase Greek letter mu (μ).

Watthours   The watt merely indicates the total amount of electric power being delivered at a given moment; it tells us nothing about the total quantity of electric power delivered over a given period. Compare this to an hourly wage earner: The fact that someone earns $15 per hr tells us nothing about that person's earnings per year unless we know how many hours are worked. Multiplying the watts by the number of hours during which this number of watts was being delivered gives us watthours, abbreviated Wh, which equals the total amount of electric power consumed during a given time. This is the amount of energy consumed. This is also the electrical equivalent of a basic concept in physics, namely that power is the rate of consumption of energy, and conversely that energy is power multiplied by the amount of time it is used. For example:

$$
\begin{aligned}
10 \text{ watts} \times 1000 \text{ h} &= 10\ 000 \text{ Wh} \\
100 \text{ watts} \times 100 \text{ h} &= 10\ 000 \text{ Wh} \\
1\ 000 \text{ watts} \times 10 \text{ h} &= 10\ 000 \text{ Wh} \\
5\ 000 \text{ watts} \times 2 \text{ h} &= 10\ 000 \text{ Wh} \\
20\ 000 \text{ watts} \times \tfrac{1}{2} \text{ h} &= 10\ 000 \text{ Wh}
\end{aligned}
$$

Kilowatthours   One kilowatthour (kWh) is 1000 Wh, 20 kWh is 20 000 Wh, etc. Electric utilities sell energy; you are, in essence, paying for electrical units of energy measured in kilowatthours. The rate schedule of a power supplier may also include a "demand factor" that allows an added charge for maximum power usage (kilowatts) under certain conditions in addition to the energy charge (kilowatthours), but the basic rate is based on the kilowatthours consumed. A demand factor is rarely involved in residential or farm installations.

READING METERS AND PAYING FOR ELECTRICITY

A meter that has a simple scale is easy to read and needs no explanation. The meter in Fig. 2–1 is a voltmeter, used to measure the potential difference in volts between two conductors or between two points. Other meters having simple scales are an ammeter, used to measure the amount of current in amperes, and an ohmmeter, used to measure the resistance in ohms of a conductor, resistor, etc. Combination meters having more than one scale, such as a combination voltmeter, ammeter, and ohmmeter, are more difficult to read than a meter with a single scale. Careful examination of the instrument and care in following the manufacturer's instructions are all that is necessary to enable you to read each of the scales properly. Always take care when using any meter to avoid damaging it by an incorrect connection. Never use a dc meter on an ac circuit, or vice versa. However, some combination meters permit using at least some of the scales on either ac or dc circuits.

A dc meter will have one terminal marked "pos." or "+," the other marked "neg." or "—." If you connect the terminals of a dc meter the wrong way, the needle of the meter will press against the zero stop peg instead of indicating a reading; if that happens just reverse the connections.

A voltmeter is always connected across a circuit or points on a circuit (that is, from one conductor to another, or from one conductor to ground) to determine the circuit voltage; or from one point on a conductor or point of a circuit to another point to determine the voltage drop between the two points. This must be done while the circuit is energized or "hot."

An ordinary ammeter is always connected in series with the load so that the entire current flows through the meter. Always turn off the circuit while making such a connection.

Fig. 2–1 Meters of this type are easy to read.

A special kind of ac ammeter is the clip-on type; two are shown in Fig. 2–2. Each has a movable member or clip that is clamped around any continuous wire carrying ac current; the meter then shows the number of amperes flowing in that wire. Each meter is also supplied with a pair of wire leads, permitting the meter to be used as an ac voltmeter also.

Although usually not quite as accurate as an ordinary ac ammeter (which requires you to disconnect a wire and connect the meter in series with it), the clip-on meter is very useful, especially in troubleshooting.

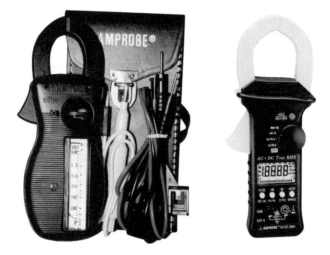

Fig. 2–2 Clip-on ammeters. They also have voltage scales. The one at the right can be used on ac circuits of widely differing frequencies, and even with dc circuits. *(Amprobe)*

Always be alert and careful when using any meter. It must be connected only to a circuit within the voltage and current range for which it was designed. The proper connections and proper setting of the selector switch must always be made on any combination meter.

Reading kilowatthour meters   Some kilowatthour meters are more difficult to read than others. Some are equipped with an easy-to-read cyclometer-type dial as shown in Fig. 2–3, and the very newest meters are all electronic, with direct numerical readouts that typically cycle between current demand and other values in addition to showing the energy usage. Others have register-type dials as in Fig. 2–4. The dials of this meter are shown enlarged to approximately full size in Figs. 2–5 and 2–6. (Four dials are shown, but you may have five dials on your meter.) Reading the four dials on the meter gives you a four-digit number indicating kilowatthours. To determine how much power has been consumed during any given period such as a month, read the dials at the beginning and at the end of that time span. Subtract the reading obtained at the beginning from the later reading to obtain the kilowatthours used.

In Figs. 2–5 and 2–6, note that the pointers on adjacent dials move in opposite directions as indicated by the numbering on the dials. Read the dials from left to right, writing down the number that each pointer has passed, as on a clock. The total of the meter in Fig. 2–5 is 1642 kWh.

◀ **Fig. 2-3** This type of register is found on some kilowatthour meters. *(Osaki Electric Co. Ltd.)*

Fig. 2-4 Ordinary meters have this type of register. ▶

Figure 2–6 demonstrates why you might find it easier to read the meter dials from right to left instead. The pointer of the third dial appears to point directly at 7. Before writing down 7, look at the pointer on the dial to the right: it has not quite reached the zero. Therefore, even though the pointer on the third dial seems to point directly to 7, it has not actually reached the 7, so write down 6 instead. The total reading is 2269 kWh.

The meter represented in Fig. 2–5 shows the position of the dial pointers at the beginning of the month (or other reading period). The same meter is shown in Fig. 2–6 at the end of the month. Subtract the first reading of 1642 from the second reading of 2269 to arrive at 627 kWh consumed during the month.

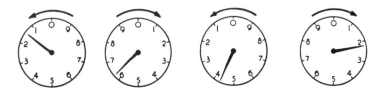

Fig. 2–5 Enlarged register of a kilowatthour meter. Read the figures that the pointer has passed. The reading above is 1642 kWh.

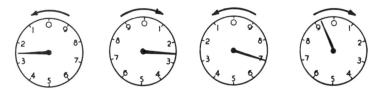

Fig. 2–6 The meter now reads 2269 kWh.

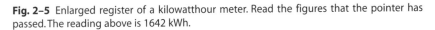

Where a large amount of power is consumed, the kilowatthour meter will have a multiplier marked on the face of the meter, such as "×10" or "Multiply by 10" or "×100." In such cases the reading obtained from the dials must be multiplied by the specified multiplier to determine the actual kilowatthours consumed.

How electricity is priced   What the customer pays for electricity varies from one locality to another depending on how much it costs the power supplier to produce and deliver the electricity. Fuel is needed to produce electricity; the type of fuel used (coal, hydro, nuclear, etc.) and its availability are major production factors affecting the cost of electricity in a given locality. Delivery cost factors include the distances and types of terrain over which the electricity is transmitted, and distribution costs such as whether lines are underground or overhead and whether the end users are in a densely populated area or spread out in a rural area.

In many cases, the commercial rate is higher than the residential rate. Usually there is a step-down rate for all users: a basic rate applies to a specific initial range of kilowatthours consumed, with increasingly lower rates for higher ranges, until the minimum rate is reached for all power over a specified amount. This policy is being rethought in many areas with tight electrical supplies. Some rate structures have higher unit costs for larger usage blocks to promote conservation and in an effort to protect average consumers from the real costs of local power generation and delivery. Table 2–3 represents a typical rate schedule, though the step-down (or up) points and the rate shown at each level will vary from one locality to another. The table will be too high in some areas with abundant hydroelectric or other inexpensive power, and far too low in others. Consider it only in terms of its calculation principles.

Table 2–3 TYPICAL ELECTRIC RATES*

First 100 kWh used per month	10.2¢ per kWh
Next 100 kWh used per month	9.1¢ per kWh
All over 200 kWh used per month	8.0¢ per kWh

Assuming you used 661 kWh during a month, your bill would be figured this way:

100 kWh at 10.2¢	$10.20
100 kWh at 9.1¢	9.10
461 kWh at 8.0¢	36.88
661 kWh total	$56.18
Average per kilowatthour	8.5 cents

*The figures above are typical of rates as this is being written. Owing to inflation, it is probable that rates will be increased in coming years. Further adding to the cost of electric power are fuel cost adjustments authorized by some rate-selling bodies, and basic monthly service charges independent of usage.

Deregulation   The electric power industry is undergoing a major transformation as technology allows the business of generating electricity to be separated from the business of delivering electricity. Customers increasingly have the freedom to choose which electricity producer they will do business with, though the power is still "wheeled" (the industry term) to their location by their local electricity distributor.

These remote suppliers may have different rate structures based on season and time of day, stimulating customers' interest in tracking their own usage patterns (through meter readings) in order to choose a supplier whose rate profile offers a cost savings. Increasingly, electricity generation is becoming a market-oriented business instead of a local monopoly. But as long as one utility pole supports only one set of power conductors, local distribution of electricity will continue to be what economists call a "natural monopoly." As such, it still attracts full government regulation of both costs and service standards.

HOW WE USE WIRES TO CARRY ELECTRICITY

If two wires are used to connect a 1½-volt lamp to the terminals of a dry-cell battery, as shown in Fig. 2–7, the lamp will light, thus indicating that electric current is flowing through the lamp filament. Since the current can reach the lamp only through the wires, there is also current flowing in the wires. Yet there is no current through the wax or other material that was poured around the battery terminals at the top of the cell, nor is there a current through the paper or other material wrapped around the cell.

Conductors and insulators   If a material will conduct an electric current, it is called a "conductor." If a material will not conduct an electric current, it is called an "insulator." Insulating materials are therefore used as insulation for conductors by enclosing conductors, such as wires, in a layer of the insulation. (The thickness of the layer depends on the kind of insulation, the purpose of the wire, and the voltage for which the wire was designed.) The insulation thus confines the electric current to the conductor and prevents it from spreading to unwanted areas. Although there are neither perfect conductors nor perfect insulators, insulating materials are available that will effectively insulate conductors.

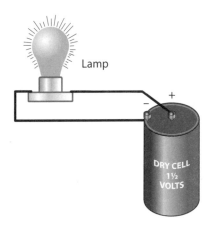

Lamp

Fig. 2–7 An ordinary dry-cell battery, no matter how large or small, develops 1½ volts when new.

Resistance of wires   If the ends of a 1000-ft length of 10 AWG copper wire are connected to a 1-volt source of electricity, 1 amp will flow through the wire, because 1000 ft of 10 AWG copper wire has a resistance of 1 ohm (abbreviated Ω, the Greek capital letter omega). If a 1000-ft length of 10 AWG aluminum wire is connected across the same 1-volt source of supply, only 6/10 amp will flow, because aluminum wire of the same length and diameter has a higher resistance than copper wire. If the aluminum wire is shortened to 600 ft, 1 amp will flow again because a 600-ft length of it has a resistance of 1 ohm. Only about 1/6 amp will flow through a 1000-ft length of iron wire of the same size, but 1 amp will flow again if the iron wire is shortened to about 167 ft; a 167-ft length of it has a resistance of 1 ohm. (The exact resistance of all types of wire depends on the chemical purity of the metal, the temperature, etc. Hence the figures above will not be precise in all cases.) This experiment shows that some electric conductors have lower resistance than others, and that the ones with the lowest resistance are the best conductors. Aluminum, for example, has a current-carrying capacity of about 78% that of copper; so if aluminum wires are used, they must be larger than copper wires for the same amount of current. Of course, the load supplied—lights, motors, etc.—imposes far greater resistance in the circuit than the conductors supplying the load. Nevertheless, it is important that the resistance of the conductors be held to a minimum.

The safe current-carrying capacity (ampacity) of different sizes and kinds of wire are further discussed in Chapters 4 and 7. All references to wire in this book are to copper wire unless otherwise indicated.

OHM'S LAW: HOW CURRENT, VOLTAGE, AND RESISTANCE INTERRELATE

As already explained, 1 volt will force 1 amp through a conductor having a resistance of 1 ohm. Additional experiments will show that if the voltage is increased to 2 volts, the current will be 2 amps through 1 ohm. If the voltage is increased to 5 volts, the current will be 5 amps through 1 ohm. Additional experiments will also show that if the resistance is reduced by half, to 1/2 ohm, the current will be doubled to 2 amps if the voltage remains unchanged at 1 volt.

Doubling the cross-sectional area of a conductor of any given length will reduce the resistance of the conductor by half. An increase in conductor size (which reduces its resistance) will not reduce the resistance of the connected load. But the point to remember is that if all other factors remain constant, an increase in current will be directly proportional to an increase in voltage and inversely proportional to an increase in resistance. This is the principle of Ohm's law, which is one of the most basic laws of electricity. (The term "directly proportional" means that one factor will be increased in proportion to an increase in another factor. "Inversely proportional" means that one factor will be decreased in proportion to an increase in another factor, and increased in proportion to a decrease of the other factor. Thus the current will increase in proportion to an increase in voltage, but the current will decrease in proportion to an increase in resistance and vice versa.)

Ohm's law formulas   Ohm's law can be expressed in simple formulas, but first you must understand some abbreviations: *I, E, R*.

I = current. I is a sort of shorthand symbol meaning that we don't know at the moment how many amperes are flowing in the circuit, and until we know, we'll just call it *I*. But when we know what the current is, we say 10 amps, 200 amps, etc.

E = voltage. E is used for voltage until we know the specific number of volts involved; then we simply say 12 volts, 120 volts, etc.

R = resistance. R is used for resistance until we know the specific number of ohms involved; then we say 15 ohms, 3.5 ohms, etc.

The basic formula for Ohm's law is

$$Number\ of\ amperes\ in\ circuit = \frac{voltage\ of\ circuit}{number\ of\ ohms\ in\ circuit}$$

By substituting the basic abbreviations discussed, that formula becomes the much simpler formula $I = E/R$.

In the above formula, the current is the unknown factor that can be determined by dividing the known voltage by the known resistance. If the voltage and the current are known, the resistance can be found by changing the formula to $R = E/I$. If the resistance and current are known, the voltage can be found by changing the formula to $E = IR$.

There are also several formulas for determining the power in watts (abbreviated W), as follows: $W = EI$, or $W = I^2R$, or $W = E^2/R$. These and other related formulas are shown in Table 2–4.

Table 2–4 OHM'S LAW AND OTHER FORMULAS (FOR TWO-WIRE CIRCUITS)	
If circuit is dc, or ac having ohmic resistance only:	If circuit is ac and has both ohmic resistance and reactance:
$E = IR$ or $\dfrac{W}{I}$ or $\sqrt{WR}$	$E = IZ$
$I = \dfrac{E}{R}$ or $\dfrac{W}{E}$ or $\sqrt{\dfrac{W}{R}}$	$I = \dfrac{E}{Z}$
$R = \dfrac{E}{I}$ or $\dfrac{W}{I^2}$ or $\dfrac{E^2}{W}$	$Z = \sqrt{R^2 + X^2}$
$W = EI$ or I^2R or $\dfrac{E^2}{R}$	
In formulas above:	
E = voltage in volts	W = power in watts
I = current in amperes	Z = impedance in ohms
R = resistance in ohms	X = reactance in ohms

Corrections for ac circuits   The formulas shown for Ohm's law are always correct for dc circuits, and for ac circuits containing only what is known as "ohmic resistance," which is the kind of resistance imposed by incandescent lamps, resistance-type heating elements such as toasters, ranges, water heaters, etc., or a wire. But in most ac circuits there is an additional deterrent known as reactance (abbreviated X) also measured in ohms. While there is some degree of reactance in all ac circuits, it becomes significant mostly in equipment involving windings of wire on a steel core (motors, transformers, fluorescent lighting ballasts, electromagnets, etc.). In any piece of equipment the reactance must be combined with the ohmic resistance to determine the impedance (abbreviated Z) also measured in ohms, thus: $Z = \sqrt{R^2 + X^2}$. When using Ohm's law formulas, for example, to determine voltage drop, if impedance is involved, simply substitute Z for R in the formula. A complete explanation of reactance and impedance is beyond the scope of this book; study any good book on electrical engineering.

Factors apply to same part of circuit   In using these or other formulas as shown in Table 2–4, it is essential that you remember this: All factors must be applied to all of the circuit, or to the same part of the circuit. In other words, you cannot divide the voltage of an entire circuit by the resistance of one part of the circuit and find the current in some other part of the circuit.

Volt-amperes   A very important fact about ac circuits concerns power. Although EI (voltage × current) or VA (volts × amperes, as usually shown in power formulas) equal watts in dc circuits, this is not true with ac circuits if reactance is involved. Just about all ac circuits have some reactance, but in residential and farm wiring it is usually so low that it can be ignored. But in commercial and industrial installations, often the reactance is high enough that it must be considered. In ac circuits, volts × amperes simply equals volt-amperes (VA), which is also known as "apparent power." To determine watts (actual power) in ac circuits, the volt-amperes (VA) must be multiplied by the power factor (pf) of the circuit. The pf may be low (60% to 70% or lower), average (80% to 90%), or high (above 90%). If the pf is 100%, it is equivalent to a dc circuit.

Power factor   The power factor (pf) can be found by dividing the actual power in watts (as measured by a wattmeter) by the apparent power in volt-amperes (VA) (as measured by a voltmeter and an ammeter). The formula is $pf = W/VA$ or $pf = kW/kVA$. For three-phase circuits, the phase-to-phase voltage in formulas must be multiplied by 1.732, which is the square root of 3. (These terms are explained further in Chapter 3.)

Percent efficiency (% eff)   The percent efficiency is usually marked on the nameplate of motors, transformers, etc. The formula for finding the percent efficiency is

$$\% \ eff = \frac{output\ power}{input\ power}$$

VOLTAGE DROP: A PRICE TO BE PAID FOR USING ELECTRIC POWER

All conductors have resistance, and it takes power to force current through them. An example is two very long wires connecting a motor to a source of electricity. If you connect a voltmeter directly to the two terminals at the circuit breaker or fuse location, as shown in Fig. 2–8, the meter will probably read 120 volts (if it is a 120-volt circuit). But if you connect the same meter to the motor terminals, the meter will indicate a lower voltage, possibly 114 volts, depending on the length and size of the wires. The difference between 120 and 114 volts indicates a 6-volt or 5% drop between the two points, which is a result of the power consumed in heating the wires as a result of the resistance.

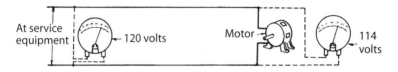

Fig. 2–8 This circuit illustrates voltage drop.

This means that less power is available to operate the motor; when operated at 95% of its rated voltage it will deliver only about 90% as much power as it would if operated at its rated voltage; if operated at 90% of its rated voltage it will deliver only about 81% of its normal power. An incandescent lamp burned at 95% of its rated voltage will produce only about 83% of its normal light; if burned at 90% of its rated voltage it will produce about 68% of its normal light. Voltage drop therefore not only leads to power wasted in heating the wires, but also seriously affects the operation of electrical equipment.

Calculating voltage drop and power loss   Voltage drop and power loss can be computed if the resistance of the conductors in ohms and the current in amperes are known. As shown in Table 2–4, the power in watts equals I^2R. Since reactive ac power heats as much as actual power (watts), the heating effect for both ac and dc circuits equals I^2R without multiplying by the power factor for ac circuits. As an example, if the current in the wires is 7½ amps and the resistance of the wires is 2 ohms, the power consumed in heating the wires will be I^2R, or 7½ (amps) × 7½ (amps) × 2 (ohms), or 112.5 VA (or watts for dc circuits). By Ohm's law $E = IR$, so the voltage drop in the wires equals 7½ (amps) × 2 (ohms), or 15 volts.

But suppose the current should be doubled to 15 amps, using the same size wire at the same voltage. The voltage drop would then be $IR = 15 \times 2$, or 30 volts drop. Doubling the amperes doubled the voltage drop. But will the power loss also be doubled? Let's find out. $I^2R = 15 \times 15 \times 2$, or 450 VA. The power loss is therefore four times greater! Hence, the power loss for both ac and dc circuits is directly proportional to the square of the current in amperes. So if you double the ampere load on the same conductors at the same voltage, your power loss will be four times greater; and if you triple the ampere load, the power loss will be nine times greater, etc. Of course, it takes some power to force current through any conductor, but wasted power can be held to a minimum. For efficient operation, voltage drop must be held to a minimum. This can usually be done by using a wire size suitable for the current and distance involved. All this is explained in later chapters.

Operating voltage   Previous paragraphs explained the fact that the greater the current in a wire, the greater the voltage drop and the greater the power loss in the form of heat. From this it should be obvious that to carry high current without undue loss, large sizes of wire are required.

The greater the distance, the larger the wire must be. Therefore it is distinctly advantageous to keep the current as low as practical.

This, at least in theory, is simple. Any given load in watts may consist of a low voltage with high current, or of a high voltage with low current. Therefore for low current, relatively high voltages must be used.

In practice, the actual voltage depends on the amount of power to be transmitted and the distance. In an automobile, while the current is very great at times, a battery of only 12 volts is used even though the current flowing through the starting motor when it is cranking the engine is often over 250 amps (3000 watts); this is practical only because the distance is so short.

For ordinary residential lighting the voltage is usually 120 volts, but for ranges and water heaters it is 240 volts. For industrial purposes, where the power demands are great, 480 volts are usually used. The distribution lines that run down city alleys and from farm to farm are usually 2400 to 13 800 volts, but the main distribution lines are at still higher voltages until, for long-distance cross-country distribution, the voltages are usually 345 000 volts or more. In many foreign countries, especially in Europe, the lowest utilization voltage is 220, in contrast to 120 volts in the western hemisphere.

Since it is advantageous to keep the current as low as possible to reduce voltage drop and power losses in the wires and avoid the necessity of buying large-size wire when a smaller size will do, and since this can be done by making the voltage higher, it would appear entirely logical to use high voltages for all purposes. You might well ask why not use, for ordinary house wiring, 240 volts or 500 volts or higher.

First of all, the higher voltages require heavier insulation because they are more dangerous in case of accidental contact. In Europe and other parts of the world that use 220-volt systems where we use 120 volts, additional safety provisions are required due to the higher voltage hazard. Another important consideration is that in the manufacture of devices consuming relatively low power (under 100 watts), the wire used inside the device is often of almost microscopic dimensions, even if the device is for a voltage as low as 120 volts. For example, the tungsten wire in the filament of a 60-watt, 120-volt lamp as manufactured today has a diameter of only 0.0018 in.; in a 3-watt lamp it is about 0.000 33 in. (3000 such filaments laid side by side would be required to cover a space of 1 in.). If the device were for 240 volts or for an even higher voltage, the wire would have to be still smaller, making factory production and uniformity decidedly difficult. The device would also be more fragile, and it would burn out more easily. The present common level of 120 volts is a compromise for lowest overall cost of installation, operation, and purchase of devices to be operated.

However, since the same home that has small devices consuming from 5 to 1000 watts also has appliances like electric ranges that may consume over 10 000 watts, it would be desirable to have available two different voltages—one relatively low for lighting and small appliances, and one relatively high for large appliances. Fortunately, this is practical.

THREE-WIRE SYSTEMS: TWO VOLTAGES FOR THE PRICE OF ONE

The three-wire system in common use in homes today provides both 120 and 240 volts. Only three incoming wires are used and only a single meter. The three-wire 120/240-volt system constitutes the ordinary system as installed in practically all houses and farms and many other installations. The higher voltage is usually used for any single appliance consuming 1800 watts or more.

Figure 2–9 shows two generators, each delivering 120 volts; the two combined deliver 240 volts. (Two generators are not actually used. This will become clear in the discussion on transformers in the next chapter.) Any load connected either to wires *A* and *B* or to wires *B* and *C* will be connected to 120 volts. Any load connected to wires *A* and *C* will be connected to 240 volts. In actual wiring the central or neutral wire *B* is white; the outer two or "hot" wires are black or some other color, but never white or green. Connect any device operating on 120 volts to one black and one white wire, and any device operating on 240 volts to the two black wires. (These colors are correct only on grounded neutral systems as explained in Chapter 9 on grounding. Practically all installations today have a grounded neutral.)

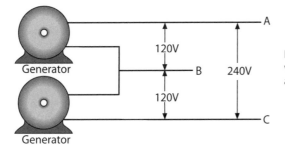

Fig. 2–9 With only three wires, two separate voltages are available.

CHAPTER 3
AC and DC; Power Factor; Transformers

YOU WILL FREQUENTLY ENCOUNTER the terms "direct current," "alternating current cycles," "cycles per second" (or "hertz," abbreviated Hz), "single phase," "two phase," "three phase," and "polyphase." This chapter provides basic explanations for these and other important terms. You will also find several brief discussions of topics that are largely beyond the scope of this book to cover in detail, but serve here as introductions to some important contemporary issues in electrical production and distribution.

DIRECT CURRENT, ALTERNATING CURRENT, AND NONLINEAR LOADS

Electric current is either direct or alternating. Definitions of these and other related terms basic to electrical work are provided here. The discussions of power factor and nonlinear loads are offered simply as introductions to important subjects that can be further explored through other resources.

Direct current (dc)   If an ordinary direct-current voltmeter is connected to a battery, the pointer will swing either to the right or to the left, depending on how the two terminals on the meter are connected to the corresponding terminals of the battery. The two terminals of the meter are marked "+" and "—," "P" and "N," or "pos." and "neg.," indicating positive and negative; the battery terminals are similarly marked. Only when the positive terminal of the meter is connected to the positive terminal of the battery, and the negative terminal of the meter is connected to the negative terminal of the battery, will the meter pointer swing in the right direction. For any source of electricity, whether battery, generator, or other apparatus, if one terminal is positive and the other is negative and they never change, the current is known as "direct current," or "dc." Current from any type of battery is always direct current.

Instead of an ordinary voltmeter with the zero at one end of the scale, a zero-center voltmeter of the type shown in Fig. 3–1 may be used. This meter is the same as that described in the previous paragraph except the terminals are not marked positive and negative. Connect the terminals of this meter to the two terminals of a battery and note which way the needle swings. Then reverse the two battery leads and the needle will swing in the opposite direction. The meter is equally easy to read whether the pointer swings to the right or left. And it provides the additional convenience that it is not necessary, before connection is made, to investigate carefully which is the positive and which is the negative terminal.

Fig. 3-1 This voltmeter is the same type as shown in Fig. 2-1 except that the zero is in the center of the scale.

Alternating current (ac)   Suppose, as a mental exercise, we take this zero-center voltmeter and connect its two terminals to a source of electricity, the type of current being unknown. Suppose the pointer of the voltmeter performs in a peculiar fashion. It never comes to rest, but keeps swinging from one end of the scale to the other with great regularity. Watch that pointer carefully, starting from zero in the center. It starts swinging toward the right, first rapidly, then more slowly, until it reaches a maximum of about 170 volts in exactly 15 seconds. Then it starts dropping back toward 0, first slowly, then rapidly, until in 15 seconds more it is back at 0. It does not stay there but continues swinging toward the left, and in 15 seconds more it reaches the extreme left at 170 volts, the same relative position as it originally had at the right. Again it swings back toward the right, and in 15 seconds more, 1 minute from the starting point, it is back where it started at 0. It repeats this pattern every minute indefinitely. From observing the pointer it is evident that each wire is first positive, then negative, then positive, then negative, alternating between positive and negative continuously. The voltage is never constant, is always changing from 0 to a maximum of 170 volts, first positive, then negative. Current in which any given wire regularly changes from positive to negative, not suddenly but gradually as described in this exercise, is known as "alternating current," or "ac." If the actual voltage is plotted against time, it will produce a chart such as shown in Fig. 3–2. This portrays one cycle of alternating current.

Alternating current may be defined as a current of regularly fluctuating voltage and regularly reversing polarity.

The majority of electric power in commercial use is produced in rotating generators that spin a coil of wire (armature) through a stationary magnetic field. At one point in the rotation the coil is moving parallel to the lines of magnetic flux. At this point it isn't cutting any lines of flux, and the voltage induced is zero. As the prime mover (which could be a water turbine, or a steam turbine powered by oil, coal, or nuclear power plant, or an internal combustion engine) keeps the coil turning, it becomes perpendicular to the lines of flux, cutting them rapidly. At this point, the voltage is at its maximum. The effect of this circular motion is to describe a trigonometric function where the voltage is proportional to the sine of the degrees of angular movement in the armature.

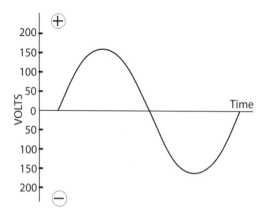

This book, intentionally, does not require an understanding of trigonometry. However, the term "sine wave" with respect to current, voltage, and power quality discussions is so widely used that you should be aware of where it comes from. If Fig. 3–2 were

Fig. 3–2 This shows one cycle of alternating current. The voltage fluctuates regularly and continuously from zero to maximum to zero, and each wire alternates regularly between positive and negative.

plotted from a 120-volt household generator (as we've just seen, the maximum voltage "e_m" would be 170 volts), it would show the instantaneous voltages (e_i) resulting from one complete revolution of the armature. The armature would turn through a complete circle, and you can easily imagine a rotation angle (θ) starting at 0 deg, passing through 90 deg at the top of the curve, 180 deg at the middle where the voltage returns to zero, 270 deg at the maximum negative value, and then finishing at zero once again at 360 deg. In this case the sine of the angle θ is the ratio of instantaneous voltage to maximum voltage. Therefore, based on the general formula $e_i = e_m \sin \theta$, in this case $e_I = 170 \sin \theta$. When you hear about sine waves and electricity, this is what's being discussed.

Note that the sine of 45 deg is 0.707, or $\frac{1}{2}\sqrt{2}$, and 170 volts $\times$ 0.707 = 120 volts. By using calculus, we can determine that this is the value that produces the same amount of heat loss through a resistance as a dc circuit of equivalent voltage, which is defined as the "effective voltage."[1] For example, an ac circuit with a maximum voltage of 170 delivers the same power as an unvarying 120-volt dc circuit.

Usually these discussions go the other direction, but it's easier to understand the first time if we start with the maximum voltage. Now you can see, for example, that a 120-volt ac circuit has a peak voltage of 170 volts. It's called a 120-volt circuit because that is what it is equivalent to if it were connected to a battery. The maximum voltage of any ac circuit is its effective voltage multiplied by $\sqrt{2}$. The rest of this book uses ac effective voltages unless specifically stated otherwise. A

---

1. This number is also the square root of the average square of all heating effects, leading to the synonymous term "root-mean-square" voltage (and current), usually abbreviated "rms." This is another term you'll encounter very frequently.

120-volt ac source will light an incandescent 120-volt lamp (a purely resistive load) to the same brilliancy as a 120-volt dc source.

Frequency   Alternating current that takes a full minute to go through the entire cycle (from no voltage to maximum voltage on the positive side, back to 0, to maximum voltage on the negative side, back to 0) would have a frequency of 1 cycle per minute. There is no such current in actual use. The ordinary 60-cycle ac used in the United States goes through the changes described above at the rate of 60 times per second, much too fast to be observed by an ordinary voltmeter. Such current has a frequency of 60 cycles, the "per second" being understood. The correct terminology for regularly repeating phenomena like this is the hertz, abbreviated Hz (named after the 19th-century German scientist who discovered the cyclical nature of electrical phenomena). Thus, we refer to the alternating current just described as "60 Hz."

In the curve of Fig. 3–2 the voltages range between 0 and 170 volts. If a 120-volt lamp is connected to an ac circuit of 1 cycle per min, it will burn far more brightly than normal while the voltage is above 120 volts, less brightly than normal while under 120, and not at all a part of the time because the voltage is very low, even zero twice during the cycle. Flickering would be extreme and unendurable. However, in the case of the ordinary 60 Hz, all this change of voltage takes place twice per cycle, 120 times every second. The filament of a lamp does not have time to cool off during the very short periods when no voltage is impressed on it, which is the reason for the lack of observable flicker. Very small lamps with very thin filaments that can cool off quickly will have an annoying flicker if operated on 25 Hz.

In the United States all standard commercial current is 60 Hz, as it is in Canada and most of Mexico. In other foreign countries most installations are 50 Hz, but there are many at 60 Hz, with the trend toward 60 Hz. In the United States 180 Hz ac is in use for equipment in a few industries, and fluorescent lighting installations operating at 400-Hz ac are in experimental use. Military equipment often operates at 400-Hz ac. The advantage of the higher frequencies is that motors, transformers, and similar equipment can be smaller in physical size as the frequency increases.

Remember that "kilo-" means "thousand." When your radio receiver is tuned to a station operating at 1250 kilocycles, it means that the signal coming into the receiver is 1 250 000 (1.25×10^6) Hz. If the receiver is tuned to an FM station operating at 90 megacycles ("mega" has been adopted to designate "millions"), it means that the signal is 90 000 000 (9×10^7) Hz (90 MHz). Recalling the brief discussion of electric fields in Chapter 2, 60-Hz electric current is simply extremely low frequency electromagnetic energy, occurring at the very lowest end of a very wide spectrum. The 90 MHz FM signal frequency is much higher, but far lower than visible light, which is roughly 500 000 000 000 000 (5×10^{14}) Hz for yellow light, with red light being a little lower and blue a little higher, which in turn is far lower than x-rays, which are on the order of 1 000 000 000 000 000 000 or 1×10^{18} Hz.

Alternating current and motors   Alternating current as discussed up to this point is "single phase" alternating current. (The word "phase" is usually abbreviated ϕ, the Greek letter phi.) Remember that when applied to a motor, alternating current magnetizes the steel poles of the motor every time it builds up from zero to peak voltage, or in other words 120 times per second as shown in Fig. 3–3, which shows three consecutive cycles of 60 Hz. At the top is indicated the time between cycles, or ⅟60 sec. At the bottom is indicated the time between alternations, or ⅟120 sec. You might say that a motor is given a push 120 times a second, just as a gasoline engine is given a push every time there is an explosion in the cylinder. Offhand, 120 times per second may seem fast enough for any purpose, but remember that an ordinary motor runs at 1800 r/min, which means that the rotor (the rotating part) makes 30 revolutions every second. In turn this means that the 120 pushes per second become only 4 pushes per revolution; if the rotor or armature of the motor has a diameter, for example, of 12 in. (circumference over 36 in.), any given point on the rotor has to turn about 9 in. between pushes. These pushes are not abrupt sudden impacts. They are gradual pushes that start slowly and build up to a maximum as the voltage builds up to a maximum value.

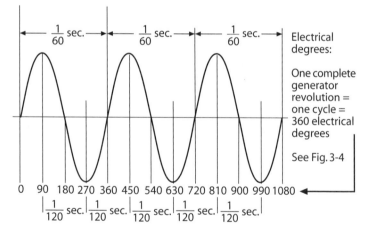

Fig. 3–3 Three cycles of 60-Hz ac.

In an ordinary one-cylinder four-cycle gasoline engine running at 1800 r/min, there is an explosion in the cylinder every other revolution, or 900 times every minute, or 15 times every second. The crankshaft gets a push 15 times every second. If more pushes are needed every second to secure smoother operation or more power, more cylinders are used: two or four or as many as needed. How is this to be done in the case of an electric motor? Use three-phase alternating current as explained in the next section.

Three-phase alternating current   The number of pushes per second on the rotor can be increased by putting into the motor three separate windings not connected to each other in any way, but each connected to one of three separate sources of single-phase alternating current. The three separate sources of current must be designed so the peak voltage of one does not coincide with the peak voltage of another.

The voltages of the three sources reach their peak in regular fashion, one after the other. The motor receives three times as many pushes as before. Figure 3–4 shows the voltage curves of the three separate sources. At the top is indicated the time between cycles in each separate source: ¹⁄₆₀ sec. At the bottom is indicated the time between pushes from the three separate sources combined: ¹⁄₃₆₀ sec. So three-phase alternating current consists simply of three separate sources of single-phase alternating current arranged so the voltage peaks follow each other in a regular, repeating pattern.

Note that for each phase the duration of a cycle is ¹⁄₆₀ sec, but there are two pushes per cycle, so that the time between pushes is ¹⁄₁₂₀ sec. Look at the bottom of the diagram and you will see that the pushes from the three phases combined are

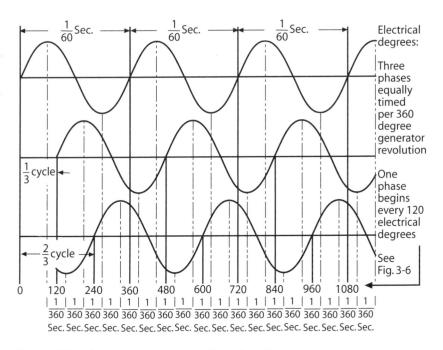

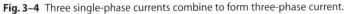

Fig. 3–4 Three single-phase currents combine to form three-phase current.

only $\frac{1}{360}$ sec apart. The pushes are imparted by each of the three windings in turn, as shown by the dashed lines from the peaks to the bottom of the diagram.

Figure 3–5 shows this diagrammatically: generator *A* and (inside the motor) winding *A*, generator *B* and winding *B*, and generator *C* and winding *C*. In practice it would be impossible to make three separate generators run in such precisely uniform fashion that the voltage peaks would come at precisely the right times; it would also be uneconomical. Instead a single generator is used with three separate windings so that the peak and the zero voltage of each winding come at precisely the right times. This is illustrated in Fig. 3–6, which also shows how the six wires in Fig. 3–5 become only three wires in actual practice.

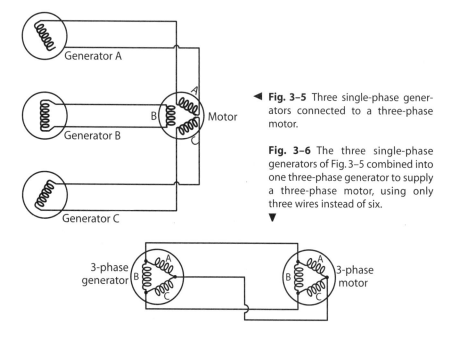

◀ **Fig. 3–5** Three single-phase generators connected to a three-phase motor.

Fig. 3–6 The three single-phase generators of Fig. 3–5 combined into one three-phase generator to supply a three-phase motor, using only three wires instead of six.

▼

Two-phase alternating current   A two-phase system, which is similar in principle but inferior to a three-phase system, is used in so few localities that it does not warrant space in this book.

Polyphase current   Both two- and three-phase currents are called "polyphase" current (from the Greek word *poly* meaning "many"). With a few exceptions, three-phase current is not used in homes, but it is widely used in commercial and industrial establishments for operating motors and similar equipment. Even in those establishments, two-wire circuits are used for lighting and small miscellaneous loads. These two-wire circuits are often referred to as "single phase," but that isn't strictly accurate. They originate in a three-phase distribution, and they behave differently with respect to how they influence the distribution of loads on the conductors feeding them, since they reach their current maximums and minimums at different moments in time.

Volt-amperes   The previous chapter contained the formula *volts × amperes = watts*. That formula is always correct for direct current; but for alternating current it is not correct because of reactance, which results in a power factor (abbreviated pf) of other than 100%. This will be discussed after volt-amperes (abbreviated VA) has been further explained.

For single-phase: *Volt-amperes = volts × amperes.*

For three-phase: *Volt-amperes = 1.732 × volts × amperes.*

In three-phase work you will frequently encounter the number 1.732, which equals $\sqrt{3}$.

Kilovolt-amperes   One thousand volt-amperes is one kilovolt-ampere (abbreviated kVA). If the power factor of the load is 100%, then and only then is one kilovolt-ampere the same as one kilowatt (kW).

Power factor   A detailed explanation of power factor is beyond the scope of this book. However, an explanation of how to measure it and a general idea of its importance are covered here.

For a dc circuit consisting of the load and an ammeter, a voltmeter, and a wattmeter, the product of the volts and the amperes is without exception equal to the reading of the wattmeter. But if this experiment is made with an ac circuit, the same is sometimes true and sometimes it is not.

For an ac circuit, if measurements show that the product of the volts and the amperes is exactly equal to the wattmeter reading, the load is said to have a power factor of 100%. Loads of this type include incandescent lamps, heating appliances of the type that have ordinary resistance-type heating elements, and in general all noninductive loads, that is, equipment that does not include windings or coils of wires, such as transformers, motors, and fluorescent lighting fixture ballasts.

If the product of volts multiplied by amperes is greater than the reading of the wattmeter, then the power factor is less than 100%.

Power factor, which is also referred to as the "cos θ" (cosine of theta) is defined as the ratio of the real or measured watts (also known as "effective power") and the volt-amperes (also known as "apparent watts"). The formula is

$$power\ factor = \frac{watts}{volt\text{-}amperes}$$

Measuring power factor   To measure power factor you need only a voltmeter, ammeter, and wattmeter. Assume a small, single-phase motor on a 120-volt circuit consuming 5 amps as indicated by an ammeter, and 360 watts as indicated by a wattmeter. The calculation is

$$power\ factor = \frac{360}{5 \times 120} \ or \ \frac{360}{600} \ or\ 0.60\ or\ 60\%$$

For a three-phase 240-volt motor consuming 12 amps, the volt-amperes are $1.732 \times 240 \times 12$, or 4988. If the watts as indicated by a wattmeter are 3950,

$$\text{power factor} = \frac{3950}{4988} \text{ or } 0.79 \text{ or } 79\%$$

Generally speaking, the power factor of a motor improves (increases in percentage) with an increase in horsepower of the motor. It also varies considerably with the type and quality of the motor in question. It may be as low as 50% for small fractional-horsepower motors, and over 90% for a 25-hp motor.

Watts in alternating-current work   The formula for ac power in watts is

$$watts = volt\text{-}amperes \times power\ factor$$

For three-phase power, multiply the volts by 1.732.

Desirability of high power factor   Assume that a small factory is using 100 amps at 240 volts, single-phase, or a total of 24 000 VA or 24 kVA. If the power factor is 100%, this is equivalent to 24 kW. At 5 cents per kWh, the power supplier receives $1.20 per h for the power.

Now assume a second factory also using 100 amps at 240 volts, but with a power factor of only 50%. That is still 24 kVA but only 12 kW, and at 5 cents per kWh, the power supplier now receives only 60 cents per h.

Since it is the kilovolt-ampere load that determines wire size, transformer and generator size, and similar factors, and since each factory uses the same 24 kVA, the power supplier must furnish wires just as big for the factory where they are paid 60 cents per h as for the one where they are paid $1.20 per h. They tie up just as much transformer capacity, generator capacity, and all other equipment for the one as they do for the other.

It is natural, therefore, that power suppliers, when furnishing power to establishments where the power factor is low, not only charge for the kilowatthours consumed but also make an extra charge based on the kilovolt-amperes used during the month or period in question, as compared with the kilowatthours used. Because the loss in wasted power decreases as the power factor increases, it is definitely worthwhile to improve the power factor. Few installations attain 100% power factor, and rarely does one fall as low as 50% The overall power factor in an industrial establishment is generally determined by the electric motors in use, although other equipment also contributes its share.

Power-factor correction   The theory of power-factor correction is entirely beyond the scope of this book, but the actual correction is accomplished by means of capacitors or synchronous motors; the required calculations should be made by one thoroughly familiar with the subject. Correcting the power factor not only reduces the charges for power consumed but also has many other advantages, including higher efficiency of electric equipment because of reduced voltage drop, and reduced heating and power loss.

Nonlinear loads   Much of today's electrical loads involve some form of electronic switching. In fact the Electric Power Research Institute (EPRI), which is the research arm of the electric utilities, made a prediction a few years ago that by this time 60% of electric power would be processed electronically, and that estimate may well be low. Much of this equipment uses switching devices that are completely turned off until the voltage wave reaches a predetermined value. Then the device turns on until the voltage wave passes through its maximum and reaches a predetermined minimum, when it turns off again. You can see that a current wave supplying a power supply controlled in this way won't have the same shape as the voltage wave supplying the same load.

The *NEC* defines loads where the wave shape of the steady-state current does not follow the wave shape of the applied voltage as "nonlinear." As in the case of power factor correction, a full explanation of the theory of nonlinear loading is beyond the scope of this book. However, given the dramatic impact electronic controls have on today's society, and the fact that such controls almost invariably involve nonlinear load problems, this general overview is offered as an introduction to the subject.

If you plot a typical nonlinear load, the curve looks hopelessly irregular. However, it has been determined that no matter how irregular the appearance of such a curve, it can always be determined to be an arithmetic sum of various sine (or cosine) curves, as shown in Fig. 3–7. These component curves have two remarkable properties. First, they can be individually plotted, and when plotted they look very similar to Fig. 3–2. Second, although the frequencies differ, every component frequency is an integral multiple of the 60-Hz fundamental. This latter property means they are harmonics of the basic 60-Hz power. For example, fluorescent ballasts generate significant third order harmonics (180 Hz), and variable speed drives with six-pulse rectifiers generate significant fifth and

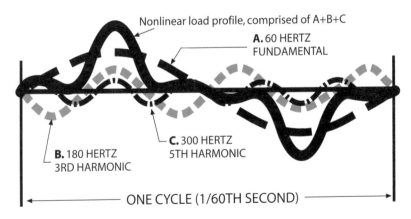

Nonlinear load profile, comprised of A+B+C

A. 60 HERTZ FUNDAMENTAL

B. 180 HERTZ 3RD HARMONIC

C. 300 HERTZ 5TH HARMONIC

ONE CYCLE (1/60TH SECOND)

Fig. 3–7 This irregular curve results from a summation of harmonic currents associated with a nonlinear load.

seventh order (300 Hz and 420 Hz) harmonics. Further, not all electric metering devices used in the trade properly measure the higher order harmonic currents. Read the documentation carefully before you decide which to buy.

When engineers discuss "harmonic loading," this is what they are talking about. Harmonic currents are currents just the same, and they contribute heat when they pass over electrical wiring and components. Just as in the case of conventional power factor deficiencies, an electrical system configured only in terms of conventional electric power (watts) may be inadequate to carry the resulting loads. And, just as in the case of poor conventional power factor, the serving utility may penalize installations that have a rich harmonic profile, since compensation based on watthours alone doesn't allow them to recover the associated costs.

To make matters worse, some facilities have both problems. Because of phase shifting (current peaks not occurring at the same instant), apparent power associated with an unloaded motor and apparent power resulting from harmonics, both of which add to the fundamental, may not add together. In this world 1 + 1 + 1 might equal 2. That might sound okay, but consider that if you install a power factor correcting capacitor to fix the motor, you might still end up with, for example, 2 when you expected something much closer to 1. A great deal of today's electrical design research money is tied up in addressing these issues.

HOW TRANSFORMERS WORK AND WHAT THEY DO FOR US

Where it is necessary to transmit thousands of kilowatts over a considerable distance, wire large enough to transmit it at 120 or even 240 volts would have to be so big that the cost would be prohibitive. If a relatively small wire and a much higher voltage are used, the voltage would be so high that it would be dangerous to use with ordinary equipment.

It would be convenient, therefore, to have a way of changing from one voltage to another as needed. For direct current there is no simple, efficient means of doing this for everyday needs, but for alternating current there is a simple and efficient means. A transformer changes one ac voltage to a different ac voltage. It has no moving parts. Basically it consists of a core made of many thin sheets of silicon steel, with two coils of wire or windings, as shown in Fig. 3–8. The windings are electrically insulated from each other and from the core. Electric energy is supplied to one winding called the primary winding. The other winding, called the secondary

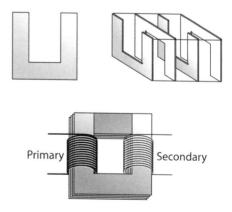

Fig. 3–8 Basic construction of a transformer.

winding, delivers the energy at a different voltage. (For some special purposes, the voltage can be the same in both windings.)

A transformer can be a *step-up* transformer that receives energy at one voltage and delivers it at a higher voltage, or a *step-down* transformer that receives energy at one voltage and delivers it at a lower voltage.

When the primary winding is energized by an ac source (which may be an ac generator or another transformer), an alternating magnetic field, called "flux," is established in the transformer core. The alternating flux surrounds both windings and induces a voltage in both. So once a produced electromotive force (emf) is available, an alternating flux that rapidly fluctuates (rises to maximum and falls to zero 120 times a second for 60-Hz ac) can be used to cut through the windings and induce a voltage in each of them.

Ratio of number of turns to voltage   Since the same flux cuts both windings, the same voltage is induced in each turn of each winding. Hence, the induced voltage is proportional to the total number of turns in each winding. The voltage in the primary will be the same as that received from the source. But the voltage in the secondary will depend on the number of turns in the secondary winding in proportion to the number of turns in the primary winding. If the secondary has the same number of turns as the primary, the voltage will be the same in both windings; if the secondary has twice as many turns as the primary, the secondary voltage will be twice as high as the primary voltage; and if the secondary has half the number of turns as the primary, the secondary voltage will be half as high as the primary voltage, etc. So the formula is

$$E_P / E_S = N_P / N_S$$

where E_P = primary voltage, E_S = secondary voltage, N_P = number of primary turns, and N_S = number of secondary turns. Therefore,

$$(E_P)(N_S) = (N_P)(E_S), \text{ and } [(E_P)(N_S)]/ N_P = E_S, \text{ etc.}$$

Transformer current and loads   Where there is no connected load (where the switch in the secondary conductors is open), a transformer has only a very small no-load current, which is called the "exciting" or "magnetizing" current. The magnetizing current produces the magnetomotive force that produces the transformer-core flux. But when the switch is closed and a secondary load (lamps, motors, etc.) is connected, then just as much current will flow in the primary as is required to deliver the required power to the secondary, but no more (assuming, of course, that the capacity of the transformer is adequate for the load connected to it).

Experiment shows that if the primary and secondary have the same number of turns, the voltage of the primary and secondary will be the same, but the current flowing in the primary from the power line adjusts itself to the current demanded of the secondary by the nature of the particular load connected to it.

If the secondary has twice as many turns as the primary, the voltage of the secondary will be twice that of the primary, but the current will be only half as great. If the secondary has ten times as many turns as the primary, the voltage in the secondary will be ten times that in the primary, but the current will be only one-tenth as great. By reversing the proportions and having fewer turns in the secondary than in the primary, it is equally simple to step the voltage down instead of up; but the current will then go up as the voltage goes down. The volt-amperes in the primary are always equal to the volt-amperes in the secondary, plus a small percentage, depending on the efficiency of the transformer.

The minimum number of turns must be kept within the limits that good engineering has shown lead to the greatest efficiency, and wire sizes in both the primary and the secondary must be adequate to carry the current involved. The smallest transformer usually found is the ordinary doorbell type, which steps 120 volts ac down to about 8 volts for operating a small doorbell and similar equipment. Chimes and some doorbells operate at a slightly higher voltage. The largest transformers are so big that it is hard to find a railway car sturdy enough to transport a single transformer.

Well-built transformers are very efficient, and, generally speaking, the larger the transformer, the greater the efficiency. In very large transformers, it is possible to obtain from the secondary over 99% of the power supplied to the primary.

In a large generating station (power house), generators produce various voltages, 13 800 volts being typical. Transformers step up this generated voltage to a much higher voltage for transmitting great distances. Transmission voltages range from 46 to 345 kV and even higher, depending on the transmission distance. At various distribution substations, the transmission voltage is stepped down to distribution voltages, which may range from 13 800 to 34 500 volts (or 13.8 to 34.5 kV). At strategic points, it is again stepped down to utilization voltages, which range from 120 to 4160 volts and sometimes higher. See Fig. 3–9. (Single-phase transformers are shown here for simplicity, but three-phase transformers are actually used for most high voltages.) Sometimes more than one distribution system is used in a single locality. For instance, 34.5 kV may be distributed throughout a general

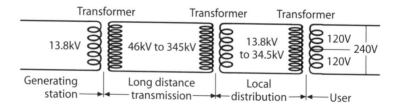

Fig. 3–9 Power is generated at a relatively low voltage, then stepped up to a much higher voltage for transmission over long distances, then stepped down to lower distribution and utilization voltages.

area, after which it is again stepped down at various substations for local or plant distributions at 4160 volts, and then stepped down again for various utilization voltages, such as 480Y/277 volts and 120/240 volts.

The voltage at which power is transmitted depends on many factors, but the principal factors are the distance and the amount of power involved. Several 500- and 765-kV transmission lines are in use. Still higher voltages are being developed.

Series-parallel connections   Power transformers for power and lighting usually have two identical coils in each primary winding, and two identical coils in each secondary winding. If the two primary coils are connected in series as shown in A of Fig. 3–10, the primary will be suitable for connection to a 4800-volt line. If connected in parallel as shown in B, the primary is suitable for connection to a 2400-volt line. Similarly, the two secondary coils can be connected in parallel to deliver 120 volts as shown in A, or in series as shown in B to deliver 240 volts. Usually, the two secondary coils are connected in series with a tap at the midpoint, forming the common three-wire, 120/240-volt system, as shown in B. If connected for 120 volts, the available current in amperes will be twice that available if connected for 240 volts; but the secondary volt-amperes will be the same regardless of which way it is connected. However, the available power [(VA)(pf)] at the connected loads will usually be slightly higher at 240 volts than at 120 volts because of lower voltage drop in the conductors. In addition, this type of connection provides a dual-voltage, three-wire system for supplying both 120- and 240-volt circuits.

Three-phase transformers   For a three-phase system, sometimes a three-phase transformer is used, and at other times, three single-phase transformers are interconnected to form a three-phase bank of transformers. A three-phase transformer consists of three separate single-phase transformers that are interconnected within a single enclosure.

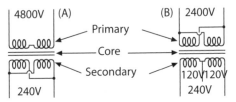

Fig. 3–10 The windings can be connected in series or in parallel for two different voltages.

Figures 3–11 and 3–12 show only the secondaries of the transformers. A three-phase transformer (or transformer bank) may be connected in the delta configuration (the name is derived from the Greek capital letter delta: Δ) as shown in Fig. 3–11, or in the wye configuration (sometimes called "star") as shown in Fig. 3–12. (The primary of a delta-connected secondary may be connected either as delta or wye, and the primary of a wye-connected secondary may be either wye or delta.)

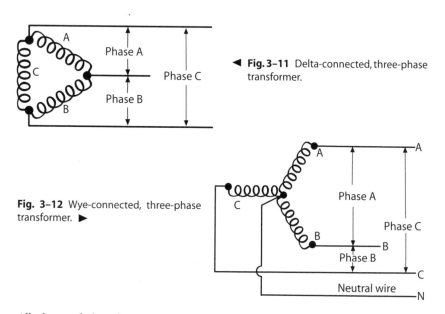

Fig. 3–11 Delta-connected, three-phase transformer.

Fig. 3–12 Wye-connected, three-phase transformer. ▶

All three of the phase wires must run to any three-phase load, such as a three-phase motor, but where only two (any two) of the phase wires of a three-phase system are used, they can supply only a single-phase load.

In the delta system of Fig. 3–11, the power will usually be at 240 or 480 volts. The single-phase power then available is one of those voltages. Single-phase power at only 240 volts isn't very practical for most purposes; we want 120/240-volt, single-phase power. Figure 3–13 shows the same transformers as in Fig. 3–11, but with one of the secondaries tapped at the midpoint. Assuming the basic voltage is 240 volts, the secondary that is tapped at the midpoint in Fig. 3–13 will deliver 120/240-volt, single-phase power, which may be used at the same time as the three-phase power. Although not shown in Fig. 3–13, the midpoint where the tap is made is also grounded to a grounding electrode. This creates a grounded system, the virtues of which are explained in the discussion relating to Fig. 9–2.

In the wye-connected system shown in Fig. 3–12, a neutral wire is run from the junction of the three secondaries. The three-phase voltage of any circuit

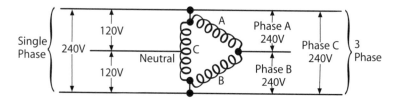

Fig. 3–13 A three-phase, delta-connected transformer can deliver both three-phase and dual-voltage single-phase power.

connected to all three wires, A, B, and C, is usually 208 volts. The single-phase voltage between wires A and B (or B and C, or A and C) is 208 volts. You would then expect the voltage between the neutral wire N and either A, B, or C to be half of 208, or 104 volts, but that is a wrong conclusion. The voltage between the neutral and any hot wire of a wye-connected, 208-volt, three-phase system is 120 volts.

At first glance this may seem all wrong, because if the voltage between wires A and B in Fig. 3–12 is 208 volts, the voltage between the neutral wire and either A or B might be expected to be half of 208, or 104 volts, instead of 120 volts as previously stated. Remember, however, that in a three-phase circuit, the voltage comes to a peak or maximum at a different time in each phase. At the instant the voltage in secondary A is 120 volts, that in B is 88 volts, so that across wires A and B there is a voltage of 120 + 88, or 208 volts. (Note that 208 = 120 $\sqrt{3}$.) The system therefore has the advantage of making it possible to transmit over only four wires (including a grounded neutral) three-phase power at 208 volts, single-phase power at 208 volts, and single-phase power at 120 volts. Occasionally in a home, instead of the usual 120/240-volt, three-wire system being provided, three wires of the wye-connected system (the grounded conductor and any other two wires of Fig. 3–12) are provided, thus furnishing 120 volts for lighting and 208 volts (instead of the usual 240 volts) for water heaters and similar large loads. In larger installations, all four wires (the three phase wires and the neutral) are used. This is a three-phase, four-wire, 208Y/120-volt system. Although not shown in Fig. 3–12, the neutral point of the windings (where all three windings are connected together) is grounded. As in the case of a center-tapped delta system, this creates a grounded system, covered in Chapter 9.

Instead of 208Y/120 volts, newer installations in commercial and industrial establishments provide power at 480Y/277 volts. More will be said about this later. European distributions, and other parts of the world using 220 volts as the base voltage, use 380Y/220 volts for these locations.

Autotransformers   An autotransformer can be defined as a transformer in which a portion of the turns of a single winding is common to both primary and secondary (see Fig. 3–14). Let there be a tap at the midpoint of the coil so that although there are, for example, 1000 turns of wire between A and C, there are only 500 between B and C. If the voltage is to be stepped down, the entire coil A and C will constitute the primary, and those turns from B to C will constitute the secondary. Whatever the voltage across A to C, the voltage across B to C will be exactly half. The tap may be at any point in the coil. The voltage across B to C, as compared with the total voltage across A to C, will always be proportional to the number of turns from B to C, as compared with the total number of turns from A to C. If the voltage is to be stepped up (instead of down), a part of the winding is used for the primary and the entire winding is used for the secondary. This type of transformer, being far less expensive than the two-coil type, is often used for transmission lines and similar purposes, but it is seldom used in buildings because *NEC* 210.9 has restrictions on its use with branch circuits. So, except as used in some types of lighting fixture ballasts and for some types of large motor starters, an autotransformer is usually used in buildings only to boost the voltage of an existing 208-volt branch circuit for supplying a 240-volt appliance, or to reduce the voltage of an existing 240-volt branch circuit for supplying a 208-volt appliance. Autotransformers are also allowed to be used in industrial occupancies for 600–volt by 480–volt transformations in either direction.

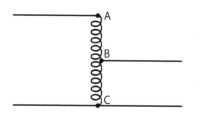

Fig. 3–14 An autotransformer has only one winding.

High-voltage dc transmission Transformers operate only on alternating current, yet you may have read about high-voltage, direct-current transmission lines, which are in use mostly in foreign countries, at voltages up to 1 000 000 volts (1000 kV) or more. How is this possible without transformers? Ac generators are used, the generated voltage is stepped up to the desired transmission voltage by transformers, and then "rectifiers" are used to change the alternating current to direct current. (A rectifier is a device that permits current to flow in one direction, but not in the other. Four of them in combination convert single-phase alternating current to direct current. More complicated combinations of rectifiers convert three-phase alternating current to direct current.) After transmission, converters are used to change the direct current back to alternating current. Then transformers are used to step the ac voltage down again. The procedures for doing this are beyond the scope of this book.

There are many advantages to long-distance transmission of dc power, but the method is practical only if very large amounts of power are transmitted from one single point to another single distant point, without need for tapping off any of the power at intermediate points. Although dc transmission is not yet in common use in the United States, there are a number of these lines in use, typically running at 1000 kV or higher. Therefore dc transmission may become more common in the future.

CHAPTER 4
Basic Electrical Power Utilization Systems

THIS CHAPTER OUTLINES basic concepts about electrical systems as a whole that you need to understand in order to use electricity safely and productively. Subsequent chapters look more closely at the various components of electrical systems. Having an understanding of the overall system will help you appreciate how those components function within it.

If electric current is to be useful, we need to control where it flows, limit how much of it flows, and prevent it from posing a shock hazard. The flow of electric current usually represents a transfer of electric energy from a source to some electrical load, over a defined path called a circuit. Circuits under our control usually, but not necessarily, begin directly or indirectly at the service.

THE SERVICE IS THE USUAL ENERGY SOURCE

The most common source of electric energy in today's buildings is an electric utility wiring system. Because these systems operate under the exclusive control of highly trained utility line crews, their wiring procedures follow an entirely different code—the National Electrical Safety Code. That code is not designed for wiring that will be used by the general public, and building wiring installed under that code would pose a substantial safety hazard. The *NEC* has special rules that govern the transition from utility wiring practice to premises wiring, which is the wiring you will be working with. That transition always occurs at the service. In fact, although a building might have other sources of power, such as on-site generation, the *NEC* defines the term "service" in a way that excludes it from applying to anything other than a utility-to-premises wiring transition. *NEC* Article 230 is devoted in its entirety to services.

A service consists of wires and control equipment that comprise the means to control the delivery of electrical energy from the supplying utility to the premises wiring system. The service has several basic components:

- *Service point*—This is the point, generally defined by government regulations, at which the utility jurisdiction over the wiring system stops and the owner's responsibility begins. It may differ from jurisdiction to jurisdiction, and even from place to place within a jurisdiction depending on local regulations. Before beginning a wiring system, be sure to get a copy of your local utility's wiring rules.

■ *Service conductors*—These are the wires running from the service point to the service disconnecting means. If they run overhead we call them collectively the *service drop,* and if they run underground they go by the term *service lateral.* After a service drop or lateral arrives at a building, it customarily supplies *service-entrance conductors,* which run from that connection to service equipment.

■ *Service equipment*—This is the main control of the electric energy supply coming to the building from the utility supply system. It always sits at the load end of the service-entrance conductors, and it always includes a switching mechanism that you can operate to disconnect the building wiring from the supply of electricity. It also includes overcurrent protective devices that prevent excessive electrical loading in the building from overheating the service conductors. In many applications, the two functions are combined in a single device. An electric meter may be placed either ahead of or after the service equipment (the local utility wiring rules govern this). Since a meter doesn't control the flow of electricity, the *NEC* doesn't consider it part of the service equipment.

OVERLOAD PROTECTION PROTECTS AGAINST SUSTAINED OVERHEATING

It is impossible for an electric current to flow through a wire without heating the wire. As the number of amperes increases, the temperature of the wire also increases. For any size of wire, the heat produced is proportional to the square of the current. Doubling the current increases the heat four times; tripling it increases the heat nine times, and so on. As the temperature of a wire increases, its insulation may become damaged by the heat, leading to ultimate breakdown. With sufficient current, the conductor itself may get hot enough to start a fire. The wiring system must include equipment, referred to as *overload protective devices,* that prevents excessive electrical loads from overheating wires.

SHORT-CIRCUIT AND GROUND-FAULT PROTECTION PREVENTS DAMAGE FROM ELECTRIC ARCS

Overload protection alone is not sufficient. The electrical system also needs to protect against fault currents, which are cases where current flows outside its intended path due to a misconnection or an insulation failure. There are two types of fault currents.

A *short circuit* results from one wire of a circuit being in direct contact with another with no load in between. With no load to limit current flow, a very high current flow results that is capable of causing extreme heating and fire within a fraction of a second.

A *ground fault* results from unwanted electrical contact between an ungrounded circuit conductor and a grounded object or wiring component. Depending on the quality of the resulting circuit, ground faults may produce currents comparable to short circuits. In addition, because ground faults often result from insulation failures that don't produce a solid electrical connection, they are usually accompanied by sparks. Sustained sparking is called an *electrical arc*, and it is one of the hottest energy releases on the surface of the earth, frequently involving temperatures over 10 000°F. In addition, when an electrical connection takes place between an ungrounded conductor and a grounded object, that object is raised to some voltage above ground, possibly to the point of being a shock hazard.

The wiring system must de-energize or "clear" an overloaded or a faulted conductor before it can cause damage. The *NEC* defines all these hazards as differing forms of *overcurrent*, and devices that provide full protection are *overcurrent protective devices*. Overcurrent protective devices today consist of either fuses or circuit breakers, both covered in Chapter 6.

Although overcurrent protection is essential to a safe electrical installation, it doesn't do everything. A conventional overcurrent protective device would never interrupt a circuit just because some unlucky person happened to be getting a shock from a faulty power tool or other source. Electrical shocks capable of killing a person (75 milliamps or so) are far below 1% of the nominal tripping point of even the smallest overcurrent device. There are special devices designed to provide shock protection. *Ground-fault circuit interrupters* trip at about 5 milliamps, and they have saved thousands of lives over the last few decades. Chapter 9 covers them in more detail.

SERVICE CONDUCTORS INVOLVE UNIQUE OVERCURRENT PROTECTION PROBLEMS

Service conductors are on the line (supply) side of all overload protective devices. Service equipment can and usually does prevent the electrical system from imposing excessive loads on the service conductors. However, in the event of a ground fault or a short circuit, there isn't any protection. The fault has to burn clear; there just isn't any practical alternative. Although electric utility systems have some form of protection on their lines, it isn't such that a fault would be cleared before it caused extensive damage within a building. This is why service protective devices (both circuit breakers and fuses), although fully qualified for use as overcurrent protective devices throughout an electrical system, must be referred to as overload protection only. They will never see the effects of a short circuit in the conductors they protect.

The *NEC* requires that service conductors stay outside of buildings, or terminate at a readily accessible point nearest their point of entry if the service equipment is indoors. This requirement minimizes the building's exposure to unprotected conductors. Although specific maximum distances are proposed every Code cycle, the Code-making panel has always rejected them, preferring to leave the distance to the field inspector. This provides some room for negotiation based on actual conditions. Some local jurisdictions amend the *NEC* and set maximum distances, so check local rules on this point. Note that for service disconnects located outdoors there is no specific distance restriction. However, many inspectors view any stand-alone remote service equipment as an independent structure. Where so interpreted, the run to the building becomes a feeder and the building then requires its own building disconnect, also at a readily accessible point nearest the point of conductor entry.

If you must bring the service equipment farther into the building, the *NEC* has special allowances for service conductors running in a raceway under concrete slabs or at least 18 in. of earth, or through fireproof vaults meeting the requirements of Part III of Article 450 for transformer vault construction, or within raceways encased in 2 in. of concrete. If the conductors fault, the arc damage will be contained and will not cause a fire in the building. These allowances also apply if you need to locate a building disconnect (supplied from a feeder from another building or structure) some distance into the interior of the building.

SERVICE DISCONNECTS—"READILY ACCESSIBLE" AND OTHER *NEC* REQUIREMENTS

"Readily accessible" is an item of *NEC* terminology that you must understand thoroughly. Defined in *NEC* Article 100, it means "capable of being reached quickly for operation, renewal, or inspections, without requiring those to whom ready access is requisite to climb over or remove obstacles or to resort to portable ladders, etc." Note that this definition doesn't preclude arrangements such as locked electrical rooms, as long as the qualified personnel who have business in such rooms ("to whom ready access is requisite") have the keys, and can get to the rooms without climbing over obstructions, etc.

SERVICE AND BUILDING DISCONNECTS

In an emergency, electrical systems have to be able to be disconnected quickly, and in an orderly way that doesn't create additional hazards in the process. The service disconnect provides the principal means of accomplishing this function. It must be marked to identify its function, and if composed of two or more disconnecting means, each must be marked to show the load served. There are additional requirements aimed at minimizing confusion while still allowing necessary flexibility in the design of electrical systems, as follows:

■ *Number of services, more than one by special permission*—There must be no more services to a building than necessary. For homes and stand-alone light commercial applications, that number is one. Additional services are permitted for multiple occupancy buildings without any common area for equipment accessible to all occupants, and for a very large building, for example, that could not be properly served from just one point. These two cases require *special permission* from the inspector. The *NEC* defines this permission as the written consent of the authority having jurisdiction. Written consent must be in specific reference to that which requires permission; a generic completion signature on a building inspection weather card could constitute an approval (review Chapter 1) but is not the required special permission. Consider for a moment how parents might handle the arrival of their child home after an established curfew. The next morning they may elect to tacitly approve the behavior by saying nothing, but the action remains one that lacked permission.

■ *Number of services, more than one by right*—There are other cases where additional services are allowed by right. Multiple single-phase services are permitted when the capacity requirements exceed that which the serving utility normally provides, and in general, multiple services are permitted for load requirements in excess of 2000 amps at 600 volts or less regardless of utility service policies. Multiple services are also permitted for different characteristics of supply, as in the case of different system voltages, or for different uses including different rate schedules. Different supply character-istics probably provide the most common basis for multiple services at a building. Additional services are also permitted when the second service provides a standby power source function, or when it supplies a fire pump (discussed at the end of Chapter 30).

■ *Number of disconnects*—Each "service disconnect" may actually be composed of up to six disconnecting means. By long precedent extending back many generations, the rule of six operations of the hand has been considered a reasonable limit to complexity. With the exception of the fire pump, the six disconnecting means must be grouped so all are capable of operation in one location. Grouping does not mean two in the building basement and one outside the building, for example, even if those disconnects back up to one another. With each disconnecting means comes overload protection, each of which must be able to carry the load assigned to it.

■ *Number of services, reciprocal labeling of disconnecting means*—If multiple power sources serve a building, permanent plaques or directories must be placed at every service (and building) disconnect showing the area served and the locations of all the other power source disconnects plus the area served by each. In an emergency situation there can be significant potential hazard unless those charged with disconnection in fact disconnect all power. Consider that each service may consist of up to six disconnecting means, and there is no rule that requires multiple services to be collectively grouped. In addition, there are cases where the power sources to a building consist of a utility service in addition to a supply from another building. The effort invested in creating the required signage is well justified by the need.

■ *Many of the same principles apply to building disconnects*—*NEC* 225.30 imposes the same limitations on the number of supplies to a second building as apply for services, with one exception. For a second building, an additional supply is permitted "for different uses" such as control of outside lighting. *NEC* 225.32 requires a local disconnecting means for all separate buildings and structures fed from another building or structure on the property, even an accessory building to a single-family dwelling. It is a fundamental principle of the *NEC* that in an emergency you should never have to go into another building to disconnect power to the site of that emergency. The only exceptions are for lighting standards and signs, and for highly restricted heavy industrial and institutional applications beyond the scope of this book. The only relaxation for residential work consists not in the requirement for a disconnect, but in an allowance for snap switches to serve for this purpose. Further, the disconnect location is the same as for service disconnects, that is, the readily accessible point nearest the point of entrance of the supply conductors, and the marking rules are the same as well. For additional details on dwelling outbuildings, refer to the discussion on wiring garages and outbuildings beginning on page 334.

■ *What is a building?*—As defined in *NEC* Article 100, a building is a structure that stands alone or is separated from any adjoining structure by a fire wall. Any opening(s) in such a fire wall must be protected by approved fire doors. In other words, if a portion of building structure is cut off from all other portions of the structure by walls that are classified by the authority having jurisdiction as fire walls, you must consider the cut-off portion as a separate building. Each such building is entitled (but not required) to be supplied by its own service drop or lateral just as though it were a separate structure. The authority having jurisdiction for classifying buildings, fire walls, fire doors, etc., is usually a building official or a fire department official. Where there is any question, the electrical inspector will usually consult with such an official.

Don't confuse fire-rated walls with fire walls. Most building codes, for example, require fire separation walls between adjoining occupancies, such as townhouse-style housing units. These walls have fire ratings but fall well short of being actual fire walls. A fire wall is comprised of masonry or concrete, and extends from the lowest level up to (and in some codes a short distance above) the roof line. It is expected to remain standing even after a complete conflagration on one side. Only an actual fire wall establishes a building separation under the Article 100 definition.

■ *And what is a structure?*—The term has recently been defined simply as that which is built or constructed. This adds little in the way of substance, but it does add support for the theory that a remote metering pedestal qualifies as a structure, requiring a second disconnect when the feeder arrives at the building (or other structure) served. Many building codes have a more specific definition of this term, so check with the inspector if you are not sure how it will be applied.

CIRCUITS

An electric circuit consists of a complete path for the electric current from the source of supply, through the connected load (equipment or apparatus), and back to the source of supply. The path must be *complete* in order for current to flow. As covered in Chapter 2, the "flow" of electricity actually involves an electric field that must be continuous from source to load and back. Any break in the complete circuit interrupts the field, and the load no longer functions. However, for code purposes we need to look at the different segments of the overall pathway according to their function in the wiring system.

That is, even though the energy that lights (see Fig. 4–1) a 150-watt floor lamp might begin at an outdoor utility transformer secondary and pass all at once through service drop, service entrance, feeder, branch-circuit conductors, cord to the lamp and the light bulb, and then back again, we wire each of those segments differently. The *NEC* considers certain major portions of an electrical installation as a separate circuit. This allows a sensible differentiation of wiring rules based on the different wires and other components that make up each portion. For example, the service conductors might be rated 200 amps, ending in a panel with a 200-amp main breaker. That panel might include a 60-amp breaker supplying a feeder to another panelboard, where our actual 20-amp branch circuit originates. The floor lamp itself has a 15-amp plug on a 10-amp cord set. Clearly each of these points in the electrical system needs its own specialized rules.

The *NEC* defines a *branch circuit* as "the circuit conductors between the final overcurrent device protecting the circuit and the outlet(s)." In our example, that includes the conductors leading to the receptacle outlet where we plugged in the floor lamp. The flexible cord to the lamp itself must comply with other requirements designed for what may be attached to a branch-circuit outlet. Moving upstream, the *NEC* defines a feeder as "all circuit conductors between the service equipment, the source of a separately derived system, or other power supply source and the final branch-circuit overcurrent device." Other rules apply to this part of the circuit, which in turn differ from the rules governing the service conductors. The ampacity topic beginning on page 103 covers the minimum sizing requirements for conductors in general.

A *separately derived system* is an energy source with no direct electrical connection or shared common conductor with another electrical system. Many transformers qualify as the source of separately derived systems, along with on-site generation facilities. Each of these systems supplies one or more feeder circuits, not services.

Source of power   In all the diagrams in this book, the word SOURCE in small capital letters means the generator, transformer, battery, or other source of power supplying the circuit. For practical purposes, you may see SOURCE indicated at the point where the wires enter the building, or the point where the particular circuit under discussion begins.

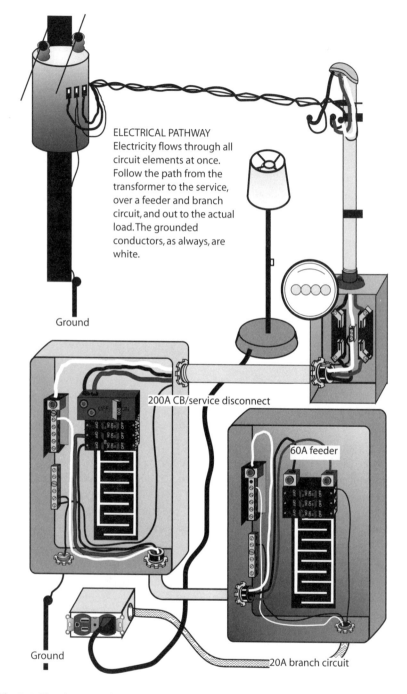

ELECTRICAL PATHWAY
Electricity flows through all circuit elements at once. Follow the path from the transformer to the service, over a feeder and branch circuit, and out to the actual load. The grounded conductors, as always, are white.

Ground

200A CB/service disconnect

60A feeder

Ground

20A branch circuit

Fig. 4–1 The elements of a complete circuit from the utility transformer, through the service equipment, and ending at a floor lamp.

Basic circuit   Figure 4–2 shows a wire running from the SOURCE to the socket with a lamp, and another wire running from the lamp back to the SOURCE. In all such diagrams in this book, the grounded wire is shown as a light line like this _____, and a heavy line like this _____ represents the ungrounded (hot) wire. This is done just to make it easier for you to follow the diagram and does not mean that one wire is larger than the other; both are the same size. Not all circuits use a grounded return conductor. Chapter 9 covers the concept of grounded conductors in more detail, because most simple residential and commercial circuits use them. Many industrial occupancies use ungrounded distributions (refer to discussion of delta distributions in industrial applications beginning on page 532).

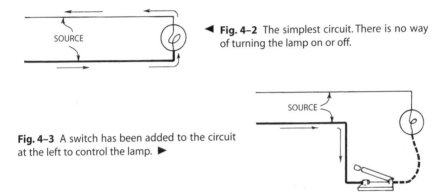

◀ **Fig. 4–2** The simplest circuit. There is no way of turning the lamp on or off.

Fig. 4–3 A switch has been added to the circuit at the left to control the lamp. ▶

The current flows outward through one wire (the heavy line), through the lamp, and back through the other wire (the light line). This makes a complete circuit, and as long as the SOURCE furnishes power, the lamp will light. This is not a practical circuit since it is necessary to disconnect one wire from the socket or unscrew the lamp to turn the light off. A switch must be included, as has been done in Fig. 4–3. An unenclosed switch is shown so you can see how it works: opening the switch is the same as disconnecting or cutting a wire. Of course, unenclosed switches are unsafe and not permitted due to the hazards of the exposed live parts; instead use the snap switches covered in the next chapter.

PANELBOARDS (AND SWITCHBOARDS)

For reasons of workmanship and convenience, most overcurrent protective devices today are installed in circuit groups within enclosures designed for this purpose. Usually the overcurrent devices will be installed on a single buswork structure contained in a common enclosure with a hinged door set in an outer trim and designed to mount to a wall or similar surface. This is a *panelboard* (see Fig. 4–4), and it differs from a *switchboard* (see Fig. 4–5) which is stand-alone

equipment, usually with rear access, that accommodates large overcurrent devices and switches (see photos). Panelboards typically run up to about 1200 amps and nearly all dwellings and small to medium sized commercial occupancies use them exclusively. The smaller versions for residential and light commercial use are referred to in some catalogs as "load centers," but that term has no meaning in the *NEC*. All "load centers" are forms of panelboards, and must comply with all rules pertaining to this equipment. Switchboards run up to 6000 amps or more, and usually power major commercial, institutional, or industrial facilities.

In the wiring of ordinary residences or other small buildings, the overcurrent devices (circuit breakers or fuses) are installed at the service equipment, and all

Fig. 4–4 This is a panelboard. The cabinet mounts to the wall. *(Square D Company)*

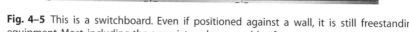

Fig. 4–5 This is a switchboard. Even if positioned against a wall, it is still freestanding equipment. Most, including the one pictured, are capable of rear access. *(Square D Company)*

branch circuits begin at that point; thus there are no feeders other than the trivial case of the buses themselves, which are technically a form of feeder conductor. In larger buildings it is not practical to begin all the branch circuits at the point of the service equipment because the wires would be impractically long. Instead, overcurrent devices are installed at the service equipment to protect feeders. The feeders are heavy wires running to other locations where panelboards containing the overcurrent devices for the smaller branch-circuit wires are installed. If the building is very large, those feeders may supply still other smaller feeders, and so on, until you reach the final branch-circuit overcurrent protective device.

Be sure to identify every overcurrent device in every panelboard or switchboard by its function, so those who come after you will know what device controls which load. Circuit directories should not only be complete, they should convey useful, accurate information. There are far too many electrical installations out in the field today for which the circuit directory was clearly an afterthought. For example, a 200-amp dwelling panelboard might have eight 15-amp circuits marked "lights and plugs," four 20-amp circuits marked "kitchen," and six 2-pole, 20-amp circuits marked "electric heat." The 2005 *NEC* requires circuit directories to show "the clear, evident, and specific purpose or use" of each of the circuits, and further that the description must include sufficient detail so "each circuit [can be] distinguished from all others."

GROUNDING

For reasons outlined in the system grounding topic beginning on page 132, electrical systems almost always involve components connected to the earth, and many systems also involve one of the circuit conductors being deliberately connected to earth as well. That means you will be making a grounding connection to your electrical system. Figure 4–6 shows a typical residential grounding connection at the service equipment.

WIRING METHODS

It isn't enough to control how much current flows through a wire. That wire has to be installed in a wiring method appropriate for the location. Otherwise the insulation on the energized wires may become damaged to the point of exposing the public to the shock hazard of live wires. In addition, a wiring method degraded by physical abuse due to inadequate consideration of environmental conditions could lead to an arcing failure. Furthermore, the owner will certainly want a reliable system even to the extent that safety issues are addressed in the wiring design. That may involve upgrading your installation plans based on probable usage patterns. The basic wiring methods are covered in Chapter 11, and less common methods, especially those used more in industrial and heavy commercial locations, are covered in Chapter 27.

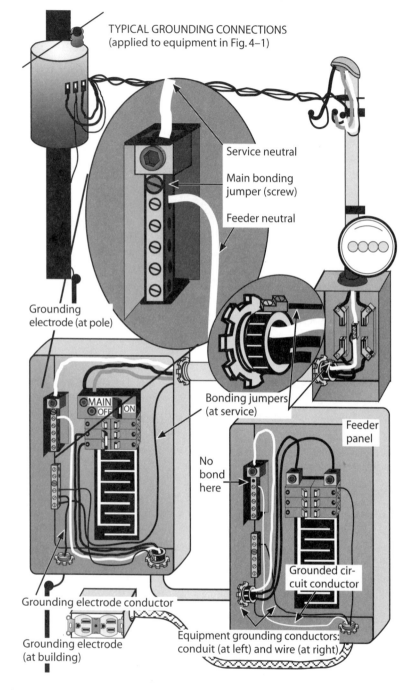

TYPICAL GROUNDING CONNECTIONS
(applied to equipment in Fig. 4–1)

Service neutral

Main bonding
jumper (screw)

Feeder neutral

Grounding
electrode (at pole)

Bonding jumpers
(at service)

Feeder
panel

No
bond
here

MAIN
OFF   ON

Grounded cir-
cuit conductor

Grounding electrode conductor

Grounding electrode
(at building)

Equipment grounding conductors:
conduit (at left) and wire (at right)

Fig. 4–6 The elements of a basic residential distribution including a feeder and remote panel.

CHAPTER 5
Basic Devices and Equipment

BEFORE YOU CAN WIRE A BUILDING, you must learn how switches, receptacles, sockets, and other devices are properly connected to each other with wire to make a complete electrical system. The wiring diagrams in this chapter and in other parts of this book are mostly parts of branch circuits rather than complete circuits. Only the basic devices and equipment necessary to make the circuit work are discussed in this chapter. There are dozens of other devices. Other materials, fittings, raceways, boxes, etc., are covered in other chapters.

DEVICES, FITTINGS, AND BOXES—DEFINITIONS

You will often see the term "wiring device," which is defined in *NEC* Article 100 as a component that carries current but does not consume it. Examples are sockets, switches, overcurrent devices, push buttons, etc. Receptacles are wiring devices in that they do not consume power but serve only as a contact point for power-consuming loads such as lamps and toasters that are plugged into them. Anything that consumes power is utilization equipment and constitutes the load on the circuit.

The *NEC* defines a "fitting" as "an accessory such as a locknut, bushing, or other part that is intended primarily to perform a mechanical rather than an electrical function."

In installing wiring, all connections of one wire to another, or to a terminal, are made inside of boxes. The boxes are often metal, though use of nonmetallic boxes is increasing, especially in residential applications. Many kinds of boxes are described in Chapter 10.

LAMPS FOR INCANDESCENT LIGHTING

Probably the most common electrical equipment is what is often called a "light bulb." The correct name is "lamp"—the glass part of the lamp is the bulb.

Lamp design principle   The lamp consists essentially of a filament, which is a wire made of tungsten, a metal that has a very high resistance and melting point making it possible to heat the filament to a very high temperature (over 4000°F in ordinary lamps) without its burning out. The filament is suspended on supports in a bulb from which the air has been exhausted and into which, in most sizes, some inert gas such as argon has been introduced to prolong the life of the filament. The ends of the filament are brought out to a convenient base, which makes replacement simple. In the base, the center contact is insulated from the outer metal part of the base, thus producing two terminals for the two wires leading up to the filament.

Any lamp that operates by means of a heated filament is called an incandescent lamp to distinguish it from a fluorescent lamp.

Sockets (lampholders)   Lamps operate in what the *NEC* calls lampholders, but which most people just call sockets. The simplest socket is the cleat lampholder shown in Fig. 5–1; a cross section is shown in Fig. 5–2. One terminal *A* is connected to the center contact corresponding to the center contact of the lamp base; the other terminal *B* is connected to the screw-shell terminal (which is carefully insulated from the center contact and terminal *A*), corresponding to the outer screw shell of the base on the lamp. When a lamp is screwed into such a socket, the current flows in at *A*, through the filament, and out at *B*. Such sockets with exposed terminals are used with open wiring, and are seldom seen today.

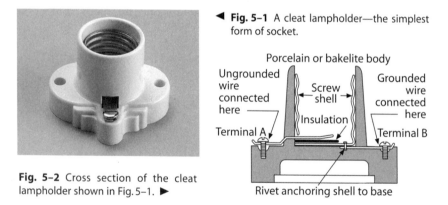

◀ **Fig. 5–1** A cleat lampholder—the simplest form of socket.

Fig. 5–2 Cross section of the cleat lampholder shown in Fig. 5–1. ▶

Porcelain or bakelite body

Ungrounded wire connected here

Screw shell

Grounded wire connected here

Insulation

Terminal A

Terminal B

Rivet anchoring shell to base

All sockets are lampholders, but not all lampholders are the type with a screw shell to engage a corresponding screw-shell base on a lamp. Lampholders (sockets) are available in great variety. The cleat lampholder in Fig. 5–1 is not used in house wiring. The commonly used porcelain or phenolic socket may be either keyless or have a pull-chain switching mechanism to turn the lamp on and off. Figure 5–3 shows a pull-chain lampholder used at outlet boxes. Other types are shown in Chapters 14 and 29.

RECEPTACLE OUTLETS FOR CONNECTING LOADS

Every point where power is taken from wires and consumed is an outlet. *Receptacle outlets* are outlets where one or more receptacles are to be installed.

The receptacle itself consumes no power, but whatever is plugged into it (toaster, radio, etc.) does consume power. The distinction between an outlet and some other opening or junction point, such as a switch or thermostat, in an electrical system is important because the *NEC* applies branch-circuit wiring rules to the supply side of outlets, but not to the load side. This is why contractors frequently estimate jobs on a "per opening" (and not "per outlet") basis.

Fig. 5–3 The socket shown fits directly on top of an outlet box. *(Pass & Seymour/Legrand)*

The principle of a plug-in receptacle outlet is shown in Fig. 5–4: a pair of metal contacts, one connected to each of the two wires from the SOURCE, a plug with two corresponding contacts that can be brought into contact with the first pair, and a pair of wires running to the appliance.

Duplex receptacles—most common   Figure 5–5 shows a duplex receptacle, called that because it has two pairs of openings to accommodate two plugs at the same time.

Single receptacles—special uses   The single receptacle of Fig. 5–6 is seldom used in new installations, but watch for special conditions. By *NEC* definition, a receptacle is a contact device installed at the outlet for the connection of an attachment plug. A duplex receptacle, shown in Fig. 5–5, is two receptacles under this definition. Some provisions in the *NEC* specifically call or allow for a single receptacle. For example, perimeter and counter receptacles in a dwelling kitchen generally must be on a special kind of 20-amp branch circuit for small appliances (covered in Chapter 13); however, a special exception allows a refrigerator (or freezer) to be on a 15-amp circuit, provided it is an *individual branch circuit*. Only a single receptacle can be used at the load end of an individual branch circuit.

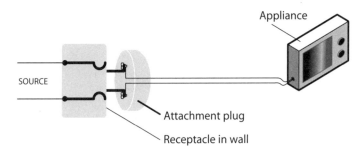

Fig. 5–4 The principle of a plug-in receptacle outlet.

Connecting switch-controlled receptacles   In any wiring diagram a receptacle can always be substituted for a socket; but if the socket is controlled by a wall
switch, then whatever is plugged into the receptacle substituted for a socket will
also be turned on and off by the switch. In any diagram or circuit, connect the
receptacle so that if it were a lamp it would always be on. (In actual wiring, it is
often desirable to connect a receptacle so that it is not permanently on, but
instead is controlled by a wall switch. Switched receptacles are discussed on page
327.) If in doubt, trace the path of the current to see if it can go from the SOURCE
to the receptacle and back again to the SOURCE even if all switches are in the open
(off) position.

Fig. 5–5 A duplex receptacle permits two
different appliances to be plugged in at the
same time. *(Pass & Seymour/Legrand)*

Fig. 5–6 Single receptacles are little used
today. *(Pass & Seymour/Legrand)*

SWITCHES FOR CONTROLLING OUTLETS

To control loads with a simple on and off function, use what the *NEC* calls a snap
switch of the type shown in Fig. 5–7. Designed for safety, it is concealed in the
wall with only the handle showing. It has two terminals just like the knife switch
in Fig. 4–3. The snap switch is a small and compact mechanism that does exactly
what the knife switch does: in one position of the handle the switch is open, in
the other it is closed.

Faceplates cover switches and other devices   For safety reasons, switches and receptacles mounted in walls cannot be left with openings around them, nor can terminals be left exposed. Instead they are covered by faceplates after installation. Figure 5–8 shows several faceplates. The smaller ones are used for single devices. Sometimes it is necessary to install two or three or more devices side by side, requiring wider plates known as "two gang," "three gang," or wider depending on how many devices the plate covers. They are also available in combinations so that switches, receptacles, and other devices can be mounted side by side, as the same figure shows.

Faceplates are made of a variety of materials such as plastic in various colors, ceramic, or brass and other metals in many different finishes to suit the user.

◀ **Fig. 5–7** A snap (toggle) switch. The mechanism is completely enclosed. It does exactly what the switch in Fig. 4–3 does—it opens one wire. *(Pass & Seymour/Legrand)*

Fig. 5–8 Faceplates must cover all switches, receptacles, and similar devices.
▼

Grounding requirements anticipate metal faceplates   You have to be sure that a switch yoke has a grounding connection even if you're using a nonmetallic faceplate. At one time most boxes were metal. Since the *NEC* generally requires the enclosing exterior metal parts of a wiring system to be grounded, most boxes including those housing switches were grounded as well. That meant that most switch yokes were grounded through their mounting screws. Wiring practices have changed over time, and now many boxes are nonmetallic. In addition, there are snap switches available with nonmetallic yokes.

If you install a metal faceplate, its mounting screws thread into and through the device yoke. If there is an ungrounded branch-circuit conductor resting against the inside of the device yoke, and the mounting screw penetrates the wire insulation, the result will be a voltage on the metal faceplate. (The same is true of a metal screw head in a nonmetallic plate, but the surface area is so small that no special rule need be made for this; the natural startle reflex is sufficient.) The *NEC* now requires that if you mount a switch in a nonmetallic box, or if you use one with a nonmetallic yoke even in a metal box, you must select a switch with an additional terminal on the yoke that provides a grounding connection. You can still use switches with conventional metal yokes, but only in grounded metal boxes. The *NEC* assumes that the person changing a faceplate to metal in the future will likely be someone with no understanding of the importance of assuring that such a faceplate is grounded. Grounding is thoroughly covered in Chapter 9.

Observe switch ratings   The maximum number of amperes that a switch is capable of handling, and the maximum voltage at which it may be used, are stamped into the metal mounting yoke of the switch. This might be a single rating, such as "15A 277V" indicating that the switch may be used to control up to 15 amps at not over 277 volts. Another common rating is "10A 125V-5A 250V" indicating that the switch may be used to control up to 10 amps at not over 125 volts, or up to 5 amps at not over 250 volts. Of course switches with higher ampere and/or voltage ratings are available, but switches for over 277 volts are usually of a different type.

Match switches to type of current (ac or dc)   There are two kinds of snap switches based on the type of current on the circuit: *ac general-use snap switches* (commonly referred to as ac-only switches), and *ac-dc general-use snap switches*. The first may be used only on ac circuits, and the second on either ac or dc circuits. The ac-only type can be identified by the letters "AC" at the end of the rating stamped on the mounting yoke of the switch; the ac-dc type does not have the letters "AC-DC" on the yoke. If the letters "AC" do not appear, the switch is the ac-dc type. Study the *NEC* on the subject of switches: Articles 100 ("Definitions") and 404.14 ("Rating and Use of Snap Switches").

Ac-only switches   In any location (except for dc circuits), it makes sense to use ac-only switches. Quiet in operation without the click of the ac-dc type, ac-only switches are the most common type switch installed. Ac-only switches have a minimum rating of 15 amps with some at 120 volts and others at any voltage up to 277 volts. On ac circuits, they may be used anywhere to control any type of load up to their full ampere and voltage rating, with two exceptions:

- They may not be used to control incandescent lamps at a voltage above 120 volts (even if the switches are rated 277 volts).

- If used to control a motor load they must have an ampere rating of at least 125% of the ampere rating of the load.

Ac-dc switches   At one time this was the only kind of snap switch made. Direct current is now virtually nonexistent for residential branch circuits. But many ac-dc switches installed in the past are still in use. As they fail, they can be (and usually are) replaced by the newer ac type, except for those on dc circuits. Switches capable of interrupting dc current must have specially designed operating mechanisms; these are usually quite loud, and consumers often object to the noise. Indeed, this feature appealed to one homeowner who specified them for his children's rooms just so he could hear if they turned the light on after bedtime.

Ordinary ac-dc switches are usually rated at 10 amps at not over 125 volts, or 5 amps at not over 250 volts. However, there are two subtypes: those that are *T-rated* and those that are not T-rated. If they are T-rated, the letter T appears at the end of the ampere and voltage rating stamped on the mounting yoke. The "T" stands for tungsten, the material used in the filaments of the incandescent lamps. The special rating addresses the fact that when an incandescent filament is cold (not lit) its resistance is far less than when it is hot (lit). When the lamp is first energized, it draws about ten times the current than when it is lit, and the special "T" rating means the switch has been designed to carry (and must be used on) incandescent lighting loads.

T-rated ac-dc switches may be used in any location to control lamps or other loads up to the full ampere ratings of the switches, unless the loads are inductive, in which case the switches must be rated at 200% of the ampere rating of the load involved. Ac-dc switches that are not T-rated may be used to control other (non-incandescent) resistive loads, such as heaters, up to the full rated capacity of the switches, or inductive loads up to 50% of their ampere rating.

Choose switches according to desired features   Switches are available in a wide variety to meet the requirements of safety, convenience, and comfort. These are some common types that you need to understand:

Single-pole switches   Any switch that opens only one wire is known as a single-pole switch. A single-pole snap switch is identified by its two terminals and the words ON and OFF on the handle.

Double-pole switches   Opening one of the two wires to a lamp turns the lamp on and off, but both wires can be opened if desired (and as required under some circumstances), as shown in Fig. 5–9. The porcelain-base switch shown is a double-pole or two-pole single-throw type. A corresponding flush snap switch of the type shown in Fig. 5–7 but with two poles instead of one is usually referred to simply as a double-pole or two-pole switch. It has four terminals for wires, and the words ON and OFF on the handle.

Double-pole switches are required by the *NEC* when neither of the two wires is grounded. In practice this means you must use double-pole switches for 240-volt motors or appliances. Grounding is discussed in Chapter 9. Be aware of two cautions regarding use of double-pole switches:

■ Don't assume you can use a two-pole (or a three-pole) switch to control two (or three) individual circuits. The listing instructions require an additional manufacturer's marking on the switch indicating "two-circuit" (or "three-circuit"), and as of this writing no such switches are on the market.

■ For a 240-volt circuit connected to a system that has no grounded system conductor that fixes the voltage to ground of the ungrounded conductors at not over 125 volts, do not use a two-pole snap switch unless its 240-volt rating carries a double underline: 240.

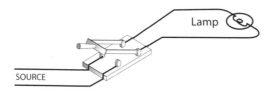

Fig. 5–9 Both wires are disconnected when a lamp is turned off with a double-pole switch.

Three-way switches   Often it is convenient to be able to turn a light on or off from two different places, for example, a hall light from upstairs or downstairs, or a garage light from either the garage door or from inside the house. For this purpose, choose three-way switches. The connections for this type of switch are described later in this chapter.

Four-way switches   For a light that must be controlled from more than two locations, a four-way switch is used in combination with three-way switches. The connections are described later in this chapter.

Special-use switches   Switches are available in many types besides those described in this chapter. The lock type in Fig. 5–10 can be operated only by someone having keys to fit. Some momentary-contact switches look like ordinary switches, but the handle is held in one position by a spring, returning to its original position when the operator releases the handle. Old-time switches had two push buttons instead of a handle. The surface type of Fig. 5–11 is used mostly in surface wiring, such as in some garages and farm buildings.

Dimmer switches   To control the brightness of the lights in part of the home, such as the dining room or family room, use special dimming switches, usually called dimmers. They are available for either incandescent or fluorescent lighting; buy the right kind because they are not interchangeable. Remove the ordinary switch and replace it with a dimmer. See Fig. 5–12. Most installation directions caution you to shut the circuit off before doing this. An inadvertent fault to a grounded object in the process of completing the installation will put enough current across the dimmer to ruin it.

Fig. 5–10 This type of switch can be turned on or off only by using a special key.

One type of dimmer for incandescent lights is an inexpensive device that has only HIGH-OFF-LOW positions, controls only up to 300 watts, and is not available in a three-way type. Others control up to 1000 watts continuously from off to full brightness, and are available in both ordinary and three-way type. Fluorescent fixtures can be dimmed, but usually only with special ballasts. Dimming fluorescent fixtures today is big business because of the potential energy savings in large commercial applications. Generally today's dimming systems involve limited-energy controls arranged to interface directly with the internal wiring of electronic ballasts. (This is discussed on page 565 in the energy efficient fluorescent lighting topic.) Some of the less expensive dimmers may cause radio or TV interference. Better ones have noise-eliminating devices built into them; the cartons should state whether they do.

Interchangeable devices   Sometimes there is not enough space at the desired location to install a ganged box for multiple devices. Or perhaps after a single-gang box has been installed, it becomes desirable to use two or more devices. There are available various types of devices physically small enough so that two or three can be installed in a single-gang box. The basic devices, such as receptacles, switches, and pilot lights, are stocked separately; typical pieces are shown in Fig. 5–13. They are mounted on a skeleton strap or yoke that comes with the faceplate, as shown in Fig. 5–14. Faceplates with just one or two openings are also available to preserve continuity of appearance throughout an installation.

Fig. 5–11 Surface-type switches are occasionally used.

Fig. 5–12 A dimmer controls the brightness of lights; it also serves as a switch. *(Pass & Seymour/Legrand)*

◀ **Fig. 5–13** Separate interchangeable devices are assembled on the job into any desired combination. Three of them occupy no more space than one ordinary device.

Fig. 5–14 Three devices are assembled on a skeleton strap, as shown in successive steps. ▶

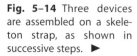

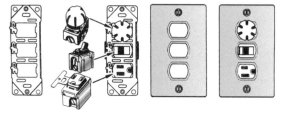

Although a single yoke with three devices can be installed in a single-gang box, or two yokes with six devices in a two-gang box, this is likely to result in overcrowding of boxes (more wires than permitted in a box of any given size). Use deep switch boxes or, better yet, where possible use 4-in.-square boxes with plaster rings as shown in Figs. 10–4, 10–5, and 10–6.

How snap switches work in electrical installations   The following material shows how common snap switches function. To see how to actually install these devices using customary wiring methods, refer to Chapter 17.

Using one switch to control multiple outlets connected in series   Often a switch must control two or more lamps. In drawing a diagram for this, many beginners will connect several sockets as in Fig. 5–15. The current can be traced along one wire from the SOURCE to the first lamp, to the second, to the third, to the fourth, to the fifth, and then through the switch and the other wire back to the SOURCE; consequently the lamps should light. They will light if not too many are used and if all are the same size and of the proper voltage. However, assume that each lamp is a different size, which means each will have a different resistance. Review Ohm's law (covered in Chapter 2); because all the current that flows through one must also flow through each of the others (making I, the current, equal across each lamp in the formula $E=IR$), the voltage dropped across each lamp must vary directly with the resistance. Therefore, the smallest lamps will see the most voltage and burn more brightly than normal. The biggest ones will see less voltage and burn less brightly than they should, while medium-size lamps may burn at normal brilliancy. Burning out one lamp or removing it from its socket, as shown in Fig. 5–16, is equivalent to opening a switch in the circuit—all the lamps go out. This arrangement is known as *series wiring* and is impractical for ordinary purposes. The series circuit was used in very old-style Christmas tree outfits, where eight identical lamps were used and consequently all burned at the same brilliancy. Each lamp was rated at 15 volts; they could be used on 120-volt circuits because each lamp received one-eighth of the total of 120 volts, or about 15 volts. At one time street-lighting lamps were also connected in series. With the development of midget fixtures and lamps that complete the circuit even after the bulb burns out, series lighting has returned to the holiday lighting scene with 50-lamp 2.5-volt lamp strings and a variety of other configurations. Airport runways commonly use a more sophisticated version of the same principle on 4160-volt circuits.

Instead of a drawing of a lamp in a socket, from this point on the symbol of Fig. 5–17 is used to denote a lamp and its socket. Note also the diagrams of Fig. 5–18, indicating whether wires that cross each other in diagrams are connected to each other or not.

Fig. 5–15 This type of wiring is known as series wiring.

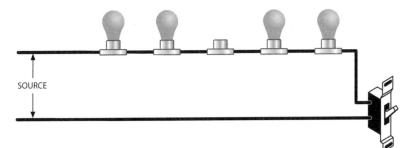

Fig. 5–16 In series wiring, when one lamp goes out, all go out.

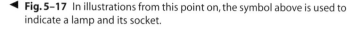

◀ **Fig. 5–17** In illustrations from this point on, the symbol above is used to indicate a lamp and its socket.

Fig. 5–18 This method is used in all diagrams to indicate whether or not crossing wires are connected to each other.

▼

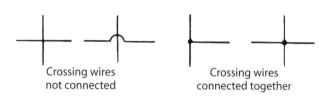

Crossing wires
not connected

Crossing wires
connected together

Parallel wiring is the scheme used in ordinary wiring; see Fig. 5–19. When one lamp burns out or is removed, the current can still be traced from the SOURCE directly to each lamp whether there are five or a dozen or more. From the other terminal of each lamp the current can be traced back along the wire, through the switch, to the SOURCE. Try it; cover one or more of the lamps with a narrow strip of paper leaving the wires exposed. The circuit will operate regardless of the number of lamps in place, and the switch will always turn the lamps on and off. This is the way the sockets in a five-light fixture (or in five separate one-lamp fixtures) are wired in order to be controlled by a single wall switch.

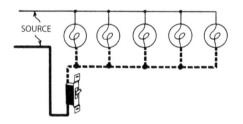

Fig. 5–19 One switch here controls five lamps.

Adding more switches to the circuit The circuits discussed so far might serve in an outbuilding on a farm, but all the lights in a house would never be controlled by a single switch. It is simple to wire a number of sockets with separate switches. Figure 5–20 is the same as Fig. 5–19 except that in place of one switch there are now five switches. With a piece of paper cover all but the first lamp and switch; you now have the simple circuit of Fig. 4–3. Cover all but the last lamp and switch and again you have the circuit of Fig. 4–3. Cover any four switches and lamps with the same result. Trace the current from the SOURCE to any lamp; it can be traced through the lamp to the switch for that lamp, and back to the SOURCE. This can be done whether one or two or all switches are on; each is independent of the others.

Turn now to Fig. 5–21, where a group of lamps has been substituted for each single lamp, so that now there are five groups of lamps and five switches, numbered *1* to *5*. With a piece of paper cover groups *2, 3, 4,* and *5* with their switches, and immediately the simple circuit of Fig. 5–19 appears—five lamps controlled by a single switch. Cover any four groups, and in each case the current

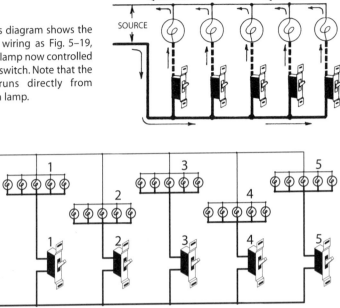

Fig. 5–20 This diagram shows the same parallel wiring as Fig. 5–19, but with each lamp now controlled by a separate switch. Note that the white wire runs directly from source to each lamp.

Fig. 5–21 This diagram is the same as Fig. 5–20 except that each switch controls five lamps.

can be traced from the SOURCE to any of the lamps and through the switch controlling the group, and back to the SOURCE.

Figure 5–21 is the basic wiring diagram for a five-room house in which each room has a five-light fixture and one switch to control it. Actually the wires would run more as shown in Fig. 5–22, which is more pictorial, with wires coming into the basement, then running to two rooms on the first floor and three rooms on the second floor.

Connecting three-way switches   The switch used for controlling a light from one point is called a single-pole switch. Three-way switches allow you to control a light from two points (not three, despite the name). With three-way switches you can turn a hall light on or off from upstairs or downstairs, or a garage light from house or garage, or a yard light from house or barn. Such switches have three different terminals for wires. Their internal construction achieves the same electrical function as the porcelain-base type (technically, a single-pole double-throw switch) shown in Fig. 5–23. If two such switches are arranged as shown in Fig. 5–24, careful study shows that if the light is on (regardless of whether the handles of the two switches are both up or both down), it can be turned off by throwing the handle of either *A* or *B* to the opposite position. The light can be controlled by either switch *A* or *B*, regardless of the position of the other switch in the pair.

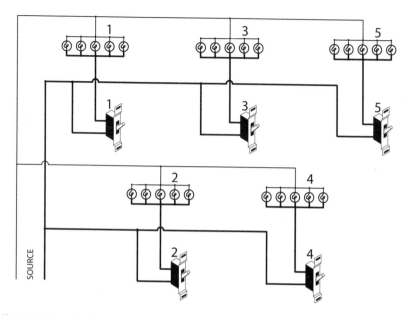

Fig. 5–22 The circuit of Fig. 5–21, but rearranged.

In actual wiring, a switch that looks like the switch in Fig. 5–7 is used, except it has three terminals instead of two, and the words ON and OFF do not appear on the handle. The term "three-way" undoubtedly refers to the number of terminals and not the number of handle positions. In one position of the handle, terminal *A* (which corresponds to the center terminal of the switches in Fig. 5–24) is connected inside the switch to terminal *C;* in the other position, terminal *A* is connected to terminal *B*. Usually, the common terminal *A* is identified by being a darker color than the other terminals which are natural brass.

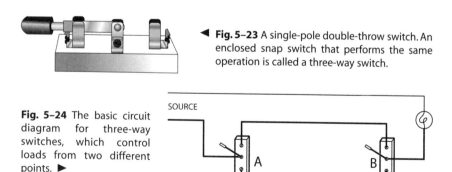

◄ **Fig. 5–23** A single-pole double-throw switch. An enclosed snap switch that performs the same operation is called a three-way switch.

Fig. 5–24 The basic circuit diagram for three-way switches, which control loads from two different points. ▶

Study the diagram of Fig. 5–25 in which the handles of both switches are down. The current can be traced from SOURCE over the black wire through *Switch 1,* through terminal *A,* out through terminal *B* and up to terminal *C* of *Switch 2,* but there it stops. The light is off.

Next see Fig. 5–26 where the handles of both switches are up. The current can be traced as before from SOURCE over the black wire through *Switch 1,* through terminal *A,* but this time out through terminal *C,* and from there up to terminal *B* of *Switch 2.* There it stops and the light is off.

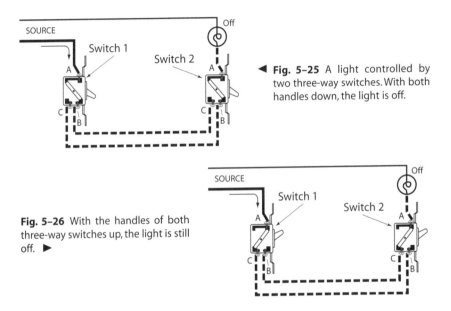

◄ **Fig. 5–25** A light controlled by two three-way switches. With both handles down, the light is off.

Fig. 5–26 With the handles of both three-way switches up, the light is still off. ►

Now examine the diagrams of Figs. 5–27 and 5–28. In both diagrams the handle of one switch is up, the other down. In each case the current can be traced from SOURCE over the black wire, through both switches, through the lamp, and back to SOURCE. The light in both cases is on. In the case of either Fig. 5–27 or Fig. 5–28, throwing either switch to the opposite position changes the diagram back to either Fig. 5–25 or Fig. 5–26, and the light is off. In other words, the light can be turned on and off from either switch. Manufacturers differ as to where the common terminal is located; in some cases the common terminal will be on one side of the switch instead of one end, so take care to find the uniquely colored terminal.

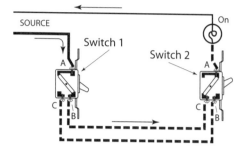

Fig. 5–27 Three-way switch: one handle down and the other up; the light is now on. Trace the current along the arrows.

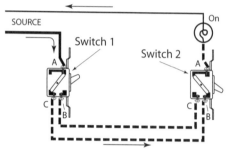

Fig. 5–28 Three-way switch: both handles are now the opposite of Fig. 5–27, but the light is still on. Either switch controls it. ▶

The wiring of three-way switches is simple, as these diagrams show. Run the white wire from SOURCE as usual to the light to be controlled. Run the black wire from SOURCE to the common or marked terminal of the first three-way switch. Run a black wire from the common or marked terminal of the second three-way switch to the light. That leaves two unused terminals on each switch; run two black wires from the terminals of the first switch to the two terminals of the second switch. It makes no difference whether you run a wire from *B* of the first switch to *C* of the second as shown, or from *B* of the first to *B* of the second. The wires that start at one switch and end at another are called runners, travelers, or jockey legs. In the case where you are using a cabled wiring method that won't easily allow for two black wires, refer to three-way switch details beginning on page 321.

Connecting four-way switches   To control a light from more than two points, use two three-way switches, one nearest the SOURCE and the other nearest the light, and four-way switches at all remaining points in between. Four-way switches can be identified by the fact that they have four terminal screws and do not have ON-OFF markings on the handles. Figs. 5–29 through 5–32 show two types of four-way switches and the wiring diagrams for circuits using each type. Depending on the manufacturer, the internal connections of four-way switches may be as shown in Fig. 5–29 or as in Fig. 5–31. In each of these figures, the four terminals are labeled *A, B, C,* and *D.*

The four-way switch in Fig. 5–29 is shown in a circuit in Fig. 5–30. The two jockey legs from one three-way switch are connected to terminals *A* and *B* of the

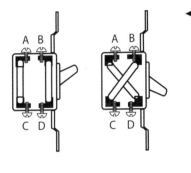

Fig. 5–29 Internal connections of one style of four-way switch with the handle in the up and the down positions. The terminals are labeled *A, B, C,* and *D.*

Fig. 5–30 Four-way switches are usually connected as shown here. With the four-way switch handles in the positions shown in the upper diagram, the light is off; in the lower diagram, the light is on. Two four-way switches are shown in the circuit, but any number may be installed depending on need. Refer to Fig. 5–29 to identify the terminals.

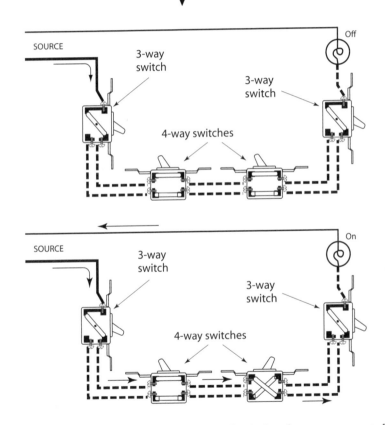

four-way switch; from the other three-way the jockey legs are connected to terminals *C* and *D* of the four-way.

The four-way switch in Fig. 5–31 is shown in a circuit in Fig. 5–32. The jockey legs from one three-way switch are connected to terminals *A* and *D* of the four-way switch, and the jockey legs from the other three-way are connected to terminals *B* and *C* of the four-way switch.

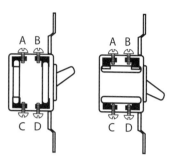

◀ **Fig. 5–31** Internal connections of the other style of four-way switch with the handle in the up and the down positions. Terminals are labeled *A, B, C,* and *D.*

Fig. 5–32 Some brands of four-way switches are connected as shown. With the four-way switch handles in the positions shown in the upper diagram, the light is off; in the lower diagram, the light is on. Refer to Fig. 5–31 to identify terminals.
▼

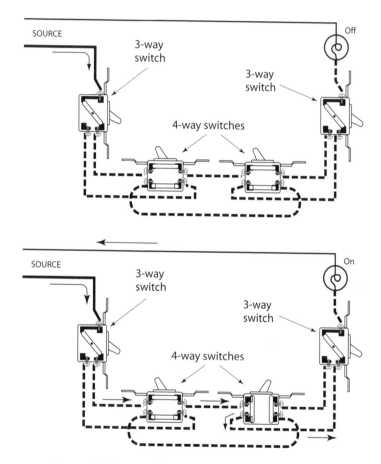

If you are not sure which type of four-way switch you have, try one diagram; if it does not work, try the other. No harm will come by making wrong connections except that the circuit will not work. Most packaging for three-way and four-way switches includes a connection diagram that should indicate which terminals do what.

CHAPTER 6
Overcurrent Devices

IT IS IMPOSSIBLE for an electric current to flow through a wire without heating the wire, and if too much current flows, the insulation will fail. With sufficient current, the conductor itself may get hot enough to start a fire. It is therefore necessary to carefully limit the current to a maximum value that is safe for a given size and type of wire.

Any device that opens the circuit when the current in a wire reaches a predetermined number of amperes is called an "overcurrent device" in the *NEC*. There are several kinds, and they all can be considered the safety valves of electrical circuits. The two types discussed in this chapter are fuses and circuit breakers. Fig. 6–1 shows how fuses and breakers are designated in wiring diagrams. The abbreviation for circuit breakers is CB (or cb), as shown in Fig. 6–12.

This chapter covers the basics of what overcurrent protective devices are and how they operate. Conductor protection is discussed in the context of very simple installations involving straightforward applications of the numbers in the basic *NEC* ampacity table. (*NEC* Table 310.16, one of the most important tables in the Code, is reproduced in the Appendix of this book as Table A–1.) The majority of residential applications and quite a few commercial and industrial applications involve only this level of detail. Chapter 7 continues the analysis with respect to wire selection, and Chapter 26 provides comprehensive coverage of this topic, one of the most important and complex in the entire *NEC*. If you're not making an exhaustive study right now, you can leave Chapter 26 for later when you need to know about heavier commercial and industrial wiring.

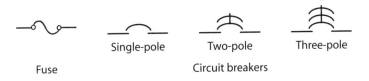

| | Single-pole | Two-pole | Three-pole |

Fuse   Circuit breakers

Fig. 6–1 At left, symbol for a fuse; at right, symbols for circuit breakers.

Besides being used to protect *wires* from too great a current, overcurrent devices are also used to protect electrical *equipment.* For example, a motor may require 15 amps to deliver the horsepower stamped on its nameplate. Motors can deliver more than their rated horsepower, but while doing so consume more than the normal number of amperes. If the overload is continued long enough, the motor will burn out. To protect the motor against such overload currents, a special protective device called an "overload device" is installed near the motor. Overload devices are explained in Chapters 15 and 30.

FUSES

A fuse is basically a short ribbon of metal alloy with a low melting point and of a size that will carry a specified current indefinitely but which will melt when a larger current flows. When the metal link melts, the fuse is said to "blow." When a fuse blows, the circuit is open just as if a wire had been cut, or a switch opened, at the fuse location. The two basic designs are the plug fuse and the cartridge fuse, with variations of each. The *NEC* limits the use of plug fuses to circuits of not over 150 volts to ground. If the premises are served by a 120/240-volt line, the voltage between either hot wire and the grounded wire is 120 volts (120 volts to ground), therefore plug fuses may be used on 240-volt circuits for water heaters, motors, and so on, although cartridge fuses are more often used.

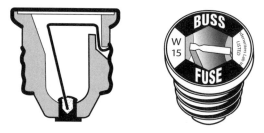

Fig. 6–2 Plug fuses are made only in ratings up to 30 amps. The type shown here is not permitted in new installations.

Plug fuses—restricted use   The common plug-type fuse is shown in Fig. 6–2. The fusible link is enclosed in a sturdy housing that prevents the molten metal from spattering when the fuse blows. There is a window that lets you see whether the fuse has blown. The largest plug fuse is rated at 30 amps; smaller standard sizes are 15, 20, and 25 amps. Sizes smaller than 15 amps are also available. The *NEC* requires that plug fuses rated at 15 amps or less be of hexagonal shape, or have a window or other prominent part of hexagonal form; those rated at more than 15 amps are round.

Traditional plug fuses, regardless of amperage, all have identical Edison (screw-shell) bases. You can use these fuses only for replacements, and then only where there isn't any evidence of overfusing (using a fuse too large for the size of wire), an obvious temptation. For new installations you must use "Type S" plug fuses.

Type S fuses—nontamperable   The *NEC* requires Type S fuses for new installations that use fuses. See Fig. 6–3. They have threads that prevent a larger size being inserted into an adapter designed for a fuse with a lower ampere rating, averting a frequent source of fires. Be sure to turn these fuses firmly so the springs on their inner shoulders flatten, allowing the center contact to connect to the mating contact in the adapter. Note that a Type S adapter cannot be removed with ordinary tools without destroying the adapter.

Fig. 6–3 A typical Type S nontamperable fuse (center and cross section at right) and its adapter. Once an adapter has been screwed into a fuseholder, it cannot be removed. This prevents the use of fuses larger than originally intended.
▼

Fig. 6–4 Cross section of a typical time-delay fuse. Time-delay fuses carry temporary overloads safely without blowing.

Time-delay fuses—convenient for motor loads   Unlike ordinary fuses, time-delay fuses take advantage of the ability of wire to carry a higher load momentarily. A typical time-delay fuse is shown in cross section in Fig. 6–4.

Consider a lighting circuit in a home, wired with 14 AWG copper wire that has an ampacity of 15 amps and is protected by a 15-amp fuse. Most of the time, the wire will be carrying less than 15 amps, and the temperature of the wire and its insulation will be well within safe limits. If the current is increased to 30 amps, the ordinary type of fuse will blow in a few seconds even though 30 amps flowing for even half a minute will not heat the wire or its insulation to the danger point, especially if the current was very low before being increased to 30 amps.

In practice, there are often conditions just as described. Perhaps 5 amps is flowing in the wire, representing 600 watts of lights. Then a motor, such as a workshop motor, is turned on. The motor consumes perhaps 30 amps for some seconds while it is starting; after that it drops to a normal of around 6 amps. Often the ordinary type of fuse blows during this starting period, although the wire and its insulation were not in any danger.

Time-delay fuses (sometimes incorrectly called time-lag) do not blow like ordinary fuses on large-but-temporary overloads, but do blow like ordinary fuses on small continuous overloads, and instantly on short circuits. The difference between the two types can be seen from Table 6–1.

All fuses, ordinary and time-delay types, are designed to carry 110% of their rated current indefinitely without blowing. But all fuses installed in very hot locations

Table 6–1 COMPARISON OF TIME-DELAY AND ORDINARY FUSES

Actual current amp	15-amp fuse		30-amp fuse	
	Time-delay, sec	Ordinary, sec	Time-delay, sec	Ordinary, sec
30	31.0	3.9		
45	10.0	0.8	140.0	22.0
60	5.0	0.3	27.0	4.4
75	1.5	0.2	11.0	1.8
90	0.5	0.1	5.4	1.0

will blow faster on any given current than they would if installed in locations having ordinary temperature conditions.

Cartridge fuses—necessary for larger loads   If the fuse is rated at more than 30 amps, there is no choice but to use a cartridge fuse. Cartridge fuses are available in all ratings, so they can be used even if currents smaller than 30 amps are involved. There are two basic types of cartridge fuses: the ferrule-contact type in Fig. 6–5 and the knife-blade contact type in Fig. 6–6. Fuses rated at 60 amps or less are of the ferrule-contact type; those rated at more than 60 amps are of the knife-blade contact type. Cartridge fuses are made both in ordinary and in time-delay type. In either time-delay or non-time-delay cartridge fuses, the fusible link melts on a short circuit. On the non-time-delay, it also melts on an overload. In the time-delay type shown in Fig. 6–7, an overload causes a bit of solder to melt, and a spring then opens the circuit, just as in time-delay plug fuses.

Fig. 6–5 The ferrule-contact type of fuse is made only in ratings up to and including 60 amps.

Fig. 6–6 The knife-blade-contact type of fuse is made only in ratings above 60 amps.

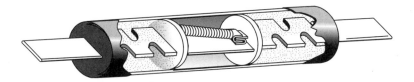

Fig. 6–7 Cartridge fuses are also made in the time-delay type. They are especially useful in protecting motors, which usually require several times as many amperes for starting as while running.

Design specialization   Cartridge fuses are made in many UL "classes," based not only on their ampere and voltage ratings, but also on their dimensions, time-delay characteristics, ability to handle high fault currents, and current-limiting abilities. Table 6–2 summarizes the more common groups of fuses, whether they are current limiting, the maximum available fault current, etc. Each type has its purpose; never substitute one for another without qualified review. In many cases, but not all, the fuse dimensions preclude improper substitutions. For example, an RK series fuse has a groove in a ferrule or a slot in a blade ("R" standing for rejection) that matches up to rejection features in fuseholders used where current limitation is required. That way a Class H fuse, which has no current limitation but uses similar physical dimensions, cannot be installed improperly.

Table 6–2 CHARACTERISTICS OF CONTROL-CIRCUIT, POWER, AND LIGHTING CARTRIDGE FUSES BY UL CLASSIFICATION

Characteristic	CC	G	H	J	K	L	RK	T
Renewable			X					
Nonrenewable	X	X	X	X	X	X	X	X
250-volt			0-600 A		0-600A		0-600A	
300-volt		0-60A						0-600A
600-volt	0-20A			0-600A	0-600A	601-6000A	0-600A	0-600A
Noninterchangeable	X	X*		X		X	X†	X
Currrent-limiting	X	X		X	‡	X	X	X
Maximum inter-rupting rating, amp	200 000	100 000	10 000	300 000	200 000	300 000	300 000	200 000
Time delay	Optional	Optional	Optional§	Optional	Optional	Optional	Optional	

* 16–20 and 21–30 amp Class G fuses are interchangeable with some miscellaneous or supplementary fuses but not with any of the other fuses in this table.

† Class RK fuses will fit in Class H fuseholders, but only Class RK fuses will fit in Class RK fuseholders.

‡ Class K fuses have current-limiting characteristics but are not permitted to be so labeled, as they are interchangeable with class H. See *NEC* 240.60(B).

§ Time delay not available in renewable type.

Renewable and nonrenewable   Class R cartridge fuses come in two types: nonrenewable or one-time type, and the renewable kind. Nonrenewable fuses once blown have no further value. Since in most cases only the fusible link is destroyed when the fuse blows, renewable fuses permit the fusible link to be replaced. Extreme care must be taken to thoroughly tighten the screws or bolts in the fuse when replacing the fuse link. Figure 6–8 shows a cross-sectional view of a renewable fuse. In external appearance there is no difference between the two types except that the renewable type can be taken apart.

Fig. 6–8 A renewable fuse can be disassembled for replacement of the fusible link.

One-time fuses usually run cooler and therefore do not overheat the panelboard clips as much as the renewable kind. They are generally considered more satisfactory and therefore are recommended. If a circuit is not overloaded, fuses don't blow very often. When one does blow, the cause for its blowing should be corrected before replacing the fuse. Many feel there is little justification for renewable fuses in properly maintained installations. The 2005 *NEC* now limits the use of these fuses to replacements in existing installations, and even there only if there is no evidence of overfusing, just like the rules for traditional plug fuses.

CIRCUIT BREAKERS

The *NEC* defines *circuit breaker* as "a device designed to open and close a circuit by nonautomatic means, and to open the circuit automatically on a predetermined overcurrent without damage to itself when properly applied within its rating." Thus a circuit breaker is a combination device composed of a manual switch and an overcurrent device. Its basic function is similar to that of a combination switch and fuse.

A circuit breaker of the type used in homes looks something like a toggle switch used to turn a light on and off. See Fig. 6–9. Essentially it consists of a calibrated bimetallic strip similar to that used in a thermostat. As current flows through the strip, heat is created and the strip bends. If enough current flows through the strip, it bends enough to release a trip that opens the contacts, interrupting the circuit just as it is interrupted when a fuse blows or a switch is opened. In addition to the bimetallic strip that operates by heat, most breakers have a magnetic arrangement that opens the breaker instantly in case of a short circuit. A circuit breaker can be considered a switch that opens itself in case of overload.

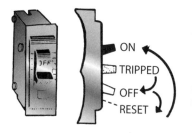

Fig. 6–9 A typical single-pole circuit breaker. If it trips, correct the reason for the overload. Then reset by pressing the handle beyond the OFF position, and then back to ON. Not all circuit breakers have an intermediate tripped position, or require movement beyond the OFF position to reset.

Circuit breakers are rated in amperes, just as fuses are rated. Like fuses, breakers will carry 110% of their rated loads indefinitely without tripping. Manufacturers of breakers furnish curves showing how long breakers of a particular variety will carry specific overloads. Most breakers will carry 150% of their rated load for perhaps a minute; 200% for about 20 seconds, and 300% for about 5 seconds, long enough to carry the heavy current required to start a motor.

ARC-FAULT CIRCUIT INTERRUPTERS

The 1999 *NEC* in 210.12 mentioned this new type of protection for the first time. It defines an *arc-fault circuit interrupter (AFCI)* as a device that can recognize the electronic characteristics of an arcing fault in progress and then open the circuit. These devices must be used for all circuits supplying outlets, for both receptacles and otherwise, in dwelling unit bedrooms. The devices have electronic circuitry capable of recognizing the chaotic waveforms characteristic of electrical arcing. Their internal calibration causes a trip based on the likely ignition energy being released in the arc. This type of low-level arcing causes many fires because it is often well below the point that would cause an ordinary overcurrent device to trip. The manufacturers and the testing laboratories conducted extensive research to determine what fault current levels will produce secondary ignition of adjacent combustible materials. At this writing, the only AFCI protective devices on the market are circuit breakers. Although there is language in the 2005 *NEC* to recognize limited applications for AFCI receptacles, the restrictions are now so severe (must be within 6 ft of the overcurrent device and wired with a metallic wiring method) that no receptacle manufacturer presently plans to actively enter this market.

Although these devices have a "push to test" button on them that looks like a GFCI, they are not GFCIs (which are covered in Chapter 9) and they have an entirely different function. However, some do incorporate GFCI sensing within the same housing.

BREAKERS *vs* FUSES

Relatively few fuse installations are being done in residential and commercial applications because breakers have many advantages. When a fuse blows, spare fuses may or may not be on hand. When a circuit breaker trips, correct the cause of the overload, then reset it as shown in Fig. 6–9. The circuit breaker provides good protection and does not trip on large but temporary overloads. Modern homes are usually equipped with circuit breakers.

However, fuses also have substantial virtues, especially in industrial settings, and they will remain an important part of the overcurrent protection scene for the foreseeable future. Fuses work very well in cases where extreme speed of operation is essential, such as where management has made the design decision to minimize the effects of a severe fault on other continuing operations at the facility. A fuse is a much simpler device than a circuit breaker. The internal moving parts of large breakers in particular have substantial inertia that affects their clearing time. In addition, because fuses are nonmechanical, they don't require exercise. It's difficult to imagine a fuse, even if ignored for 20 years, not blowing when subjected to fault current.

DETERMINING PROPER RATING OF OVERCURRENT DEVICE

Both fuses and breakers are available in standard ratings of 15, 20, 25, 30, 35, 40, 45, 50, 60, 70, 80, 90, 100, 110, 125, 150, 175, and 200 amps, and of course larger sizes (up to 6000 amps) for use where required. See *NEC* 240.6. For fuses only, the *NEC* sets additional standard ratings of 1, 3, 6, 10, and 601 amps.

The fuse must blow, or the breaker open, when the current flowing through it exceeds the number of amperes that is safe for the wire in the circuit. The larger the wire, the greater the number of amperes it can safely carry.

The *NEC* specifies the ampacity of wires—the maximum number of amperes that can be safely carried by each size and type of wire. The ampacity of any size and kind of copper (or aluminum) wire can be found in the *NEC* ampacity tables, beginning with Table 310.16. This is the table that you'll use almost exclusively; it is reproduced in the Appendix of this book as Table A–1. You may wish to memorize the allowable overcurrent protective device sizes for the smaller sizes of copper wire usually used in residential, commercial, and farm wiring. These sizes are:

14 AWG	15 amps
12 AWG	20 amps
10 AWG	30 amps
8 AWG	40 amps*
6 AWG	55 amps*

* Conductor ampacities, not overcurrent protective device size limits. See text.

When you look at *NEC* Table 310.16, you'll note that 14 and 12 AWG have higher ampacities, along with an asterisk note pointing to *NEC* 240.4(D). That section sets upper limits on overcurrent protection for these conductors regardless of the table values. The 8 and 6 AWG conductor limits come directly from the table, in the 60°C column. Even though you may have 90°C-rated conductors, you still need to observe these limits because of listing restrictions, unless the devices *at both ends of the wire* have been listed and marked for other limits. These and many more details are discussed in Chapter 7.

Available fault current   As electrical systems get larger, the total amount of power that must be available to support them increases as well. This results in a potential safety problem. *NEC* 110.9 requires that electrical devices intended to "interrupt current at fault levels" must be able to handle the total amount of fault current "available at the line terminals of the equipment."

This amount is much different than the nominal rating (trip rating on the handle) of an overcurrent device. It assumes what engineers call a "bolted fault." That is, it is the amount of current that would flow if you were to deliberately take two ungrounded conductors out of the line terminals of the overcurrent device and bolt them together, creating the worst-possible-case short circuit condition. Because you make the calculated connection on the line side of the device, you aren't allowed to count on any impedance within the overcurrent device to reduce the amount of current that might flow. Overcurrent devices must be able to clear this fault without endangering personnel or associated electrical equipment. Testing laboratories carry out actual tests under conditions that essentially duplicate these extreme conditions, in accordance with the applicable product standards.

Current limitation   The default clearing rating for circuit breakers and fuses, as given in the *NEC*, is 5000 amps in the case of circuit breakers, and 10 000 amps for fuses. Overcurrent devices suitable for use with available fault currents above these values are marked with the increased rating. In addition to the ability to handle high fault current, some overcurrent devices are "current limiting." Think back to the sine wave graphs of alternating current in Chapter 3. Fault currents rise to a current maximum just like any other alternating current, the only difference being the extreme magnitude of that current maximum.

Suppose your fuse melted or your circuit breaker opened so quickly that the fault never reached its first current maximum. Remember, that first current maximum is reached in only one quarter of a cycle, or $1/240$ sec. Such devices actually do exist, however, and as such meet the *NEC* 240.2 definition of being current limiting. Electrical equipment differs in its ability to withstand massive fault currents and return to service. As you progress, you will find installations engineered for an absolute minimum of downtime. Part of that design invariably involves confining fault damage to the point of the fault.

For example, suppose the available fault current at a motor lead was 12 000 amps. A bolted fault, such as could result from misconnecting the leads, would inevitably damage the system at that point. However, the absence of current limitation in the overcurrent devices would make that damage more extensive, besides putting the motor controller out of service by the enormous currents passing through it. UL generally tests 50-hp and smaller motor controllers for 5000-amp withstand if they incorporate running overload protection, which is the usual case. That means they will handle a fault of that magnitude without being damaged to the point of needing replacement. To address the product limitations, the system might include current-limiting overcurrent devices that prevent faults from reaching this point. If you use the wrong device, you could defeat the engineering and introduce a serious hazard.

Current limitation also figures in system coordination, which the *NEC* defines as "properly localizing a fault condition to restrict outages to the equipment affected, accomplished by the choice of selective fault-protective devices." The 12-kA fault just discussed has the potential to open every overcurrent device upstream from the fault, even to the point of putting the entire facility into darkness. Whether or not that happens depends on whether the overcurrent device closest to the fault clears quickly enough so the resulting fault current doesn't open upstream devices.

This degree of coordination requires very sophisticated engineering that is well beyond the scope of this book, and may not be practicable. Nevertheless, the 2005 *NEC* for the first time mandates this level of performance on emergency (Article 700) and legally required standby (Article 701) systems.

Actual calculations of available fault current consider capacities in the utility supply system, impedances in premises wiring and equipment, and motor contributions. (Electric motors continue to rotate while a fault is in progress. A rotating electric motor with its supply disrupted becomes a generator, adding current and thereby worsening the severity of the initial stages of a fault.) Such calculations are also beyond the scope of this book. However, you need to know why the requirements exist so you don't inadvertently create a hazard. These problems are occurring more frequently even in residential applications. Consider, for example, a house or small apartment building in a city, with a 200-amp service. If the utility goes to a network distribution in the area, the fault current available could exceed 100 kA. Utilities use network distributions, where a large group of distribution transformer secondaries are wired together in parallel, to minimize outages due to a single transformer failure. Always inquire of the local utility regarding the available fault current at the service point.

Series ratings Some systems use a tandem approach to clearing high-current faults. For example, suppose a main circuit breaker capable of clearing a 22-kA fault were put ahead of a 10-kA-rated branch-circuit breaker. Would that work with, say, 20 kA available at the branch? The answer is yes, if and only if the combination is part of a *series combination* marked on the end use equipment. Be careful to use only combinations marked on the equipment, because they are the only combinations actually tested for these extreme conditions. Note that all wiring systems have impedance, and 22 kA available at a service may be only 9 kA at a downstream panel. If a suitably trained person verifies this by appropriate calculations, then the need for the listed series combination is eliminated. If relying on a series combination listing, however, you must mark the equipment with a warning as to the system rating, for the benefit of those who come after you.

Joining of different sizes of wire If two different sizes of wire are joined together, as in Fig. 6–10, the ampere rating of the overcurrent device must be no greater than that permitted for the smaller wire. In this case the smaller 14 AWG can't be protected above 15 amps, so that is the maximum ampere rating of the overcurrent device used. In practice, a situation of this kind is sometimes found, especially in farm wiring, where 8 AWG is used for an overhead wire for

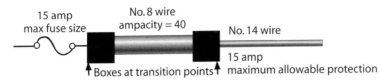

Fig. 6–10 When two different sizes of wire are connected in series, the largest fuse that may be used is one that protects the smaller wire.

mechanical strength and to avoid excessive voltage drop. Although the 8 AWG wire has an ampacity of 40, the circumstances may be that more than 15 amps is never required and the maximum rating of 15 amps for the overcurrent device is not inconvenient.

On the other hand, there may be a situation as shown in Fig. 6–11 where 8 AWG is used but where more than one 15-amp circuit is to be connected. In that case a 40-amp overcurrent device is used at the starting point, but one or more smaller overcurrent devices are installed at the point where the wire size is reduced. Each of these overcurrent devices must have an ampere rating not more than the ampacity of each of the smaller wires installed.

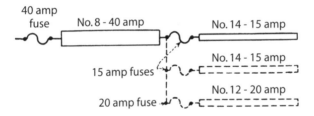

Fig. 6–11 Fuses or breakers are used to protect the smaller wires where the wire is reduced in size.

There are fifteen exceptions to these general requirements, covered in *NEC* 240.21. One covers service conductors, which by definition don't have overcurrent protection at their load end. Generators and busways have special rules in their own *NEC* articles. Motor circuits also have special rules, which this book covers in Chapter 30. Six allowances address connections involving transformers; for more detail see the discussion beginning on page 549 on installing and protecting transformers and their conductors. That leaves five rules covering the tapping of a smaller wire from larger wire, and one of those addresses a complicated industrial provision beyond the scope of this book. The remaining four are listed here. An important point, which applies to all fifteen *NEC* allowances in 240.21, is that the allowances may not be applied to one another; that is, you can't tap a tap. Once you use one of the *NEC* allowances, and end up with a wire without full overcurrent protection at its supply end, you can't tap that wire with an even smaller wire, regardless of its length.

Twenty-five-foot tap rule   The first allowance, generally referred to as the "25-ft tap rule," allows you to omit overcurrent protection at the point of supply if all four of the following conditions are met:

1. The smaller wire is not over 25 ft long.

2. The smaller wire has an ampacity at least one-third that of the rating of the overcurrent device protecting the larger wire.

3. The smaller wire ends in a single overcurrent device having an ampere rating not greater than the ampacity of the smaller wire.

4. The smaller wire is protected against physical damage.

The 25-ft tap rule is illustrated in Fig. 6–12, which shows a 1 AWG wire to which a 6 AWG wire is connected. (In this case both the 1 AWG wire and the 6 AWG wire are "feeders" and the 6 AWG wire is a "tap.") Beyond the single overcurrent device at the end of the 6 AWG wire, there may be as many circuits of smaller size as desired, but each one must be protected by an overcurrent device based on the ampacity of the wire it protects.

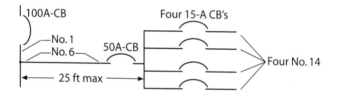

Fig. 6–12 This illustrates the "25-ft rule" which permits a wire size to be reduced without fuse or breaker at the point where reduced. See text.

Ten-foot tap rule   The second allowance is the "10-ft rule" permitted by *NEC* 240.21(B)(1). A smaller wire may be tapped to a larger one without an overcurrent device where the size is reduced, if the tap wires meet all the following conditions:

1. They must be not over 10 ft long.

2. They must have an ampacity not less than (a) the combined computed loads on the circuits supplied by the tap conductors, and (b) not less than the ampere rating of the panelboard or other device supplied by the tap wires.

3. They may not extend beyond the switchboard, panelboard, disconnecting means, or control device supplied by the tap wires.

4. They must be installed in a raceway (such as conduit) from the tap to the panelboard or other overcurrent-device enclosure.

5. In a field installation, their ampacity must equal at least one-tenth the rating of the overcurrent device protecting the larger wire. This restriction doesn't apply if the smaller wires stay entirely within a transformer vault (these generally have a three-hour fire rating), or entirely within a single enclosure.

Outdoor feeder taps   The third allowance allows outdoor field wiring to be run under similar rules as for comparable utility practice. As long as you stay completely outside the building (except at the termination), the tap can be of any length and of any relative size. At the load end, however, the smaller wire must arrive at a single overcurrent device having an ampere rating not greater than the ampacity of the smaller wire. Such taps supply electrical equipment in buildings or at structures. Remember, the rule would lose its technical validity if it were to result in conductors of unlimited length within a building or structure. For that reason, the tap must end within the building or structure disconnecting means, or immediately adjacent thereto, and the disconnect must be outdoors or inside nearest the point of conductor entry. Effective with the 2002 *NEC*, however, the point of entrance can be artificially extended into the building if the wiring method is protected in the same way as for similar extensions of service conductors. Review *NEC* 230.6 for the specific methods, which include routing the conductors in raceway below the building, or through a transformer vault (or equal), or encasing in at least 2 in. of concrete.

Branch circuit taps   The fourth allowance, covered in 240.21(A) and the exceptions to 210.19(A)(4), reflects common sense. You can make taps not over 18 in. long from circuit wires to serve an individual outlet (but not a receptacle outlet) provided that the short wire (a) has sufficient ampacity to serve its specific load and (b) has a minimum ampacity of 15 amps on circuits smaller than 40 amps, and 20 amps on 40- or 50-amp circuits.

This provision, through the reference to *NEC* 240, includes flexible cord and fixture wires. Here in tabular form are the cord and fixture wires you can supply from the branch circuits as described:

20-amp circuit	18 AWG fixture wire (up to 50 ft run length)
	18 AWG (or tinsel) cords (if part of listed appliance or portable lamp)
	16 AWG fixture wire (up to 100-ft run length)
	16 AWG cord (including listed extension cords, or cords made up with listed parts)
	14 AWG and larger fixture wire and cords
30-amp circuit	16 AWG or larger cord (if part of listed appliance)
	14 AWG fixture wire or larger
40-amp circuit	20-amp-capacity cord or larger
	12 AWG fixture wire or larger
50-amp circuit	20-amp-capacity cord or larger
	12 AWG fixture wire or larger

CHAPTER 7
Selecting Conductors

A CONDUCTOR IS ANY MATERIAL that can carry electric current. Wires are the most common form of conductors used to deliver electric power from the point where it is generated to the point where it is used, and copper is the material most commonly used, although aluminum is also widely used. The *NEC* seldom uses the word "wire" but frequently uses the word "conductor," which may be a wire, busbar, or any other form of metal suitable for carrying current. All electric wires therefore are electric conductors, but not all conductors are wires. Copper busbars, for example, are conductors but not wires. This book frequently uses the word "wire."

Insulators are materials that do not conduct electric current. Metal wire is enclosed in plastic or other insulation to help protect against stray current. Current flowing through a wire causes heat; the heat varies as the square of the current (amperes). There is a limit to the degree of heat that various types of wire insulation can safely withstand, and even bare wire must not be allowed to reach a temperature that might cause a fire. The *NEC* specifies the ampacity (the maximum current-carrying capacity in amperes) that is safe for wires of different sizes with different insulations and under different circumstances.

You must always use wire (1) that has insulation suitable for the voltage, temperature, and location (wet, dry, corrosive, direct burial, etc.); and (2) that has an ampacity rating adequate for the current. These conditions assure that your installation meets the minimum safety threshold required by the *NEC*. A responsible designer goes beyond these minimums and selects conductors of sufficient size to avoid excessive voltage drop, which is wasted power. Insulation types and ampacities are listed in *NEC* Table 310.16 (reproduced in the Appendix of this book as Table A–1). But it is up to you to determine voltage drop. This chapter shows you:

■ What you need to know about selecting insulation types for wires

■ How to minimally size a wire based on the conditions of use

■ When and how to adjust that size to avoid excessive voltage drop

The *NEC* recognizes many different types of wire that may be used in wiring buildings. This chapter covers only the more ordinary types. The term "wire" refers to a single conductor and not a cable. This chapter focuses on how to select individual conductors. Many wiring methods consist of multiple conductors within an overall assembly. These cabled wiring methods are covered in Chapter 11. You can apply the knowledge gained from these two chapters when you select the wiring method. For example, from what you learn here you might decide you need an 8 AWG wire, and after reading Chapter 11 you might decide to run Type MC cable. So when you go to the supply house, you will look for 8 AWG Type MC cable.

CHOOSING A TYPE OF WIRE INSULATION

As stated in the chapter introduction, the insulation type you choose must be suitable for the voltage, temperature, and location.

***NEC* limits on installation locations**   The *NEC* limits the use of some wires to dry locations only, some others to just dry and damp locations but not wet locations, and still others may be used in dry or wet locations, although some have a more restricted temperature range when used in wet locations. The definitions of these different locations are important. Note that the location is that of the wiring method itself. For example, if you run rigid nonmetallic conduit underground, the wiring within it must be suitable for wet locations. You cannot assume that the wiring method will forever shield the enclosed conductors from moisture in the ground. *NEC* Article 100 defines locations as follows:

- *Damp location*—Partially protected locations under canopies, marquees, roofed open porches, and like locations, and interior locations subject to moderate degrees of moisture, such as some basements, some barns, and some cold-storage warehouses.

- *Dry location*—Location not normally subject to dampness or wetness. A location classified as dry may be temporarily subject to dampness or wetness, as in the case of a building under construction.

- *Wet location*—Installations underground or in concrete slabs or masonry in direct contact with the earth, and locations subject to saturation with water or other liquids, such as vehicle washing areas, and locations exposed to weather and unprotected.

Types TW and THW   These wires are of very simple construction, consisting of thermoplastic insulation applied over the conductor, and they can be used in dry, damp, or wet locations. Type TW is rated 60°C; Type THW with better insulation is rated 75°C. These wires have been largely supplanted by Type THHN/THWN and XHHW.

Types THHN and THWN   Types THHN and THWN both have thermoplastic insulation with an outer nylon jacket. Type THWN is moisture-resistant, allowing its use in wet locations. Type THHN is suitable for higher temperatures (90°C) than Types THW and THWN (both rated 75°C), and consequently has a slightly higher ampacity. Nylon has exceptional insulating qualities and great mechanical strength, resulting in a wire that is much smaller in diameter than ordinary Types TW and THW of corresponding size. Most of this wire made today has a dual rating of THHN/THWN. Since THHN can be used only in dry or damp locations, that means the wire has the higher temperature rating only when not in wet locations.

The *NEC* limits the number of wires of any given size that may be installed in a particular size of raceway. (For calculating number of wires in conduit, see p. 170.) In any given raceway size, a greater number of small-diameter wires (such as Types THHN or THWN) are permitted than wires with thicker insulation (such as Types TW or THW), and this may be a factor in deciding which wire to use. Recently Type THW and Type THWN wires have been recognized with a "-2" suffix (as in Type "THWN-2" or "THW-2") combining moisture resistance with a higher temperature rating. These wires can be used at a full 90°C ampacity in either dry or wet locations.

Type XHHW   This wire resembles Types TW or THW in appearance, but because of a thinner layer of insulation, the overall diameter is smaller, although not as small as Type THWN/THHN in sizes 6 AWG and smaller. The insulation is cross-linked synthetic polymer, which has excellent insulating, heat-resisting, and moisture-resisting properties. It may be used in dry or wet locations. Two different temperature ratings apply depending on whether the wires are in a wet location (restricted to 75°C), with the higher 90°C ampacity value allowed only in dry or damp locations. Here again, if the wire carries the "-2" suffix it can be used in wet locations at the higher temperature rating.

Rubber-covered wire   At one time, all wire for general use had rubber insulation. Although the "R" is still used in the designation, today it no longer refers to natural rubber. This wire consists of a thermosetting polymer, commonly seen as Type RHW, which has a 75°C temperature rating and is suitable for both dry and wet locations. As in the case of thermoplastic insulation, this insulation is being manufactured in a form that withstands 90°C temperatures as well, designated with the "-2" suffix.

Fixture wire   For the internal wiring of lighting fixtures, special wire known as "fixture wire" is used. There are many types, and the particular type used depends to a great extent on the temperature that exists within the fixture (while it is in use) and therefore in the wire itself. For wiring to incandescent and high-wattage, high-intensity discharge lampholders within enclosed fixtures, those temperature requirements may be very high, requiring 150°C or higher conductors. For fluorescent ballast compartments, the 90°C fixture-wire counterpart to THHN (TFN or TFFN) is commonly used. Refer to *NEC* Table 402.3 for all the possibilities.

Fixture wires may be used only in the internal wiring of fixtures, and from the fixture up to the circuit wires in the outlet box on which the fixture is mounted. They may never be used as branch-circuit wires leading to an outlet. Fixture wires have their own ampacity table, *NEC* Table 402.5. The *NEC* permits 18 AWG fixture wire (ampacity of 6 amps) in lengths not over 50 ft to be protected by up to a 20-amp branch-circuit protective device, and similarly 16 AWG (ampacity of 8 amps) in lengths not over 100 ft.

Other types of wire   There are many other types of wires, cables, and cords. Some are rarely used in the kinds of wiring discussed in this book and will not be mentioned. Other types are used for specific purposes, such as underground wiring, and are discussed in the chapters pertaining to that kind of wiring. Cables used for limited-energy applications such as telephones and thermostats are considered separately in Chapter 19.

UNDERSTANDING WIRE SIZES

The diameter of wire is measured in mils (1 mil = $\frac{1}{1000}$ or 0.001 in.) and its area in circular mils. A circular mil (abbreviated cmil) is the area of a circle that is 0.001 in. or 1 mil in diameter. Thus a wire that is 0.001 in. or 1 mil in diameter has a cross-sectional area of 1 cmil. Since the area of a circle is always proportional to the square of its diameter, it follows that the cross-sectional area of a wire 3 mils (0.003 in.) in diameter is 9 cmil; that of a wire 10 mils in diameter is 100 cmil; that of a wire 100 mils in diameter is 10 000 cmil, etc. The cross-sectional area of any round wire in circular mils is the area of the metal only, and is found by squaring the diameter in mils or thousandths of an inch (multiplying the diameter by itself).

American Wire Gauge (AWG) numbering system   The commonly used sizes of wire have been assigned a numbering system. The gauge commonly used is the American Wire Gauge (AWG); it is the same as the Brown and Sharpe (B&S) gauge. (This gauge is not the same as that used for steel wires used for nonelectrical purposes, such as fence wires.) Although other countries use wires sized on a metric basis, typically in square millimeters, the AWG system is so thoroughly integrated throughout the entire *NEC* and product standards that there are no plans to replace it. Even foreign countries that otherwise use the metric system still stock and use AWG sizes for conductors if they are using the *NEC*. Many countries throughout the western hemisphere and in other parts of the world follow this practice.

Table 8 in Chapter 9 of the *NEC* (reproduced in the Appendix of this book as Table A–5) shows all the *NEC* recognized sizes of wire, their areas in circular mils, their resistance in ohms per thousand feet, and their dimensions in fractions of an inch. The 2002 *NEC* began the practice of also including the corresponding metric dimensions for each conductor size. At the same time, the *NEC* dropped use of the abbreviation "No." ahead of wire sizes, and now uses only the wire size followed by "AWG." This book follows that convention except at the beginning of a sentence, when "Number" is used to avoid starting a sentence with digits.

The approximate sizes of typical sizes of wire without the insulation are shown in Fig. 7–1.

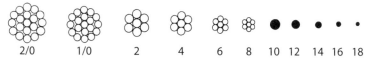

| 2/0 | 1/0 | 2 | 4 | 6 | 8 | 10 | 12 | 14 | 16 | 18 |

Fig. 7–1 Approximate diameters of typical sizes of electric wires without the insulation.

Wire size designations   Number 14 AWG wire, commonly used for ordinary house wiring, has a copper conductor 0.064 in., or 64 mils, in diameter. Wires smaller than 14 AWG are 16, 18, 20 AWG and so on. Number 40 AWG has a diameter of approximately 3 mils (0.003 in.), as small as a hair; many still smaller sizes are made. Sizes progressively larger than 14 AWG are 12, 10, 8 AWG and so on. Zero is indicated as 1/0 AWG, and is followed by 2/0, 3/0, and 4/0 AWG; these are sometimes shown as 0, 00, 000, 0000 AWG, and in either case called one-aught (or naught), two-aught, etc. Wires larger than 4/0 AWG are designated simply by their cross-sectional area in circular mils, beginning with 250 000 mil (250 kcmil) up to the largest recognized size of 2 000 000 mil (2000 kcmil or thousands of circular mils).

You may find it useful to remember that any wire that is three sizes larger than another will have a cross-sectional area twice that of the other. For example, 11 AWG has an area exactly twice that of 14 AWG, and 3 AWG has an area twice that of 6 AWG. Any wire that is six sizes larger than another has exactly twice the diameter and four times the area of the smaller wire. Number 6 AWG wire has exactly four times the area of 12 AWG.

The usual gauge used in measuring wire size is shown in Fig. 7–2. The wire is measured by the slot into which it fits, not by the hole behind the slot. You will not actually need this gauge because the *NEC* requires all building wire to be continuously marked with its size. If it is not properly marked, do not use it.

Stranded wires   Where considerable flexibility is needed, as in flexible cords, the conductors consist of many strands of fine wire twisted together. If a wire is stranded, it is slightly larger in overall diameter than a solid wire of the same size would be, but the number of circular mils in any one size is the same whether the wire is stranded or solid. The number assigned to stranded conductors is determined by the total cross-sectional areas of all the fine wires (the individual strands) added together.

Number 6 AWG and larger building wires (also 8 AWG if pulled into conduit or other raceways) must be stranded to be practical. Solid wires in larger sizes are

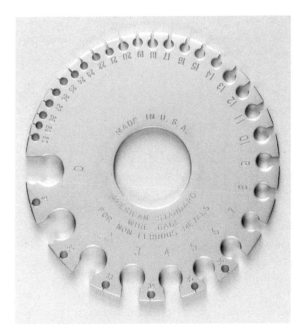

Fig. 7–2 A wire gauge shown slightly reduced in size. Measure the wire by the slot into which it fits. *(General Tools Mfg. Co., LLC)*

too stiff to handle, although this does not apply to wires such as the weatherproof type installed only overhead in free air. The stranding of each size has been standardized; the number of strands and the size of each strand can be found in Table 8 of Chapter 9 of the *NEC*.

Typical uses for various wire sizes   Wires in sizes from 50 AWG (diameter less than $1/1000$ in.) to 20 AWG are used mostly in manufacturing electrical equipment of all kinds. Numbers 18 and 16 AWG are used chiefly for flexible cords, signal systems, and similar purposes where small currents are involved, although many cables used for telephony and computer connections use even smaller sizes (24 AWG is common) because of the small currents involved in today's electronic systems. Numbers 14 to 4/0 AWG are used in ordinary residential and farm wiring, and of course in industrial and commercial work, where still larger sizes are also used. Number 14 AWG is the smallest size permitted for ordinary wiring. The even sizes, such as 18, 16, 14, 12, 8 AWG are commonly used; the odd sizes, such as 15, 13, 11, 9 AWG, with the exception of 3 and 1 AWG in service-entrance cables and other larger circuits, are seldom used in wiring. The odd sizes, however, are very commonly used in the form of magnet wire for manufacturing motors, transformers, and similar equipment, for which fractional sizes such as $15\frac{1}{2}$ AWG are not at all uncommon.

UNDERSTANDING AMPACITY AND THE *NEC* AMPACITY TABLES

Ampacity tables allow you (and the inspector) to predict what value of current will overheat and damage the insulation on a wire. Conductor insulation can be damaged by excessive heat in various ways, depending on the kind of insulation and the degree of overheating. Some insulations melt, some harden, some burn. In general, be aware that insulation loses its usefulness if overheated, leading to breakdowns and fires.

The ampacity specified in ampacity tables for any particular kind and size of wire is the current that it can carry without increasing the temperature of its insulation beyond the danger point. The insulation of Type TW wire stands the least heat, consequently this type has a lower ampacity than other kinds. High temperature fixture wire will stand far more heat, and consequently has a much higher ampacity per given wire size. The temperature of such wire carrying its rated current will be much higher than the temperature of plastic-insulated wire carrying its rated current, but the insulation of the high temperature fixture wire will not be damaged by the higher temperature.

Temperature ratings of wires   The rated ampacity of each kind and size of wire is based on an ambient temperature of 30°C, or 86°F. (Ambient temperature is the normal air temperature in an area while there is no current flowing in the wire. When current does flow, heat is created, and the surrounding air temperature as well as the conductor temperature will increase above the ambient.) Table 310.13 in your copy of the *NEC* gives information about each kind of wire, including the maximum temperature that the insulation of each kind of wire is permitted to reach. That temperature will be reached when a wire is carrying its full ampacity where the ambient temperature is 30°C, or 86°F. The maximum permitted temperature is called the *temperature rating* of the wire. Note that while the temperature rating of any wire can be found in *NEC* Table 310.13 as already mentioned, you can find the temperature ratings of wires ordinarily used in wiring buildings (but not all other types used for other purposes) at the top of the various columns in *NEC* Table 310.16.

In other words, if a wire has a temperature rating of, say, 60°C, or 140°F (the lowest temperature rating assigned to any kind of wire), it does *not* mean the wire may be used where the ambient temperature is 140°F. It means that the temperature of the wire itself must not exceed 140°F. It will reach that temperature when carrying its rated current in a room where the ambient temperature is 30°C, or 86°F. If the room ambient temperature is higher than 86°F and the wire is carrying its full rated current, its actual temperature will exceed its temperature rating of 60°C, or 140°F. Therefore if any kind of wire is installed in hot locations, its ampacity is reduced from that shown in *NEC* Tables 310.16 to 310.19. How to apply the necessary correction factors is explained later in this chapter.

Ampacity is not directly proportional to wire size   As covered earlier in this chapter, for two wires that are six sizes different from each other (for example, 6 AWG compared with 12 AWG), the larger will have twice the diameter, twice the circumference, and twice the surface area, but four times the cross-sectional area of the smaller wire. (The diameters of all sizes of wire as well as their area and resistance can be found in Table 8 of Chapter 9 of the *NEC*.)

Number 12 AWG wire has an ampacity (in the 75°C column) of 25, and you might expect 6 AWG, with four times the cross-sectional area, to have four times the ampacity of 12 AWG, or 100, but its actual ampacity is only 65. And 1/0 AWG, with four times the cross-sectional area of the 6 AWG, might be expected to have four times its ampacity, or 260, but its actual ampacity is only 150. The larger the wire size, the lower the ampacity per thousand circular mils.

A wire that has four times the cross-sectional area of another wire has only twice the surface area of the smaller wire, and the heat developed in the wire can be dissipated only from its surface. Compare 1/0 AWG with 12 AWG. The 1/0 AWG has sixteen times the cross-sectional area, but only four times the surface area from which heat can be dissipated. The ampacity of a wire, while not directly in proportion to its surface area, is more closely related to its surface area than to its cross-sectional area. Another factor, especially in the larger sizes, is the "skin effect." Alternating current flows more densely through the outer circumference of a conductor, and less densely through its center. At the usual 60 Hz this is only a factor on very large conductors above 300 kcmil, but it further degrades the efficiency of very large conductors since not all of the conductor material will carry an equal share of current.

How to select a conductor using *NEC* Table 310.16   For normal applications, first determine how many amps your wire will have to carry safely, based on *NEC* rules explained throughout this book. For example, Chapter 13 shows how to size simple services and Chapter 24 shows how to figure the load at a recreational vehicle park. Then look in the table for a wire with an ampacity at least as large as that load. The following paragraphs give an overview of the process that should be sufficient for almost all residential and farm work. The subject can be very complicated, particularly in commercial and industrial applications, which are described in detail in Chapter 26, "Sizing Conductors for All Load Conditions."

Adjusting *NEC* Table 310.16 values for actual conditions   Ampacity represents current-carrying ability under actual conditions of use; *NEC* Table 310.16 represents the same ability with no more than three current-carrying conductors in a raceway or cable assembly, and operating at not over 30°C. If your installation falls outside those parameters, then you have to adjust Table 310.16 values to fit your installation.

Correction factors for higher ambient temperature   The first adjustment occurs as part of the actual table when 30°C as a temperature limitation would be invalid. At the bottom of the table there is a series of adjustment factors to be used in this case. For example, 3/0 THHN has a normal ampacity of 225 amps. However, if it runs through a hot attic in a part of the country where a design temperature of 55°C is assumed for these conditions, that nominal ampacity must be adjusted using the correction factor 0.76 (refer to your copy of the *NEC* or the Appendix of this book to find this value). Multiplying 225 amps by the 0.76 correction factor gives the actual ampacity under these conditions of 171 amps.

Correction factors for higher number of wires   The second adjustment concerns the phenomenon of mutual conductor heating. The more wires that are confined in a given space, the more difficult it is for them to dissipate heat. *NEC* Table 310.15(B)(2)(a), reproduced in the Appendix of this book as Table A–2, requires that if the number of current-carrying conductors in a raceway or cable assembly exceeds three, the ampacities given in the table need to be adjusted by the factors in this table. For example, if you were installing seven 3/0 THHN wires in one conduit, and they all had to be considered current carrying, you would adjust the table ampacity by 70%. Multiplying 225 amps by 0.70 gives the actual ampacity under these conditions of 158 amps.

Note that the wires have to be current carrying to count. For example, a neutral conductor of a 120/240-volt, three-wire circuit, as diagrammed in Chapter 20, is never counted because it can be shown that it only carries current and contributes heat to the extent that current and heat are decreasing on one of the associated ungrounded wires. Similarly, in a three-way switch loop only one of the two travelers could carry current at any given time and thus only one of the two need be counted for this purpose.

When both conditions occur simultaneously   If both elevated temperatures and excessive numbers of wires are involved, then multiply twice, once by each factor. If our seven THHN wires run through the 55°C attic, then their ampacity in that location is 225 amps multiplied by both 0.76 and 0.70 in sequence, or 120 amps, far lower than the original 225 amps.

Termination temperatures sometimes determine minimum wire size

Although the 90°C column is perfectly valid for its insulation materials, including the common THHN and XHHW wires, it can't be used to decide how big a wire is needed to terminate at an overcurrent device or other piece of equipment. This is because test labs and *NEC* 110.14(C) limit termination temperatures. Terminations on equipment rated over 100 amps can't exceed 75°C. This means taking your assumed current and making an entirely separate calculation using the 75°C column. Using our example of 3/0 THHN, it could not be loaded beyond 200 amps (refer to the 75°C column in *NEC* Table 310.16) unless the devices connected to it, at both of its ends, were listed and marked with an allowance for 90°C terminations. In ordinary work, there are no such devices.

The 90°C column, with its higher ampacities, is still an advantageous place to start if you're facing derating penalties for high temperatures or mutual heating. For example, in the case of the seven wires in the attic, 3/0 THHN could carry 120 amps, as before. However, a 3/0 THW (75°C rating) could carry only

200 amps × 0.67 (remember the temperature correction factors change as well) × 0.7, or 94 amps. That's quite a difference, and well worth the use of the higher-temperature insulation.

The same considerations apply to devices rated 100 amps and below, except the temperature restriction drops to 60°C. In this case, however, there are a number of devices rated "60/75°C" which can use the 75°C column. Here again, remember that a wire has two ends and the restriction applies at both ends.

What to do when the ampacity doesn't match a standard overcurrent device size   In the foregoing paragraphs, the developed ampacity of 3/0 THHN has turned out to be 225, 171, 158, or 120 amps, depending on varying conditions. Only one of these is a standard size (225 amps). In the other cases, the *NEC* allows you to use the next higher standard size protective device. For example, in the case of seven wires in a raceway (ampacity = 158 amps), you could use a 175-amp circuit breaker or fuse. However, the calculated load on this wire must never exceed the 158 amps. This allowance recognizes that overcurrent devices must be standardized as a practical matter, but it does not extend to, in effect, deliberately overloading a conductor. This allowance does not apply over 800 amps.

REDUCING VOLTAGE DROP

It is impossible to prevent all voltage drop. Sometimes it is difficult to hold the voltage drop to a desired level. But you must hold it to a practical minimum. A drop of over 3% on the branch-circuit conductors at the farthest outlet, or a total of over 5% on both feeder and branch-circuit conductors, is definitely excessive and inefficient. It is usually possible and practical to reduce the voltage drop to less than that.

Advantages of reducing voltage drop   Voltage drop is simply wasted electricity. Moreover, all electrical equipment operates most efficiently at its rated voltage. If an electric motor is operated on a voltage 5% below its rated voltage, its power output drops almost 10%; if operated at a voltage 10% below normal, its power output drops 19%.

If an incandescent lamp is operated on a voltage 5% below its rated voltage, the amount of light it delivers drops by about 16%; if the voltage is 10% below normal, its light output drops by more than 30%. (Fluorescent lamps will not operate at all if the voltage is considerably below their rated voltage.) So in all cases, the output drops much faster than the reduction in voltage. It is apparent then that the voltage drop must be limited to as small a value as practical.

Excessive voltage drop creates problems in terms of wasted power and increased stress on equipment that functions best at full voltage, such as motors, but it doesn't cause a safety problem. A motor may burn out early, but as long as it is protected in accordance with all *NEC* rules, it poses no safety hazard while operating on diminished voltage. Preventing excessive voltage drop is appropriate design, but because it is not a safety issue it is completely beyond the scope of the *NEC*, which has as its function to establish the minimum requirements for safety. Although the *NEC* mentions as reasonable upper limits a 3% drop on branch circuits and 5% for the combined drop across a feeder plus a branch circuit, it does not impose this as a mandatory requirement. It is your responsibility as a designer to take it into account as you lay out an installation.

Practical voltage drops   In ordinary residential wiring, there are no feeders, or very short ones, because the branch circuits are usually connected to the service equipment panelboard that also contains the overcurrent devices (circuit breakers or fuses). In farm wiring, however, there usually are feeders, sometimes quite long, such as between the meter pole and the various buildings. The drop in the feeders then should not exceed 2%, and in the branch circuits should not exceed 3%. Achieving even less drop is desirable. A commonly accepted figure is 2% from the beginning of the branch-circuit wires to the farthest outlet, with an additional 1 to 2% in the feeders, depending on their length.

This means that on a 120-volt circuit the voltage drop from the branch-circuit panelboard (in the service entrance) to the most distant outlet should not exceed 2.4 volts; on a 240-volt circuit it should not exceed 4.8 volts. In residential wiring, if 14 AWG wire is used for ordinary branch circuits, the voltage drop will not usually exceed 2%. But lamps are getting bigger and 1000- to 1500-watt appliances are common. People are using more electric power every day, which means circuits are becoming loaded closer to their limits. There is good reason for the trend toward using 12 AWG wire for residential wiring, especially if the circuits are long. The *NEC* already requires 12 AWG for the small-appliance circuits and special laundry and bathroom circuits in dwellings (all discussed in Chapter 13 beginning on p. 224). Some local codes already require 12 AWG as the minimum, and engineers and architects often specify a minimum of 12 AWG copper. In commercial and industrial establishments, where branch circuits are longer than in dwellings, it is often essential to use 12 AWG for ordinary branch circuits to avoid excessive voltage drop.

Calculating voltage drop by Ohm's law   The actual voltage drop can be determined by Ohm's law, which is discussed in Chapter 2: $E = IR$ or *voltage drop = amperes × ohms.*

For example, assume that a 500-watt floodlight is to be operated at a distance of 500 ft from the branch-circuit circuit breaker or fuse. This requires 1000 ft of wire. At 120 volts, a 500-watt lamp draws about 4.2 amps. If 14 AWG wire is used, *NEC* Table 8 shows that it has a resistance of 2.57 ohms per 1000 ft. The voltage drop then is (4.2)(2.57) or 10.8 volts, far greater than 2%. It is 9%.

Trying other sizes, 6 AWG with 0.410 ohms per 1000 ft has a drop of 1.7 volts, less than 1½%. Number 8 AWG with 0.64 ohms per 1000 ft has a drop of 2.69 volts, or nearly 2¼%. Therefore if the floodlight is to be used a great deal, use 6 AWG wire; if it is to be used relatively little, use 8 AWG wire; if it is to be used only in emergencies, use 10 AWG wire (4.28 volts drop, about 3½%), or even 12 AWG wire with 6.8 volts drop, about 5½%. The amount of power wasted when used so rarely would be insignificant. Normally, however, a drop of 5½% should not be tolerated.

If in this example the distance had been 400 ft instead of 500 ft, the length of the wire would have been 800 ft instead of 1000 ft. The voltage drop then would have been $800/1000$ or 80% of what it is for 1000 ft.

Desirability of higher voltages   Now assume that the same floodlight is to be operated at the same distance of 500 ft but at 240 volts instead of 120 volts. The amperes then would be about 2.1 instead of 4.2. Making the same calculations, 14 AWG wire now involves a drop of only about 5.4 volts or about 2¼%. Number 12 AWG with a resistance of 1.62 ohms per 1000 ft has a drop of 3.4 volts, less than 1½%. This emphasizes the desirability of using higher voltages where a considerable distance is involved as well as where considerable power is involved.

For any given *number of watts* at any given *distance,* the voltage drop *measured in volts* on any given size of wire is always twice as great on 120 volts as it is on 240 volts. Doubling the voltage reduces the voltage drop *in volts* exactly 50% if the watts, the distance, and the wire size remain the same. It is the power in watts, and not the current in amperes, that must remain unchanged for this statement to be correct.

When the voltage drop *in percentage* is considered, remember that in the case of the 240-volt circuit the initial voltage is twice as high, but the voltage drop measured in *volts* is only half as much as in the case of the 120-volt circuit. From this you can see that the voltage drop measured in *percentage* will be only one-fourth as much for 240 volts as it is for 120 volts—if the watts, the wire size, and the distance remain the same.

The foregoing paragraph can be restated simply: Any size of wire will carry any given amount of power in watts at 240 volts *four* times as far as it will at 120 volts, with the same *percentage* of voltage drop. This statement should not be confused with the statement made in the paragraph previous to that, which is that any size wire will carry any given amount of power in watts twice as far with the *same number of volts* drop.

More practical method of calculating voltage drop   The preceding method of computing voltage drop was explained in order to emphasize the difference in drop at different voltages. A more practical and more frequently used formula is:

$$\textit{Cross-sectional area} = \textit{constant} \times \textit{amperes} \times \frac{\textit{distance in ft}}{\textit{volts drop}}$$

or $\quad CSA = \dfrac{kIL}{Ed}$

where L = distance in feet one way, I = amperes, and Ed = volts drop. You can remember this as a sentence said and easily remembered phonetically as: "casa/kill/ed."

The constant to use for a complete circuit is 22 cmil-ohms/ft.

This formula is based on the resistance in ohms per circular mil-foot of copper wire, which is 10.8 ohms at 25°C. (A "circular mil-foot" means a wire 1 ft long with a diameter of 1 mil or 0.001 in.; that is, having a cross-sectional area of 1 cmil.) The constant increases at higher temperatures, and the temperature to use is the operating temperature of the wire, not the ambient temperature. Since most devices have a termination limit of 75°C, the typical worst-case voltage drop occurs at that conductor temperature, and *NEC* Table 8 was recalculated in the 1984 edition based on 75°C instead of 25°C. For example, the constant increases to 12.3 at 60°C and 12.9 at 75°C. However, for ordinary residential and light commercial applications where the circuits aren't intensively loaded, 25°C is close enough. For more intensively loaded applications use a one-way constant of 12 or a complete circuit constant of 24. Since the wire has to go to and from a load, either the distance or the resistance has to be doubled. To make it simple, the resistance was doubled to 21.6 ohms, and to compensate for the added resistance of joints, the number was rounded off to 22. Therefore the distance *one way* is used with 22 ohms per cmil-ft in this formula. (For aluminum wire, substitute 36 for 22 in the formula, or 39 for heavier loading. Aluminum has 61% of the conductivity of copper, so its constant will always be the value for copper divided by 0.61.)

For three-phase circuits, the formula[1] is

$$CSA = \frac{(22)IL(0.866)}{Ed}$$

For aluminum wire, again substitute 36 for 22 in the formula.

Applying this more practical formula to the floodlight example, which involves a distance of 500 ft, 4.2 amps, and a drop that is to be limited to 2.4 volts (2%), the calculation becomes

$$CSA = \frac{(22)(4.2)(500)}{2.4} = \frac{46\,200}{2.4} = 19\,250 \text{ cmil}$$

In other words, to limit the drop to exactly 2.4 volts, the wire must have a cross-sectional area of 19 250 cmil. *NEC* Table 8 shows there is no wire having exactly this cross-sectional area, which falls about halfway between 6 and 8 AWG. Therefore use 6 AWG with a little under 2.4 volts drop if the floodlight is to be frequently used, or 8 AWG with a little more than 2.4 volts drop if infrequently used.

---

1. Note that 0.866 is half the square root of 3.

Voltage drop on ac circuits   Strictly speaking, the preceding discussion is valid only for dc circuits and ac circuits with nearly 100% power factor. If the power factor (review this topic in Chapter 3) drops significantly below 100%, as with motor and other inductive loads without power factor correction, then the voltage drop increases beyond that predicted by the formula, which considers only resistance. These calculations are beyond the scope of this book; however, the *NEC* now includes a revised Table 9 that provides complete tabular infor-mation on reactance and resistance of ac circuits under various installation conditions, from which impedance can be calculated. In addition, this table provides effective impedances for each of the configurations based on an assumed 85% power factor. That power factor is very practical in most commercial and industrial applications, and allows you to quickly calculate the impedance for these circuits without significant error.

Cost of voltage drop   In calculating cost of voltage drop as a factor in selecting wire size, you need to consider load, distance, circuit length, wire size, and time. Additional factors that may influence wire selection include load efficiency and need for mechanical strength.

Voltage drop on 120-volt circuit   Assume you want to operate a motor consuming 20 amps at 120 volts (2400 VA) at the end of a 100-ft circuit. Using the formula, the 20-amp load on 8 AWG wire will result in a 2% drop if the circuit is 90 ft long; at 100 ft the drop will be about 2.2%. Suppose the circuit is in use 3 hours per day, about 1000 hours per year. In 1000 hours, that circuit will consume about 2400 kWh, which at 6 cents per kWh would cost you $144.00; of that, 2.2% or about $3.16 is wasted. That does not seem an unreasonable total. But suppose again that in order to reduce your initial investment you had elected to use 12 AWG wire. Carrying 20 amps, the same formula tells you the drop will be 6.1% for a 100-ft circuit. Then 6.1% of your $72.00 or $8.78 is wasted. That is a difference of $5.62, which in five years becomes $28.10, which would have paid for the difference between the small and the larger wire. More important, your motor would perform more efficiently and deliver more horsepower with the larger wire.

Voltage drop on 240-volt circuit   Now, instead of operating the motor at 120 volts, you decide to operate it at 240 volts. As pointed out in earlier paragraphs, when the watts remain unchanged and the voltage is changed from 120 to 240 volts, without changing the wire size, the drop at the higher voltage is one-quarter, in percentage, of the drop at the lower voltage. In other words, by using 8 AWG wire, the drop changes from 2.2 to 0.55%, and by using 12 AWG wire, it changes from 6.1 to 1.53%. You can use the smaller wire and still have less voltage drop.

Consider load efficiency   In debating whether to use the smaller or larger of two sizes of wire, remember that the labor for installing the larger size is little if any more than for the smaller size. The difference is basically the difference in cost of the wire only. Consider also that with a larger wire and less voltage drop, your motor (or other load) will operate more efficiently than when using the smaller wire.

Consider mechanical strength for overhead spans   In running wires overhead, there is an additional factor that must be considered—mechanical strength. The wires must be large enough to support not only their own weight but also the strain imposed by winds, ice loads, and so on. In northern areas it is not unusual to see outdoor wires encased in an inch-thick layer of ice after a severe sleet storm. *NEC* 225.6(A) requires a minimum of 10 AWG for spans up to 50 ft, and 8 AWG for longer spans, unless the run includes a messenger wire. Messenger wires are special high-tensile strength steel wires designed for this type of duty.

In installing outdoor overhead wires in northern areas, take into consideration the expansion and contraction that take place with changes in the temperature. A 100-ft span of copper wire will be almost 2 in. shorter when it is -30°F than on a hot summer day when the temperature is 100°F. Therefore, if installing wires on a cold winter day, pull them as tight as practical. If installing on a hot day, allow considerable sag in the span so that when the wires contract in winter no damage will be done.

Minimum ampacity requirements override voltage drop calculations
You must always use a size and type of wire that has an ampacity rating at least as great as the number of amperes to be carried. For short runs, voltage-drop calculations may show a wire size smaller than is permitted for the current involved, but the minimum ampacity requirements always override any voltage drop calculations because they involve minimum provisions that are essential to a safe installation. First calculate the minimum ampacity for your wires, and then check whether voltage drop considerations merit a further increase in wire size.

CHAPTER 8
Making Wire Connections and Splices

WIRES MUST BE PROPERLY CONNECTED to switches, receptacles, and other devices. It is frequently necessary to splice (join) wires to each other. These connections and splices must be mechanically and electrically sound. The procedures are not difficult, but they are often done improperly. This chapter explains how to do these things correctly. Attention is also given to reliability issues relating to wire connections and splices, including connections made to grounding electrodes. Aluminum wire presents special problems that affect the reliability of connections and splices, and is therefore discussed in some detail.

REMOVE INSULATION BEFORE CONNECTING OR SPLICING WIRES

Figure 8–1 shows a wire stripper, a very handy tool for removing insulation from wires. Ordinary strippers are for smaller wires up to 10 AWG in size; larger strippers are needed for larger wires. Always follow the manufacturer's instructions when using a stripper. Another easy way to strip off the insulation is to use a pair of side-cutter pliers. Place the wire to be stripped between the handle grips close *behind* the cutter hinge, and squeeze the insulation enough to soften it and break it down, but not hard enough to damage the wire. Do this from the point where the insulation is to be stripped to the end of the wire. Then place the cutter jaws over the wire at the point where the insulation is to be stripped, squeeze just hard enough for the jaws to grasp the insulation (but not hard enough to touch the conductor), then slide the insulation off.

Fig. 8–1 A wire stripper saves much time in removing insulation.
(Ideal Industries, Inc.)

If you have no stripper and have the wrong kind of pliers, use a knife to cut through the insulation down to the conductor, holding the knife not at a right angle but at about 60 degrees, so that it will cut the insulation as shown in Fig. 8–2. This precaution reduces the danger of nicking the conductor, which weakens it and sometimes leads to breaks. After the insulation has been cut all around, pull it off, leaving the conductor exposed far enough to suit the purpose. Most insulation strips off easily, leaving a clean surface. If any spots of insulation remain on the stripped portion, scrape them off thoroughly for a clean connection, using the back edge of a knife blade for the purpose so that the conductor is not nicked or otherwise damaged. Stranded conductors must be specially watched to make sure each strand is clean.

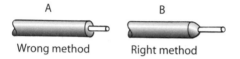

A   B

Wrong method   Right method

Fig. 8–2 In removing insulation from a wire using a knife, hold the knife at an angle of about 60 degrees.

The *NEC* requires a minimum of 6 in. of free conductor at each junction or connection point, measured from the point in the box where it emerges from its cable sheath or raceway entry. In addition, for deep boxes that you can't fit both hands into (defined as having openings less than 8 in. in any dimension), the *NEC* requires the conductors to be long enough to reach outside the box opening at least 3 in. (see Fig. 8–3). You may decide to leave the wires somewhat longer for ease in working the connections; however, after you're done you must fit all the wires into the box without overcrowding. Finally, if you decide to use the trick of looping a wire around a terminal screw (see final paragraph under the heading "Wrapping the wire under the screw"), leave an extra 4 in. of conductor to serve as a "handle" to work the loop. These are minimum requirements—with practice you'll decide what works for you.

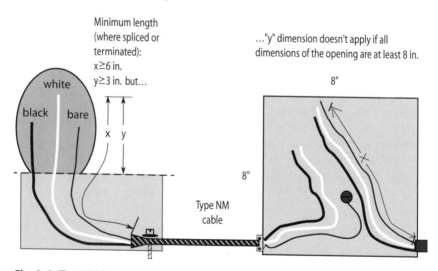

Minimum length
(where spliced or
terminated):
x≥6 in.
y≥3 in. but...

..."y" dimension doesn't apply if all
dimensions of the opening are at least 8 in.

8"

white

black   bare

x   y

8"

Type NM
cable

8"

Fig. 8–3 The *NEC* has minimum length provisions for wires in boxes. The required 6-in. minimum length is measured from the end of the cable sheath, so strip as close to the cable connector as possible to avoid overcrowding the box.

These rules apply to any conductors entering the box, even equipment grounding conductors (covered in the next chapter), which are the bare or green-colored wires that aren't used for carrying normal circuit load currents. The practice of cutting these conductors short, extending only a single pigtail (explained later in this chapter) to a device terminal, violates the *NEC*. The only conductors exempt from minimum length requirements are those that pass directly through a box, as often happens with raceway wiring methods (covered in Chapter 11) where the wires are inserted into tubes or channels after the basic wiring system is complete. Think about the person who comes after you before you use this exception, however. If someone wants to splice one of those wires, perhaps to cut in an additional device, the cut ends of the wires won't be able to meet the 6-in. rule and those conductors will need to be replaced. Generally, it's much better to leave a loop long enough to allow someone to terminate the wire later, provided the box will accommodate the wires involved. Effective with the 2005 *NEC*, uncut wires passing through a box must be counted twice in box-fill calculations if, when cut in the middle, each cut end would meet the 6-in. rule. Refer to Chapter 10, "Calculating Allowable Number of Wires in Box" for more information on this topic.

TERMINALS FOR CONNECTING WIRES TO DEVICES

Various kinds of terminals provide the means for connecting wires to devices. Terminals must be appropriate for the size and numbers of wires to be connected.

Screw-type terminals   For 10 AWG and smaller wires, screw terminals of the type shown in Fig. 8–4 are usually used. The brass base of the terminal or the surrounding insulation is shaped to prevent the wire from slipping out from under the terminal screw, which usually has its lower threads "upset" or "staked" so it cannot be entirely removed.

Fig. 8–4 A terminal for connecting 10 AWG or smaller wires.

Two wires not allowed under screw—use "pigtail"   Do not connect two wires under a single terminal screw even though it might appear logical in some situations. It is prohibited by *NEC* 110.14. Instead, connect the two ends of wire together with another short piece using a solderless connector. These connectors are described later in this chapter in the discussion on making splices. Connect the short length (called a pigtail) to the terminal screw, as shown in Fig. 8–5.

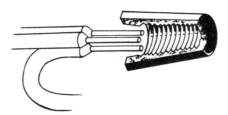

Fig. 8–5 Make a "pigtail" to connect two wires at a screw terminal. The *NEC* prohibits more than one wire under a screw terminal unless listed for that duty, and very few are.

Wrapping the wire under the screw   After about 4 in. of the wire has been cleanly stripped of its insulation, and the terminal screw has been backed out until the upset prevents further loosening, wrap the wire clockwise around the screw, as in Fig. 8–6, in such a way that (1) the insulated portion of the wire is near enough to the screw to allow the head of the screw, when tightened, to clear the insulation by not over ¼ in., as shown in Fig. 8–7; and (2) the stripped portion of the wire is brought around the screw so that it nearly closes the gap but does not overlap—not less than about three-fourths nor more than seven-eighths of a turn. Then tighten the screw so its head fits flat and snugly against the entire wraparound portion of the wire. Next, tighten the screw about another half-turn so that it is firm and tight. If you use a torque screwdriver, tighten to 12 lb-in. Finally, take the excess "tail" of bare wire and twirl it around a few times until it breaks off near the screw head.

Screw closes loop
RIGHT

Screw opens loop
WRONG

Fig. 8–6 Insert wire under a terminal screw so that tightening the screw tends to close the loop.

Insulation close to
terminal screw

Don't leave long
exposed wire

RIGHT

WRONG

Fig. 8–7 Don't leave a long bare wire next to a terminal screw.

The steps stipulated by testing labs for connecting aluminum wire to 15- and 20-amp screw terminals are shown in Fig. 8–8, and methods testing labs do not sanction are shown in Fig. 8–9. Both of these figures also apply to copper wire, with one exception: Some screw terminals are listed (for copper or copper-clad aluminum only) that permit a wire to be inserted straight, without wrapping around the screw. If this is the case, look for a groove to set the wire in. Many inspectors do not look with favor on this type of connection.

If the wire in the box is too short to allow 4 in. or so for twisting off the "tail" after the screw has been tightened, then form the end of the wire into a loop using long-nose pliers, and insert the loop under the head of the screw, closing it completely with the same pliers. Be sure the loop fits flat under the screw head. In this method sometimes it is necessary to remove a screw totally despite the staked or upset end, which can usually be done without damaging the threads. What you are doing is using the female terminal thread to rethread the screw thread, which the manufacturer deliberately cross-threaded so the screw wouldn't

come out easily. Work slowly
and deliberately, periodically
retightening the screw before
loosening it a little further with
each pass. Stripping the female
threads renders the terminal
useless. It is better to avoid
removing screws with staked
(upset) ends.

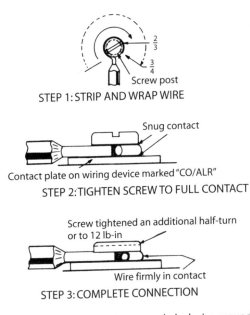

STEP 1: STRIP AND WRAP WIRE

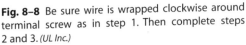

Contact plate on wiring device marked "CO/ALR"

STEP 2: TIGHTEN SCREW TO FULL CONTACT

Connecting stranded wire
While the *NEC* permits 10
AWG wire to be connected
with screw-type terminals, it is
difficult to make a good
connection if the wire is
stranded. At one time most
stranded configurations in
these sizes used a 7-strand
configuration. The individual
strands have enough thickness
to stay in place if tightly
twisted together before being

STEP 3: COMPLETE CONNECTION

Fig. 8–8 Be sure wire is wrapped clockwise around
terminal screw as in step 1. Then complete steps
2 and 3. *(UL Inc.)*

connected under a screw terminal. Today the more common arrangement is a
19-strand, and these thinner strands are almost impossible to entirely contain
under a screw terminal. The best way to terminate stranded wire is to use devices
with clamping terminals designed for this use (see Fig. 8–10). Loose strands that
escape from a screw terminal can cause short circuits or ground faults depending
on what they happen to contact. If a screw terminal is used, leave a short length
of insulation intact at the cut
end of the wire to retain the
strands on each side of the
terminal screw, or add a lug to
the cut end (see Fig. 8–11).

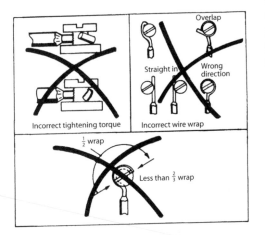

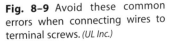

Fig. 8–9 Avoid these common
errors when connecting wires to
terminal screws. *(UL Inc.)*

Fig. 8-10 These device terminals enclose all the strands of stranded wire. They use an internal moving clamp secured by the screw. These devices are often specified for the most demanding industrial applications. Don't confuse them with the "push in" devices pictured in Fig. 8-12. *(Pass & Seymour/Legrand)*

Devices without terminal screws Some switches and receptacles are made without terminal screws. The straight end of a wire is pushed into small holes in the device, as shown in Fig. 8-12. Strip the insulation off the wire as far as indicated by a marker on the device. If it becomes necessary to remove a wire, push a small screwdriver blade into slots on the device to release the wire. These devices must not be used with all-aluminum or any composition stranded wire but are acceptable for solid copper or copper-clad aluminum. Some devices have both terminal screws and push-in terminals. Many inspectors do not favor push-in terminals, and many electricians choose not to use them. Due to concerns about the stability of these connections, recent changes in product standards now prohibit their use for larger than 14 AWG wire on a receptacle.

Connecting larger wires The terminal screws discussed up to this point must not be used for wires larger than 10 AWG. For 8 AWG and larger wires, solderless connectors or terminals of the general type shown in Fig. 8-13 are used. Simply insert the stripped and cleaned end of the conductor into the connector and tighten the nut or screw.

Fig. 8-11 Two strategies for terminating small sizes of stranded wire on a device.

MAKING SPLICES

Splices are often necessary and it is important to make them properly. (Many cases in which the *NEC* does not permit spliced wires are noted in this book in the course of discussing the related work.) A splice must be as strong mechanically and as good a conductor as a continuous piece of wire. After the splice is completed, its insulation must be equivalent to the original insulation. First, where the wires are to be spliced it is necessary to remove the insulation, as described at the beginning the chapter. Then make the mechanical joint using a solderless connector. Finally, make sure the insulation is effective.

Fig. 8–12 Some devices do not have terminal screws. Push the bare wires into openings in the device for a permanent connection. The device has a "strip gauge" showing how much insulation to remove from the wire.

Solderless connections   You must use a solderless connector that will accommodate the number and sizes of wires to be joined. The *NEC* prohibits soldered connections in the service equipment, in the ground wire, and in all grounding wires; the reasons are explained on p. 299. At one time soldering was commonly used for other splices and taps of the types discussed in here, but such soldering is rarely if ever done today. Accordingly, the art of soldering is not discussed in this book.

Insulating splices   Most solderless connectors have insulation shells that completely cover the bare ends of the wire if properly installed. Strip off only enough insulation so no bare wire is exposed after installing the connector.

If your connectors do not have insulating shells, you must insulate the splice with electrical insulating tape. Plastic electrical tape is very strong mechanically, and has a very high insulating value per mil ($1/1000$ in.) of thickness, making a comparatively thin layer sufficient. In applying tape on a splice, lay it over the original

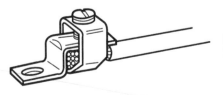

Fig. 8–13 A typical solderless connector for larger wires.

insulation at one end of the splice, then wind it spirally toward the other end letting the successive turns slightly overlap each other. Keep the tape stretched so that the overlapped portions will fuse to each other. Work back and forth in this fashion until the several layers of tape are as thick as the original insulation.

Connectors without insulating shells are also used with wires that don't require insulation, such as equipment grounding conductors (discussed in the next chapter). Connectors for this purpose are discussed later in this chapter.

Solderless, self-contained connectors For wires 8 AWG (in some cases 6 AWG) and smaller, one of the most popular splicing devices is the twist-on solderless connector, often called by the trade name "Wire Nut." Two types are shown in Fig. 8–14. One type has a removable metal insert; the ends of the wires to be spliced are pushed into this insert, the setscrew of which is then tightened and the plastic cover screwed on over the insert. The other type has no removable insert but has an internal tapered thread; lay the wires to be spliced parallel to each other (if one is much smaller than the others, let it be a bit longer than the larger wires), then screw the connector over the bare ends of the wires.

Fig. 8–14 Two types of solderless connectors for smaller wires, often called by the trade name "Wire Nut." The version with the setscrew maintains the integrity of the electrical connection even with the insulating cap removed. This makes it a good choice if the connection will be tested with diagnostic equipment while energized, such as with some motor connections.

Although many believe that the wires should always be twisted together prior to installing the twist-on wire connector, that is never required and may violate installation instructions for some of these devices. Leave the wires straight, just as you stripped them, and twist on the wire connector. Let the connector grab the wires and twist them as you tighten the connector. If you're joining both solid and stranded wire, set the stranded conductor(s) with the end(s) extending slightly farther into the connector than the solid wire. If you're joining different wire sizes, allow the smaller wire to lead the larger wire in a similar fashion. If the ends of the wires have been stripped to the proper length, no bare wire will be exposed and no taping is needed.

Recent new applications of the twist-on wire connectors have been developed that involve connectors prefilled with special compounds that make the electrical connection more reliable under adverse environmental conditions. For example, the connector in Fig. 8–15 is oversized and extremely rugged, and has a water-proofing compound that protects the splice from moisture. These connectors are even listed for direct burial applications, though not many direct burial applications allow individual conductors to be exposed in the earth. Although Type UF cable (covered in Chapter 11) comes in both multi-conductor and single-conductor versions, both listed for direct burial applications, many people are unaware that the individual conductors within multi-conductor cable are not listed for earth contact. This means that if you need to splice a damaged multi-conductor Type UF cable (its usual configuration) in the earth, you need a kit that encloses the individual conductors from cable jacket to cable jacket. Although the Fig. 8–15 connectors significantly increase reliability, they aren't adequate for underground applications unless the spliced conductors, configured as individual conductors, are suitable for direct earth contact. One form of Type UF and most Type USE cables do, however, meet this criterion.

Fig. 8–15 A twist-on wire connector prefilled with a moisture resisting compound that protects the electrical integrity of the joint.

Spring-type connectors   One type of connector uses a spring; an example is the Scotchlok shown in Fig. 8–16. Within the tough outer insulating shell there is a steel sleeve, and inside that a coiled tapered spring that unwinds as the connector is screwed onto the wires; after you release the connector, the spring compresses tightly and forms a good permanent connection. Be sure no bare wire is visible after the connector has been installed.

Crimp-type connectors   An example of a connector without an insulating shell is the crimp sleeve shown in Fig. 8–17. It is applied with a crimping tool of the sort pictured in Fig. 8–18. Although insulation is available for these connectors, their most common use is to join bare equipment grounding conductors (discussed in the next chapter).

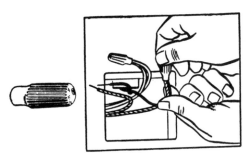

Fig. 8–16 Spring-type solderless connectors are also very popular.

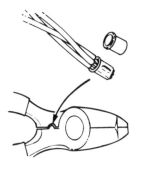

◀ **Fig. 8-17** This crimp sleeve can be installed quickly. It is suitable only for copper wires.

Fig. 8-18 This crimping tool has jaws for both insulated and uninsulated connectors.

▼

Splicing larger wires   For heavier sizes of wire, use connectors of the general type shown in Fig. 8–19. Being made of bare copper, the splice must be insulated with tape. However, such connectors are also available with insulating clamp-on shells that eliminate the need for tape.

Splicing to a continuous wire ("tap" connection)

Often it is necessary to connect a wire to another continuous wire—this is called a tap connection, or just a tap. In many cases this can be accomplished using a twist-on connector as shown in Fig. 8–15: cut the continuous wire to create two ends, with the wire to be spliced making the third wire. For larger wires this is not practical; the simplest way is to use a split-bolt solderless connector (usually referred to as a "bug") of the general type shown in Fig. 8–20. Unless the connector has a clamp-on insulating cover, the connection must be taped with electrical tape.

Fig. 8-19 Use a connector of this type to splice two large wires. The splice must be taped.

Fig. 8-20 A split-bolt connector of this type permits one wire to be tapped to another continuous wire. Some connectors have special plating and a separator bar, and can be used to join aluminum, copper, or both in combination.

SOLUTIONS TO RELIABILITY PROBLEMS OF CONNECTIONS AND SPLICES

Unreliable connections and splices pose a risk of shock, fire, and installation failure. Such connections and splices can result from inadequate workmanship, or they may be due to the inherent nature of the materials involved. Degradation of initially sound connections and splices can occur in problem environments. Following are some solutions for assuring reliability.

Observe torque specifications   Generally speaking, most electrical connections at equipment involve tightening some form of a screw either on a device directly or within the body of a lug bolted to a larger device. Common sense dictates that such connections be made tight. However, many people don't appreciate that there's such a thing as too tight. Actually, many studies have shown that the larger problem in the industry is overtightened connections. If the lug doesn't fail immediately, it may fail due to metal fatigue at some indefinite point in the future. Most electrical equipment comes with torque specifications for its connections, which must be respected. You need to have and routinely use torque wrenches and torque screwdrivers. (See Fig. 8–21.)

The upper curve of the drawing in Fig. 8–22 shows how joint resistance decreases as torque increases until the curve becomes horizontal. This is the point where any further force doesn't decrease contact resistance, but may overstress the metal in the lug. This curve represents our common-sense understanding of what happens when we tighten the set screw on a lug.

Fig. 8–21 A torque wrench in use in a panelboard. A torque screwdriver sits at the lower right.

The bottom curve may not, however, agree with common sense. It shows contact resistance plotted against *decreasing* torque. It does not simply retrace the tightening curve. Instead, it shows that resistance doesn't increase at all until the connection gets comparatively loose. At this point, resistance increases rapidly until it rejoins the initial tightening curve. This means that periodic retightening of connections may not accomplish very much. It's true that as the wire strands relax and better conform to the shape of the lug barrel under constant pressure, the connection loosens somewhat. However, that doesn't mean there's any impact on actual contact resistance.

Irreversible compression connections for larger wires   Another way to solve reliability problems at terminations involves special equipment that applies lug-type ends to larger conductors using extremely high force and chemical inhibitors (see Fig. 8–23). After the crimp is complete, the conductor and the termination become, in essence, a single mass of metal, with no opportunity for

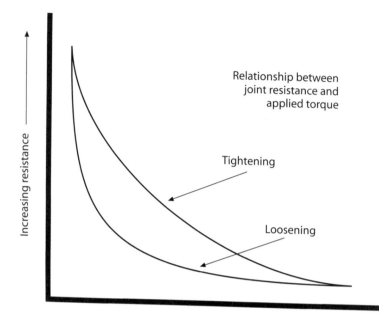

Increasing resistance

Relationship between
joint resistance and
applied torque

Tightening

Loosening

Fig. 8–22 Torque versus contact resistance. The upper curve shows the lug being tightened;
the lower curve shows it being loosened.

an oxide film to interfere with electrical conductivity. Be very certain that you use a crimping die that corresponds to the lug employed, which is generally a condition of the listing on the lug, although some (definitely not all) tools take a broad range of sizes. The worst of both worlds occurs when the irreversible crimp attaches a lug that is crimped incorrectly, resulting in an unreliable connection that appears solid.

Exothermic welding for grounding-related connections Not all electrical connections are made to electrical devices. Particularly in the area of grounding, connections must be made to various grounding electrodes and even to steel reinforcing bars and other structural components. Although there are lugs available in some cases, they may not fit a particular configuration, and they may allow air and moisture to degrade the connection over time.

There is a chemical reaction, the "thermite" reaction, that can produce various molten metals in the process of liberating enormous amounts of heat. If the right chemicals are used in the charge, the reaction generates molten copper capable of welding a copper conductor to copper or steel parts contained within the mold. Special molds are available for a wide variety of common needs, such as attaching grounding conductors to ground rods, steel reinforcing, fencing, other copper grounding conductors, and many other purposes. See Fig. 8–24.

Fig. 8–23 This photo shows an irreversible compression tool in use.

Aluminum wire—special rules address unique characteristics Because aluminum has a higher resistance per unit length than copper, a larger size is needed to carry the same current. Nevertheless, the use of aluminum wire is increasing, particularly in larger sizes. It offers less mechanical resistance to being bent, allowing for easier manipulation in crowded switchgear enclosures. In addition, it is now manufactured using different alloys than in the past, which have more dimensional stability under heat-cycling conditions.

Fig. 8–24 This graphite mold can withstand the enormous heat liberated in the thermite reaction, and the result will be a solid, welded joint capable of lasting as long as the copper wire itself. *(Erico Products, Inc.)*

Responding to a copper shortage in the 1960s, the industry began producing aluminum conductors as part of nonmetallic cables used in household wiring for ordinary branch circuits. Two problems—cold flow and oxidation—began showing up in the years following the first uses of these cables. Both problems are related to the metallurgy of aluminum.

Cold flow   First, any wire under load develops heat. When an aluminum wire heats up, it expands faster than its copper equivalent. If it expands against the underside of an unyielding screw head, it may deform slightly, depending on the screw head design. As it cools it contracts, but rather than returning to its original dimensions it retains its slightly changed shape (called "cold flow"). With the repeated heating and cooling cycles that electrical circuits undergo in use, the connections loosen. Since electrical circuits never stay at the same temperature over time, this outcome would repeat continuously.

Oxidation   The second problem compounds the first. When copper is exposed to air, it forms cuprous oxide, which is still conductive. When aluminum is exposed to air it immediately forms aluminum oxide, which is an excellent insulator. Therefore when these terminals cooled, the aluminum not only loosened, but it then oxidized. That put an insulator in series with the connection, which proceeded to generate heat in proportion to the square of the current as shown in the power formula presented in the second chapter. The connection now being much hotter in use led to increasing deformity and loosening of the aluminum wire when the circuit was no longer under load. This allowed for more oxidation, which produced even more heat the next time the circuit carried load, etc. In a disturbing number of documented cases this phenomenon progressed to the point of causing fires.

Industry response   In response to these problems the industry developed aluminum alloys that weren't as likely to be affected by cold flow, and the device manufacturers developed special terminations that better accommodate aluminum connections. In addition, copper-clad aluminum conductors, having a thin layer of copper bonded to the outer surface of aluminum, came into production in the small branch-circuit sizes. At present, the shortage of copper is long behind us, and the concerns that arose years ago have driven the smaller sizes of aluminum conductors off the market. However, if those economic issues recur and smaller sized aluminum cables make a comeback, it will be necessary to know how to use them. In addition, there are untold thousands of homes still wired with aluminum, and you'll almost certainly come across it.

Before September 1971, switches, receptacles, and other devices rated at 15 or 20 amps and intended only for copper were not marked. Those intended for copper or aluminum were marked "CU-AL" (CU is the chemical abbreviation for copper, and AL is for aluminum). It gradually became evident, however, that the CU-AL devices were not suitable for aluminum in the 15- and 20-amp ratings. Since September 1971, product standards have required still different terminals (in the 15- and 20-amp sizes) and these are marked "CO-ALR" (CO became an arbitrary mark in place of the usual symbol CU for copper, and ALR means aluminum "revised").

Look for the appropriate markings   Devices rated at 15- and 20 amps with unmarked terminals should be installed only when using copper wire. If the terminals are marked CU-AL they should now only be installed using copper wire. If marked CO-ALR (Fig. 8–25), they may be installed using copper, copper-clad aluminum, or all-aluminum wire. But note that devices with push-in or screwless terminals, as shown in Fig. 8–12, must not be used with all-aluminum wire, but are acceptable for copper and, if so marked, for copper-clad aluminum.

Fig. 8–25 A "CO-ALR" marked receptacle.

Note also that devices rated at 30 amps or higher (whether made before or after September 1971), if *not* marked may be used with either copper or copper-clad aluminum. If marked CU-AL they may be used with copper, copper-clad aluminum, or all-aluminum. Devices rated at 30 amps or more are *not* marked CO-ALR. Aluminum conductors used on 30-amp or higher circuits are usually at least 8 AWG, and 8 AWG and larger conductors have to be stranded. A stranded conductor is more forgiving in terms of stability at a termination.

Terminals and splicing devices for aluminum wire   At one time twist-on wire connectors were routinely listed for use with copper and aluminum conductors, but in use they showed some of the same problems that had affected device terminations, and the test labs rewrote the product standard to include severe heat cycling tests. At the time, no such connectors could pass the test when used on aluminum wire, and recognition for use with aluminum conductors was withdrawn in all cases. This created a major problem when electricians needed to replace devices or extend branch circuits in homes that had aluminum branch-circuit conductors.

For a long time the only listed method involved a compression connection with a special tool that, for liability reasons, no electrician could buy. If you wanted to use the tool, you had to obtain it by agreement with its manufacturer, which always included a requirement for factory training in its use, with the manufacturer retaining ownership. Recently, however, certain twist-on wire connectors (see Fig. 8–26), factory prefilled with special oxide inhibitor, have been listed for use with specified common combinations of aluminum and copper wire. Although smaller, these connectors are similar to the one in Fig. 8–15. These connectors aren't inexpensive, but they often spell the difference between being able to work on this old wiring and having to rewire entire circuits.

Fig. 8–26 This twist-on wire connector is prefilled with oxide inhibitor and listed for use with copper and aluminum wire combinations. *(Ideal Industries, Inc.)*

There are other strategies for handling aluminum terminations. Some connectors have a thin, self-healing polymeric coating that pushes out of the way and then seals up to (on the molecular level) the contact point between an aluminum conductor and the lug body, preventing any chance for the aluminum contact surfaces to oxidize. Other connectors rely on field-applied compounds applied to the raw aluminum contact surfaces (see Fig. 8–27). As always, follow any directions that accompany the equipment, because the test lab will have reviewed the termination provisions in those instructions.

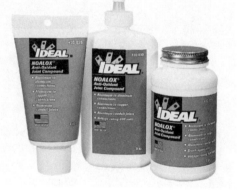

Fig. 8–27 One of several compounds available for treating aluminum connections. Apply it immediately after stripping the wire because raw aluminum develops an oxide skin almost immediately after exposure to air. *(Ideal Industries, Inc.)*

CHAPTER 9
Grounding for Safety

IN ALL DISCUSSIONS of electrical wiring, you will regularly meet the terms *ground, grounded,* and *grounding.* They all refer to deliberately connecting parts of a wiring installation to the earth. Actually the connection is made to something called a "grounding electrode" that is in contact with the earth—such as the buried metal piping of a water system, or the grounded metal frame of a building, or a ground rod driven into the earth. This chapter covers only the purpose and functionality of grounding. It assumes an *NEC*-appropriate grounding electrode system is in place, and explains how ground pathways back to that point function and promote safety. The mechanics of installing a grounding electrode system are covered in Chapter 16. See Figs. 16–21 to 16–23 and related text.

THREE TYPES OF GROUNDING
Grounding falls into three categories: (1) system grounding, or grounding one of the current-carrying wires of the installation; (2) equipment grounding, or grounding non-current-carrying parts of the installation, such as service equipment cabinet, the frames of ranges or motors, the metal conduit or metal armor of armored cable; and (3) bonding, or permanently joining conductive parts together to assure continuity and the ability to safely carry any fault currents likely to be imposed. Although all three concepts involve safety, they protect against distinctly different but sometimes interrelated hazards. System grounding stabilizes voltages to ground so equipment works properly, but it also allows for automatic clearing of ground faults (discussed on page 57), and in so doing automatically removes elevated voltages from surfaces that may be touched. Equipment grounding primarily removes hazardous voltages from touchable surfaces, but the local ground reference it provides is essential to the functioning of some equipment. Bonding is the process of assuring the integrity of the grounding path so that both system and equipment grounding systems function as intended. When you finish this chapter, be sure you understand both the similarities and differences among these grounding concepts, neatly summarized in *NEC* 250.4. And if you ever get confused about a grounding rule, always start by sorting out in your mind which of these concepts is addressed.

The purpose of grounding is safety. A wiring installation that is properly grounded allows excess electrical current, such as from lightning strikes, to travel into the earth without causing serious injury to people or damage to the wiring system. Grounding facilitates the proper functioning of fuses and circuit breakers, limiting the risk of shock from defective equipment. An installation that is not properly grounded can present an extreme danger of shocks, fires, and damage to appliances and motors. Proper grounding throughout the system reduces such dangers and minimizes danger from lightning, a problem especially on farms.

Grounding is an important subject. It is so important that many points are repeated throughout this book for emphasis. Study the subject thoroughly and understand it. The *NEC* rules for grounding are extensive and at times appear ambiguous. However, for installations in homes and light commercial buildings the rules are relatively simple. Recent *NEC* changes regarding agricultural buildings have made grounding very complicated for many aspects of farm wiring. Grounding requirements on farms are discussed throughout Chapter 22, "Farm Wiring."

GROUNDING TERMINOLOGY

Ground   In this book the term "the ground" means what the *NEC* calls the "grounding electrode system"—the metal pipe of an underground water system, or effectively grounded building steel, or a driven pipe or rod that is used where there is no metal underground piping and no other grounded component available. When it is said that something is "grounded" it means that it is connected to the ground. The remainder of the coverage in this chapter covers grounded systems, which are the only ones permitted for general use. Ungrounded systems are covered at some length in Chapter 28, particularly under the topic "Delta distributions are still important for industrial applications."

Voltage to ground   This term is often used in the *NEC* and is defined in Article 100. If one of the current-carrying wires in the circuit is grounded, the *voltage to ground* is the maximum voltage that exists between that grounded wire and any hot wire in the circuit. If no wire is grounded, then the voltage to ground is the maximum voltage that exists between any two wires in the circuit.

Ground wire   In the *NEC* the "ground wire" is called the "grounding electrode conductor"—the wire that runs from the service equipment to the ground. The ground wire can be insulated, covered, or bare. Because its function is reasonably obvious, the *NEC* generally allows you to use any color insulation, although the most common method is to use bare wire. The literal text of the *NEC* does reserve the colors white and green, however, for grounded and equipment grounding conductors respectively (in 200.7 and 250.119), so you may have a quibble with those colors, although few inspectors would actually cite them. In the service equipment enclosure (cabinet) the ground wire is connected to the neutral wire, thus grounding it. The metal cabinet is grounded (bonded) to the grounded neutral wire; thus the cabinet is also grounded, and so is the conduit, or armor of armored cable, which are connected (bonded) to the cabinet.

Equipment grounding wire   The "equipment grounding wire" does not carry current at all during normal operation; it is connected to non-current-carrying parts of the installation, such as the frames of motors or clothes washers, and the outlet boxes in which switches or receptacles are installed. The grounding wire is in the same cable or conduit and runs with the current-carrying wires. In other words, it is connected to parts that normally do not carry current, but do carry current in case of damage to or defect in the wiring system or the connected appliances. When using metal conduit or cable with metal armor, it may not be necessary to install a separate grounding wire because often the metal conduit or armor can serve as the grounding conductor.

The grounding wire may be green, or green with one or more yellow stripes (to enable American appliances to be exported without modification for the European market), or in many cases bare uninsulated wire. Green wire may not be used for any purpose other than the grounding wire. Other chapters discuss when a separate grounding wire must be installed. If metal conduit or cable with metal armor is used, the conduit or armor may serve as the grounding wire in many (not all) instances.

Grounded neutral wire   In ordinary residential or light commercial wiring, the power comes into the premises from the utility over three wires. See Fig. 9–1 in which the wires are marked N, A, and B. Wire N is grounded (both at the service equipment and at the transformer supplying the power) and is called the grounded neutral wire, and wires A and B are ungrounded and are called "ungrounded wires," "phase wires," or usually just "hot" wires. Note that the voltage between the neutral N and either A or B is 120 volts; between the two hot wires A and B it is 240 volts. Somewhat larger commercial enterprises use a three-phase, four-wire wye system as shown in Fig. 3–12. For instructional purposes, however, it's simpler to look at the single-phase, three-wire system for now.

If you touch both A and B, you will receive a 240-volt shock. If you touch either A or B while also touching N, you will receive a 120-volt shock. But note that N is connected to the ground; so to receive a 120-volt shock you don't have to actually touch N. Touching A or B while standing on the ground is the same as touching A or B while touching N—you will receive the 120-volt shock.

The two hot wires may be any color except white, gray, green, or green with one or more yellow stripes. One is usually black and the other red, and sometimes both are black.

Is it a neutral wire?   Where there are three service wires, as in Fig. 9–1, the grounded wire is definitely a neutral wire. What about the wires that begin at the service equipment and run to various circuits in the building? If the circuit is a three-wire, 120/240-volt circuit (explained in Chapter 20, "Wiring for Multiple

Circuits and Specialized Loads"),
the grounded wire is definitely a
neutral wire. But if the circuit is a
two-wire, 120-volt circuit, the
grounded wire, even if it does
connect to the neutral wire in the
service equipment, is not a neutral
wire; it is called just the grounded
wire. Many people call the
grounded wire in a two-wire
circuit a "neutral wire," but there
cannot be a neutral in a two-wire
circuit.

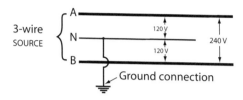

Fig. 9–1 Using only three wires, two different
voltages are available. The neutral wire *N* is
grounded.

SYSTEM GROUNDING

As mentioned in the opening to this chapter, "system grounding" refers to intentionally connecting a current-carrying conductor of the electrical system to earth
(or to some conducting body that serves in place of the earth). System grounding
does not directly eliminate potential shock hazards since we don't build electrical
systems with the idea that bare circuit conductors, even grounded ones, should be
safe to touch, though there are exceptions, such as service equipment bonded to
a grounded service conductor. The two principal reasons for system grounding
are related to stabilizing system voltages to ground, partially covered in Fig. 9–2,
and allowing for automatic fault clearing, as covered later in this chapter. Figure
9–2 illustrates one reason why you have to ground certain systems and one of the
most important benefits of doing so.

Stabilizing system voltages to ground   Figure 9–2 illustrates a basic utility
three-phase, four-wire, wye-connected distribution system (there are others) as
diagrammed in Chapter 3, but the voltages between phases are 13 800, and the
voltage to ground is 8000. Depending on load, each transformer is connected to
only one phase conductor, and the grounded conductor is common to both the
primary and secondary sides of the transformer. In the event of an accidental
cross as shown, the primary current flows through the secondary winding.
Usually the impedance is low enough to trip the protective devices at the utility
substation within a reasonable time. Until that happens, the premises voltage
stays within reasonable limits. If the system were ungrounded, the inside wiring
would have to carry the full 8000 volts to ground, and the result would be major
repairs and possible injuries.

Main bonding jumper—a critical connection   The other reason for
providing a grounded system is that it allows wiring faults to usually clear
automatically and immediately. This is covered in great detail later in this chapter.
For now, review Fig. 4–6 to satisfy yourself that any voltage applied to a

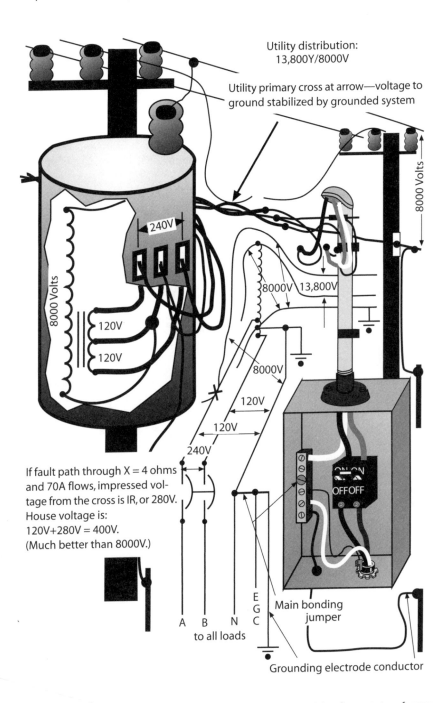

Utility distribution:
13,800Y/8000V

Utility primary cross at arrow—voltage to
ground stabilized by grounded system

8000 Volts

240V

8000 Volts

120V

120V

8000V 13,800V

8000V

120V

120V

240V

If fault path through X = 4 ohms
and 70A flows, impressed voltage from the cross is IR, or 280V.
House voltage is:
120V+280V = 400V.
(Much better than 8000V.)

ON ON
OFF OFF

A   B   N E Main bonding
              G   jumper
              C
to all loads

Grounding electrode conductor

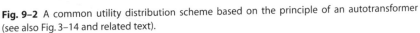

Fig. 9–2 A common utility distribution scheme based on the principle of an autotransformer
(see also Fig. 3–14 and related text).

grounding wire will immediately become part of a complete circuit back to the wiring system source. The *NEC* defines this interconnection as the *main bonding jumper* (in the case of a service) or *system bonding jumper* (in the case of a non-utility, separately derived source). For grounded systems they are the most important connections in the entire electrical system.

White (grounded) wire   In this book, the grounded wire is called just the "white wire" or "grounded wire" unless it is in fact a neutral wire, in which case it is called that. The *NEC* calls it the "identified conductor" in either case. *NEC* 200.6 requires a grounded wire to be identified by a white finish throughout its length (or by three longitudinal white stripes along the entire length) if it is 6 AWG or smaller. If larger than 6 AWG, it must (1) have the same white color, or (2) be painted white at each terminal, or (3) be taped with white tape at each terminal. The *NEC*, with two exceptions (see the white wire exception topic on p. 314) requires that a white building wire must never be used for any purpose other than the grounded wire.

The *NEC* generally permits either "white" or "gray" as the identifying color for wire insulation. Take care to avoid confusion regarding the old term "natural gray," which was used from the 1923 *NEC* until the 1999 *NEC* as an equal identifier to white. This term referred to the color of uncolored latex and the unbleached muslin applied around it for mechanical resiliency. With the advent of modern insulations, true natural gray insulation has not been in production for over 50 years. Many jurisdictions, however, recognized the controlled color gray as a neutral, particularly for 480Y/277 systems in order to easily distinguish this neutral from a 208Y/120V system neutral (usually white) on the same premises. The 2002 *NEC* implicitly recognized this practice by removing the term "natural" from the Code. This means ordinary gray wire now is accepted on the same footing as white wire. However, the *NEC* still mandates no connection between system voltage levels and grounded conductor color; gray can be used on the 208Y/120V system, and white on the higher voltage system. Some installations use color to distinguish between systems from different sources even if they are at the same voltage.

In general the choice of white or gray is up to the installer unless there is a local requirement. However, if two systems are represented in the same enclosure, *NEC* 200.6(D) requires that those systems be distinguishable, and that the means of system identification be posted at each branch-circuit panelboard. White and gray are one logical basis for making this distinction, with white (or gray) wire with colored tracers to be used if more than two systems are present. Be aware that white wire with a colored tracer is available but usually only on special order and with large minimum lengths required. Meanwhile, based on the literal text of 310.12(C) in prior editions of the *NEC*, gray wire could have been used in the past as an *ungrounded* conductor because gray is not "natural gray." Discuss the local practice with those who know, and test all connections, especially those involving gray wire, for voltage before assuming conductors are grounded. For simplicity this book refers only to the color "white" and ignores the term "gray."

The grounded wire is never interrupted by a circuit breaker, fuse, switch, or other device. This simple fundamental requirement of wiring must at all times be kept firmly in mind. An exception would be a device designed so that in opening the grounded wire it will at the same time open all the ungrounded wires, but such devices are not commonly used. So the white wire, with the exception noted, always runs from its source to the equipment where the current is finally consumed, without being switched and without overcurrent protection. However, it may be spliced when necessary, or connected to terminals.

The grounded wire and one hot wire must run to equipment operating at 120 volts. But two hot wires without a grounded white wire must run to 240-volt loads. (Anything connected to a circuit and consuming power, such as a lamp, toaster, or motor, constitutes a "load" on the circuit. Switches and receptacles do not consume power and therefore are not loads, but anything plugged into a receptacle is a load.) Some appliances such as ranges have both 120- and 240-volt loads, so the white wire and both hot wires must run to the appliance. A separate grounding wire runs to the non-current-carrying parts of most loads (unless metal conduit or the armor of cable serves as the grounding conductor). In the case of 120/240-volt loads (appliances such as ranges having both 120- and 240-volt loads) the grounded neutral wire under some circumstances is permitted to serve also as a grounding wire. These cases, which now involve only existing installations, are explained under "Special provisions for ranges, ovens, cooking units, and clothes dryers" beginning on page 384. The white wire, whether it is a neutral or not, is grounded to the local grounding electrode system at the service equipment cabinet. The utility also grounds the neutral service wire at the transformer serving the premises.

If the neutral wire is properly grounded both at the transformer and at the service equipment at the building, it follows that if you touch an exposed white wire at a terminal or splice, no harm follows, no shock, any more than if you touch a water pipe or a faucet, because the white wire and the piping are connected to each other. Any time you touch a pipe, you are in effect touching the white grounded wire.

Since the white wire is actually grounded and there is no possible danger in touching it, why put insulation on that wire? The white wire is a current-carrying wire; if it is uninsulated it will touch piping and the like, and part of the current that the wire normally carries will flow through the piping and the like, to travel all over the building and possibly start fires. Furthermore, when a wire carries current there will always be voltage drop across the length of the wire. This means that even a grounded conductor always operates at some voltage above ground. Unless otherwise stated, it is necessary to use the same kind of insulation and the same careful splices for the white wire as for the hot wires.

EQUIPMENT GROUNDING HELPS PREVENT ELECTRIC SHOCK

Figures 9–3 through 9–10 illustrate correct installations as well as faulty conditions in which shock can occur. The 115- and 230-volt motors represented could be either freestanding or part of an appliance. (Instead of the usual 120 and 240 volts, the NEC still classifies motors as 115- and 230-volt equipment.) The coiled portion of the line on the right side of each motor drawing represents the motor's winding.

Grounding and fuse/breaker placement for 120-volt circuit   See Fig. 9–3, which shows a 115-volt motor with the white wire grounded, and a fuse in the hot wire. (Actually the fuse and the ground connection might be a considerable distance from the motor, although there might be an additional fuse near the motor). If the fuse blows (which is the same as cutting a wire at the fuse location), the motor stops. What happens if, in inspecting the motor, you accidentally touch one of the wires at a terminal? Nothing happens because the circuit is hot only up to the blown fuse. Between the fuse and the motor the wire is now dead just as if the wire had been cut at the fuse location. The other wire to the motor is the grounded wire, so it is harmless. You are protected. But if the fuse is *not* blown and you touch the hot wire, you will receive a 120-volt shock because the current will flow through your body to the earth, and through the earth back to the grounded wire, and then to its source of supply.

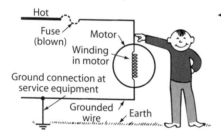

◀ **Fig. 9–3** A 115V motor properly installed except for a grounding wire. With no grounding wire, the installation is safe *only as long as the motor remains in perfect condition.*

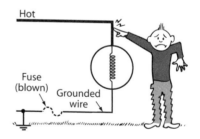

Fig. 9–4 A 115V motor with a fuse wrongly placed in the grounded wire. It is a dangerous installation. ▶

Fuse/breaker placement error   The same motor is shown in Fig. 9–4 except the fuse, instead of being placed in the hot wire, is wrongly placed in the grounded wire. The motor will operate properly, and it will stop if the fuse blows—but the circuit will still be energized through the motor up to the blown fuse. If you touch one of the wires at the motor, you complete the circuit through your body, through the earth, to the grounded wire ahead of the fuse; you are directly connected across 120 volts and are likely to receive a nasty shock. The degree of shock and danger will depend on the surface on which you are standing. If you are on an absolutely dry and nonconducting surface, you will notice little shock; if you are on a damp surface as in a basement, you will experience a severe shock; if you are standing in water, you will undergo maximum shock, often leading to death.

Accidental internal ground—unprotected   Now see Fig. 9–5, which again shows the same 115-volt motor as in Fig. 9–3. But suppose the motor is defective so that at the point marked *G* the winding inside the motor accidentally comes into electrical contact with the frame of the motor; the winding becomes "grounded" to the frame. That does not prevent the motor from operating. But suppose you touch just the frame of the motor. What happens? You will receive a shock up to 120 volts because you will be completing the circuit through your body, providing a path back to the grounded wire at the service. The exact voltage depends on where the motor winding insulation failed. If it failed near the hot wire connection, the voltage would approach 120 volts; if it failed near the grounded wire connection the voltage would be much less. It is a potentially dangerous situation; shocks of much less than 120 volts can be fatal.

In fact, it is not uncommon for breakdowns in the internal insulation of a motor to result in an accidental electrical connection between the winding and the frame of the motor. The entire frame of the motor becomes energized. The same situation develops if the motor is fed by a cord that becomes defective where it enters the junction box on the motor so that one of the bare wires in the cord touches the frame. If there is no cord but the motor is fed directly by the circuit wires, a sloppy splice between the circuit wires and the wires in the junction box of the motor can lead to the same result: the frame of the motor becomes energized.

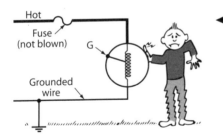

Fig. 9–6 The same defective motor as Fig. 9–5, but now a grounding wire has been installed from the frame of the motor to ground. Even though the winding is accidentally grounded to the frame, as represented by *G*, there is no shock hazard. ▶

◀ **Fig. 9–5** The same motor as in Fig. 9–3, but the motor is defective. *G* represents an accidental grounding of the winding to the frame. The grounding wire has not been installed. This is a dangerous installation.

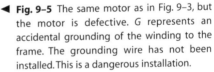

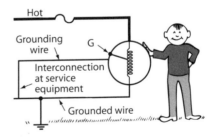

Accidental internal ground—protected   Now see Fig. 9–6, which shows the same motor as in Fig. 9–5 with the same accidental ground between winding (or cord) and frame, but protected by a grounding wire connected to the frame of the motor and running back to the ground connection at the service equipment. When the internal ground occurs, current will flow over the grounding wire, the amount depending on the degree of insulation breakdown within the motor housing. It will sometimes but not always blow the fuse. But even if the fuse does not blow, the grounding wire will protect you because it reduces the voltage between the frame of the motor to substantially zero as compared with the ground you are standing on; you will not receive a dangerous shock provided that a really good job of grounding and bonding was done at the service equipment. If there is a poor bonding connection between the equipment grounding wire and the grounded wire in the circuit, you will receive a shock, which may be dangerous, but still less hazardous than if there were no connection to a local grounding electrode.

Grounding and fuse/breaker placement for 240-volt circuit Now refer to Fig. 9–7, which shows a 230-volt motor installed with each hot wire protected as required by a fuse or circuit breaker. Remember that in 240-volt installations the white wire does not run to the motor, but the white wire is nevertheless grounded at the service equipment. If you touch both hot wires, you will be completing the circuit through your body, and you will receive a 240-volt shock. But if you touch only one of the hot wires, you will be completing the circuit through your body, through the earth, back to the grounded neutral at the service equipment, and you will receive a shock of only 120 volts: the same as touching the white wire and one of the hot wires of Fig. 9–1. The difference between 120- and 240-volt shocks may be the difference between life and death.

◀ Fig. 9–7 A 230V motor with each hot wire correctly protected by a fuse/breaker. The grounded wire does not run to 240V loads. Because the grounding wire has not been installed, shock hazard is avoided *only as long as the motor remains in perfect condition.*

Fig. 9–8 The same installation as Fig. 9–7, but the motor is now defective. The winding is accidentally grounded to the frame, as represented by G. This is a dangerous condition. ▶

Accidental internal ground—protected But assume that the motor becomes defective, that the winding in the motor becomes accidentally grounded to the frame, as shown in Fig. 9–8. This is the same as Fig. 9–5, except that the motor is supplied by 240 volts instead of 120 volts. Touching the frame will produce a 120-volt shock. But if the frame has been properly grounded, as shown in Fig. 9–9, one of the fuses will probably blow; even if it does not blow, touching the frame will not produce as dangerous a shock because the frame is grounded, again assuming that a really good grounding and bonding job was done at the service equipment. In any of the illustrations of Figs. 9–3 to 9–9, if there is an accidental contact between any two wires of the circuit, a fuse will blow regardless of whether the contact is between two hot wires or between a hot wire and the grounded wire. For this to happen, there must be bare places on two different wires touching each other, which doesn't occur very often in properly installed jobs.

Metal conduit or armored cable   Consider a system in which all the wires are installed in a metal raceway such as steel pipe called conduit, or in a cable with metal armor such as Type AC cable (often called by the trade name "BX"). The raceway or armor is grounded at the service equipment. It is also connected to the motor itself. (If the motor is connected by a three-wire cord and three-prong plug, the metal conduit or cable armor is connected to the motor frame through the grounding wire in the cord.) No separate grounding wire is required inside the metal conduit or metal armor. If there is now an accidental ground from winding to frame, or if a hot wire at a defective bare spot comes in contact with the metal raceway or armor, it has the same effect as a short between the hot wire and the grounded wire because the grounded wire and the raceway or armor are connected to each other at the service equipment. A fuse then will blow whether the motor is supplied by 120 or 240 volts. This is shown in Fig. 9–10.

◀ **Fig. 9–9** The same defective motor as Fig. 9–8, but a grounding wire has been installed from the frame of the motor to ground. Even though the winding is accidentally grounded to the frame, as represented by G, the shock hazard is minimized.

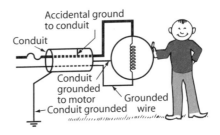

Fig. 9–10 Wires in metallic conduit or metallic armor are used to install the motor. The conduit or armor is grounded both at the service equipment cabinet and to the motor. It is a safe installation. ▶

Separate grounding conductors   Many wiring methods use nonmetallic cable sheaths or raceways. In most such cases you install (or the cable includes) a separate grounding conductor that performs the same equipment grounding function as the steel raceway or cable armor. In addition, some designers insist on a separate equipment grounding conductor even within a steel raceway in order to add additional reliability. The *NEC* actually requires that arrangement for branch circuits serving patient care areas of hospitals. Although hospital wiring is beyond the scope of this book, that rule points to the many instances where you're likely to encounter separate grounding conductors.

EQUIPMENT GROUNDING CONDUCTORS MUST HAVE LOW IMPEDANCE

The NEC and the electrical product standards go to great lengths to assure that line voltages don't reach conductive surfaces where they would create a shock hazard. However, since nothing is perfect, the NEC also provides a place for current to return in the case of an insulation failure. This is the equipment grounding system, as just described. To work properly, this return path must have as low an impedance as possible, and it must be comprehensive, reaching all conductive surfaces capable of presenting a shock hazard.

Continuous grounds   At every location where there is an electrical connection you can expect to find an equipment grounding connection as well. It may be a metal raceway, such as a run of conduit, or it may be a separate grounding conductor, or both. Then, where a grounded neutral wiring system is used (which is nearly 100% of the time for residential and farm wiring as well as for most commercial occupancies) you must ground not only the white wire but also the conduit, cable armor, or separate grounding conductor. The white wire is grounded only at the service equipment, but the conduit, metal armor, or separate grounding conductor must be securely bonded to every metal box or cabinet, and to grounding terminals, yokes, and fixture support bars as required in the case of nonmetallic enclosures. When fixtures and devices are mounted on these enclosures they become grounded. When you connect your electrical system to motors or appliances, they become grounded. This is also the case where a motor or appliance is connected by means of a flexible cord that includes a grounding wire, a three-prong plug, and a grounding receptacle, all explained later in this chapter.

It is of the utmost importance that you solidly drive down the locknuts on the conduit (or on the connectors of armored cable) so that the locknut bites down into the metal of the box. In that way you provide a continuous metallic circuit entirely independent of any of the wires installed, which is an essential part of grounding. Such continuous grounding is necessary even if there is an additional separate grounding conductor within the wiring method. This is because fault return currents on ac systems, due to skin effect (described on p. 305), attempt first to flow on the outer margin of the wiring method. If this path is open, they will flow over the enclosed grounding conductor, but only at a cost of increased impedance. As explained under the next heading, this small but definite increase in impedance could directly result in a fire.

Similarly, take care to make solid electrical connections when you terminate a separate grounding conductor. Don't use sheet metal screws, for example, because their coarse threads don't have the mechanical advantage of a machine screw, and the NEC prohibits their use for this purpose. Regardless of the wiring method used, a continuous ground all the way back to the source (usually the service) is of the utmost importance.

Return path—run with or enclose circuit conductors   In addition to continuity, one of the most important concepts in proper grounding involves routing equipment grounding return paths so they always run with or enclose the circuit conductors that potentially supply the fault current that could be required to return. Understanding this requires a review of some basic principles that apply to ac circuits and their protective devices.

AC circuits involve an alternating magnetic field associated with the alternations of current and voltage. Each second that magnetic field collapses 120 times, and each time it collapses it induces a voltage counter to the principal system voltage, a potential source of significant impedance. However, when the return current is in the same physical location, its magnetic field is opposite of that generated by the current leaving the source. This self-cancellation of magnetic fields reduces the overall impedance in the circuit path.

As discussed at the beginning of Chapter 4, overcurrent protective devices protect circuits from overloads and fault currents. These devices clear overload conditions on an inverse current basis, with a deliberate time delay arranged so that the amount of delay varies inversely with the amount of current. If the current exceeds overload levels, those faults are cleared essentially instantaneously. This threshold might be on the order of ten or more times the nominal trip rating.

Using Ohm's law, the current flowing in a fault equals the voltage divided by the impedance. Suppose we want to be sure a 100-amp breaker trips immediately on a ground fault. If the circuit is 120 volts to ground, the impedance has to be on the order of $\frac{1}{10}$ ohm or lower. If the impedance exceeds this amount by even a fraction of an ohm, the overcurrent protective device will "see" the fault as an overload such as a motor trying to start. Accordingly, the fault will be allowed to continue until the overcurrent device, in effect, decides that the high current isn't just a nuisance and it clears.

Putting these principles together, in order to be sure that overcurrent protective devices operate in their instantaneous range, we have to do everything possible to reduce or eliminate sources of impedance in a fault return path. When an equipment grounding return path doesn't run with or enclose the supply circuit, a small but definite impedance degradation results. That additional impedance frequently will drop the potential fault current below the instantaneous response range of the overcurrent device.

When an arcing fault progresses for any length of time, such as when an overcurrent device "thinks" it is dealing with a nuisance overload that needn't be immediately cleared, the amount of damage is likely to be catastrophic. As noted in Chapter 4, electrical arcs are among the hottest phenomena known on the surface of the earth. The result is almost certain to be a major fire if there are any combustibles in the vicinity.

This is why the *NEC* in 250.134(B) and other locations emphasizes the principle that fault current supply and return paths must be confined to the same physical location over the entire length of the path. And this is why every aspect of workmanship on equipment grounding issues has such a crucial bearing on the overall safety of the electrical system. Take the time to do this part of the job right.

Grounding of fixed equipment   In this chapter we have discussed the equipment grounding wire running to a motor. But many other items must have an equipment grounding wire (unless the wiring is in rigid metal conduit, EMT, or armored cable is used, any of which serve the purpose of a separate equipment grounding wire). *NEC* 250.110 generally requires grounding of exposed non-current-carrying parts of fixed equipment that is "likely to become energized." This terminology is often used; as defined in product standards, something is likely to become energized if the failure of even a single element of insulation, such as that on a wire, could energize a conductive surface. For example, a metal box must be grounded because a nick in the insulation on any ungrounded conductor within it could put a hazardous voltage on the surface of the box. Careful reading of *NEC* 250.110 reveals that it requires grounding in almost every case. One requirement is that any equipment supplied by a wiring method that provides an equipment ground must be grounded. Since all wiring methods in common use do provide for equipment grounding, that provision alone covers most any installation.

Note that the universal equipment grounding principle includes metal faceplates. For snap switches installed in metal switch boxes, their metal faceplates will be properly grounded if the boxes are properly grounded. But if you install a metal faceplate on a nonmetallic box, you must take steps to ground it as required in *NEC* 404.9(B). Snap switches with a green grounding screw terminal are available and commonly used for this purpose. They are wired exactly the same as the grounding terminals on receptacles, shown in Fig. 9–16. The faceplate is automatically grounded when it is screwed into the grounded device yoke.

GROUNDING-TYPE RECEPTACLES INCREASE SAFETY

The original receptacle having only two parallel openings for the blades of the plug, as shown in Fig. 9–11, hasn't been allowed in new construction for almost 40 years, but there are still millions in everyday use. They continue to be manufactured because they are allowed by the *NEC* in 406.3(D)(3)(a) as replacements on the millions of branch circuits still running on obsolete wiring methods that don't include equipment grounding provisions. Anything plugged into such a receptacle duplicates the condition of Fig. 9–3. If the appliance is defective, anyone handling it could receive a shock, as shown in Fig. 9–5. This danger led to the development of the "grounding receptacle," which has the usual two parallel slots for two blades of a plug, plus a third round or U-shaped opening for a third prong on the corresponding plug (see Fig. 9–12).

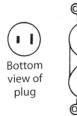

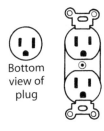

Fig. 9–11 An ordinary receptacle and plug. Only plugs with two blades will fit it. Although nongrounding, it is polarized. The larger slot and blade connect to the grounded circuit conductor.

Fig. 9–12 A grounding receptacle and plug. Plugs with either two blades or two blades plus a grounding prong will fit it. Not all grounding plugs have different-sized blades since the ground pin polarizes the plug (the blade to the left, shown as smaller, is the ungrounded conductor). The receptacle slots, however, are polarized, because nongrounding plugs may be inserted and their polarity of the plugs must be *enforced*.

In use, the third prong of the plug is connected to a separate grounding wire in the cord; this grounding wire, which is green (sometimes green with one or more yellow stripes) runs to and is connected to the frame of a motor or appliance. On the receptacle, the round or U-shaped opening leads to a special green terminal screw that is connected to the yoke or mounting strap of the receptacle. When installing the receptacle, you must connect this green terminal to ground unless the yoke screws serve that purpose, as more fully covered in Chapter 18. If conduit or cable with armor is used, run a bonding jumper, of the same size as would be required for a separate grounding wire, from the grounding terminal to the outlet or switch box, and the conduit or the metal armor becomes the grounding conductor. Cable without armor (and even some with metal armor) and all nonmetallic raceways will contain an extra wire, bare or green, in addition to the insulated conductors; this wire must be connected to the green terminal of the receptacle, and grounded to the boxes. In this way the frame of the motor or appliance is effectively grounded, leading to extra safety as shown and discussed in connection with Fig. 9–6.

The *NEC* takes no position on the orientation of the ground hole on a receptacle, up or down (or left or right). Some prefer to mount the ground hole up, based on the ground pin providing some protection against a metal object dropping against a partially inserted plug and potentially contacting the hot blade. Others prefer the ground pin down, based on the orientation of multitudes of angle cord caps that prevent the cord from hanging straight down unless the ground pin is in the lower position. Many receptacles are mounted horizontally, and many cord caps orient at a 45-deg angle from the ground pin so they can be used in either vertical or horizontal receptacles. It is impossible to make receptacle orientation into a rule, so let the owner decide.

For residential occupancies, *NEC* 250.114(3) requires a three-wire cord and three-prong plug for all cord-connected refrigerators, freezers, air conditioners, clothes washers, clothes dryers, dishwashers, information technology equipment (computers), aquarium equipment, sump pumps, and hand-held motor-driven tools such as drills, saws, sanders, hedge trimmers, lawn mowers and similar items. The three-wire cord and three-prong plug requirement does not apply to ordinary household appliances such as toasters, irons, radios, TV sets, razors, lamps, and similar items. Grounding receptacles are designed to allow either grounding or nongrounding plugs to fit (see Fig. 9–12).

What about double insulation?   In the case of ordinary portable tools, the *NEC* does not require grounding if the tool is constructed with what is called *double insulation*—as opposed to only functional insulation, which is only as much as required for proper functioning of the tool, including the insulation on the wires used to wind the motor, an insulated switch, and so on. Double insulation goes much further: usually the case of the tool is made of insulating material rather than metal; the shaft is insulated from the steel lamination of the rotor of the motor; switches have handles of an especially tough insulating material; and there are many other points of specially durable insulation.

NEC 406.3(A) requires that for new installations all receptacles on 15- and 20-amp circuits must be of the grounding type. If you replace a defective receptacle in an existing installation, the new one must be of the grounding type provided you can effectively ground it. If that is difficult or impossible, the replacement receptacle must *not* be of the grounding type unless it has GFCI protection as covered later in this chapter.

Three-to-two adapters   What if you have only nongrounding two-wire receptacles and you want to use an appliance with a three-wire cord and a three-prong plug? The best procedure is to replace the old receptacles with new grounding receptacles if you can ground them properly. An alternative is to use a "three-to-two" adapter as shown in Fig. 9–13. First, temporarily remove the screw holding the faceplate, and then plug in the adapter so the grounding lug aligns with the screw hole. The adapter will be polarized, so you can only do this on one of the two receptacles at the outlet. Then reattach the faceplate screw, grounding the

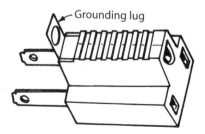
Grounding lug

Fig. 9–13 This adapter permits a three-prong grounding plug to be used in an ordinary two-slot receptacle. The use of the adapter, however, does not provide the same protection as a properly installed grounding receptacle.

adapter. But be warned: Unless the existing circuit provides a grounding connection to the screw in the mounting yoke of the receptacle (or the yoke has an otherwise acceptable grounding connection as covered in the page 344 topic on grounding receptacles) you will *not* have grounding protection when the adapter is used.

How dangerous are shocks? Most people think it is a high voltage that causes fatal shocks, which is not necessarily accurate. The amount of current flowing through the body determines the effect of a shock. A milliampere (mA) is one-thousandth of an ampere. A current of 1 mA through the body is just barely perceptible. One to eight milliamperes causes mild to strong surprise. Currents from 8 to 15 mA are unpleasant, but usually victims are able to let go of the object that is causing the shock. Currents over 15 mA are likely to lead to "muscular freeze" which prevents the victim from letting go and often is fatal. Currents over 75 mA are likely to be fatal. Much depends on the individual involved.

The higher the voltage, the higher the number of milliamperes that would flow through the body under any given set of circumstances. A shock from a relatively high voltage while the victim is standing on a completely dry nonconducting surface will result in fewer milliamperes than a shock from a much lower voltage while standing in water. Many deaths have been caused by shock on circuits considerably below 120 volts; many people have survived shock from circuits of 600 volts and more.

Another determining factor in danger is the number of milliamperes a source of power can produce. When you walk on the dry carpeting of a house in the winter when the humidity is low, you pick up a static charge and feel a mild shock when you touch a grounded object. The voltage is high, but the current is infinitesimal. If you touch the spark plug of a car while the motor is running, you receive a shock of at least 20 000 volts, but the current is exceedingly small, so no harm is done. The same principle applies to electric fences used on farms—a voltage high enough to be uncomfortable, but a very small current for a very short period of time that does no harm.

It should be noted that farm animals are much less able to withstand shocks than are human beings. Many cattle have been killed by shocks that would be only uncomfortable to a person.

GROUND-FAULT CIRCUIT INTERRUPTERS (GFCI)

A ground-fault circuit interrupter (GFCI or GFI) is a device that can be installed to protect either a complete 120-volt, two-wire circuit or a single receptacle against ground-fault currents. If there is a fault between the hot wire and the frame of a defective tool or appliance held by a person, the GFCI will open the circuit quickly enough to keep the shock from being dangerous, even if it will be felt. Present product standards require a GFCI to trip the circuit if the fault current reaches 4 to 6 mA. The allowable tripping time varies inversely with the magnitude of the current.[1] The principle of operation is shown in Fig. 9–14. In a properly installed two-wire, 120-volt circuit, the amount of current flowing in the grounded wire is precisely the same as in the hot (ungrounded) wire. But if there is a fault in a tool or appliance, part of the current (the fault current) will flow through the grounding conductor, if present. Some will flow through the body of the operator. Therefore the amount of current in the grounded conductor will be less than in the hot conductor. The GFCI senses this difference and opens the entire circuit or the ungrounded wire to the receptacle that it protects.

An ordinary GFCI will not protect a three-wire circuit, or receptacles connected to two of the wires of a three-wire circuit (three-wire circuits are discussed at the

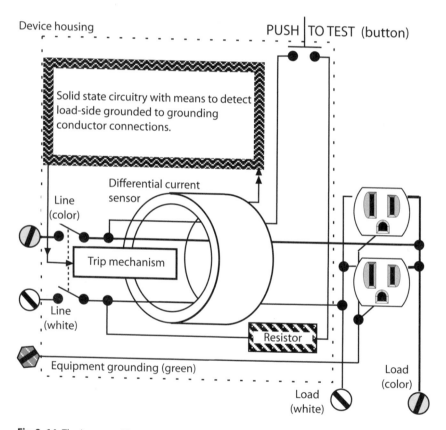

Fig. 9–14 The inner workings of a GFCI protective device.

1. The actual allowable tripping time in seconds cannot exceed $(20/I)^{1.43}$, where I is the current in milliamperes. To provide a little perspective, since 1 raised to any power is 1, a 20 mA fault must be cleared in not more than one second. The typical device test circuit creates about a 7 mA current differential, and even at that lower amount most devices trip much faster than one second.

beginning of Chapter 20). But if a three-wire circuit is divided into two 2-wire circuits, a GFCI can be used to protect one of those two-wire circuits, or to protect one or more receptacles on one of those circuits. Be sure the GFCI has the same ampere rating as the circuit, and be sure it is listed for the type of protection for which you are using it. Two-pole GFCI circuit breakers are also available to protect an entire three-wire circuit, or a straight 240-volt, two-wire circuit with no neutral required.

GFCI types   Ordinary GFCIs are available in various types. One form is a separate device installed to protect a receptacle or circuit. More popular are combination circuit breakers and GFCIs, which are installed in a panelboard to protect an entire 120-volt circuit (see Fig. 9–15). To install the device, run the coiled white wire to the neutral bus bar in the panel, and terminate both the ungrounded and grounded circuit conductors at the appropriate points on the breaker. The breaker needs to sense current in both sides of the line in order for its GFCI function to operate.

The most common type consists of GFCI protection built into a receptacle, as shown in Fig. 9–16. If you only need the GFCI protection and not the receptacle (or code rules prohibit installing a receptacle on the circuit you need to protect), you can use the device in Fig. 9–17.

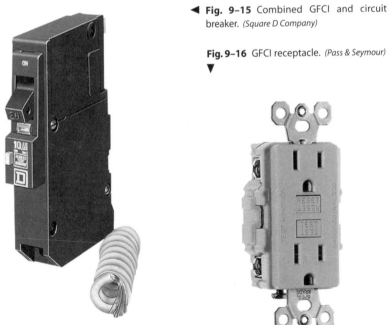

◄ **Fig. 9–15** Combined GFCI and circuit breaker. *(Square D Company)*

Fig. 9–16 GFCI receptacle. *(Pass & Seymour)*
▼

The receptacle in Fig. 9–16 is a "feed through" device, which means it has LINE and LOAD markings. It protects downstream loads connected to its load terminals. *Do not mix up the line and load sides of these devices.* If you wire the device backwards, downstream loads will be protected, but appliances plugged into the receptacle *will have no protection.* The test button will trip, apparently functioning normally, but the ungrounded receptacle slots will still be hot.

Figure 9–17 is essentially a GFCI feed-through receptacle as depicted in Fig. 9–16, but without any receptacle slots. If the test and reset buttons have additional OFF and ON markings, the device has been evaluated as a manual motor controller and can be used as a switch.

Fig. 9–17 A "master trip" GFCI device. *(Pass & Seymour)*

Where GFCIs are required   The *NEC* requires GFCI protection for 15- or 20-amp, 125-volt receptacles as follows (there are many other required locations, but this list covers most of the common examples):

- In residential properties, for *all* receptacles in the following locations: outdoors including balconies (except those for rooftop snow melting equipment, for which special rules apply); bathrooms; garages except for receptacles for dedicated appliance space or on the ceiling ("not readily accessible"); serving kitchen countertops; within 6 ft of a laundry, utility, or wet bar sink; and in crawl spaces and unfinished basements. The GFCI may be installed to protect individual receptacles, or the entire branch circuit supplying the receptacles.

- In other occupancies, for all receptacles in bathrooms and on rooftops, again with an exception for rooftop de-icing equipment for which special rules apply. The requirement also applies to all nonresidential kitchen receptacles ("kitchen" being defined for this purpose as an area with a sink and permanent facilities for food preparation). The 2005 *NEC* extends the requirement to most outdoor receptacles, excepting only those not accessible to the public. The requirement also applies to all outdoor receptacles installed to facilitate servicing of heating, air-conditioning, and refrigeration equipment, whether publicly accessible or not.

■ Formerly just for construction sites, now for all 15-, 20-, and 30-amp, 125-volt, single-phase receptacles that are used for "construction, remodeling, maintenance, repair, and demolition" activities, whether or not on a construction site. An exception applies for some industrial occupancies with expert supervision, but only in cases where a random nuisance trip would create a greater hazard, or in cases where the powered equipment can be shown to be incompatible with GFCI devices.

■ Swimming pools: There are very stringent requirements that are discussed separately in the special applications topic beginning on page 389.

Replacement receptacles   If you replace a receptacle, and it is in a location where the *NEC* now requires GFCI protection for that receptacle, even if no GFCI protection was required at the time the original receptacle was installed, you must provide GFCI protection for the replacement. If you replace a receptacle and there is an equipment grounding conductor available at the outlet, the replacement receptacle must be of a grounding configuration and properly grounded whether or not the receptacle needed to be grounded at the time it was originally installed. However, if you replace a receptacle at an outlet with no equipment grounding conductor present, you have three options, assuming you don't plan to rewire the branch circuit. You could (1) use a new, nongrounding receptacle, or (2) use a GFCI receptacle (they only come in grounding configurations) with a NO EQUIPMENT GROUND marking on the faceplate, or (3) use a conventional grounding receptacle fed from a GFCI protective device and mark the receptacle location with both NO EQUIPMENT GROUND and GFCI PROTECTED.

Ground-fault protection of equipment (GFPE)   There is a form of protection closely related to the GFCI that is used to prevent very low-level arcing faults from causing fires. It is typically available in specialized circuit breakers. Imagine circuitry identical to that of a GFCI, but with the trip set in the neighborhood of 30 mA. This would not provide adequate shock protection, but the *NEC* does require this protection for frost protection systems run for snow melting purposes (*NEC* 426.28) and along piping systems (*NEC* 427.22). Experience has shown that these cables have so much resistance that faults will not trip an overcurrent device, but they will sputter for long periods, eventually starting fires. Although GFPE in this sense has a far different "feel" than the industrial systems covered in Chapter 28, with their current imbalance trigger points set for hundreds of amperes (see Fig. 28–14 and associated text), the principles are the same. Arcing faults usually produce a current flow outside of normal circuit paths; that difference can be detected, and the circuit automatically opened. Be sure you understand the differences so you can order and apply the appropriate protective devices.

CHAPTER 10

Outlet and Switch Boxes

IN THE INTEREST OF SAFETY from both fire and shock hazards, switch and outlet boxes or equivalent enclosures must be installed at every point where wires are spliced or connected to terminals of electrical equipment. This chapter covers only outlet and device boxes, generally not over 100 cu in.; large pull boxes are discussed in the topic on junction and pull boxes beginning on page 499.

BOXES SERVE DUAL PURPOSE

Boxes house the ends or splices in wires at all points where the raceway (or cable armor, shield, or jacket) has been terminated or interrupted, and where the insulation of the wire has been removed. A continuous piece of wire properly protected against physical damage is substantially safe, but a poorly made splice or terminal connection can lead to short circuits, grounds, or overheating. Electric arcs are hotter than gas flames and can readily ignite dust, cobwebs, many kinds of thermal insulation or its wrapping, and many types of construction or finishing materials in the hollow spaces of walls and ceilings. Furnishings and stored materials are similarly vulnerable. Therefore there is some danger of fire at splices and terminals, but enclosing them in boxes practically eliminates this danger. Moreover, metal boxes provide a continuity of ground for a metal wiring system, as explained in the preceding chapter. The *NEC* requires that boxes be supported according to definite standards that provide mechanical strength for supporting lighting fixtures, switches, and other equipment, eliminating the danger of mechanical breakdown.

KNOCKOUTS ALLOW CONDUIT OR CABLE ENTRY

Around the sides and in the bottoms of boxes there are *knockouts,* which are sections of metal, or thin sections of plastic in the case of nonmetallic boxes, that can be knocked out easily to form openings for conduit or cable to enter.

Removing knockouts Metal knockouts are completely severed around these sections except at one small point that serves to anchor the metal until it is to be removed. It is a simple matter to remove these knockouts—usually a stiff blow with a pair of pliers on the end of a heavy screwdriver held against the knockout will start it, and the metal disk is then easily removed with the pliers. If the knockout is near the edge of a box, a pair of pliers is the only tool needed. Some brands of metal boxes have the pry-out type of knockout. Insert a screwdriver into the small slot near or in the knockout and pry out the disk. Remove only as many knockouts as actually needed for proper installation. If you accidentally remove a knockout that won't be used, you must close the opening with a knockout seal. The most common type, shown on the left in Fig. 10–1, is a metal disk with tension clips around the edge on one side. Push the clips into the unused opening with your thumb, then tap it in with a pair of pliers or similar tool. For larger sizes of knockouts, use the type shown on the right in Fig. 10–1; this consists of two disks, one for the inside and the other for the outside of the box. A screw through the disks holds the seal in place.

Concentric knockouts   The cabinets of service switches and breakers do not have room for separate knockouts of all the different sizes that may be required in various installations, so *concentric knockouts* are provided, as shown in Fig 10–2. In addition, many box manufacturers have decided to reduce their inventory

Fig. 10–1 Any unused opening must be sealed with a knockout closure.

problems by making boxes with concentric knockouts; that way a single box knockout configuration accommodates two or three different raceway trade sizes. Be especially careful to remove only as many sections as required for the size you need. Remove just the center section to provide the smallest size; remove two sections for the next larger size, and so on. If you make a mistake and remove too many sections, you'll need to install a set of reducing washers (Fig. 10–3) to correct the resulting oversized opening.

Fig. 10–2 A concentric knockout. Remove as many sections as necessary to provide the size of opening required.

Fig. 10–3 Reducing washers are available in almost every conceivable combination of sizes.

There is a trick to removing intermediate sections that have two points of support. Twisting from side to side often results in distorting and weakening more sections than you intend to remove. Instead of twisting, try pushing the middle of the section you want to remove halfway between the supporting tabs, all the way back until it is roughly perpendicular to the side of the enclosure. Then do the same on the other side. The two halves will be almost touching each other. Take your diagonal cutting pliers ("dikes" in the trade) and cut through both halves of the knockout where they almost touch. Now you can move each half of the section independently in a back-and-forth motion that doesn't weaken the other portions of the concentric knockout that you want to leave in place.

TYPES OF BOXES

NEC 314.24 requires that all boxes be at least ½ in. deep, and boxes that contain flush devices such as receptacles must be at least $^{15}\!/_{16}$ in. deep. The kind and size of the box you use depends on the purpose of the box and, more important, on the number of wires that will enter the box, as discussed in this chapter. All boxes must be covered.

Steel outlet boxes   Figure 10–4 shows the common 4-in. octagonal outlet box. These boxes are still available in smaller sizes but even the 4-in. box is not very roomy. It is generally more practical to use square boxes as shown. They are made in both 4- and 4$^{11}\!/_{16}$-in sizes, and in 1½- and 2⅛-in depths. Either type can be used for either one or two switches or similar devices. Today every metal outlet box must be grounded. If you need to add a box to an existing wiring method that doesn't include an equipment grounding conductor, you must either arrange for equipment grounding or use a nonmetallic box.

Fig. 10–4 Typical octagonal and square outlet boxes.

Sometimes it is necessary to install three or more switches or similar devices side by side. Boxes similar to square boxes are made in wider sizes to permit this. Figure 10–5 shows a 4-in. box with two switches, and a wider four-gang box and the cover to permit mounting four devices.

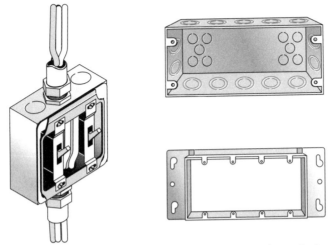

Fig. 10–5 A square box with two switches, and a four-gang box and cover for it.

Nonmetallic boxes   Figure 10–6 shows a representative assortment of nonmetallic boxes. Many types are available, including those with mounting brackets for installation on timbers of buildings. Nonmetallic boxes are constructed in one piece and cannot be ganged, but multigang boxes are available.

Fig. 10–6 Nonmetallic boxes have become the predominant device enclosure in residential construction today. In addition to the ones shown here, which are for cabled wiring methods, there are nonmetallic boxes with conduit-sized knockouts designed for nonmetallic raceways.

Fig. 10–7 With square boxes, use covers of the type shown in the center and on the left unless the wiring is permanently exposed. Although the use of plaster finish has declined dramatically, "plaster ring" is still the usual term for these covers. If the wiring is permanently exposed, use covers of the type shown on the right, usually referred to as "raised covers." *(Hubbell Electrical Products)*

Install nonmetallic boxes as you would metal boxes, except that per *NEC* 314.17(C) connectors or clamps are not required at a single-gang device box. If you use this provision, which applies only to single-gang device boxes, secure the cable within 8 in. of every box, measured along the cable, and the cable sheath must enter the box at least ¼ in. For all other boxes the support distance is 12 in., just as for metal boxes.

Covers A fixture mounted on a box and completely concealing it is considered a cover. For square boxes installed in walls or ceilings, use covers of the type shown at the left in Fig. 10–7; the fronts of the covers will be flush with the wall or ceiling surface. Use covers of the type shown to the right in the same photo for boxes installed on the surface of a wall in permanently exposed wiring.

With octagonal boxes, use the covers shown in Fig. 10–8. At *A* is shown a blank cover used only on outlet boxes that serve only as pull or junction points. At *B* is a drop-cord cover used with pendant lights or fixtures; the hole is bushed to eliminate sharp edges that might otherwise injure the cord's insulation. At *C* is a "spider cover" sometimes used to mount surface-type switches. At *D* is a cover with a duplex receptacle used mostly in basements, workshops, and similar locations. At *E* and *F* are lampholders used as inexpensive lighting fixtures in closets, attics, basements, farm buildings, and similar locations; one is unswitched and

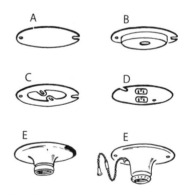

Fig. 10–8 Covers for octagonal boxes are made in dozens of types.

the other has a pull-chain switch. There are many other types of covers. All metal covers must be grounded. Where mounted on grounded metal boxes, the mounting screws take care of the grounding connection. If the box is nonmetallic, then you need to add a grounding bracket to the box (available in some brands), or independently ground the cover. The grounding clip shown in Fig. 18–5 does the job nicely.

Figure 10–9 shows how box, cover, device, and final faceplate all fit together. The distances between mounting holes are standardized.

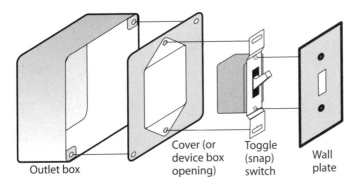

Outlet box   Cover (or   Toggle
                      device box   (snap)   Wall
                      opening)   switch   plate

Fig. 10–9 All parts of an electrical outlet are standardized in size to fit properly and easily.

Switch boxes   The *NEC* uses the term "device box" for what is commonly called a "switch box." The *NEC* term is more accurate because switch boxes are also used for receptacles and other devices and not just switches, but the term "device box" is heard less often.

Figure 10–10 shows the ordinary switch box. The removable sides allow you to create a double-size ("two gang") box out of two single boxes as shown in Fig. 10–11. Discard one side of each box and join the boxes together as shown in the picture. No additional parts are needed. In similar fashion you can join several boxes together to mount three or more devices. In most cases, however, a single box made for the purpose is sturdier and easier to support.

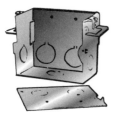

◀ **Fig. 10–10** A typical single-gang switch box. The sides are removable.

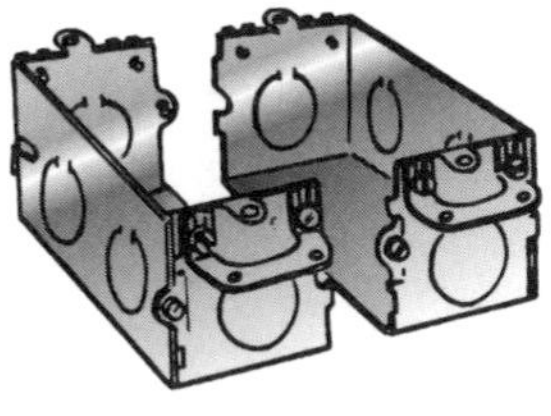

Fig. 10–11 Two single-gang boxes joined to form a two-gang box. Any number of boxes may be ganged to form a box of any size needed. ▶

Switch boxes range in depth from $1\frac{1}{2}$ to $3\frac{1}{2}$ in. Under today's rules the shallower ones seldom have a practical application. Review the sizing rules at the end of this chapter.

INSTALLING BOXES

To provide a good, safe continuous ground throughout an installation, it is absolutely necessary that every length of metal conduit or armored cable entering a box be firmly and solidly anchored to the box.

Securing conduit to boxes Conduit is anchored to boxes by means of a locknut and bushing, both shown in Fig. 10–12. Note that the locknut is not flat, but is dished so the lugs around the circumference are actually teeth that will dig into the metal surface of the box. In the case of bushings, the smallest internal diameter is slightly less than the internal diameter of the conduit. This causes the wires, where they emerge from the conduit, to rest on the rounded surface of the bushing, preventing damage to the insulation of the wires that might occur if the wires were to rest against the sharp edge of the cut end of the conduit.

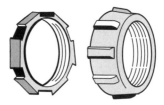

Fig. 10–12 Locknuts and bushings are used at all ends of conduit. They anchor the conduit to the box and provide a continuous grounded raceway.

Screw a locknut on the threaded end of the conduit with the teeth facing the box. Slip the conduit into the knockout. Then screw the metal bushing on the conduit inside the box as far as it will go. Only then tighten up the locknut on the outside of the box; drive it home solidly so the teeth bite into the metal of the box, making a good grounding connection. See Fig. 10–13. Detailed instructions for cutting and using conduit are covered near the beginning of Chapter 11.

Where the wires are 4 AWG or larger, the bushing must be of the insulating type shown in Fig. 10–14. Such bushings may be made of metal with an insulated

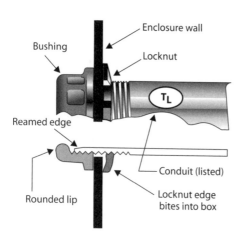

Enclosure wall

Bushing

Locknut

Reamed edge

TL

Rounded lip

Conduit (listed)

Locknut edge bites into box

◀ **Fig. 10–13** Cross section showing how a locknut and metal bushing are used to anchor conduit to any box or cabinet.

Fig. 10–14 Insulated bushings are used with larger sizes of conduit. *(Hubbell Electrical Products)*
▼

throat. Bushings made entirely of insulating material are available, but if they are used, a locknut must be installed between the box and the bushing in addition to the locknut outside the box. The double-locknut method is actually required for circuits running over 250 volts to ground unless other steps are taken to ensure grounding continuity. This is covered in Chapter 11 (see Fig. 11–43) and Chapter 27 (see Fig. 27–15).

Securing cable to boxes   If the box does not have the built-in clamps for cable as described in the next paragraph, use connectors of the type shown in Fig. 10–15. After the connector is solidly anchored to the cable, slip the connector into a knockout, install the locknut inside the box, driving it home tightly as in the case of conduit so the teeth bite into the metal of the box. The shoulder of the connector acts as a locknut on the outside of the box. Many nonmetallic cable connectors today, and even some for metallic cables, skip the locknut in favor of snap-in catches or other designs. Make certain the connector you use has been evaluated for the cable to be terminated. Since this information is quite detailed, it is spelled out on the smallest unit shipping carton. Make sure you review this information if you have any questions about a connector's suitability.

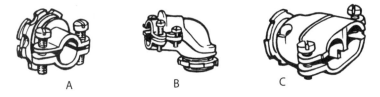

A   B   C

Fig. 10–15 Cables are anchored to boxes with connectors of this general type.

It is common to use boxes that have built-in clamps, which eliminate the need for separate cable connectors. Typical boxes of this kind are shown in Fig. 10–16. The ends of the wires would be much longer than shown.

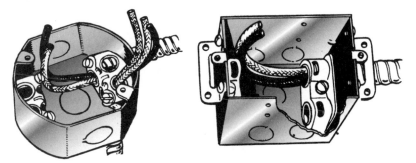

Fig. 10–16 Boxes are available with clamps for cable, making the use of separate connectors unnecessary.

Boxes in ceilings or walls If the wall or ceiling is combustible (meaning if it can burn), the front edge of the box (or the cover installed on the box) must be flush with the surface. If the wall material is noncombustible, the front edge may be as much as ¼ in. below the surface, but it is better to install the box flush. You will have to find out the thickness of the ceiling or wall from the builder. If the box is installed in an unfinished area that may later be finished, mount the box so its front edge will be flush with the future wall or ceiling.

Outlet boxes used for load support Outlet boxes cannot normally support static loads, such as heavy chandeliers, that weigh more than 50 lb unless the box is specifically listed for heavier duty. In addition, outlet boxes cannot support dynamic loads, such as paddle fans, unless specifically listed for that duty regardless of the weight of the load. Boxes used to support fixtures must be "designed for the purpose" and switch boxes don't fall within that category, although a special exception allows switch boxes to support small fixtures. Chapter 18 provides detailed information for hanging fixtures of any weight at outlets, and for supporting paddle fans.

Supporting outlet boxes The usual method of supporting a ceiling box for a lighting fixture is by means of a bar hanger as shown in Fig. 10–17. Such hangers consist of two telescoping parts, and thus are adjustable in length. Remove the center knockout in the bottom of a box, slip the fixture stud that is part of the hanger (as shown in the upper part of the illustration) into the opening, adjust the length of the hanger to fit between the two joists, and install the locknut on the inside of the box. Then nail the hanger to the joists as shown in Fig. 10–18. The fixture stud can later be used to support a lighting fixture. If the box is not to be used for a fixture, use a hanger without a fixture stud as shown in the lower part of Fig. 10–17.

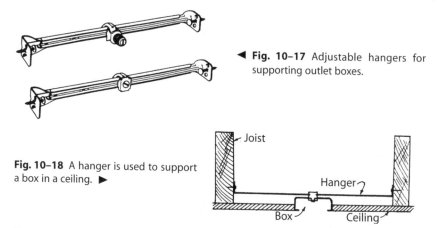

◀ **Fig. 10–17** Adjustable hangers for supporting outlet boxes.

Fig. 10–18 A hanger is used to support a box in a ceiling. ▶

Factory-assembled combinations of boxes with hangers, as shown in Fig. 10–19, are time savers.

Boxes may be installed in walls in the same way as in ceilings, but using boxes with mounting brackets will usually save much time; some types are shown in Fig. 10–20. Nail the brackets to the studs and the installation is finished. The same bracket boxes may be used in ceilings if a location next to a joist is suitable.

Fig. 10–19 Preassembled boxes with hangers save time in installation.

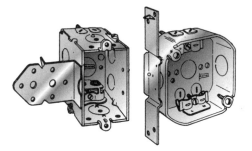

Fig. 10–20 Outlet and switch boxes with mounting brackets. Nail the brackets to studs.

Supporting switch boxes   Ordinary switch boxes as shown in Fig. 10–10 can be nailed directly to studs at the proper depth to be flush with the final surface. Some boxes of this type have nails already installed in the boxes, staked so they can't fall out. But it saves a great deal of time to use boxes with mounting brackets as shown in Fig. 10–20. The brackets can be nailed directly to the studs. Be sure to select boxes with the brackets at the proper location considering the thickness of the particular wall materials that will be used. Instead of wooden studs in buildings, metal studs and door bucks are becoming popular even for residential applications. Special brackets are available for mounting standard boxes to such metal structural members without requiring drilling or tapping of metal members. Fig. 10–21 shows a practical application of boxes and their supporting fittings in metal framing.

Surface boxes   When the wiring is on the surface of a wall and will be permanently exposed, ordinary boxes are impractical because of the sharp corners both on the boxes and on covers for switches or

Fig. 10–21 These boxes have brackets that snap into metal framing members. The sheet metal bracket is arranged to bottom on the wallboard on the back side of the partition, providing additional stiffness to the box. This feature is particularly helpful at receptacle outlets.

similar devices. Use a special box known as a utility box or "handy box"; both box and covers are shown in Fig. 10–22. Square boxes and appropriate raised covers are also often used for surface wiring.

Fig. 10–22 The "handy box" and covers shown are used for surface wiring. *(Hubbell Electrical Products)*

CALCULATING ALLOWABLE NUMBER OF WIRES IN BOX

NEC 314.16 and Table 314.16(A) limit the number of wires that may enter a box, depending on the cubic-inch capacity of the box and the sizes of the wires. Besides being unsafe because of possible damage to the insulation, crowding a box with wires makes it difficult to install devices in the box. Shorts and grounds may develop as a result of forcing a switch or receptacle into a box already crowded with wires. Use a box of adequate size.

The *NEC* requirements are reproduced here in Table 10–1, in slightly abbreviated and modified form. The last three boxes shown are the handy boxes of Fig. 10–22. To calculate minimum box size, compare the sizes of possible boxes with the minimum fill requirements. The minimum fill requirement consists of the sum of the following volumes as they apply in any given instance. In each case, apply Table 10–2 values per each allowance required:

1. For each conductor that originates outside the box and terminates within it, one allowance based on the conductor size. In the case of fixture wires, this principle is modified by item 5 below.

2. For each conductor that passes through the box without joint or splice, one allowance, based on the conductor size. If the length of a free conductor equals or exceeds double the minimum length required for a terminating conductor (see Figure 8–3 and associated text), it must be counted twice.

3. For each conductor that begins and ends within the same box, no allowances required.

4. For each set of equipment grounding conductors entering the box, however many wires in the set, one allowance based on the largest equipment grounding conductor contained in the set. If there are two sets of such conductors (for an example, refer to the discussion of isolated ground receptacles and equipment on page 505), make one allowance for each set. If the only separate equipment grounding conductor, whether one or more, is

smaller than 14 AWG and enters from a domed fixture or similar canopy, it can be ignored.

5. For each fixture wire smaller than 14 AWG (not exceeding four) entering the box from a domed fixture or similar canopy such as a paddle fan canopy, no allowances required. If the fixture wires exceed four, add the excess into the volume allowance calculations.

6. For one or more fixture studs or hickeys, one allowance for each type, based on the largest conductor entering the box.

7. For internal cable clamps, such as those shown in Fig. 10–16, one allowance (however many clamps are present) based on the largest conductor entering the box. Cable clamps with the clamping mechanism outside the box need not be counted.

8. For each device strap (see left-hand side of Fig. 10–5 for examples of two such straps), a double allowance based on the largest conductor terminating on the strap.

The volume of any box can be calculated based on the volumes of its assembled sections, provided each section is either a standard box listed in the table, or is marked with its volume by the manufacturer. If the box (or one of its sections) meets both conditions, use the marked volume. If the box or section meets neither condition (to their competitive disadvantage not all manufacturers bother marking their plaster rings), then its volume contribution must be

Table 10–1 VOLUMES OF SELECTED STANDARD METAL BOXES

KIND OF BOX	SIZE IN INCHES		INTERIOR VOLUME IN CUBIC INCHES
Outlet box	4 x 1¼	ROUND	12.5
	4 x 1½	OR	15.5
	4 x 2⅛	OCTAGONAL	21.5
	4 x 1¼ SQUARE		18.0
	4 x 1½ SQUARE		21.0
	4 x 2⅛ SQUARE		30.3
	4¹¹/₁₆ x 1½ SQUARE		29.5
	4¹¹/₁₆ x 2⅛ SQUARE		42.0
Switch box	3 x 2 x 1½		7.5
	3 x 2 x 2		10.0
	3 x 2 x 2¼		10.5
	3 x 2 x 2½		12.5
	3 x 2 x 2¾		14.0
	3 x 2 x 3½		18.0
Handy box	4 x 2⅛ x 1½		10.3
	4 x 2⅛ x 1⅞		13.0
	4 x 2⅛ x 2⅛		14.5

Table 10–2 VOLUME REQUIRED PER CONDUCTOR

WIRE SIZE	VOLUME IN CUBIC INCHES PER WIRE
18	1.50
16	1.75
14	2.00
12	2.25
10	2.50
8	3.00
6	5.00

ignored. All nonmetallic boxes must be marked because their sizes have never been standardized. For example if you take a standard 4-in.-sq. metal box (1½ in. deep) and put a plaster ring on it with a marked 6-cu-in. capacity, the total box volume is 21 cu in. + 6 cu in., for a total of 27 cu in.

Let's look at an example of calculating a box size. Suppose a kitchen box will have a switch and a receptacle. The receptacle connects to a 20-amp small-appliance circuit (12 AWG wire), and the switch for the overhead light connects to a 15-amp lighting circuit (14 AWG wire). For both circuits, the feed from the panel comes into the box, which has internal cable clamps. The run to the overhead light eventually supplies other lighting outlets, so it is three-wire cable. In addition, there will be other receptacles fed from the receptacle in this box, so another 12-2 cable leaves the box. The wiring method is Type NM cable, which has a separate equipment ground within it. How big a box must you provide?

There are two current-carrying 12 AWG conductors entering the box, and two leaving to supply other receptacles. There are two current-carrying 14 AWG conductors entering the box, and three leaving: a wire that is always hot to supply other lighting outlets, a wire controlled by the switch for the overhead light, and a grounded circuit (white) wire. There is one set of equipment grounding conductors consisting of two 12 AWG and two 14 AWG conductors, one in each Type NM cable connected to the box. There are two device straps, one with 14 AWG conductors on it for the switch, and one with 12 AWG conductors on it for the receptacle. Here is the calculation:

12 AWG wires	4×2.25 cu in. = 9.00 cu in.
14 AWG wires	5×2.00 cu in. = 10.00 cu in.
Grounding wires	1×2.25 cu in. = 2.25 cu in.
Devices w/12 AWG ($\times 2$)	2×2.25 cu in. = 4.50 cu in.
Devices w/14 AWG ($\times 2$)	2×2.00 cu in. = 4.00 cu in.
Internal clamps	1×2.25 cu in. = 2.25 cu in.
Total volume required	32.00 cu in.

There are a number of possibilities for meeting this requirement. You could find a two-gang plastic box with a marked capacity of at least 32 cu in. You could gang together two 3½ in. deep metal device boxes (18 cu in. each). You could use a 4-in.-sq × 2⅛-in. deep metal box (30.3 cu in.) because almost any two-gang plaster ring is likely to provide the missing 1.7 cu in. Be sure the manufacturer marked the volume on the plaster ring—the inspector has every right to turn it down if it isn't marked.

CHAPTER 11
Wiring Methods

THE *NEC* RECOGNIZES many different wiring methods. Some of these are used only in large buildings and are not discussed here. Except for outdoor overhead spans, the most commonly used wiring methods use conductors installed in some form of raceway or assembled into some type of cable.

Article 100 of the *NEC* defines a raceway as "an enclosed channel of metal or nonmetallic materials designed expressly for holding wires, cables or busbars, with additional functions as permitted in this Code." Although not stated in this *NEC* definition, raceways are installed so that wires can be placed in them or removed from them after installation. The metal armor of armored cable is not a raceway because it does not permit removal of the wires contained in it. Raceways may be made of metal as exemplified by rigid metal conduit, electrical metallic tubing, surface raceways, and many others, or of nonmetallic materials as exemplified by nonmetallic conduit. Although the designed purpose of a raceway is to contain electric conductors, a metal raceway may usually also be used as an equipment grounding conductor.

The word "cable," not defined in the *NEC*, usually refers to an assembly of two or more wires within an outer metal or nonmetallic enclosure such as metal armor or a nonmetallic sheath. Some types of cable for direct burial in the earth (refer to the page 284 underground service wire topic and the page 121 discussion about Fig. 8–15) contain a single conductor in a tough outer nonmetallic jacket.

Each type of raceway or cable and its use is further defined in an *NEC* article covering that type of wiring method. Some of those methods are discussed in Chapter 27 of this book. In this chapter, only the methods most commonly used in the wiring of residential, farm, and commercial buildings are covered. The chapter concludes with a note on using flexible cord.

WIRING METHODS USING TUBULAR RACEWAYS (CONDUIT AND TUBING)

There are many types of circular raceways, only some of which, technically, are conduit. Generally the heavier-walled methods are conduits, and the thinner-walled methods have the word "tubing" in their article titles. Since various code provisions depend on knowing which is which, this book uses the word "conduit" to refer only to wiring methods with that word in their article titles. This section covers the more usual raceway methods, including rigid metal conduit (RMC), intermediate metal conduit (IMC), rigid nonmetallic conduit (RNC), flexible metal conduit (FMC), liquidtight flexible metal and nonmetallic conduits (LFMC and LFNC), electrical metallic tubing (EMT), and electrical nonmetallic tubing (ENT).

The *NEC* now requires all tubular raceways to be listed, with the exception of brass rigid conduit used for swimming pool lighting. Because this material is necessary in cases where metal raceways are needed for underwater lighting, and because there is no longer any manufacturer now producing listed brass conduit, "approved" brass conduit can be used for these applications. Suitable heavy-wall brass piping is available from plumbing supply houses. Review this with your local inspector, who must grant such approval.

A discussion of rigid metal conduit serves here as a basis for covering the general issues in using circular raceways. This doesn't imply that this wiring method is primary in any way, though it is the oldest of the circular raceway methods, having evolved from piping used for gas lighting. Actually, the use of rigid metal conduit has declined significantly over the years in favor of wiring methods that are less labor intensive. Descriptions of other wiring methods in this category follow the RMC discussion, with any differences from general installation procedures noted. Special mention is made of the importance of making installation accommodations for the effects of earth movement and thermal expansion.

Rigid metal conduit (RMC)   In this method, all wires are enclosed in pipe called conduit. The pipe is usually steel, sometimes aluminum, or brass (especially for swimming pool lighting). Conduit differs from water pipe in that the interior surface is carefully prepared so wires can be pulled into it with a minimum of effort and without damage to their insulation. Also, the chemical composition of the steel in conduit is carefully controlled so it will bend easily. Steel conduit usually has a galvanized or similar finish inside and outside. Some galvanized steel conduit and some aluminum conduit have an additional enamel or plastic coating for additional protection against corrosion. Note, however, that test labs routinely refuse to speculate on the suitability of such supplementary coatings for corrosion protection. Be sure to discuss with the local inspector the suitability of various supplementary coatings in any particular application. Steel conduit with only a galvanized finish may be used indoors or outdoors, above ground or underground in ordinary soils. In highly acid or otherwise corrosive soil conditions, steel conduit should have the additional coating already mentioned, or asphalt paint, or some other treatment acceptable to the inspector. In addition, supplementary protection is always required where a conduit length extends from poured concrete directly into the earth. This juncture is judged by the test labs as severely corrosive, and it is a major safety problem. Many otherwise intact conduit runs end up rusted completely open at the below-grade interface just outside the building wall. Aluminum conduit also, by test lab restrictions, always needs an additional protective coating where installed in contact with the earth or where embedded in concrete. Never use steel or aluminum conduit in cinder fill or cinder concrete unless the conduit has an additional protective coating.

Conduit comes in 10-ft lengths with a coupling on one end (length is measured to include that coupling); each length bears a listing mark. See Fig. 11–1. Conduit can be ordered in longer lengths for special applications. The ½-in. size is the smallest used in ordinary wiring. All sizes are identical in dimensions with the corresponding size of water pipe, but as in the case of water pipe the nominal size indicates the internal diameter— the actual diameter is larger than the indicated diameter. For example, "½ in." conduit has an internal diameter of nearly ⅝ in. and an external diameter of about ⅞ in.

Fig. 11–1 Rigid metal conduit looks like water pipe but differs in several important ways.

Because these sizes are approximate, the *NEC* refers to them without dimensions, in either the metric or the English system. For example, if you look for 3-in. conduit in the *NEC*, you won't find it, the reference being "metric designator 78 (trade size 3)." This book will, at least for this edition, continue the customary practice of referring to (in this case) "3 in. conduit."

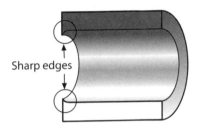

Sharp edges

Fig. 11–2 When pipe is cut, a sharp edge is likely to be left; this must be removed before conduit is installed.

Cutting   A hacksaw blade with 18 teeth to the inch will do a good job of cutting RMC. Avoid using a metal-parting tool such as a plumber's pipe cutter because it will leave an expanded sharp edge that will tear the insulation of wires as they are pulled into the conduit; see Fig. 11–2. If you do use a pipe cutter, you will have to ream the end extensively until the inner diameter is completely restored. A reamer of the type shown in Fig. 11–3, which fits the brace of a brace-and-bit set,

Fig. 11–3 A pipe reamer is handy for removing burrs.

is ideal for this purpose. The edge left by a hacksaw will be much less sharp than that left by a pipe cutter and is easily and quickly reamed with the tool in Fig. 11–3, or a file or even a knife blade. For 1-in. or smaller conduit, the handle of a pair of side-cutting pliers or the shaft of a large squareshaft screwdriver can be used—wring the

Fig. 11–4 If many cuts must be made, a power portable bandsaw is a good investment.

handle of the pliers or the shaft of the screwdriver round and round a few times inside the cut end. Always inspect the cut end after you have reamed it to make sure it is absolutely smooth.

If you do a considerable amount of conduit work, a power saw of the general type shown in Fig. 11–4, with a metal-removing blade, is a good investment.

Threading   The tools used for threading conduit are the same as for threading water pipe, but with special dies having a taper of ¾ in. to the foot. The dies for threading water pipe have a greater taper, and if used for conduit fewer threads on the conduit will engage with the threads of couplings or fittings. This results in poor connections and poor grounding continuity. For the same reason, never put running threads on a conduit end, that is, threads running the full length of a coupling to allow the coupling to be unthreaded from one conduit while being threaded onto an adjoining one. The coupling will never be tight on the first conduit because the remaining threads won't have any taper. Figure 11–5 shows a ratchet-type hand threader with interchangeable die heads.

Fig. 11–5 This threader can be used for all conduit sizes 2 in. and smaller. Larger sizes require special geared threaders.

Bending   Because the wires are pulled into conduit after the conduit is installed, all bending must be done carefully to avoid substantially reducing the internal diameter in the bending process. Make the bends uniform and gradual, using a conduit bender. RMC has a heavy enough wall to resist kinking. At one time electricians used conduit hickeys, as shown in Fig. 11–6, to bend conduits routinely. Inch the hickey gradually along the bend, taking care to keep the hickey always in the same plane so the bend lies flat. In the hands of an experienced journeyman the hickey can be used to produce concentric bends; in a family of conduits changing direction in the same plane, the bend radius increases from innermost to outermost conduit in the sequence and all the conduits maintain equidistant spacing throughout. See Fig. 11–7.

Fig. 11–6 A conduit hickey.

Today, however, a fixed-radius hand bender of the general type shown in Fig. 11–8 is far more widely used, especially for sizes 1 in. and smaller. Select a bender suitable for the sizes of conduit to be bent. Some will handle several sizes. To create a right-angle (90-deg) bend so that the end of the conduit will have a "rise" of 13 in. above the floor, start by determining how far from the end to begin the bend. From the rise, subtract 5 in. for the ½-in. conduit, 6 in. for ¾-in. conduit, and 8 in. for 1-in. conduit. If you forget the calculation, check the bender head— the take-up dimensions (which may vary by brand) are usually imprinted there. In the example of Fig. 11–9, assuming ½-in conduit, subtracting 5 in. from the rise of 13 in. leaves 8 in. Hook the bender over the conduit so the arrow on the bender points to a spot 8 in. from the end of the conduit. Most benders also have another point marked that "predicts" where the bend will end up in terms of horizontal travel. You'll need this feature instead of the take-up mark when you're positioning the bender to bring the conduit tight against an adjacent wall or up within a partition. See Fig. 11–10. Full instructions for many types of bends usually come with a bender as purchased.

Factory-made elbows are usually used for bends in 1¼-in. and larger conduit, and in many cases for all sizes. For bends other than 90 deg (and in many cases all bends), hand benders are used for the smaller sizes and power benders for the larger sizes. Power benders are available in a variety of types, hydraulically or electrically operated. Most power benders are the one-shot type, although they can be used for segment bends similar to those made by a hickey, for concentric pipe layouts. Bend radii can't be less than those specified in Table 2 in Chapter 9 of the *NEC*.

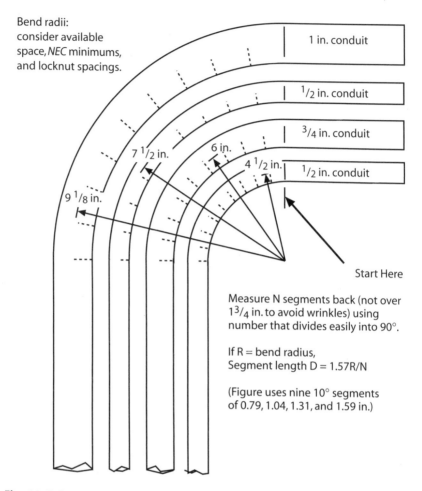

Bend radii:
consider available
space, *NEC* minimums,
and locknut spacings.

1 in. conduit

$^1/_2$ in. conduit

$^3/_4$ in. conduit

7 $^1/_2$ in.   6 in.

4 $^1/_2$ in. $^1/_2$ in. conduit

9 $^1/_8$ in.

Start Here

Measure N segments back (not over
$1^3/_4$ in. to avoid wrinkles) using
number that divides easily into 90°.

If R = bend radius,
Segment length D = 1.57R/N

(Figure uses nine 10° segments
of 0.79, 1.04, 1.31, and 1.59 in.)

Fig. 11-7 Sequential segmented bends laid out to produce a concentric conduit
arrangement.

Bending conduit is an art that requires a great deal of practice. Take pride in your
work, always making sure you don't kink the conduit, which dangerously reduces
its inner diameter.

NEC 344.26 (and comparable sections in other tubular raceway articles) prohibits
more than four quarter-bends or their equivalent (360 deg) in one "run" of
conduit. A run of conduit is that portion between any two openings, such as a
cabinet, a box, or conduit fitting with a removable cover. The fewer the bends, the
easier it is to pull wire into the conduit. This dimension limit includes all bends
between pull points no matter how minor or gradual, and it includes the typical
opposing pair of 10-deg bends at the end of a run (the "box kick") that brings the
conduit ½ in. off the wall and into a knockout in a surface-mounted enclosure.

Fig. 11–8 A conduit bender, and method of use. This one bends 1-in. rigid or intermediate metal conduit, or 1¹/₄-in. EMT. For best results, use the heaviest foot pressure you can manage. This bender has a hinged, two-position foot pad allowing heavy, straight-down foot pressure throughout the bending process.

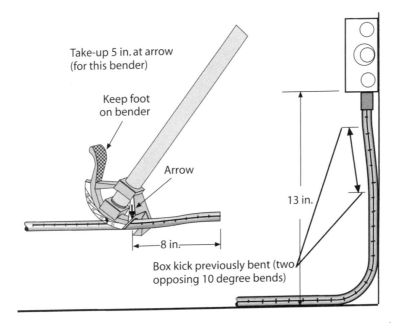

Take-up 5 in. at arrow
(for this bender)

Keep foot
on bender

Arrow

13 in.

8 in.

Box kick previously bent (two
opposing 10 degree bends)

Fig. 11–9 This shows how to figure a 90-deg bend. Note the box kick at the end of the 90-deg angle that was bent first. There are one-and-one-fourth 90-deg bends so far.

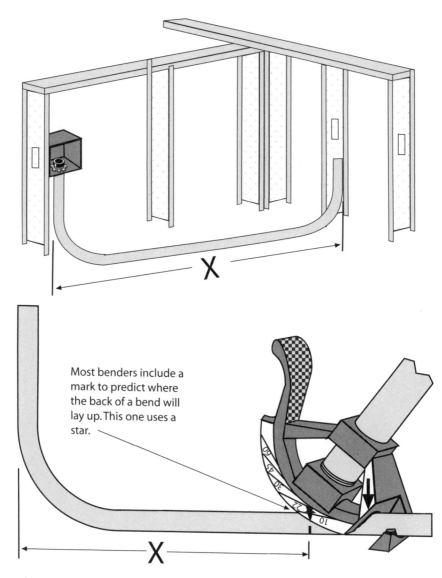

Most benders include a
mark to predict where
the back of a bend will
lay up. This one uses a
star.

Fig. 11–10 Laying out a back-to-back bend by predicting where the outer edge of the conduit will fall.

Calculating number of wires in conduit   At one time the *NEC* applied the same wire fill limit to all circular raceways of a given trade diameter. For example, the allowable fill for 3-in. EMT was the same as for 3-in. RMC, even though the actual cross-sectional area for the two products differs significantly. The *NEC* now bases allowable wire fill on the actual dimensions of the individual circular raceway. The basic rules, however, haven't changed. If more than two wires run in a conduit, their total cross-sectional area can't exceed 40% of the conduit cross section. (For two wires the percentage is 31%, and for one wire, 53%.)

Selecting a conduit size is a two-step process. First, in your copy of the *NEC* go to Table 5 in Chapter 9 and add up the actual cross-sectional areas of the wires you plan to install. Be sure to distinguish correctly among the various insulation types, each of which has a different physical dimension. Then from Table 4 select a conduit size that accommodates the required fill.

Here's an example: Suppose you need to install a new three-phase, 30-hp, 460-volt motor near an existing 50-hp motor, and you need to know if the existing conduit is large enough for both circuits. Remember, if you use a single conduit you have to adjust wire sizes to accommodate mutual heating in the conduit run (review how to adjust *NEC* Table 310.16 values on page 105). In this case the existing motor would need 3 AWG wire, and the new motor would need 6 AWG wire. Suppose you add a 6 AWG bare conductor as a supplementary grounding conductor. How big does the conduit need to be if you use THHN for the ungrounded conductors?

three 3 AWG THHN (from Table 5)	$= 3 \times 0.0973$ in.^{2}	$= 0.2919$ in.^{2}
three 6 AWG THHN (from Table 5)	$= 3 \times 0.0507$ in.^{2}	$= 0.1521$ in.^{2}
one 6 AWG bare (from Table 8)	$= 1 \times 0.027$ in.^{2}	$= 0.0270$ in.^{2}
Total wire fill:		$= 0.4710$ in.^{2}

Now review *NEC* Chapter 9 Table 4, Rigid Metal Conduit. Use the 40% fill column because there are three or more (in this case seven) conductors. A 1-in. conduit, with an allowable fill of 0.355 in.^{2} is too small. The smallest conduit you can use is a 1¼-in. conduit, with an allowable fill of 0.610 in.^{2}. If all your wires are the same size and insulation type, the *NEC* has done these calculations for you in Appendix C. For example, suppose both motors are 50 hp, and there is no supplementary grounding conductor. *NEC* Table C8 in Appendix C shows the capacity of 1¼-in. RMC as six 3 AWG THHN, so that would be the answer. Note that if you used THW insulation instead, only five conductors would fit, and the minimum conduit size would go up to 1½ in.

There are a few other general rules to keep in mind regarding wire fill. For instance, where all the wires are the same size, the *NEC* allows a slightly increased fill. To determine the allowable fill, you would normally divide the 40% area by the area of your proposed wires. As an example, suppose the 40% limit is 0.98 in.^{2}, and the wire cross section amounts to 0.1 in.^{2} each. Dividing the first number by the second gives 9.8 conductors. Although the general rule prohibits rounding up (meaning this result normally restricts the fill to 9 wires), a special allowance in cases where all the wires are the same size permits rounding up (in this example to 10 wires) if the decimal fraction is at least 0.8.

Another variation on allowable wire fill concerns short raceway sections, generically described as conduit or tubing nipples, which the *NEC* effectively defines as not over 24 in. long. The allowable fill in this case rises to 60% of the raceway cross section. In addition, since heat can easily escape from so short a length, derating for mutual conductor heating need not be applied to nipples.

Occasionally raceways enclose nonstandard conductors or even cable assemblies. In these cases, the actual dimensions of the conductors, as provided by the manufacturer or other industry sources, must be used in the calculations. For example, you may encounter metric conductors on imported equipment. In the case of cable assemblies, they often aren't round. For example, two-conductor Type NM cable is almost always elliptical in cross section, since the two conductors aren't twisted. If you sleeve this through a circular raceway, the *NEC* requires that you calculate its area as though it were circular in cross section, with a diameter equal to the major axis of the ellipse.

Pulling wires into conduit   Wires in any kind of conduit must be continuous and without splice in the conduit itself. See *NEC* 300.15. Remember that splices are permitted only in switch or outlet boxes, junction boxes, or where the splice is otherwise permanently accessible. For short runs of conduit the wires can be pushed through from one outlet to the next. But in most cases, especially if the conduit contains bends, "fish tape" is used. For occasional work, a length of ordinary galvanized steel wire may be used, but a fish tape made of thin tempered steel about ⅛ in. wide is usually used. Figure 11–11 shows a fish tape on a plastic reel. If the loop breaks off, heat the end of the tape with a blowtorch (which removes the temper) and form a new loop with your pliers. If you attempt to make a new loop without heating, the tape will often break. Some steel fish tapes are about ¼ in. wide and are particularly useful on very large conduits where smaller tapes can fold back on themselves over a long run. There are also fiberglass types that work well for pulling additional wires into a partially filled conduit because they are less likely to damage existing insulation.

Fig. 11–11 Fish tape in its reel, with a convenient rewinding handle.

Both steel and fiberglass tapes come on convenient reels that provide a "handle" for pulling in the wires after they have been attached to the eye or loop on the end of the tape. Push the tape into the conduit through which the wires are to be pulled, and attach the wires that are to be pulled into the conduit to the tape. This is done by removing the insulation from about 4 in. of each wire, threading the wires through the eye or loop of the

Tape optional

Fig. 11–12 How to attach wires to fish tape for pulling.

tape, and folding the wires back on themselves. Then wrap a few turns of friction or plastic tape over the wires to hold them fast. If there are existing wires, be sure to tape the fish-tape loop as well, to prevent the tape from tangling. See Fig. 11–12. A tape in use is shown in Fig. 11–13. Another strategy is to use a pulling basket, as shown in Fig. 11–14. It is made of steel mesh that grabs tighter the harder it is pulled.

Pulling the wires into the conduit usually requires one person pulling at one end and another feeding the wires in at the other end to make sure there will be no snarls or damage to the wires. For difficult pulls, apply a listed wire lubricant, one certain not to degrade insulation over time.

Conduit systems must always be installed complete between access points prior to pulling in any wires. That rule provides some rough assurance that the raceway will always be able to function as intended, namely, as a wiring method capable of having its wires replaced as required. Don't pull wires through a length of

Fig. 11–13 Using fish tape to pull wires into conduit. *(Klein Tools, Inc.)*

Fig. 11–14 This steel-mesh grip is reusable. Put some tape on the end of the wire bundle so you don't lose a conductor, and to assist getting the basket over all the conductors. Add a little more tape over the end of the basket so it won't release when the pressure is off between pulling efforts.

conduit at a time, eventually accumulating so much friction over the entire length that pulling between access points results in insulation damage. Power assistance may be required for large conductors and for very long runs. The longest conventional fish tape is not much over 200 ft. For a longer run, there are blower systems that propel a lightweight plug and twine over great distances; assuming the run is free of obstructions, the twine serves as a codline to pull in a heavier rope. To pull the conductors, heavy equipment is available, capable of applying many tons of force. For more modest requirements, consider the setup in Fig. 11–15.

Fig. 11–15 The driving motor for this portable pulling setup is the same heavy-duty drill shown in Fig. 21–15. *(Greenlee Textron)*

Supporting rigid metal conduit   Conduit in all sizes is required by *NEC* 344.30 to be supported within 3 ft of each box, fitting, or cabinet, and in addition at intervals of not over 10 ft. However, if the conduit is in straight runs with threaded-type couplings, an exception to these support rules allows for greater support distances, provided the installation assures that stresses don't get transferred to the conduit terminations. Figure 11–16 shows three of the many types of

Fig. 11–16 A one-hole and a two-hole clip for supporting conduit on a surface, next to a "Minerallac" clip that keeps it off the surface. "Minerallac" is a brand name used generically in the trade, but most major conduit hardware manufacturers make equivalent clips.

conduit support hardware. Another exception allows up to 20 ft between supports on a vertical drop to stationary equipment, provided the conduit is securely anchored at both the top and bottom of the run. This is very useful in industrial areas with high ceilings, and it applies in nonindustrial occupancies as well.

Intermediate metal conduit (IMC)   This is similar to ordinary rigid metal conduit and, size for size, has the same outside diameter but thinner walls. That makes the internal area in square inches a little more than in ordinary conduit, and with the 1999 *NEC* changes in Chapter 9, Table 4, that size difference often makes a difference in allowable wire fill. IMC is installed in the same way as ordinary RMC. The threads are identical and RMC fittings can be used without modification. It is more likely to kink due to the reduced wall thickness, making it less tolerant of conduit hickeys. Conduit benders work very well. The only real application difference between the two products is that RMC is made in all trade sizes up to 6 in., but IMC goes only up to 4 in.

Both RMC and IMC can support boxes that have threaded hubs, or conventional sheet metal boxes to which hubs identified for this purpose were added. (Refer to the topic on raceway support of boxes beginning on page 495.) A "hub" in this case is a heavy, female-threaded connection provision intended to directly receive threaded conduit ends. *No other wiring methods are allowed to do this.* In all other cases, independently support the box, and then add the wiring method. In addition, both wiring methods can have their first point of support (normally 3 ft) as far as 5 ft from the end, provided there aren't any structural elements available nearer to the termination.

Electrical metallic tubing (EMT)   A raceway that is similar to rigid metal conduit but of thinwall construction is called in the *NEC* "electrical metallic tubing" (abbreviated EMT).

A length of it is shown in Fig. 11–17. It is made of galvanized steel or occasionally of aluminum, either of which may have an additional plastic or other protective coating. The rules that apply to rigid metal conduit—such as number of wires permitted, bending, supports—apply also to EMT. It is made only in sizes through 4 in. It must be supported within 3 ft of every box, cabinet, or fitting, and additionally at intervals not exceeding 10 ft, regardless of size. If there isn't any structural element available for support, the first point of support can be as far as 5 ft from the termination, provided the intervening tubing consists of an uncut length of EMT. In addition, an unbroken length of EMT can be fished. This may seem odd, but there are cases where you may be able to poke a length of tubing down into a wall cavity.

Fig. 11–17 Electrical metallic tubing ("EMT" or thinwall conduit) is never threaded in the field. It is lighter and easier to use than rigid metal conduit.

The internal diameter in the smaller sizes is the same as in rigid conduit, but in the larger sizes it is a little larger. However, because the walls are so thin, EMT must never be threaded in the field (there is a patent in existence for a factory-threaded product, but it isn't presently in production). Instead, all joints and connections are made with threadless fittings that hold the material through pressure.

Figure 11–18 shows both a coupling and a connector; note that these fittings do have external threads. Each coupling and connector consists of a body and a split ring through which tremendous pressure is exerted on the EMT when the nut is forced home tightly. In installing these fittings, the compression nut is first loosened until the fitting will slip over the end of the EMT. Be sure the end of the EMT goes past the inner ring to the shoulder inside the fitting. Then tighten the nut. Setscrew fittings are commonly used as well. There are other methods, such as a straight female steel sleeve used with a special indenting tool that permanently connects the fitting to the tubing, but setscrew and compression fittings are the two principal methods in use today.

Fig. 11–18 A compression connector and coupling (left and center) used with thinwall conduit. The connector on the right, by comparison, uses a setscrew instead of a compression ring to secure the tubing.

Always cut EMT with a hacksaw having a blade with 32 teeth to the inch, or with a metal-removing power saw. After the cut, ream the ends to remove burrs or sharp edges.

Occasionally it will be necessary to join a length of EMT to a length of rigid conduit or to a fitting with threads for rigid conduit. There are changeover fittings designed for this purpose. In addition, you can always use a connector of the type shown in Fig. 11–18. The threaded portion of the connector will always fit the internal thread of any rigid metal conduit coupling, box, or other fitting designed for the corresponding size of rigid conduit.

Although the *NEC* seems to suggest EMT can go in the earth, never do so unless you provide additional corrosion resistance beyond the galvanizing, such as asphalt paint. The test lab guide card restrictions on this product generally require this additional protection for direct burial. Further, the same test labs refuse to evaluate supplementary coatings for this function, just as in the case of rigid conduit. This means that a decision on suitability of supplementary protection is between you and the inspector, and many inspectors refuse to recognize such coatings for direct-burial applications. Be sure to ask rather than have to redo the job.

Flexible metal conduit (FMC)   This material is covered by *NEC* Article 348 and is shown in Fig. 11–19. It is often called "flex" or "greenfield." It is similar in appearance to armored cable but is larger in diameter since it is an empty raceway and wires must be pulled into it after it is installed. It is available in steel or aluminum in both normal and reduced wall thickness. The aluminum and reduced-wall steel products require care to prevent separating the convolutions, and they require use of appropriate terminal fittings evaluated with these forms. Try to use connectors that clamp the entire circumference of the raceway in order to more reliably distribute pull-out forces, particularly in areas where the flex may be disturbed after installation.

Fig. 11–19 Flexible metal conduit. The wires are pulled into place after the conduit is installed.

The smallest size used in ordinary wiring is the ½-in. trade size, which actually has an outside diameter of nearly ⅞ in., similar to RMC. Although it looks like armored cable, it is a true raceway, with the wires pulled into it after installation just as with rigid conduit or EMT. Unlike those wiring methods, however, the *NEC* allows this wiring method in the smaller ⅜-in. trade size to be used in lengths up to 6 ft to general utilization equipment. *NEC* 348.22 includes a special wire fill table for these instances.

Flexible metal conduit can be fished, which is one of its principal applications. Where not fished, it must be supported within 12 in. of every box, cabinet, or fitting and also at intervals not over 4½ ft, with two exceptions: (1) a free length not over 36 in. is allowed at terminals where flexibility is required, such as at a motor where vibration is a factor or where the motor location must be slightly changed for proper belt tension, and (2) a 6-ft length is allowed between a recessed lighting fixture and a junction box. The latter exception is now not often used, as explained in the page 332 discussion of high temperatures relating to recessed fixtures.

Occasionally a short length of flexible metal conduit is used in a run of rigid metal conduit (or even EMT) where it would otherwise be extremely difficult to bend the rigid conduit or tubing into the necessary configuration. This should not be done at all if it can be avoided (as it usually can) since this introduces additional resistance into the equipment grounding path.

Metal conduit or tubing may be used as the equipment grounding conductor. Flexible metal conduit has a very much higher resistance per foot than the pipe types, and it requires a separate equipment grounding wire in the conduit. So when using flexible metal conduit, a separate equipment grounding wire must be installed. The wire may be bare or insulated (if insulated it must be green) and must be installed in the conduit along with the circuit wires. At each end, connect it as you would the bare equipment grounding wire of nonmetallic-sheathed cable as discussed later in this chapter. The size depends on the rating of the overcurrent protection in the circuit involved. Alternatively, if the run of flex isn't over 6 ft long, you can run the wire along the outside, per *NEC* 250.102(E). With 15-amp protection use 14 AWG wire; with 20-amp protection use 12 AWG; with 30- to 60-amp protection use 10 AWG; with 100-amp protection use 8 AWG; with protection over 100 amps but not over 200 amps use 6 AWG. All these values are from *NEC* Table 250.122 and are for copper wire only.

There is one exception, covered in *NEC* 250.118(5). In a 6-ft (or shorter) length, ordinary flexible conduit may be used without the equipment grounding wire, but only if the fittings are listed for the purpose and the circuit is protected by a breaker or fuse rated not over 20 amps. The 6-ft restriction applies to the total of all flexible wiring methods in the grounding return path. That is, if you have 4 ft of liquidtight flexible metal conduit somewhere else in the home run, and you install another 3 ft of flex, you can't avoid the separate ground wire regardless of the circuit rating. Finally, this allowance does not apply if the flex is installed "where flexibility is necessary after installation." This means you can use the allowance where you installed the flex to solve a problem where you couldn't bend a rigid raceway in a particular case, or where you fished a short section (not over 6 ft) in a wall. However, if you installed it to supply a swinging sign, or a movable motor, or for vibration isolation, you must add an equipment grounding conductor even for 15- or 20-amp circuits.

Liquidtight flexible metal conduit (LFMC)   This material is similar to ordinary flexible metal conduit, plus it has an outer liquidtight nonmetallic sunlight-resistant jacket. It is commonly called "Sealtite," which is the trade name of one manufacturer. It is covered by *NEC* Article 350. Figure 11–20 shows both the material and the special connector that must be used with it. In order to properly seat the connector, take care to cut the conduit squarely. Part of the connector goes inside the conduit, making a good connection for grounding continuity; part of it goes over the outside, forming a watertight seal. The plastic content in the outer jacket limits the amount of heat it will stand. In general, size your wires so the current they will carry does not exceed the 60°C ampacity column limits (in *NEC* Table 310.16) because higher temperatures will soften the jacket. You can exceed those limits if the product is marked accordingly. The product is made and can be used in the smaller ⅜-in. trade size for applications similar to those where ⅜-in. flexible metal conduit can be used (and with the same wire fill).

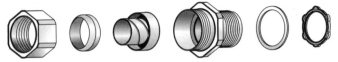

Fig. 11-20 Liquidtight flexible conduit and the special fittings used with it.

In general, as in the case of flexible metal conduit, you need to install a separate equipment grounding conductor when you use this product. Similarly, there is a limited exception for small circuits running with not over 6 ft of flexible wiring in the total grounding return path, and with a similar restriction against use where flexibility is required—however in this case the exception is considerably more complex. The limit is 20 amps in the ½- and ⅜-in. sizes, and 60 amps in the ¾-, 1-, and 1¼-in. trade sizes. Larger sizes cannot be used for this purpose regardless of circuit size. There is another complication. For many decades this wiring method was widely—and in many locations exclusively—available as an unlisted product. The unlisted versions of this product do not provide the grounding continuity of the listed products (along with other deficiencies). The listed varieties have a strip of copper wound into the convolutions, and only those varieties qualify for the limited grounding path allowances. Use of the listed product is now an *NEC* requirement; make sure that what you are being sold is actually listed.

Liquidtight flexible nonmetallic conduit (LFNC)   This material has the same function as its metallic precursor, but has a completely nonmetallic wall. It is available in three forms, two of which are relatively uncommon—the "A" type with reinforcement between the core and cover, and the "C" type having a corrugated wall without additional reinforcement. The "B" type, with integral reinforcement within the conduit wall, has become a very popular wiring method. This type does not have the restriction to 6 ft (unless a longer length is required for flexibility, a relatively unusual circumstance) that the other types share. As in the case of the metallic version, you have to be sure the enclosed wires don't run above 60°C, unless the product is marked accordingly.

Both this product and the metallic version can be used outdoors, and even directly buried if listed and marked for this duty. Where exposed to sunlight, it must be marked for this duty as well.

Rigid nonmetallic conduit (RNC)   The *NEC* covers this nonmetallic counterpart to RMC in Article 352. Rigid nonmetallic conduit is immune to conventional corrosive influences, although some industrial chemicals will attack it. There are many forms of this product for use underground. One form of underground conduit, a black, high-density polyethylene, can be supplied in continuous lengths on a reel (now with its own Article 353); this form can also be shipped with preinstalled conductors, in which case it comes under *NEC* Article 354. This form will burn and can't be used in exposed locations or in buildings, so it is not covered further in this book. However, the polyvinyl chloride (PVC) version can be used above and below grade, and is the predominant type used. Recently a form of fiberglass has been successfully listed for use above grade, but it is comparatively difficult to terminate and its use is only a tiny fraction of PVC. The remainder of this discussion, to the extent it doesn't apply generally, concerns PVC.

Terminations   A hacksaw cuts RNC easily, and many tool manufacturers make cutting tools for RNC if you will be using this wiring method all the time. As in the case of the steel product, you need to ream the cut end, but you can do it easily with a pocket knife. To bring the conduit into an ordinary enclosure, add male threads to the end using a terminal adapter. Terminal adapters have a female socket that accepts the cut RNC on one end and male conduit threads of the same trade size on the other end, allowing for connections with a simple locknut. To attach, use PVC cement designed for the materials you're using. Clean the surfaces, preferably with PVC cleaning solvent. Then apply a thin coat of glue to the mating surfaces. Don't use too much cement, or it will overflow and ball up at the end, resulting in a partially obstructed passage. Push the conduit all the way in, and then turn the adapter one-quarter to one-half turn (see Fig. 11–21). This assures the glue is uniformly applied within the adapter. Hold the adapter and pipe in place for a few moments until it sets up permanently. Each 10-ft length comes with a coupling provision on one end, usually just an expanded female bell

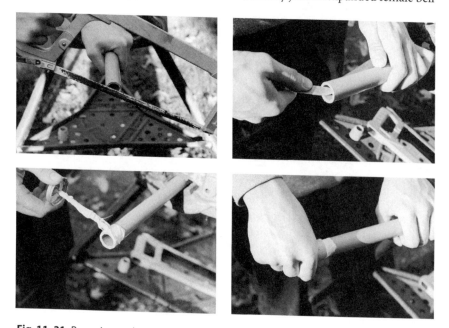

Fig. 11–21 Preparing and applying a terminating fitting to a length of PVC conduit.

end that will accept the next length of conduit. Some nonmetallic enclosures come with female hubs ready to receive a length of conduit, eliminating the need to prepare a terminal adapter. In addition, couplings are readily available where necessary to join nonstandard lengths.

Bends   As with all circular raceways, you must never exceed 360 deg in bends between access points no matter how gradual the bends, and the minimum bending radius is the same as for RMC. Bending PVC raceways involves heating the raceway until it becomes pliable, completing the bend, and then cooling it off to set the bend. The *NEC* requires "bending equipment identified for the purpose." Never use a blowtorch for this, because the flame temperature is far too hot. For small conduit sizes there are electrically heated blankets (Fig. 11–22), and for any size there are bending boxes. Be sure not to kink the conduit. Larger sizes need internal sidewall support to prevent that from happening. Support is easily provided by using conduit plugs in each end, applied before the conduit is heated. As it heats up, the air inside becomes pressurized, and that's all you need to support the inside walls (Fig. 11–22). Wear gloves when you handle the hot conduit, especially in larger sizes.

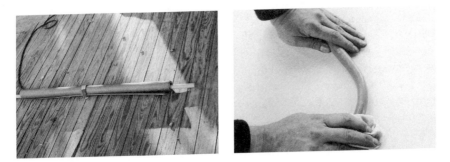

Fig. 11–22 A heating blanket for nonmetallic conduit. After making the bend, set it with a wet rag, as at right.

Factory-made bends are readily available in both 90-deg and 45-deg forms for almost every trade size of RNC (and some 30-deg segments as well). In a pinch, if you need a different bend pattern and have no bending equipment, you can make almost any bend you need in a regular household oven. Set the thermostat for about 300°F, and put a factory bend into the oven, watching carefully. When the PVC heats up, the conduit will remember the form it was made from, namely, a straight length, and begin unbending. Take the conduit out of the oven when it relaxes to the bend angle you want, and freeze it in place with a wet rag. This won't damage the conduit, and since it only amounts to unbending a piece of conduit, the sidewalls don't have to be supported. You cannot, however, go very far the other way on the larger sizes; attempting to create a more severe bend will only collapse the sidewall.

Supports   Where used above grade, RNC requires far greater support than its steel counterpart. Although both products require support within 3 ft of a termination, there isn't any allowance for the nonmetallic variety to extend that distance to 5 ft if no intervening structural support is available. In addition, the nonmetallic version must be supported every 3 ft in the smaller sizes (not 10 ft!). Larger sizes have that distance relaxed somewhat depending on the size, as covered in *NEC* Table 352.30. However, even for the very largest size (6 in.) the allowable distance (8 ft) is still less than the customary 10 ft for steel products. The outer diameter of PVC raceways is approximately the same as their steel counterparts, which means steel support fittings (one-hole clips, Minerallac clips, etc.) will work with PVC. In cases subject to thermal expansion, your support arrangements must accommodate such movement.

Environmental forces on raceways must be accommodated   Earth movement and temperature changes are two environmental phenomena that can cause significant damage to wiring installations. Both nonmetallic and metallic raceways are vulnerable to the effects of earth movement. For nonmetallic installations in particular, temperature changes must be anticipated.

Earth movement—frost action and fill settlement   You must address the issue of earth movement in your wiring design. In colder areas of the country, the seasonal freezing and thawing of the ground creates problems with raceway movement (usually referred to as frost-heaving). Nor are southern areas immune to the effects of earth movement on raceway, because any time construction takes place on fill, that material frequently settles over time. The *NEC* now requires that earth movement, whether due to frost conditions or settlement, be taken into account. The problem (and the *NEC* rule addressing it) affects nonmetallic and metallic raceways equally. Using an expansion fitting of the type shown in Fig. 11–23 in a vertical riser would prevent damage from occurring. A house with an underground service lateral terminating in an above-ground meter socket offers a classic example of the problem. Where comparatively small movements are anticipated, the inspector may allow the conduit to enter a vertical elliptical hole capable of accommodating ups and downs.

Temperature changes   PVC, like all substances, changes its physical dimensions as its temperature changes. The amount of change per degree of temperature change per unit length is referred to as the "coefficient of thermal expansion." PVC has a coefficient roughly four to five times that of concrete or wood or steel or most other building surfaces. One of the leading PVC conduit makers recommends allowing 140°F as a design temperature differential when working in areas with direct sunlight exposure, because sunlight heats the conduit more than the ambient temperature would predict. Referring to *NEC* Table 352.44(A), this would cause almost 6 in. of possible movement over a 100 ft run. How much of that appears at any given installation depends on the time of year it was first installed. For example, if it was installed in a moderate

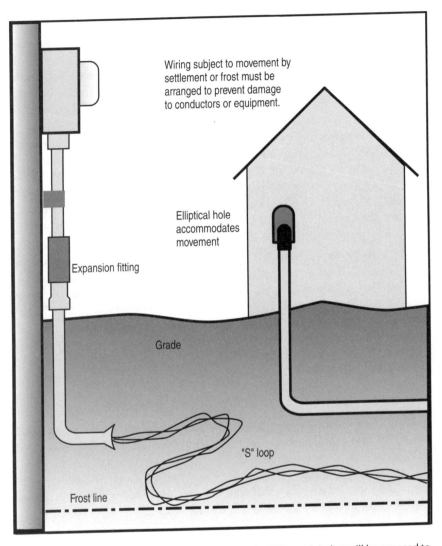

Wiring subject to movement by settlement or frost must be arranged to prevent damage to conductors or equipment.

Elliptical hole accommodates movement

Expansion fitting

Grade

"S" loop

Frost line

Fig. 11–23 Support hardware and expansion fitting for PVC conduit that will be exposed to outdoor conditions.

temperature on a cloudy day, it probably was in the middle of its size range, and might be expected to contract 3 in. during the coldest night and expand 3 in. during the hottest afternoon.

The expansion coefficient of steel, on the other hand, is approximately the same as most common construction materials on which it would likely be supported. Expansion fittings are not needed for steel unless special circumstances are involved, such as very long travel distances or crossing another expansion joint as occurs in some buildings or on bridges and similar structures. For PVC, however, the movement resulting from ordinary outdoor temperature changes cannot be safely accommodated with conventional fittings over even comparatively short distances. Such movement is enough to break out concentric knockouts, break enclosures free of their supports, and do other damage. The *NEC* insists on expansion fittings for this material if you expect total expansion/contraction between fixed points to exceed ¼ in.—that amount of movement would occur over less than a single 10 ft conduit length in most climates. Plan on installing these fittings routinely on all outdoor installations unless the basic arrangement accommodates the movement in other ways. For example, for a straight PVC service riser (assuming the service head doesn't butt against the roof—see drawings in Chapter 16, Fig. 16–10) no expansion fitting would be required. However, the appropriate mounting hardware must be used so this movement can occur safely. Refer to Fig. 11–23.

Electrical nonmetallic tubing (ENT)   As RNC is the nonmetallic counterpart to RMC, electrical nonmetallic tubing is the counterpart to EMT. It is made of the identical PVC in trade sizes up to 2 in., but in a corrugated wall construction that allows it to be bent by hand without the application of heat. It can be supplied in continuous lengths from a reel. It is even available as a prewired assembly with specified conductor combinations already pulled in place. However, it is not a cable, and it is subject to all the normal restrictions for raceways, including the 360-deg bend rule. It must be supported every 3 ft, and within 3 ft of terminations. It cannot be used outdoors or for direct burial, however, it can be used in cases where it runs in concrete, even if the concrete is below grade.

Since the outside diameter and chemical composition of this product is the same as for PVC rigid nonmetallic conduit, you can use the same solvent-welded fittings. But there are snap-on fittings that are much quicker to apply (Fig. 11–24). For commercial wiring use either PVC or plastic boxes and plaster rings, although metal boxes are acceptable as long as you don't forget to ground them.

You can use the ENT wiring method either exposed or concealed in low-rise construction. However, in buildings that exceed three floors above grade, it must

Fig. 11–24 This shows the corrugations on ENT, which allow it to be bent by hand (a "pliable" raceway according to the *NEC* definition). A snap-on connector has been added, which will lock in place when it goes into one of the knockouts in the box. *(Carlon, a Lamson & Sessions company)*

never be exposed, even in the first three floors. Instead, in other than fully sprinkled buildings, it needs to be behind a thermal barrier that has at least a 15-minute finish rating as defined in listings of fire-rated assemblies. In the case of walls, this is fairly easy to arrange, since most ½-in. drywall used in commercial construction carries this rating. The same holds true above a drywall ceiling. However, if there is a suspended ceiling (common in commercial occupancies), check with the building inspector. The support grid and the ceiling panels need to be identified as a combination for this duty. For example, having 15-minute panels would do no good if the T-bars dumped those panels onto the floor after 11 minutes of fire exposure.

The first floor is defined as the one with at least half its exterior wall area at or above grade level; one additional floor level at the base is allowed for vehicle parking or storage, provided it is not designed for human habitation. Beginning with the 2002 *NEC*, ENT can also be used in high-rise buildings (those over three floors above grade) without the use of a thermal barrier if the entire building has a complete fire sprinkler system in full compliance with *NFPA 13, Standard for the Installation of Sprinkler Systems*. The sprinkler system must cover all floors, not just the area where the use of ENT is being considered.

CABLED WIRING METHODS

For many purposes it is desirable to have two or more wires grouped together in the form of a cable. This is easy to install in any case, but especially in rewiring a building because the cable lends itself well to being fished through hollow wall spaces. A cable that contains two 14 AWG wires is known as "14-2" (fourteen-two); if it contains three 12 AWG it is called "12-3" (twelve-three), etc. If a cable has, for example, two insulated 14 AWG wires and a bare uninsulated grounding wire, it is called "14-2 with ground." If a cable contains two insulated wires, one is white and the other black. If it contains three, the third is red. A few cables are single-conductor type, such as Type USE and some Type UF applications for direct burial in the earth. Since cables never require access to their individual conductors except at terminations, they have no restrictions on the number of bends in the run, although the allowable radii of those bends will be limited to prevent damage to the cable. Splices are not normally permitted in cable except in outlet or switch boxes, where they are made as in any other wiring method, although they are permitted underground without an enclosure if a listed splicing device is used for this purpose..

Nonmetallic-sheathed cable   This cable is available in two principal types—Type NM and Type NMC—and contains two, three, or four insulated conductors in sizes not larger than 2 AWG, bundled together, in addition to a bare grounding conductor. It costs less than other kinds of cable, is lightweight, and easy to install. It is very popular and widely used. It has become the overwhelming choice for residential wiring, and in many areas for light commercial wiring as well.

Type NM is the ordinary kind that has been available for many years; it may be used only in permanently dry locations. It is often called "Romex," which is the trademark of one manufacturer. Fig. 11–25 shows the construction. The outer jacket consists of moisture- and flame-resistant thermoplastic. The individual conductors must now carry a 90°C temperature rating, which you can recognize by the suffix "-B" on its marking (thus, "Type NM-B"). There may or may not be an additional wrap over the conductors. Although most constructions used that additional layer in the past, recent changes in the product standard allow it to be eliminated if the cable passes a very strenuous test designed to show that under extremely cold conditions it can be pulled without damage through a series of nonaligned bored holes in wood joists. But the bare equipment grounding conductor must be wrapped.

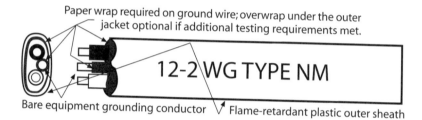

Fig. 11–25 Nonmetallic-sheathed cable is popular for ordinary wiring. This is *NEC* Type NM and may be used only in dry locations.

There is a recent subform of this type, designated Type NMS. The "S" stands for signaling, and refers to an additional set of signaling conductors for use in applications covered in *NEC* Article 780, the so-called "smart house" technology. As originally conceived, this technology anticipated a futuristic generation of household appliances featuring addressable computer chips for central control. The wiring involves uniquely configured receptacle outlets that remain de-energized unless the appropriate appliance is installed and is drawing its appropriate current.

This technology has proven premature, and other than a few prototypes, no such houses have come on the general market. However, there are a number of wiring schemes that provide for computerized control of specific outlets. There is a continuing controversy over whether Type NMS cabling can be used for this other control method. As a general rule, control conductors of the type anticipated in Type NMS cabling aren't allowed within the same cable assembly (or raceway) as power conductors. Be aware of this controversy, and don't plan on installing this wiring without getting a ruling in advance from your local inspector.

Type NMC is shown in Fig. 11–26. The individually insulated wires are embedded in solid plastic. Therefore it may be used in dry, damp or wet, or corrosive locations (such as barns with corrosive materials from the excreta of animals), but it must not be run directly in the earth.

Solid plastic

12-2 WG TYPE NMC

Bare equipment grounding conductor

Fig. 11–26 Type NMC nonmetallic-sheathed cable may be used in dry or wet locations.

Building restrictions   The NFPA Standards Council made a dramatic change in the allowable uses of Type NM cable when it released the 2002 *NEC*. The Council agreed that Type NM cable could now be used in any building permitted to be of Types III, IV, or V construction, even if actually constructed as Type I or II, provided (for nonresidential uses) the cable is located behind the same sort of thermal barrier normally required for ENT in high-rise buildings. There is no *NEC* waiver for buildings with a sprinkler system, however, building codes generally permit more extensive buildings to be constructed as Type III, IV, or V if a full sprinkler system is in place. The Council action was based on the entire record before it, including NFPA fire statistics that showed no association between Type NM cable and fire prevalence, and the report of the NFPA Toxicity Advisory Committee to the effect that in a fire the contribution of Type NM cable jacketing materials is a negligible fraction of the total smoke load.

This is the first time the *NEC* has put building construction types at the center of a rule governing the use of a wiring method, although some rules on transformer installations include these references. Refer to *NFPA 220, Standard on Types of Building Construction* for the complete descriptions. They are summarized as follows:

- *Type I*—All structural members are noncombustible (or limited-combustible) and have fire ratings generally of three or four hours (less in some cases) depending on the specific usage.

- *Type II*—All structural members are as in Type I but the fire resistance generally drops to two or one hours (less in some cases) depending on the specific usage.

- *Type III*—All exterior bearing walls are noncombustible (or limited combustible) and have fire ratings of at least two hours, but interior structural elements can be of approved combustible material.

■ *Type IV*—All exterior and interior bearing walls are noncombustible (or limited combustible) and interior columns, beams, girders, arches, trusses, floors and roofs are of heavy timber construction without concealed spaces.

■ *Type V*—Buildings constructed of approved combustible material, that for some structural elements is subject to a minimum one-hour fire resistance rating.

For about thirty years nonmetallic-sheathed cable could not be used anywhere in any building that exceeded three floors above grade; in contrast to ENT, this restriction applied whether or not the wiring was concealed behind a fire finish. The only exception was in the case of a one- or two-family building, where it could be used regardless of total height. This restriction became the subject of intense and continuing national debate. Some jurisdictions imposed further restrictions, and some others removed the height limitations entirely. Be sure to review your local code because this provision may have been amended.

Ampacity restrictions Although the individual conductors have a 90°C temperature rating, the final allowable ampacity must not exceed that given in the 60°C column (refer to page 105 discussion on adjusting *NEC* Table 310.16 values for actual conditions). This applies after adjustments for ambient temperature and for mutual conductor heating have been factored. Although mutual conductor heating may seem odd, you must apply this to all the conductors in a group of cables if those cables are "bundled" for longer than 24 in. This includes routing more than one cable through a succession of single bored holes. You need to consider this requirement in a basement, for example, where you might be tempted to run large numbers of cables back to the panelboard through a set of holes lined up through the floor joists and ending at the panel. The derating penalties in *NEC* 310.15(B)(2)(a) apply to this condition. In addition, a special restriction (which also applies to any wiring method) prevents making use of the general permission to round up to the next higher overcurrent protective device if a branch circuit supplies multiple receptacle outlets. These two conditions mean that for any branch circuits supplying multiple receptacle outlets, if the resulting derated ampacity drops below the branch circuit size even by a single ampere, then you must separate the conductors. For example, the ampacity of 12 AWG in the 90°C column is 30 amps, and for 14 AWG it is 25 amps. Suppose just five 2-wire cables run through a common set of bored holes. Since the derating penalty is 50%, the resulting diminished ampacity condemns any 14 AWG cables on 15-amp circuits, and any 12 AWG cables on 20-amp circuits— the usual assumption. You could, of course, increase the wires by a cable size, but normally it is better to bore a parallel set of holes.

Removing the outer cover The outer cover must be removed at the ends for a distance of about 10 in., or whatever distance you decide based on the rules for free conductor length in boxes, as covered at the beginning of Chapter 8 and the end of Chapter 10. For Type NM this can be done by slitting the jacket; there is a wide variety of cable rippers on the market today (Fig. 11–27). Cut off the dangling cover with a knife or pair of side-cutter pliers, being careful not to damage the individual wires.

Fig. 11–27 This cable ripper saves much time in installing nonmetallic-sheathed cable.

Anchoring cable to boxes   The cable is anchored to switch and outlet boxes with connectors, several types of which are shown in Fig. 11–28. Some of these connectors just snap in without requiring locknuts. As mentioned in Chapter 10 (see Fig. 10–6 for an example), many boxes have built-in cable clamps that eliminate the need for separate connectors. Today most NM cable never sees a metallic enclosure once it leaves the panelboard, and even some panelboards are nonmetallic. Connectors undergo testing for specific cable configurations. Never use a connector for a cable out of its listed size range, and never use a connector for multiple cables unless it is listed for that duty.

Fig. 11–28 An assortment of connectors for anchoring cable to outlet and switch boxes. The two-screw metal one near the center is also listed for two 12-2 or 14-2 Type NM cables.

Mechanical installation   If installed while a building is under construction, nonmetallic-sheathed cable is placed inside the wall and ceiling spaces where it will be concealed after the walls and ceilings are installed. *NEC* 334.30 requires that the cable be supported at least every 4½ ft and in any event within 12 in. of any box. This support requirement applies in addition to the general requirement in *NEC* 300.4(D) specifying that all cables must run at least 1¼ in. from the face of a stud or framing member when the wiring will be concealed afterward. This spacing requirement is intended to prevent errant drywall screws or picture nails from hitting the cable, but it doesn't leave much room on the face of a 2 × 4 stud, so clamps of the type shown in Fig. 11–29 have been developed that space the cable away from the stud. Alternatively, you can provide a shield consisting of a steel plate at least 1/16 in. thick, but that is seldom practical when running the long way along a framing member. But steel plates are very practical when you have to run a cable through a stud at less than the prescribed distance from the wall finish. The *NEC* also allows thinner but harder plates listed as providing comparable protection.

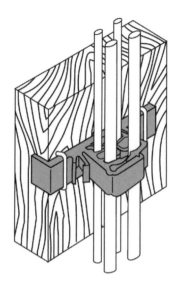

If you do anchor directly to the stud, use staples with rounded shoulders that won't cut into the cable; don't use ones designed for armored cables. Don't staple two-wire cables on edge. Some jurisdictions require insulated staples (Fig. 11–30), and even if not required they add considerable reliability to the installation. There are a variety of other fittings on the market designed to support this cable. The

Fig. 11–29 This anchoring device keeps cables away from picture nails and misdriven screws. Remember that the spacing rule applies to all cable wiring methods and even some raceways such as flex and ENT. *(Caddy)*

outer jacket of Type NM cable is plasticized, which means it will flow under constant pressure. An overdriven steel staple can cause problems years later. There are also listed staples designed for some sizes of NM cable that are driven from a staple gun. In old work where the cable is fished through wall spaces, the cable does not need to be supported except where exposed; of course connectors or built-in cable clamps must be used at all boxes.

Type NM cable is commonly installed in light commercial work, and when that entails routing the cable through steel framing make sure you use suitable grommets that encircle the entire punched opening in the framing member. The grommets protect the cable from being damaged by the sharp edges of the opening. All bends in cable, wherever made, must be gradual to prevent damaging the cable. According to *NEC* 334.24, if a bend were continued to form a complete circle, the diameter of the circle must be at least ten times the diameter of the cable.

Where exposed, as in basements, attics, barns, etc., cable must be protected against physical damage. This protection can be provided in a variety of ways; see Fig. 11–31. If the cable is run along the side of a joist, rafter, or stud as at *A*, or along the bottom edge of a timber as at *B*, no further protection is required. If it is run at an angle to the timbers, the cable may be run through bored holes as at *C* and no further protection is required. The holes should be bored in the approximate center of the timbers. If the cable is run across the bottoms of joists, then it must be run on substantial running boards, as shown at *D*, but the running boards are not required if the cable contains at least three 8 AWG or two 6 AWG or larger conductors. Cables smaller than that, if not run through bored holes or on running boards, are required by *NEC* 334.15(A) to follow the surface of the building structure, as shown at *E*.

Therefore, to save material and reduce voltage drop, cables crossing timbers should be run through holes or on running boards. If a ceiling is to be added later, run the cable through bored holes. Do not run this cable above a suspended ceiling in a nonresidential occupancy. The 2005 *NEC*

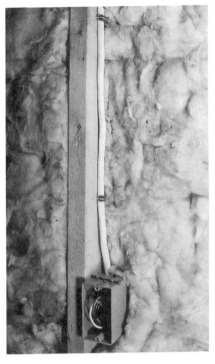

Fig. 11–30 Insulated staples help prevent damage to Type NM cable.

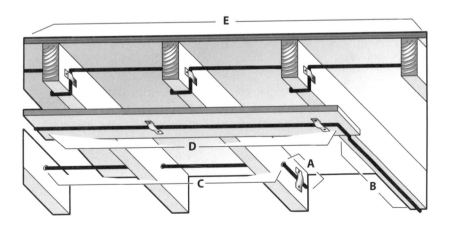

Fig. 11–31 Five ways in which cable may be run on an open ceiling. Various methods are used to protect the cable, depending on the method of installation.

prohibits exposed use in these locations, and the definition of "exposed" in Article 100 is such that all wiring above hung ceilings qualifies as exposed

In accessible attics, if the cable is run across the top of floor joists instead of through bored holes, it must be protected by guard strips at least as high as the cable, as shown in Fig. 11–32. If run at right angles to studs or rafters, it must be protected in the same way at all points where it is within 7 ft of floor joists. No protection is required under other conditions. If the attic is not accessible by means of stairs or a permanent ladder, this protection is required only for a distance of 6 ft from the attic opening.

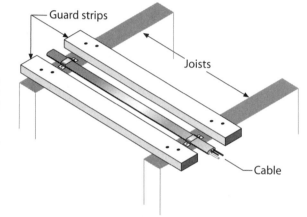

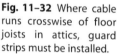

Fig. 11–32 Where cable runs crosswise of floor joists in attics, guard strips must be installed.

In all cases the cable must follow a surface of the building or of a running board, unless run through bored holes in timbers not over 4½ ft apart. It must never be run across open space, either unsupported or suspended by wires or strings.

Type UF cable   In some localities Type NMC cable is hard to find, while Type UF (Underground Feeder) is available. Type UF is very similar in appearance to Type NMC and costs very little more. It may be used wherever Type NMC may be used. In addition, it may be buried directly in the earth provided it has overcurrent protection at its starting point. Thus, it may not be used in a service entrance. However, on farms, if it starts at the yard pole and runs to various buildings, it may be used but again only if it has overcurrent protection at the pole. Where it enters a building, it may continue into the building and be used for interior wiring. When this occurs, follow all the requirements for Type NM cable, including the 90°C conductor ampacity rule. Type UF cable styled "UF-B" is available for this purpose.

Service-entrance cable, Type SE   Refer to Fig. 16–15 and related text for an illustration of how Type SE cable is constructed and for more information on its use in service entrances. It can also be used for ordinary feeders and branch circuits, provided the grounded circuit conductor is insulated. Normally this entails use of a cable construction with an additional insulated wire, reserving the bare conductor for equipment grounding purposes. Such cables are round (often referred to as "SER" cables, but that isn't an *NEC* designation) and follow the normal rules for installing Type NM cable within buildings with the exception that the cable need not be derated to 60°C ampacity limits for this purpose. Pay close attention, however, to the discussion of the effects of thermal insulation on conductor ampacity toward the end of this chapter.

Armored cable   The ordinary kind of armored cable, called "BX" by most people, is shown in Fig. 11–33. The *NEC*, in Article 320, refers to it as Type AC cable.

Fig. 11–33 Armored cable. Steel armor protects the wires. A bare grounding strip runs inside the armor for better equipment grounding purposes.

In armored cable, the insulated wires are wrapped in a spiral layer of tough kraft paper, with a spiral outer steel armor. Between the paper and the armor there is a narrow aluminum (or copper) bonding strip that reduces the resistance of the armor itself by short circuiting the convolutions, thus providing better continuity of ground. Without the bonding strip, fault currents would tend to spiral back along the tape armor instead of moving in a straight line parallel to the faulted line conductor.

Cutting cable   A hacksaw can be used to cut armored cable. Do not hold the saw at a right angle to the cable, but rather at a right angle to the strip of armor, as shown in Fig. 11–34. You must use extreme care to avoid cutting past the armor into the conductor insulation or the bonding strip. After the cut is made through the armor, grasp the armor on each side of the cut and give a twist as shown in Fig. 11–35. The end of the armor can then be pulled off. The cut should be made about a foot from the end of the cable.

Instead of a hacksaw, a time-saving BX cutter can be used; one type is shown in Fig. 11–36. It has a built-in automatic stop that allows the armor to be cut without damage to the insulation.

Insulating bushings   When armored cable is cut, a sharp edge is left at the cut end, with burrs extending in toward the insulation on the wires. To eliminate the danger of such burrs damaging the insulation, you must insert a plastic (or fiber)

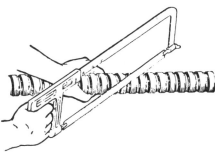

◀ **Fig. 11–34** In sawing armored cable, hold the hacksaw blade at an angle as shown.

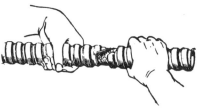

Fig. 11–35 After sawing, twist the cut end of armor to remove it. ▶

◀ **Fig. 11–36** The "Roto-Split" cutter provides clean cut ends of armor and saves much time compared with using a hacksaw. It is also available in larger sizes, which also cut flexible metal conduit. *(Seatek Co., Inc.)*

bushing between the armor and the wires. One is shown in Fig. 11–37. Such bushings are called antishort bushings or sometimes just "redheads." Since there is little room between the armor and the paper, space is provided by removing the paper. The steps shown in Figs. 11–38 and 11–39 demonstrate how to unwind the paper beneath the armor, and then with a sharp yank tear it off some distance inside the armor, creating space for the bushing which is inserted as shown in Fig. 11–40. The final assembly is shown in Fig. 11–41.

Connectors Connectors for armored cable are shown in Fig. 11–42. They differ from those for nonmetallic cables due to different cross-sectional dimensions and the fact that for armored cable the connector is a vital link in the equipment grounding return path. As such you must take care to install it properly. Armored cable connectors have openings or peepholes through which the inspector can see

Fig. 11–37 Insert a bushing between the armor and the wires to protect them against danger of their insulation being punctured by cut ends of armor.

◄ **Fig. 11–38** Unwrap the paper over the wires as far as you can; do not tear off.

Fig. 11–39 Remove paper from inside the armor as far as you can; then with a sharp yank tear it off inside the armor. ▶

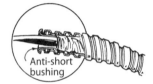

◄ **Fig. 11–40** Insert the fiber bushing.

Fig. 11–41 Cross section of cable, showing paper removed inside the armor and protective bushing in place. ▶

Anti-short bushing

the red antishort bushing from the inside of the box. Because of these peepholes such connectors are known as the "visible type."

To install the connector properly, first insert the antishort bushing under the armor and then bend the bonding strip back over the armor so it will hold the bushing in place. Push the cable into the connector as far as it will go so the fiber bushing cannot slip out of place and can be seen through the peepholes in the connector. Then tighten the screw(s) on the connector to anchor it solidly on the cable. Remove the locknut, slip the connector into the knockout of the box, and drive the locknut solidly home inside the box, as with other cable; see Figs. 11–42 and 11–43. If this is carefully done, with the teeth of the locknut biting into the

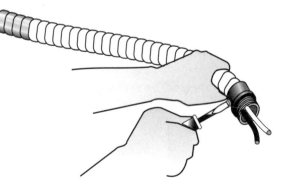

Fig. 11–42 Installing connector on cable.

metal of the box, all the outlets on the circuit are then tied together through the armor of the cable and the bonding strip, providing the continuity of ground discussed in Chapter 9.

Using the bonding tape to help retain the antishort bushing is strictly a convenience. Because there isn't any way to know where that tape lies when cutting the armor, and because it must be in intimate contact with the armor in order to do its job, it often ends up being cut. It doesn't matter if that happens. Again, the only function of this

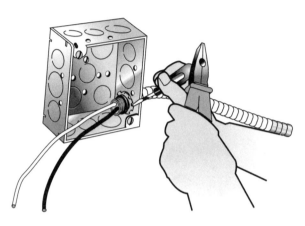

Fig. 11–43 Drive the locknut of the connector down tightly so the teeth of the locknut bite down into the metal of the box.

bonding tape is to short-circuit the spiral convolutions. It isn't a separate equipment grounding conductor as in the case of Type NM cable, and no special termination rules apply to it.

Bends Avoid sharp bends. The *NEC* requirement is that if a bend were continued into a complete circle, it would have a diameter not less than ten times the diameter of the cable. In the case of armored cable, any sharper bend is likely to break the convolutions apart.

Supporting armored cable Except where fished through existing wall spaces, armored cable must be supported within 12 in. of each box and cabinet, and at intervals of not less than 4½ ft. However, a free length not over 24 in. long is permitted at terminals where flexibility is needed, such as a motor that must be adjusted for belt tension. Staples of the type shown in Fig. 11–44 are more convenient than two-hole straps, but such staples must have a rustproof finish as required by *NEC* 300.6(A).

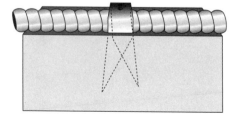

Fig. 11–44 Staples of this kind are used to support armored cable on wooden surfaces. Do not use them with nonmetallic-sheathed cable.

Where used   Like Type NM nonmetallic-sheathed cable, armored cable may be used only in permanently dry locations. For residential wiring it is usually acceptable, but it must not be used in barns or other locations where corrosive conditions are the norm. Some local codes have restrictions in addition to those imposed by the *NEC*. Check your local code for any additional restrictions.

Metal-clad cable, Type MC   At one time designers considered this wiring method only for large feeders in major industrial projects. Today it ranks as one of the leading cable wiring methods, with configurations running from the traditional industrial applications all the way down to widely used 15- and 20-amp branch circuits in commercial and even residential occupancies of all sizes.

The reason for this adaptability is that Type MC cable includes three different categories. The most common type looks like Type AC cable, with a spiral metal armor (see Fig. 11–45). However, there isn't a bonding strip, nor a paper wrap to keep such a strip tight against the armor. The armor jacket, therefore, is not an equipment grounding path. Instead, the cable comes with an equipment grounding conductor, usually with green insulation. Although the jacket does have to be grounded at terminations, the separate equipment grounding conductor carries the principal grounding responsibility. Recently one manufacturer solved the grounding continuity problem across interlocking MC cable armor by using a bare, fully sized aluminum equipment grounding conductor against the armor and on the outside of the plastic wrap that goes over the circuit conductors. This product can be used similarly to Type AC cable, without a separate grounding conductor entering the box.

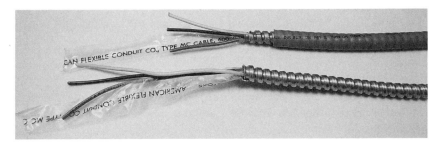

Fig. 11–45 Interlocking armor Type MC cable. The version at the top has an outer nonmetallic jacket and is listed for direct burial.

The second type of Type MC cable uses a seamless corrugated aluminum armor. See Fig. 11–46. In most (but not all) cases the total cross-sectional area of the armor is at least equivalent to the size of the required grounding conductor for the enclosed circuit conductors. For example, typical 12-2 corrugated MC cable would be used on a 20-amp circuit; if that circuit were wired with aluminum, it would have a 10 AWG aluminum equipment ground. In actuality, the cross-sectional area of the aluminum armor roughly equals the size of an 8 AWG conductor. One form of this cable comes in a rectangular cross section just large enough to enclose the two circuit conductors, and relies on the armor as the equipment grounding return path. In this sense these cables are electrically closer to AC cable than the interlocking form, which is far closer in outward appearance. The third type, used primarily in heavy industrial applications that are beyond the scope of this book, has a smooth outer sheath that is far less flexible than the others. Type MC cable, in any of its forms, can be manufactured with an overall nonmetallic jacket. In this form it may carry a listing indicating it is suitable for direct burial or other uses in wet locations.

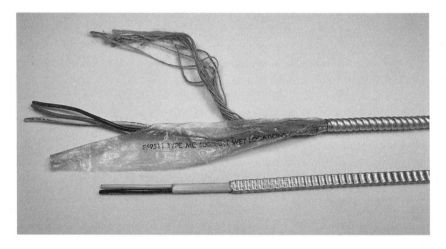

Fig. 11–46 Corrugated Type MC cable comes in several configurations. Both of the cables in this photo have outer armor that is a fully qualified equipment grounding conductor. The upper cable has an additional green wire, however, for use when a separate grounding conductor is needed.

Bends   Since the three cable types differ widely in flexibility, the *NEC* has different rules regarding bend sharpness. Bends in interlocking-armor and corrugated types must be no sharper than the point at which the bend, if continued into a circle, would have a diameter 14 times the cable diameter. The smooth types are much less tolerant of sharp bends, and the *NEC* sets much larger minimum bend requirements for these cables, with comparable circle diameters running from 20 to 24 to 30 times the cable diameter based on cable size ranges. Refer to *NEC* 330.24 for the exact requirements.

Supports   Type MC cable can be fished between access points. Otherwise, support it at least every 6 ft, a more liberal distance than for armored cable. The small branch-circuit sizes, consisting of four or fewer conductors and not larger than 10 AWG, carry an additional requirement for support within 12 in. of a termination. As in the case of cables generally, when you run Type MC cable through holes in structural framing, whether joists or studs, the *NEC* considered those runs supported. However, at the final support point where the cable enters a box, take care to do more than just cradle the cable in the nearest hole through a framing member. At terminations cables need to be "secured" and not just "supported." That means using a clip or staple that actually prevents the cable from moving in any direction.

Connectors   Type MC cable must terminate in connectors designed for the particular cable involved. At the outset of this book the close interplay between product standards and the *NEC* was described, and MC cable connectors illustrate the point very well. The *NEC* doesn't require anti-short bushings for Type MC cable, although one leading manufacturer does make them available for that purpose, and many contractors choose to use them anyway. They are not required because the throat designs of listed connectors keep the conductors away from the cut edges of the armor. In addition, since these connectors may have to handle ground-fault currents, they are tested with their designed cable types for this duty. Those tests have no validity beyond the cable types actually tested.

Before you run MC (or any other cable), look closely at the box or shipping carton for the connectors. Particularly in the case of Type MC cable, you'll find very specific size ranges given. For example, the box might indicate suitability for smooth Type MC in a particular range of cable diameters, corrugated in a range of diameters, and interlocking with a range of conductor configurations. If you don't find your particular application, select a different connector. Only some connectors, for example, are suitable for the corrugated, rectangular cross-section type of cable shown in Fig. 11–46. The same principle applies to internal box connectors, as illustrated in Fig. 10–16. Look at the shipping carton for the box, because the label (for a listed box) will always tell you what cables the internal clamps are designed for.

Knob-and-tube wiring   In the early days of electric wiring, open wires were installed supported on porcelain insulators and run through porcelain tubes in holes through timbers. As explained in the discussion of knob-and-tube wiring that begins on page 392, this wiring method is so rarely used today that it does not warrant detailed coverage in this book.

Changing from conduit to cable   At times you will find it necessary to make a transition from conduit wiring to cable wiring, which presents no problem. Make the change in an outlet box as shown in Fig. 11–47. Connect all the black wires together, connect all white wires together, and connect the bare equipment grounding wire of the cable to the box using a screw used for no other purpose. Use a metal box, if metal wiring methods are involved. Use a box large enough to avoid crowding of the wires. Leave the wires long enough so the splices can be made easily, being sure to observe the 6-in. minimum length rule covered at the beginning of Chapter 8. The box must be permanently accessible and covered with a blank cover.

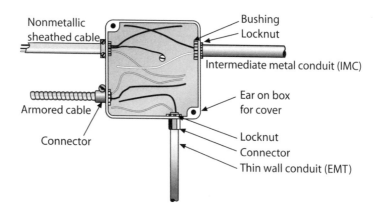

Fig. 11–47 How to change from one wiring system to another. A cover must be installed on the box.

THERMAL INSULATION DEGRADES WIRING METHOD AMPACITIES

Thermal insulation not only impedes the transfer of heat to and from conditioned spaces, it impedes the transfer of heat away from raceways and cables. For example, a special rule applies to Type AC cable when you run it through thermal insulation. The individual conductors must be rated 90°C, and the cable must not be loaded beyond the 60°C ampacity limits. This rule also applies to Type NM cable, in fact, the NM cable rule applies whether or not the run is in insulation. These rules provide some headroom in the temperature ratings in reflection of the fact that no ampacity table comes anywhere near the actual current-carrying capacity of conductors embedded in thermal insulation. Actual tests using what would normally be 100-amp Type SE cable (with 90°C conductor insulation), showed excessive heating at 67 amps when embedded in cellulose insulation, and actual incineration of the XHHW insulation at 100 amps.

These results vary by wire size, tending to get worse as the size increases. This seemingly paradoxical result follows from the fact that most installations of larger conductors reflect an expectation that more current will be carried. The amount of heat loss varies by the square of the current, and inversely by the first power of the wire resistance (I^2R). This means that the heating effects from higher currents quickly overwhelm the benefits of decreased resistance. Although we cover the issue here because these wiring methods have specific rules hinged on the use of thermal insulation, it is a general problem. It can affect all cables and all raceways, especially those carrying significant current. Never run large feeders in locations where they will be embedded in thermal insulation any distance much greater than straight through a wall.

FLEXIBLE CORDS ARE NOT WIRING METHODS

Where wires are installed permanently, they need only to be sufficiently flexible to permit reasonably easy installation and, for some flexible connections, to allow for slight movement or normal vibrations. If the wires must be moved about, as on a floor lamp, vacuum cleaner, or portable tool, they must be very flexible. This is necessary for convenience and to prevent the conductors from breaking, which would be more likely if they were solid copper, especially if of considerable diameter. Flexible wires of this kind are called "flexible cords" in the *NEC*. There are a great many different kinds, only a very few of which are described here. Others are covered by *NEC* Table 400.4. They have special ampacities not covered in *NEC* Table 310.16; use Tables 400.5A or 400.5B as applicable.

Cords are covered here at the end of the wiring method chapter only because of their usual multiconductor character. Although there is a superficial resemblance between a flexible cord and some cabled wiring methods, the *NEC* places flexible cords at the beginning of Chapter 4, which covers equipment. With very few exceptions, flexible cord is something to be supplied from an outlet, and not to be used to supply an outlet. Flexible cords are intended primarily for use as an integral part of a portable tool, appliance, or light. Cords are not intended to be used as extension cords except for temporary duty. Do not use a flexible cord as a permanent extension of the fixed wiring. Never use cord to connect an appliance unless it is one with fasteners and mechanical connections obviously designed for ease of removal for maintenance. Equipment that is hard wired or cord- and plug-connected frequently comes with a knockout, and you make the decision whether to connect a length of cord, or to come all the way to the equipment with per-manent wiring. If you decide to use cord, be sure to select the appro-priate type. Never permanently attach cord to a building surface, although there is an exception for a strain-relieved take-up as illus-trated in Fig. 11–48.

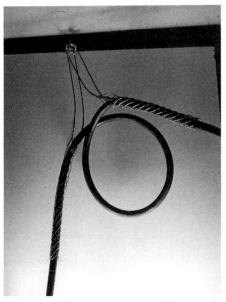

Where an extension cord is used temporarily, be sure to unplug it when you have finished using the portable equipment. Do not go off and leave it plugged in if in so doing you are using the cord as a substitute for the permanent wiring in the building, which the *NEC* prohibits. Use only extension cords with a suitable female member of a separable connector. Be sure that all portable lights (as

Fig. 11–48 This arrangement, with a single attachment to the building, is often used to complete the wiring to frequently repositioned equipment.

distinct from extension cords) are equipped with a suitable handle and lamp guard; some portable lights may also need a glass globe for additional protection.

Types SP and SPT   This is the cord commonly used on lamps, clocks, radios, and similar appliances. As shown in Fig. 11–49, the wires are embedded in a solid mass of insulation, so no outer jacket is required. Type SP cords have rubber insulation; Type SPT have plastic insulation. Often the cord is made with a depression between the two conductors for ease in separating them to make connections.

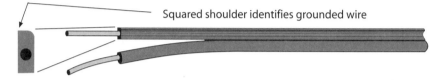

Squared shoulder identifies grounded wire

Fig. 11–49 Type SP and SPT cords are used mostly on floor and table lamps, clocks, and similar equipment. Note the square corner on the cross section of one of the two wires. This is the identified conductor and must be connected to the grounded (white) branch circuit wire.

Types S, SJ, SV   The cords described in the preceding paragraphs are designed for ordinary household devices that are generally moved about very little once they are plugged into a receptacle. They will not withstand a great deal of wear and tear. A cord that has a sturdier construction is needed for portable appliances and tools, such as vacuum cleaners, washing machines, and electric tools.

A Type S cord is shown in Fig. 11–50. It consists of two or more stranded conductors with a serving of cotton between the copper and the insulation to prevent the fine strands from sticking to the insulation. Jute or similar fillers are twisted together with the conductors to make a round assembly that is held together by a fabric overbraid. This assembly has an outer jacket of high quality rubber. Type SJ is similar in construction but has a lighter jacket. Type SJ is for ordinary hard usage, while Type S is for extra-hard service such as in commercial garages. Type SV has a still lighter outer jacket and is used only on vacuum cleaners.

Fig. 11–50 Type S cord is very tough and durable.

Types ST, SJT, SVT   If the outer jacket is made of plastic materials instead of rubber, Types S, SJ, and SV become Types ST, SJT, and SVT.

Oil-resistant cords   When cords using ordinary rubber insulation are exposed to oil, the oil attacks the outer rubber or plastic jacket, causing it to swell, deteriorate, and fall apart. For that reason ordinary cords cannot be used where exposed to oil or gasoline, such as in commercial garages. Cords made with a special oil-resistant jacket of neoprene or similar material have the letter O added to the type designation; Type S, for example, becomes Type SO. For additional protection, you can get cord with both the outer jacket and the individual conductor insulation resistant to oil, designated SOO.

Cords in wet locations   When you use a cord in an outdoor wet location, test lab restrictions require a "W-A" or "Outdoor" designation on the cord. For example, you might see a marking "SO W-A" which would mean a cord suitable for extra-hard usage in wet locations and which also has an outer jacket suitable for exposure to oils. If the cord will be used for immersion in water outdoors, as in the case of fountains, look for a "W" marking.

Flexible cord has different overcurrent protection rules   The ampacities of flexible cord are given in *NEC* Tables 400.5(A) or 400.5(B) as applicable. However, regardless of the ampacities in those tables, flexible cord used in an extension cord set can be as small as 16 AWG. The *NEC* now defers to the testing laboratory review of extension cords for ampacity limitations, having deleted the specific size rules formerly in 240.5, so the associated instructions with such cords must be carefully followed.

CHAPTER 12
Planning Residential Installations

AN ADEQUATELY WIRED HOME is one that has been wired to be completely safe, and to provide the occupants with maximum convenience and utility from the use of electric power. There must be sufficient light where needed, from permanently installed fixtures or portable lamps. The occupants must be able to plug in lamps, radios, TV sets, etc., where they please, with a variety of furniture arrangements, and without resorting to extension cords. They must be able to turn lights on and off in any room without stumbling through darkness to find a switch, and to move from basement to attic with plenty of light but without leaving unneeded lights turned on behind them. They must be able to plug in needed appliances without unplugging others; motors must get full power and heating appliances must heat quickly, and lights must not dim excessively as an appliance is turned on. There should be no reason to overload circuits, causing fuses to blow or circuit breakers to trip.

As explained in Chapter 1, the *National Electrical Code* is concerned only with safety. *NEC* 90.1(B) states plainly that compliance with the *NEC* and proper maintenance will result in a reasonably safe installation, "but not necessarily efficient, convenient, or adequate for good service or future expansion of electrical use." Many houses are being wired with entirely too little thought about adequacy, even though today's norm requires much more light and a greater variety of appliances than ever before. In addition, the electronic revolution is creating increasing demand for permanent wiring that accommodates Internet access and other signaling circuit interconnections throughout the house. Adequate wiring today is accomplished only through careful planning. This chapter covers the essential *NEC* minimum requirements, and notes areas where good design dictates going beyond the minimum. Review, if necessary, the basic concepts in Chapter 4.

FACTORS IN ADEQUATE WIRING

Designing a house wiring installation involves making decisions about the size of the service, the size of the wires, and the number of branch circuits. Always take into account both present and future needs of the occupants.

■ *Install a large enough service.* The components of the service are described in detail in Chapter 4. The maximum load in use at one time should neither overload the service-entrance wires, causing excessive voltage drop and wasted electricity, nor trip breakers or blow fuses. Today the *NEC* minimum service size for a one-family dwelling is 100 amps. Future household needs will likely be met more adequately by a 150- or 200-amp service. In some areas, local codes are already requiring services larger than 100 amps. The larger service costs relatively little more than the 100-amp size. Service installation details are covered in the next chapter.

■ *Consider the benefits of larger wires.* Remember that the *NEC* specifies only minimum wire sizes. For long runs or heavy loads, be sure to use wires large enough to prevent excessive voltage drop (discussed on page 107). For house wiring, the minimum circuit wire size permitted by the *NEC* is 14 AWG protected by a 15-amp fuse or circuit breaker, except for door chimes and other limited energy wiring. Some homes are designed with 12 AWG as the minimum branch-circuit size (protected at 20 amps), and in a few localities it is required as a minimum.

■ *Plan a generous number of circuits.* The greater the number of branch circuits, the greater the flexibility and the less likelihood of breakers tripping or fuses blowing due to circuit overload, and the lower the voltage drop, thus making for brighter lights and greater efficiency of appliances. In addition, more branch circuits make it easier to arrange overlapping coverage, so if a fuse or breaker does open, the entire area doesn't go dark. The next chapter explains how to calculate the minimum number of branch circuits based on the size of the house. The rest of this chapter focuses on the room-by-room requirements and recommendations for all the receptacle (plug-in) outlets, lighting fixtures, wall switches, and other outlets and devices that are characteristic of a well-designed home wiring installation.

NEC EMPHASIZES RECEPTACLE PLACEMENT RULES IN DWELLINGS

Receptacles in sufficient number eliminate the need for extension cords, which are not only unsightly and inconvenient, but hazardous—people can trip on them and receive injuries from falling, and cords that are defective or damaged can lead to shock or fire. *NEC* 210.52(A) requires that:

> In every kitchen, family room, dining room, living room, parlor, den, sunroom, bedroom, recreation room, or similar room or area of dwelling units, receptacle outlets shall be installed in accordance with the general provisions specified in (1) and (2).

> (1) *Spacing.* Receptacles shall be installed so that no point measured horizontally along the floor line in any wall space is more than 1.8 m (6 ft) from a receptacle outlet in that space.

(2) *Wall space.* As used in this section, a wall space shall include the following:

(a) Any space 600 mm (2 ft) or more in width (including space measured around corners) and unbroken along the floor line by doorways, fireplaces, and similar openings.

(b) The space occupied by fixed panels in exterior walls, excluding sliding panels.

(c) The space afforded by fixed room dividers, such as free-standing bar-type counters or railings

The receptacles required are in addition to any receptacles that are over 5½ ft above the floor, or are inside a cabinet, or part of an appliance or lighting fixture. Note that the 6-ft and 2-ft rules measure all wall space, including that behind a door swing or in a narrow room-entry area. The Code-making panel understands that these areas won't be used for furniture, but wants to increase the likelihood of unobstructed access to some receptacles, such as for vacuum cleaners.

Receptacle access from countertops   In kitchens and dining rooms the *NEC* also requires a receptacle at each counter space 12 in. or wider, and for kitchen wall counters a very generous receptacle spacing in consideration of the multitude of today's kitchen appliances. No point measured along the wall line can be more than 2 ft from a receptacle outlet, and countertops interrupted by sinks, range tops, and refrigerators must be considered as separate counter spaces. Counter receptacles must be above the counter, but not more than 20 in. above. This rule follows changes in product standards that shortened appliance cords so they won't hang over a counter edge within reach of toddlers, reducing the likelihood of pulling hot appliances over on top of themselves. See Fig. 12–1.

Island and peninsula countertops must have at least one receptacle on each. These countertops present a challenge. A receptacle on the side leads to the toddler-pulling-on-cord problem. A receptacle on the top requires some sort of structure or a "tombstone" configuration, the oval-shaped pedestals with squat bases generally seen in commercial office space mounted to the floor, since another rule prohibits mounting receptacles in a face-up position in all dwelling unit work surfaces. Many kitchen peninsulas and islands today amount to permanently anchored tables routinely used for meals. Owner resistance to a mandated receptacle placement in the middle of their "table" has usually been fierce. The present rule is a reasonable compromise. If the island or peninsula has two levels, such as a backsplash around a sink area, or if there is a suspended cabinet or other structure that could carry a receptacle within 20 in. of the work surface, then the normal rule applies. But if the surface is entirely flat and there isn't any structure close above, then the receptacle can go on the side of the island or peninsula (though no *NEC* rule prohibits a "tombstone" on the countertop if the owner wants one). A receptacle on the side must be within 12 in. of the countertop and

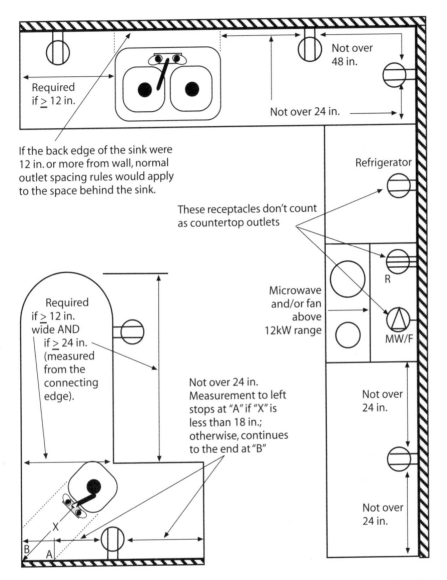

Fig. 12–1 This kitchen floor plan illustrates most of the receptacle placement rules. The 2005 *NEC* has new rules for the space behind ranges and sinks (as shown).

located in an area where the counter doesn't extend more than 6 in. beyond the support base. This assumes that parents, in their role as responsible adults, won't use or will police the use of these receptacles in the presence of toddlers. An additional exception grants complete relief from the above-the-counter rule in construction for people with physical handicaps. In this case, receptacles are often located under the front counter lip, easily accessed by a person in a wheelchair.

Many appliances consume 1000 watts or more, and often several are in use at the same time. The *NEC* requires at least two special 20-amp small-appliance circuits for portable appliances. No lighting outlets may be connected to these circuits. Discussion of small-appliance circuits begins on page 222.

Special-function receptacles on ordinary lighting circuits   Although the *NEC* normally covers dwelling-unit receptacle placements through perimeter spacing rules, the general rule in 210.50(B) still holds. Receptacles must be provided at every point that flexible cords with attachment plugs are used. Two examples are line-powered electric clocks, and loads that can't be served from wall receptacles and must use floor receptacles to avoid cords running across traveled areas.

Clock receptacles   Although many clocks are battery powered today, some owners want a clock connected to the power system. Although not an *NEC* requirement, the most workmanlike way to connect a line-powered clock is to install, where the clock is to be placed, a clock-hanger type of receptacle of the kind shown in Fig. 12–2. The receptacle itself is located at the bottom of a well in the device; the outlet supports the clock. Cut the cord of the clock to a few inches in length; the cord and plug will be completely concealed behind the clock. This clock-hanger outlet may be installed either on a lighting circuit, or on a 20-amp small-appliance circuit if it will be in a room served by one of these circuits.

Floor receptacles   The *NEC* classifies the movable portion of a sliding door as just that, a door. If multiple door units are mulled together, the result will usually be nonmovable panels located between "doors." Assuming the fixed portions are at least 2 ft wide, the result is one or more glass walls that require receptacle support, and clearly a conventional wall receptacle won't work. For these and similar applications, such as adjacent to an open railing that defines the boundary of a room area, use a floor receptacle. The *NEC* counts these receptacles as perimeter receptacles as long as they are within 18 in. of the wall.

The *NEC* requires floor receptacles to be installed only in boxes that have been specifically listed for floor duty. Because of the way the listing process works in practice, you have to get a box and receptacle as a combined kit that is evaluated for

Fig. 12–2 This outlet supports a clock. The receptacle is in a "well" that completely conceals the cord and plug of the clock. *(Pass & Seymour/Legrand)*

floor use. To prevent damage to floor receptacles, especially in frame construction, pay careful attention to locating the receptacles where they will get a minimum of floor traffic. Also, be sure you are ordering the correct type of device. The *NEC* waives the special box requirement if the receptacle isn't likely to see floor action, such as in a commercial display window. These receptacles are sold separately, identified as "display receptacles." They look like floor receptacles but have no box and haven't been evaluated with respect to the amount of walk-over they will withstand. Don't use them as general floor receptacles.

Service receptacles for heating, air conditioning, and refrigeration equipment   To correlate with mechanical codes, the *NEC* requires service receptacles within 25 ft of this equipment wherever located, even outdoors at grade, and at all dwelling units. The former limitations applying this rule only within attics or crawl spaces and, for other than one- or two-family dwellings, on rooftops, have been lifted. This outlet can be on any local lighting circuit; the only restraint is that it not be connected to the equipment circuit. The service receptacle requirement applies only to furnaces or equipment with compressors; if all you have is some form of blower, there isn't any requirement. For dwelling units, the outdoor convenience receptacle (see requirements for outdoor receptacles, covered below) may serve for this purpose, but only if located within the required 25 ft. For all equipment, the receptacle must be on the same level as the equipment. Be careful to follow this rule on multilevel rooftops.

NEC REQUIRES LIGHTING, USUALLY WITH SWITCH CONTROL

Many rooms require one or more permanently installed lighting fixtures for general lighting. Additional lighting in most rooms can be and usually is provided by floor or table lamps. The lighting fixtures themselves can be whatever the owner wants; fixture selection is discussed in detail in Chapter 14.

A permanently installed lighting fixture controlled by a pull chain or by a switch on the fixture itself is inexcusable today, except perhaps in a closet so small that it is impossible to miss a pull chain. With this exception, every fixture should be controlled by a wall switch. There must be lighting controlled by a wall switch in every habitable room and bathroom, and in hallways, stairways, attached garages, detached garages with electric power, and on the exterior side of all outdoor entrances. In kitchens and bathrooms the switch-controlled lighting must be provided by permanently installed fixtures; other interior rooms (other than bathrooms and kitchens) may have switch-controlled receptacle outlets into which floor or table lamps can be plugged.

The *NEC* does not specify where switches must be located, although for stairways of six or more risers there must be switch control somewhere on both levels. In a room that has only one door, the logical location for a switch is near that door. If there are two entrances, a pair of three-way switches allows the light to be controlled by a switch at each entrance. For three entrances, a switch at each is a touch of luxury the occupant will appreciate—add a four-way switch to the pair of three-way switches. (Installation of three- and four-way switches is explained in two chapters: one discussion begins on page 79 and the other on p. 320.) In an adequately wired house, you can enter by any outdoor entrance and move from basement to attic without ever being in darkness, yet never having to retrace your steps to turn off lights. Making it easy to turn off unused lights helps conserve energy.

There are other options to elaborate three-way and four-way switch loops. You can use relays (see the low-voltage remote-control switching topic on page 366), and you can use occupancy sensors. If an occupancy sensor is located at a normal switch location and has an override switch that allows it to function as a wall switch, the NEC allows it to substitute for normal wall switch control. Occupancy sensors sense movement in a room and turn the lights off when everyone has left.

ROOM-BY-ROOM WIRING NEEDS

Some rooms need much more lighting than others; some need more receptacles than others. Provide each room with what it needs.

Living room   At one time lamps bigger than 100 watts were seldom used in floor or table lamps; today 300-watt halogen torchiere lamps are common. However, not all floor and table lamps provide illumination beyond local areas. Often they leave dark places throughout much of the room, therefore general illumination is needed. It can be provided by ceiling fixtures, but not all owners like them. Cove lighting, installed on the wall some distance below the ceiling, provides good indirect lighting if properly designed, but may not be compatible with the owner's tastes or the overall architecture. Consider all the possibilities, which are more fully explored in Chapter 14.

Be generous with receptacle outlets in the living room. The NEC minimum of receptacles placed so that no point along the wall will be more than 6 ft from a receptacle is rarely adequate in a living room. Cords on floor and table lamps are not always 6 ft long. Install enough receptacles so that floor and table lamps can be placed where you want them without using extension cords. Locate one receptacle where it will always be readily available for a vacuum cleaner no matter how the furniture is arranged.

The ordinary duplex receptacle is really two receptacles in a single housing allowing two things to be plugged in at the same time. In this book, each set of openings is referred to as one of the "halves" of the duplex receptacle. Most duplex receptacles come from the manufacturer with both halves either on or off at the same time. However, as explained in the two-circuit duplex receptacle topic on page 329, usually you can split the two halves in the field so one half is permanently live and the other half can be controlled by a wall switch. With several two-circuit receptacles in a room, a group of lamps can be turned off at one time by a single wall switch. Alternatively, the room furniture can be rearranged in the knowledge that a floor lamp has access to a switched receptacle whatever the furniture arrangement.

Although the *NEC* does permit wall-switch control of a receptacle to substitute for a permanent fixture in most rooms, always split the receptacle or find some other way to assure a permanent source of power at every required receptacle outlet in rooms with switched receptacles. Otherwise owners who want a television or other appliance plugged into that location will either leave the switch constantly in the ON position, entering a dark room to reach the turn knob on the floor lamp, or will resort to an extension cord plugged into an unswitched receptacle. Either outcome is poor design and frustrates *NEC* objectives.

Sunroom, den, and similar spaces   Provide a generous number of receptacle outlets for lamps and appliances. A room used for a home office would benefit from many receptacles, perhaps arranged as double-duplex outlets (two duplex receptacles in adjacent positions in a two-gang box) to minimize the need for extension cords feeding multi-outlet assemblies. If you're sure the room will be used for office purposes, placing such receptacles above desk height is a great convenience for computers and peripheral equipment, but many owners will not want a room hard wired in that way. The *NEC* does not restrict the height of perimeter receptacle outlets; as long as the outlet is not over $5\frac{1}{2}$ ft above the floor it may be counted as meeting perimeter outlet spacing requirements. Provide general illumination (probably a ceiling fixture) depending on how the room is to be used.

Dining room   Be sure to provide a ceiling outlet for a lighting fixture, controlled by three-way switches located at both entrances to the room. Visualize the arrangement of the furniture and locate the ceiling light over the center of the dining table rather than in the center of the room. Lots of receptacle outlets should be provided, taking into consideration the probable location of the furniture. Too many dining-room outlets are located where it is impossible to get at them easily for a vacuum cleaner and table appliances.

It is possible to use switched receptacles in a dining room instead of overhead or wall-mounted fixtures, but special rules apply. The receptacles must be served by special small-appliance circuits that don't supply lighting loads, and in this case the *NEC* presumes a switched receptacle to be a lighting load. The only way to solve this problem is to make certain that the appliance circuits indeed supply receptacles at every perimeter location required under the 6-ft and 2-ft spacing rules. Then, in addition, provide a switched receptacle. You could even use an ordinary duplex receptacle, completely split on both sides. Connect one of the halves to an appliance circuit, and the other half to the switched lighting circuit. Take care not to mix up the grounded (white) wires from the two different circuits. Beware of the new common disconnect rule (see page 331).

Kitchen   Kitchens vary from modest to deluxe, but whatever the nature of a particular kitchen it is likely to be a center of household activity. Kitchens require abundant wiring, and good lighting is essential. For general lighting there should be one or more ceiling outlets controlled by switches at each entrance to the kitchen. A light over the sink, controlled by a wall switch, and another light over the stove are necessary because without these the occupant will be standing in his or her own shadow when working at these points. Special fixtures and very thin-profile, limited-load lighting track assemblies are available for use under cabinetry to minimize glare and shadows on the counter surface.

The NEC allows but does not require trash compactors, dishwashers, and kitchen waste disposers to be cord and plug connected. This is a very useful design choice—if the appliance needs to be removed for service, the owner doesn't have to involve an electrician with a mechanical problem. The cords are limited to between 1½ and 3 ft for waste disposers and between 3 and 4 ft in length for the other appliances (measured from the back of the appliance). Review the installation instructions, because the NEC requires the use of cord assembly kits identified in those instructions for this purpose. Other appliances may be cord-and plug-connected as well, provided "the fastening means and mechanical connections are specifically designed to permit ready removal for maintenance or repair, and the appliance is intended or identified for flexible cord connection."

Hallway   Every hallway should have a ceiling light and a receptacle. The rather long hallways common in many houses, with rooms opening off either side, should have two ceiling lights controlled by three-way switches at either end. A receptacle for a vacuum cleaner is a necessity, and the NEC requires a receptacle in any hallway 10 ft or longer.

Bedrooms   Every bedroom should be provided with general illumination controlled by a wall switch; some people prefer ceiling fixtures, others switched receptacles. It may be wise to anticipate the use of paddle fans by installing boxes of appropriate design. (Discussion of paddle fan requirements begins on page 356.) As in the case of the living room, try to avoid having the switch control the entire duplex receptacle. If possible, have the switch control more than one receptacle to allow for furniture rearrangement. One of the required general purpose receptacles, or an additional one, should be located where always accessible for the vacuum cleaner. Be sure there are enough for table lamps, heating pad, radio, and so on. If the house isn't provided with central air conditioning, consider whether to add a separate circuit for a room-size unit. Smaller models can be plugged into an ordinary circuit if it is not already loaded, but a larger one might require a circuit of its own.

Clothes closets   It is exasperating to grope around trying to find something in a dark closet. This has caused people to use a match or lighter for illumination with tragic results. A hot lamp (bulb) can also have tragic results if it comes in contact with or is left turned on near combustible material in a clothes closet. The glass bulb of a lamp often reaches a temperature over 400°F. NEC 410.8 has specific rules about the placement and type of lights in a clothes closet.

A pendant light (a light socket on the end of a drop cord) is not permitted at all because it can be moved around and laid on or against clothing. An incandescent fixture is not permitted unless the bulb is completely enclosed. Fluorescent fixtures have cooler lamp temperatures and can be installed without an additional enclosure over the bulbs. Either type of fixture, if used, must be on the wall above the door, or on the ceiling, and nowhere else. Within those locations, an assembled surface-mounted incandescent fixture must be at least 12 in. from a storage area, and an assembled fluorescent fixture (or recessed incandescent fixture) must be at least 6 in. from a storage area. For the purposes of applying those distances to fixtures above the door or on the ceiling, the *NEC* defines "storage area" as 12 in. (or the width of the shelf, if greater) from the back or side walls of the closet and extending from 6 ft (or the highest closet rod, if higher) to the ceiling. This means that the minimum depth of a closet with a surface incandescent fixture would be about 30 in., given the storage depth of 12 in., the clearance of 12 in., and the likely minimum fixture dimension (including globe) of 6 in. Study *NEC* 410.8 carefully—many closets are too shallow to have any internal illumination under these rules, and others could only accept a recessed fixture over the door. You may, of course, always install fixtures just outside the closet door. Shallow closets and some walk-in closets are usually of such size that a pull-chain switch can be found without having to grope for it, though a wall switch (preferably with a pilot light) is more satisfactory. For a deluxe installation, use an automatic door switch, as shown in Fig. 12–3, which automatically turns the closet light on when the door is opened and off when the door is closed. A door switch requires a special narrow box, often available with the switch.

Bathrooms To provide enough light for putting on makeup or shaving, a lighting fixture (such as a fluorescent bracket or one that mounts a string of small globular incandescent bulbs) on each side of the mirror is essential; of course they should be controlled by a wall switch. A ceiling fixture is needed in a large bathroom, although for the average bathroom the lights at the mirror may be sufficient. The *NEC* requires a wall switch to control some permanently installed lighting in the bathroom.

A touch of luxury is an infrared heating lamp fixture installed in the ceiling, or a permanently wired electric wall heater installed near the floor well away from the tub, shower, and basin. Either

Fig. 12–3 A door switch turns a closet light on when the door is opened, off when it is closed. *(Pass & Seymour/Legrand)*

one should be controlled by a wall switch. An exhaust fan is desirable in all cases, and often is required by the building code. *NEC* 210.52(D) requires a receptacle within 36 in. of each bathroom basin, on an adjacent wall (or in the face of the basin cabinet, but not over 12 in. below the top), for electric shavers, hair dryers, and similar appliances; it must be protected by a GFCI. There are special circuiting rules as well; see the bathroom-circuit discussion on page 224.

Porches, decks, and patios—outdoors   If the porch is a simple stoop, a ceiling or wall light lighting the floor and steps is sufficient; of course it must be controlled by a wall switch inside the house. An illuminated house number is a touch the owners and their visitors will appreciate.

If the porch is larger, or if there is a deck or patio used more or less as an outdoor living room in summer, provide a number of receptacles for radio or CD player, floor lamps, appliances, or yard tools. *NEC* 210.8(A)(3) requires that such receptacles be protected by GFCIs. In general, the *NEC* now requires grade-accessible outdoor receptacles at both the front and back of every one-family dwelling, and the front and back of each dwelling in a duplex or in a multifamily house that is at grade level.

Basement   First of all there should be a light that illuminates every step of the stairs, controlled by three-way switches at the top and bottom of the stairs. If one switch is in the kitchen or some other location from which the basement light cannot be seen, use a switch with a pilot light to indicate when the basement light is on. Beyond this, the requirements vary greatly depending on how elaborate the basement.

If there is an all-purpose finished space for family use, which might be anything from a children's playroom to a second living room, wire it as you would the living room. If the ceilings are relatively low, consider recessed fixtures that fit flush or semi-flush with the ceiling. One such fixture is shown in Fig. 12–4. Recessed fixtures don't provide light over as wide an area as conventional fixtures, but they do give good light directly below for cards, table tennis, and other games. Provide receptacles generously, and provide wall switches for the lights. The *NEC* requires a receptacle for general maintenance in the unfinished portion of a basement; the receptacles you have to provide in the finished portion won't meet this requirement.

The *NEC* also requires at least one receptacle for laundry facilities, except in a multifamily application with central facilities, or in other than single-family dwellings where management policy precludes the use of laundry facilities. As in the case of bathrooms, there are special circuiting rules for laundry outlets; these are discussed toward the bottom of page 223. Even where the laundry is in an unfinished portion of the basement, the laundry receptacle won't qualify as the service receptacle previously mentioned. Since the clothes washer requires one receptacle, at least one more may be required for an iron. Many people like to

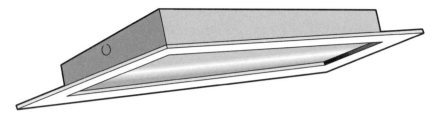

Fig. 12-4 Recessed ceiling fixtures are convenient where ceilings are low. Additional lighting from floor lamps is usually needed.

iron in the basement; a good ceiling light, preferably with a reflector, is essential in that area.

At other points in the basement, install ceiling lights as required. For basements, utility rooms, and crawl spaces, the *NEC* requires at least one wall-switch-controlled lighting outlet if the area is used for storage or contains equipment requiring service. In the case of serviceable equipment, the outlet logically must be located near that equipment. Nearly every basement has at least a corner that sooner or later becomes a workshop, and a light plus a receptacle, usually several, should be considered for that area.

In unfinished basements without a plastered or other reflective ceiling finish, it is wise to install reflectors of the general type shown in Fig. 12–5. Such reflectors greatly increase the amount of useful light. Dark ceilings absorb light; reflectors, if kept clean, throw the light downward where it is needed.

Fig. 12-5 A reflector greatly increases the amount of useful light from a lamp.

Attic   If the attic is used mostly for storage and has a permanent stair (or even a retractable stair), a single light placed to light the stairs may be sufficient. It should be controlled by wall switches on both levels, the lower level switch preferably with a pilot light. If the access is only through a scuttle hole, one wall switch would be enough. If it is a large attic, install additional lights as required; they can all be controlled by the same switch. If the attic is really an additional floor, unfinished but capable of being finished later into completed rooms, do not make the mistake of providing only a single outlet from which cable will later radiate in octopus fashion to a large number of additional outlets. Anticipate possible future construction and provide an extra circuit, terminating it in a box in some convenient location from which additional wiring can be started when needed. You could also install a raceway leading from the panel, or even just an empty cable sleeve originating in an unfinished area of the basement, and running to a convenient location in the attic.

Additional features   Every house needs a doorbell or chimes with a push button at each door. Pilot lights are desirable at switches that control lights that cannot be seen from the switch location, to indicate whether the lights are on; common locations for pilot lights are switches for basement, attic, and garage lights. Careful consideration of such details will make a home much more livable. Other suggestions are offered in Chapter 17.

PUTTING YOUR WIRING PLAN ON PAPER

The plans for an electrical installation begin with an outline "map" of the rooms involved, showing their relative locations and proportions. Standardized graphic symbols representing fixtures, receptacles, and other equipment are placed to show where the various outlets are to be located. A great many symbols are required to cover the wiring of all kinds of buildings from the smallest to the very largest. Symbols likely to be encountered in the wiring of the kinds of buildings discussed in this book are shown in Figs. 12–6 and 12–7, excerpted by permission from American National Standard "Y32.9–1972, Graphic Electrical Wiring Symbols for Architectural and Electrical Layout Drawings," copyright by the Institute for Electrical and Electronics Engineers, Inc.

Symbols of this kind are included in most wiring plans. You must understand them in order to be able to read and follow plans. Study the symbols of Figs. 12–6 and 12–7 until you can recognize them easily.

Sometimes a symbol is supplemented by additional letters near the symbol to more fully define the outlet. Examples:

WP	weatherproof	EP	explosionproof	UNG	ungrounded
RT	raintight	PS	pull switch	R	recessed
DT	dusttight	G	grounded	DW	dishwasher

The wiring plans for a split-level house are shown in Figs. 12–8 and 12–9, showing how symbols are used to indicate outlet locations. Typically blueprint plans show lines from switches to the particular outlet(s) to be controlled, but they do not show how various outlets are to be connected to each other. Plan drawings are supplemented by detailed written specifications giving size and type of service entrance, number of circuits, types of material to be used, and similar data.

Make some plans of a similar nature of other installations, for example: (1) your own home as it is wired, (2) your own home as you would like to see it wired, probably with many more appliances that you do not have now, and (3) an "ideal" home in which you would like to live. Practice in this way until you are familiar with using wiring symbols.

Often the electrician may be called upon to make the plans for a job, instead of finding plans ready-made. In that case include all those details found in plans of the general type shown in Figs. 12–8 and 12–9. Include in the plans such things as size and location of service-entrance wires and service equipment, number of circuits, materials to be used, and similar details.

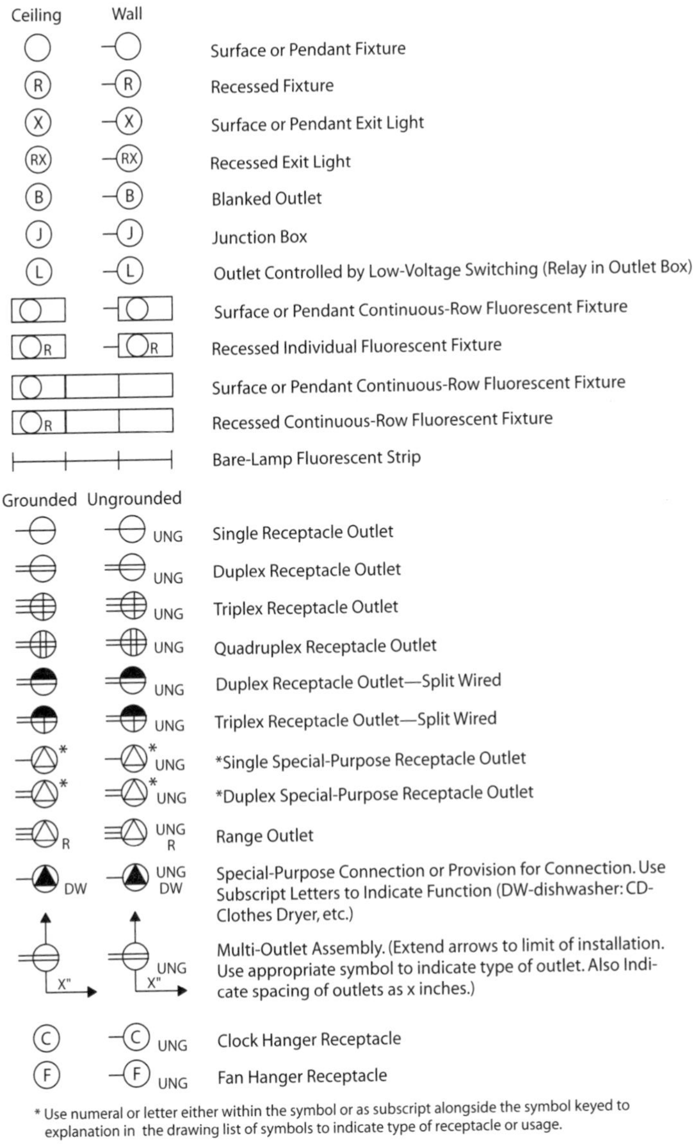

Ceiling	Wall	
◯	─◯	Surface or Pendant Fixture
Ⓡ	─Ⓡ	Recessed Fixture
Ⓧ	─Ⓧ	Surface or Pendant Exit Light
ⓇⓍ	─ⓇⓍ	Recessed Exit Light
Ⓑ	─Ⓑ	Blanked Outlet
Ⓙ	─Ⓙ	Junction Box
Ⓛ	─Ⓛ	Outlet Controlled by Low-Voltage Switching (Relay in Outlet Box)
▭Ⓞ▭	─▭Ⓞ▭	Surface or Pendant Continuous-Row Fluorescent Fixture
▭⒪R	─▭⒪R	Recessed Individual Fluorescent Fixture
▭Ⓞ▭▭		Surface or Pendant Continuous-Row Fluorescent Fixture
▭⒪R▭		Recessed Continuous-Row Fluorescent Fixture
├──┼──┼──┤		Bare-Lamp Fluorescent Strip

Grounded Ungrounded

─⊖	─⊖ UNG	Single Receptacle Outlet
═⊖	═⊖ UNG	Duplex Receptacle Outlet
⊞	═⊕ UNG	Triplex Receptacle Outlet
⊞	═⊕ UNG	Quadruplex Receptacle Outlet
═◒	═◒ UNG	Duplex Receptacle Outlet—Split Wired
⊟	═⊕ UNG	Triplex Receptacle Outlet—Split Wired
─△*	─△* UNG	*Single Special-Purpose Receptacle Outlet
═△*	═△* UNG	*Duplex Special-Purpose Receptacle Outlet
═△R	═△ UNG R	Range Outlet
─▲ DW	─▲ UNG DW	Special-Purpose Connection or Provision for Connection. Use Subscript Letters to Indicate Function (DW-dishwasher; CD-Clothes Dryer, etc.)
↑⊖ ⌐X"→	↑⊖ UNG ⌐X"→	Multi-Outlet Assembly. (Extend arrows to limit of installation. Use appropriate symbol to indicate type of outlet. Also Indicate spacing of outlets as x inches.)
Ⓒ	─Ⓒ UNG	Clock Hanger Receptacle
Ⓕ	─Ⓕ UNG	Fan Hanger Receptacle

* Use numeral or letter either within the symbol or as subscript alongside the symbol keyed to
  explanation in the drawing list of symbols to indicate type of receptacle or usage.

Fig. 12–6 Graphic symbols used in architectural drawings to designate electric outlets.

It is your job to make sure the installation you are planning will meet *NEC* requirements. The average homeowners know little about such things, so their suggestions and ideas might result in an installation that is far from adequate, and possibly not even safe. Normally you will have no choice about meeting *NEC* and local code minimums anyway, due to local enforcement of applicable law and

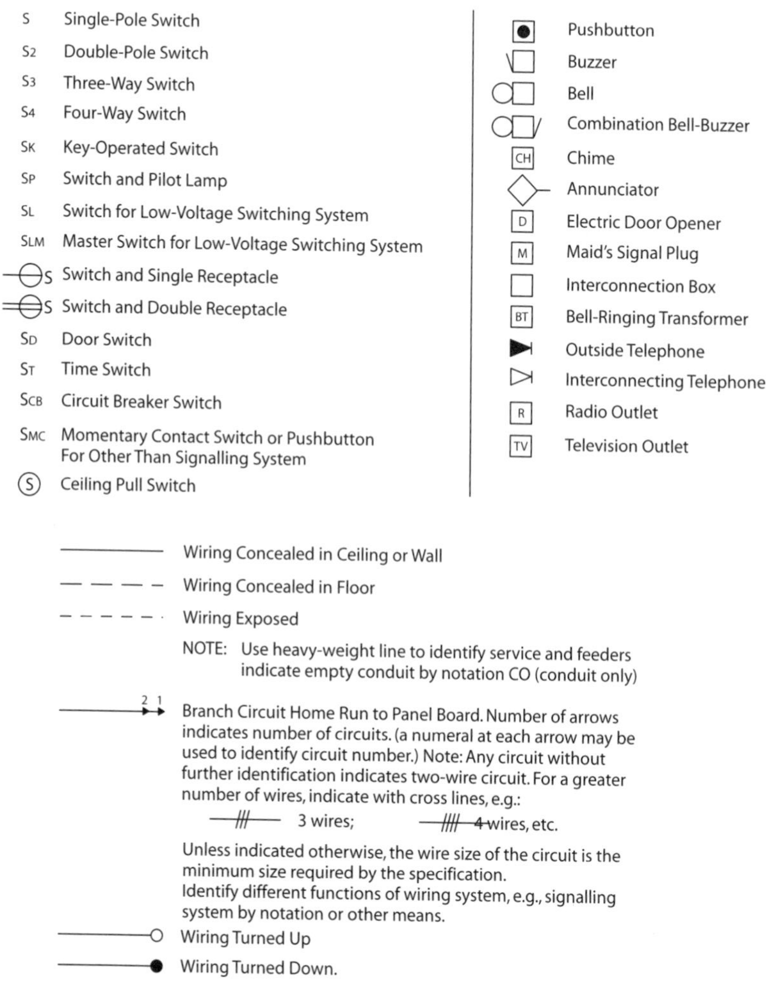

S	Single-Pole Switch		⬛	Pushbutton
S2	Double-Pole Switch			Buzzer
S3	Three-Way Switch			Bell
S4	Four-Way Switch			Combination Bell-Buzzer
SK	Key-Operated Switch		CH	Chime
SP	Switch and Pilot Lamp		◇	Annunciator
SL	Switch for Low-Voltage Switching System		D	Electric Door Opener
SLM	Master Switch for Low-Voltage Switching System		M	Maid's Signal Plug
—⊖S	Switch and Single Receptacle			Interconnection Box
═⊖S	Switch and Double Receptacle		BT	Bell-Ringing Transformer
SD	Door Switch		▶	Outside Telephone
ST	Time Switch		▷	Interconnecting Telephone
SCB	Circuit Breaker Switch		R	Radio Outlet
SMC	Momentary Contact Switch or Pushbutton For Other Than Signalling System		TV	Television Outlet
Ⓢ	Ceiling Pull Switch			

—————————— Wiring Concealed in Ceiling or Wall

— — — — Wiring Concealed in Floor

– – – – – Wiring Exposed

NOTE: Use heavy-weight line to identify service and feeders indicate empty conduit by notation CO (conduit only)

————————↠ 2 1 Branch Circuit Home Run to Panel Board. Number of arrows indicates number of circuits. (a numeral at each arrow may be used to identify circuit number.) Note: Any circuit without further identification indicates two-wire circuit. For a greater number of wires, indicate with cross lines, e.g.:

—⫫— 3 wires;   —⫫—4 wires, etc.

Unless indicated otherwise, the wire size of the circuit is the minimum size required by the specification.
Identify different functions of wiring system, e.g., signalling system by notation or other means.

————————○ Wiring Turned Up

————————● Wiring Turned Down.

Fig. 12–7 More symbols. All are from American National Standards Institute (ANSI) "Standard Y-32.8–1972."

licensing regulations. Explain to your customers the benefits of an adequate installation; it will mean a larger sale for you and greater satisfaction for your customers. If you are bidding competitively, consider a code-minimum bid that lets the customer compare apples with apples, and a second bid with additional design features that the owner would appreciate, or offer a list of optional additions with the first bid.

Fig. 12–8 Wiring diagram of first and second levels of split-level house.

Fig. 12–9 Wiring diagram for basement of house shown in Fig. 12–8.

CHAPTER 13

Residential Electrical Distribution

THIS CHAPTER DISCUSSES residential load calculations, which determine the number of branch circuits to install, the size of wire to use for each circuit, the size of the service-entrance wires, and similar details. The chapter also covers the basic type of service equipment that must be included. Selection and installation of field-wiring components are discussed in Chapter 16.

The *NEC* recognizes two types of branch circuits. One type is a circuit serving a single current-consuming appliance or similar load, such as a range or water heater. The other type is the ordinary circuit serving two or more outlets, consisting of permanently installed lighting fixtures or appliances, and receptacles for portable loads such as lamps, vacuum cleaners, and similar small appliances. Both types of circuits are covered in this chapter.

There is a discussion of 3-wire circuits at the beginning of Chapter 20. A 3-wire circuit is equivalent to two 2-wire circuits; it allows use of the same amount of power but with less material and with less voltage drop. When installing receptacles on a 3-wire circuit, it is practical to connect the upper halves of all the receptacles to the neutral and the black wire of the 3-wire (multiwire) circuit, and all the bottom halves to the neutral and the red wire. If protected by a GFCI, as required for all kitchen countertop outlets, the GFCI must be the special circuit-breaker type designed for multiwire circuits, and must have a 20-amp rating. Note that formerly in dwellings and now in all occupancies, any split-wired equipment supplied by a multiwire circuit must have means provided that opens both circuits simultaneously, and the two-pole GFCI breaker does exactly that.

In this chapter there are many references to overcurrent devices, which may be either circuit breakers or fuses. Since circuit breakers ("breakers") have largely supplanted fuses in residential applications, the references are mostly to breakers but apply equally to circuit breakers or fuses unless otherwise stated.

INSTALL ENOUGH BRANCH CIRCUITS FOR CONVENIENCE AND SAFETY

When groups of outlets are sensibly distributed on plenty of separate circuits, an entire building need never be in complete darkness on account of a tripped breaker except on rare occasions when a main breaker trips. Most of the time each breaker carries only a portion of its maximum capacity. But there are times during each day when many lights and other loads are put into service at the same time—a family might start the day using an iron, hair dryer, maybe even the washing machine at the same time as various cooking appliances. With a sufficient number of circuits in a properly designed installation, preferably with at least two circuits entering each room, the load will be fairly well divided among the various breakers with the result that none is overloaded and none trips. Where there are only a few circuits, each breaker will carry a heavier load and trip far more frequently. Breakers that trip frequently tempt the owner to change to a breaker larger than permitted for the size of wire used, or perhaps to use substitutes that defeat the purpose of overcurrent protection. Both are dangerous practices that can lead to fires. Another practical consideration is that the greater the number of circuits, the lower the voltage drop will be in each circuit, with less wasted power and greater efficiency—brighter lights, toasters that heat quicker, percolators that make coffee faster, and so on. Except for circuits restricted to serving a single outlet for special equipment as described in this chapter, the NEC places no limits on the number of outlets that may be connected on one residential branch circuit. But if you have provided the required number of circuits, you will have no need to overload any circuit supplying lighting and receptacle outlets.

Square footage determines number of circuits   NEC 210.11 and 220.12 specify the minimum number of circuits that may be installed. The note to 220.12 advises: "The unit values herein are based on minimum load conditions . . . and may not provide enough capacity for the installation contemplated." The starting point in determining the minimum number of circuits is to calculate the floor area according to the guidelines in NEC 220.12:

> The floor area for each floor shall be computed from the outside dimensions of the building, dwelling unit, or other area involved. For dwelling units, the computed floor area shall not include open porches, garages, or unused or unfinished spaces not adaptable for future use.

Note the description "not adaptable for future use." Many houses are being built with unfinished spaces that are not at first used for living purposes, but are intended to be completed later by the owners at their discretion. Too often this future use of space is disregarded in planning the original electrical installation. One outlet might be installed, and when the space is later finished many more outlets are added, branching off from the one outlet originally installed. That overloads the existing circuit. Instead run a separate circuit to the unfinished space.

Requirements for including basement space in minimum circuit calculations are somewhat vague. If the basement space is to be used only for ordinary household utility purposes, it can be safely disregarded in your volt-amperes per square foot calculations for lighting circuits. But if any part of a basement can be finished off into living space, such as for a recreation area, add that area to the total computed floor area. Do the same for any unfinished upper floor areas that can be finished.

How many lighting circuits? NEC Table 220.12 specifies a minimum of 3 volt-amperes per sq ft of floor area in determining the minimum number of lighting branch circuits. A lighting branch circuit includes not only permanently installed lighting fixtures, but also receptacle outlets for floor and table lamps, radios, TVs, computers, vacuum cleaners, and similar portable appliances, all of which consume comparatively small amounts of power. But this does not include larger appliances such as kitchen appliances and larger room air conditioners.

A house that is 25 × 36 ft has an area of 900 sq ft per floor, or 1800 sq ft for two floors. Assume that it has a space for a future finished room in the basement, 12½ × 16 ft, or 200 sq ft. This makes a total area of 2000 sq ft and will require a minimum of 3 × 2000 or 6000 volt-amperes for the lighting circuits.

The usual lighting circuit is wired with 14 AWG wire that has an ampacity of 15 amps, which at 120 volts is 1800 volt-amperes (15 × 120 = 1800). For 6000 volt-amperes, 6000 divided by 1800, or 3.3 lighting circuits would be required. That is more than three circuits, so four circuits must be provided because circuits must not be overloaded. Divide the loads as evenly as practical among the four circuits.

You can reach the same answer another way. Since each circuit can carry 1800 volt-amperes and 3 volt-amperes must be provided for each square foot, each circuit can therefore serve 1800 divided by 3, or 600 sq ft. An area of 2000 sq ft would require 2000 divided by 600, or 3.3 circuits as before.

Remember that the NEC is concerned primarily with safety, not with convenience or adequacy. Many people feel that one 15-amp lighting circuit should be provided for every 500 sq ft of area rather than every 600 sq ft. In the example of the preceding paragraphs, 2000 divided by 500 would lead to an answer of exactly four circuits, the answer we already reached. If instead the area had been 2200 sq ft, the answer on the basis of 600 sq ft per circuit would be four lighting circuits, but on the basis of 500 sq ft it would be five circuits. The extra circuit would provide for more flexibility and allow some provision for future needs.

How many small-appliance circuits? Many years ago kitchen appliances consisted of a toaster and an iron, each consuming about 600 watts (for these resistive loads, watts and volt-amperes are the same). They were plugged into ordinary lighting circuits and were rarely both used at the same time. A modern kitchen is equipped with many additional appliances: coffeemaker, mixer, blender, food processor, roaster, frying pan, deep-fat fryer, bread machine, microwave oven, toaster oven, and more. Even the smallest kitchens seem to have many of the same appliances, just more crowded together in the smaller area. Many of today's small appliances consume more than 1000 volt-amperes, and often several are used at the same time. Ordinary lighting circuits cannot handle such loads.

The *NEC* requires special circuits called "small-appliance branch circuits" for such appliances. *NEC* 210.11(C)(1) and 210.52(B)(1) require two or more 20-amp small-appliance branch circuits for all small-appliance loads including refrigeration equipment in the kitchen, pantry, breakfast room, dining room, or similar area of dwelling occupancies. These small-appliance circuits must not have other outlets. An exception does permit a clock-hanger outlet as discussed in Chapter 12 (see Fig. 12–2) and the electronic ignition, lights, and controls of gas ranges to be connected to one of the two small-appliance circuits; no lighting outlets may be connected to them.

Both these special circuits must serve the kitchen so that some of the countertop receptacles in the kitchen are connected to each of the circuits, thus increasing the number of kitchen appliances that can be operated at one time. Either (or both) of the circuits may extend to the other rooms mentioned, and to other kitchen receptacles. Other than the exceptions noted, the *NEC* reserves the full 20-amp capacity in both circuits to support portable appliance loads; the circuits may supply *no other outlets.* Do not connect a receptacle for a permanently installed microwave oven that's located in a cabinet over the cooktop (although a portable microwave oven may be plugged into any kitchen outlet). Do not connect a receptacle in a sink base for a kitchen waste disposer, and do not connect a range hood exhaust fan to one of these circuits either.

The *NEC* sets no limit on the number of small-appliance branch circuits other than the minimum of two. However, since only small-appliance branch circuits may supply the specified receptacle outlets, it follows that every circuit supplying such outlets must be a 20-amp circuit supplying nothing else. The owner is entitled to believe that every receptacle in his/her kitchen or dining room (other than a switched receptacle as covered in Chapter 12) has 20 amps of capacity behind it undiminished by other than small-appliance loads. If you bring three or four circuits to the kitchen counters, more than the two-circuit minimum, you still can't use these circuits to connect other loads such as the range hood. Keep track of the number of circuits you choose to install (and larger kitchens should have more than two) because each of these circuits has to be counted in the total load calculation as covered later in this chapter. A special exception allows refrigeration equipment to be on an "individual branch circuit rated 15 amperes or greater." This allows a cost effective way to add capacity in the kitchen—by running a 15-amp (14 AWG wire) circuit to a single 15-amp receptacle for the refrigerator.

NEC 210.11(B) also requires one 20-amp circuit for the laundry area, where it may serve one or more receptacle outlets. Like the small-appliance branch circuit, it too may serve no other outlets and it must also be tracked in the total load calculation.

Note that on a 20-amp circuit, receptacles rated at either 15 or 20 amps may be installed, unless the 20-amp circuit is an individual branch circuit—a circuit supplying only a single receptacle; such a receptacle must have a 20-amp configuration. If you supply a duplex receptacle, then you are supplying two receptacles, which is not an individual branch circuit even though only a single outlet is involved. Also note that a plug designed specifically for a 20-amp receptacle will not fit a 15-amp receptacle, but the ordinary plug designed for 15-amp receptacles will fit the 20-amp receptacle.

How many bathroom circuits?   The *NEC* now recognizes the transitory but heavy loading imposed by today's hair dryers. *NEC* 210.11(C) sets a 20-amp minimum circuiting rule for bathroom receptacle outlets. This rule differs in key respects from the other 20-amp circuit rules. First, although the basic rule requires at least one 20-amp circuit to supply the bathroom receptacles collectively, with no other outlets, a crucial exception allows such a circuit to supply other load as long as the circuit is confined to a single bathroom. In other words, the *NEC* allows a trade-off of other load in a single bathroom for receptacle loads in multiple bathrooms. Since some ceiling vent fans have a listing restriction requiring GFCI protection, that trade-off makes them easy to wire because you can connect them to the load side of the GFCI receptacle. But don't forget that as soon as you supply any nonreceptacle load on a bathroom receptacle circuit, that circuit may no longer leave that bathroom. Finally, given the transitory nature of this load, you don't have to include any additional allowance in your load calculation for these bathroom circuits.

Branch circuits serving single outlets   It is customary to provide a separate circuit for each of the appliances listed below. Although some of these loads, if less than 50% of the branch circuit rating, might share their circuit with additional load, that seldom works out well in practice; it's better to provide separate circuits.

- Range (or two circuits for separate oven and counter unit)
- Water heater
- Clothes dryer
- Kitchen waste disposer
- Dishwasher
- Each permanently connected appliance rated at 1000 volt-amperes or more (example: a bathroom heater)
- Each permanently connected motor rated at $\frac{1}{2}$ hp or more, (example: motor on a water pump)
- Central heating equipment (other than fixed electric space-heating equipment), which may share that circuit with other loads directly related to that equipment, including zone valves, air cleaners, or circulators or other related pumps. *NEC* 422.12 mandates the separate circuit in these central heating applications, although the circuit can supply air conditioning equipment since it is a noncoincident load. Some oil burner regulations even

require the oil burner motor to have a suitable switch at the head of the basement stairs (or other highly visible location), identified by a red faceplate.

The circuit for the appliance may be either 120 or 240 volts depending on the voltage rating of the appliance. The wire used must have sufficient ampacity for the current rating of the load it is to serve. The ampere rating of the overcurrent device in the circuit depends on the ampere rating of the specific appliance or motor served. For a discussion of motor-operated appliances refer to Chapter 15, because the installation rules for these appliances essentially follow the rules for motor circuits, which differ in fundamental ways from other circuits. For continuous loads (seldom a factor in residential applications) refer to page 476. For a branch circuit serving a single appliance and nothing else, up to 20-amp overcurrent protection is acceptable for any appliance rated 13.3 amps or less. If the appliance consumes 13.4 amps or more, the overcurrent protection may be up to but not exceeding 150% of the ampere rating of the appliance. If that turns out to be a nonstandard size, the next larger standard size may be used. If, however, the appliance has a marked upper size limit on overcurrent protection, then you must observe that marking.

Sizes of branch circuits   For most locations, there are five sizes of circuits that serve two or more outlets: 15, 20, 30, 40, and 50 amps, based on the ampacities of 14, 12, 10, 8, and 6 AWG wire. Of course circuits 60 amps and larger may be installed but must serve only one outlet.

The ampere rating of a circuit depends on the rating of the breaker or fuse protecting the wire, not on the size of the wire in the circuit. While normally the rating of the breaker or fuse is matched to the ampacity of the wire in the circuit, this is not always the case. A 20-amp circuit is normally wired with 12 AWG wire with ampacity of 20, but if you used a 20-amp breaker or fuse and used 10 AWG wire (for example, for reduced voltage drop or for mechanical strength in an overhead run), it would still be a 20-amp circuit.

Fifteen-ampere branch circuits   This is the size of circuit usually used for ordinary lighting circuits, which include general-use receptacle outlets other than those in cooking, dining, and laundry areas. It is wired with 14 AWG wire and protected by a 15-amp breaker or fuse. The receptacles connected to it may be rated at no more than 15 amps, which means that only the ordinary household variety of receptacle may be used. Any type of socket for lighting may be connected to it. No cord- and plug-connected appliance used on the circuit may exceed 12 amps (1440 watts) in rating. (Appliance classifications are defined on page 378.) If the circuit serves lighting outlets and/or portable appliances, as is usually the case, as well as fixed (permanently connected) appliances, the total of all the fixed appliances may not exceed 7½ amps (900 VA).

Twenty-ampere branch circuits   The special small-appliance, bathroom, and laundry circuits described earlier in this chapter are required to be 20-amp circuits. Also, 20-amp circuits are permitted to be used for lighting and general-use receptacles if 12 AWG wire is used.

Any circuit wired with 12 AWG wire and protected by a 20-amp overcurrent device is a 20-amp circuit as defined in the *NEC*. It may serve any combination of lighting outlets, receptacles, and permanently connected appliances. Any kind of socket for lighting may be connected to it. Receptacles may have either a 15- or 20-amp rating. No single portable appliance may exceed 16 amps (1920 VA) in rating. If fixed (permanently connected) appliances are also on the circuit, the total of all such fixed appliances may not exceed 10 amps (1200 VA).

Thirty-, forty-, and fifty-ampere branch circuits   Circuits of these ratings, each serving a single outlet for a range, clothes dryer, or similar appliance, are commonly used. (The appliance circuit discussion begins on page 382.) However, such circuits serving lighting outlets are not permitted in dwellings, and they are rarely used in dwellings to serve more than one outlet for appliances. But such circuits are commonly used in nonresidential wiring; forty- and fifty-ampere branch circuits are discussed on page 555.

Balancing circuits   In a house served by a 120/240-volt service (which of course contains a neutral wire), the white wire of each 120-volt branch circuit is connected to the incoming neutral at the neutral busbar in the service equipment. Great care must be used in connecting the hot wires of the branch circuits so they are divided approximately equally between the two incoming hot wires. If this is not done, practically all the load may be thrown on only one of the two hot wires; unbalanced conditions lead to frequent tripping of breakers or blowing of fuses.

Location of branch-circuit overcurrent protection   The service equipment, which consists of the service disconnecting means and service overcurrent devices—circuit breaker(s) or fuses on pullout block(s)—is required by *NEC* 230.70(A) to be located near the point where the wires enter the building. In most houses, the cabinet of the service equipment also contains the branch-circuit breakers or fuses.

In many houses, most of the power is consumed in the kitchen, while others have large appliances such as a water heater and clothes dryer in the basement, but in either case there are only relatively small loads in the remainder of the house. For minimum cost, plan your layout so the wires enter the house at a point from which the relatively large wires to the range, water heater, clothes dryer, and kitchen appliances can be short; let the smaller, less expensive wires to the remainder of the building be the long ones.

In houses with basements, the service equipment is usually located in the basement. For houses built slab-on-grade, the service equipment often ends up on the wall of an attached garage, or at some other convenient point. Sometimes it is located in the kitchen where it is more accessible, though in an adequately wired house immediate accessibility becomes less important because breakers will rarely trip. However, it is good design to put the panel, whenever possible, in a location that allows future access to the outside of the panelboard enclosure if circuits need to be added. Don't confuse this design principle with the *NEC* rule requiring "ready access" as discussed on page 58. If such a location is physically impossible, at least consider running a spare raceway out of the panel and extending to a box in an accessible location such as an attic space. The branch-circuit panelboard must also be located to provide adequate workspace and to prevent the possibility of the installation being compromised by leaking fluids from a failed piping system. Adequate workspace is discussed in detail on page 298.

Branch-circuit schemes   The typical scheme of locating all branch-circuit breakers (or fuses) at one point is shown in Fig. 13–1. As far as protection is concerned, the wires beyond the branch-circuit breakers may be as long as you wish, consistent with keeping voltage drop to a reasonable level. Note that 240-volt circuits may be run from any point by simply tapping off the two ungrounded wires. For example, the ungrounded wire of circuit no. 1 and the ungrounded wire of circuit no. 2 together would make one 240-volt circuit, the grounded wire of course being disregarded. One such 240-volt circuit is shown in dashed lines.

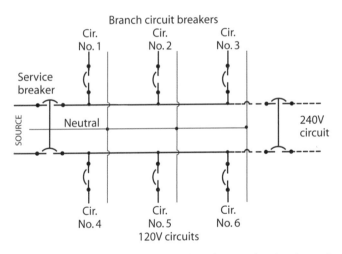

Fig. 13–1 The most common scheme: Locate all branch-circuit breakers in one location.

Occasionally, in a very large house the owner may prefer to locate the branch-circuit breakers in various locations throughout the house—the breakers protecting the basement circuits, for example, being placed in the basement, those protecting the first floor placed in the kitchen, and so on, as shown in Fig. 13–2. Where the service equipment and the branch-circuit fuses are in the same cabinet, as would be the case in an installation made in accordance with

Fig. 13–1, the breakers (other than the service or main breakers) are branch-circuit breakers, and the wires beyond these breakers are branch-circuit wires. If the installation is made in accordance with Fig. 13–2, the same wires (but now larger and running to another "panelboard") become feeders, and the multipole breaker ahead of the feeder wires and the downstream panelboard (but now of a higher ampere rating) becomes the feeder overcurrent protective device.

In Fig. 13–2, the wires from the service equipment to each location where a panelboard is installed are feeders, as already explained. The wires of any feeder must be big enough to carry the maximum load that will be imposed on the feeder by the branch circuits at any one time. The ampere rating of the overcurrent device for the feeder must not be greater than the ampacity of the feeder, subject to all the complicating factors fully discussed on page 105 under "How to select a conductor using NEC Table 310.16" and the paragraphs following here. (See Appendix in this book for full NEC Table 310.16.) The NEC in 215.2(A)(2) requires that the ampacity of the feeder must not be lower than the ampacity of the service-entrance wires if the feeder carries all the current supplied by service-entrance conductors that have an ampacity of 55 amps or less.

NEC 215.2(A)(3) correlates these provisions with 310.15(B)(6). For individual dwelling unit feeders, whether they are within a house or run to an apartment in

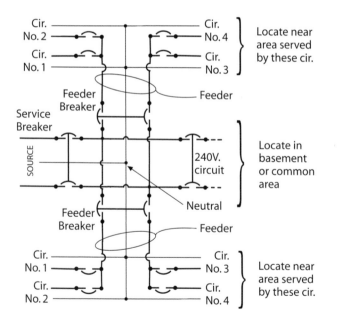

Fig. 13–2 In very large houses, and very commonly in nonresidential installations, feeders distribute power to groups of branch-circuit breakers near points of load concentration.

a multifamily dwelling, the *NEC* liberalizes its ampacity rules significantly. For example, 2 AWG aluminum (assuming 75°C terminations; refer to termination temperature discussion on page 106) is normally a 90-amp wire. However for a dwelling unit feeder it is a 100-amp wire. For that reason, 2 AWG aluminum XHHW service entrance cable is widely used for residential 100-amp services— but only on true single-phase, three-wire distributions. If, for example, a three-wire feeder to an apartment originates in a three-phase, four-wire service, you would have to increase that 100-amp feeder to 1 AWG aluminum. The relaxation of the ampacity rules discussed in this paragraph applies only to dwelling units. Don't attempt to use it in a commercial setting.

If the scheme of Fig. 13–2 is used, an important point relates to the grounding busbar. The feeder panels at the end of feeders downstream from the service equipment panelboard will of course contain a grounded busbar to which all the white grounded wires of the branch circuits will be connected. In the service equipment, this busbar must be bonded to the cabinet of the panelboard. In a branch-circuit panelboard downstream from the service equipment this not only is not required, it is prohibited. To repeat: the grounded busbar in the service equipment *must* be bonded to the cabinet; in a downstream panelboard it must be *insulated* from the cabinet. If you are using nonmetallic-sheathed cable, with a bare equipment grounding wire in addition to the insulated wires, the grounded conductor busbar (which you have just carefully kept insulated from the cabinet) is used only as a connection point for all the grounded (white) wires of the branch circuits. The bare equipment grounding wires of all the cables must be connected to a separate grounding busbar, and that busbar *must be bonded* to the cabinet.

SIZE THE SERVICE FOR PRESENT AND FUTURE NEEDS

The service must always be large enough for the connected load. In some cases the *NEC* also imposes absolute minimums: for one circuit, such as to a stand-alone telephone booth, 15 amps; for not more than two 2-wire branch circuits, 30 amps; for other nonresidential applications, 60 amps; for a one-family dwelling, 100 amps. Don't confuse service size with feeder size. In a multifamily dwelling with gas heat and cooking, a 30-amp feeder to each dwelling unit might easily supply the computed load. The 100-amp rule applies only to the actual service (utility interface as discussed at the beginning of Chapter 4) for a one-family dwelling.

For houses specifically, even if a 100-amp service may be permissible, remember that the minimum specified by the *NEC* is based on safety, not practicability or convenience. Look ahead—instead of the acceptable 100-amp minimum, install 150- or 200-amp equipment. Some local codes already require a service larger than the 100-amp minimum.

Selecting service wire size   As explained in Chapter 7, the ampacity of a given size of wire depends not only on its size but also on its insulation, and how the wire is installed. The ampacity of each size and kind of wire can be found in *NEC* Table 310.16, which is reproduced in the Appendix of this book. In addition, for dwelling units connected to single-phase distributions, don't forget to use the special ampacity table in *NEC* 310.15(B)(6) when the entire dwelling unit load is connected to a feeder or service conductor. The reason for this special table is explained in the ampacity allowance discussion on page 296.

Service switches for fused equipment are rated at 30, 60, 100, 200, and 400 amps (and of course higher ratings for larger buildings) with no in-between ratings. When using fused service equipment of a size larger than the minimum required (because a switch rated at the specific minimum is not available), it is not contrary to the *NEC* to use service wires with a smaller ampacity rating than the ampere rating of the switch, provided the fuses in the switch do not have a higher ampere rating than the ampacity of the wires. But if you are using a wire that has an ampacity for which there is no standard fuse of that ampere rating, you may use a fuse of the next larger standard rating, as covered in more detail in the page 105 discussion on adjusting *NEC* Table 310.16. The same principle holds for circuit breakers.

Applying demand factors in calculating service load   The service equipment and conductors do not need to have an ampere rating equal to the total ampere ratings of all the individual branch circuits. There never will be a time when every circuit in the house is loaded to its maximum capacity. Therefore "demand factors" have been established, based on many tests and past experiences, that represent the maximum part of various types of loads that are likely to be in use at any one time. These demand factors and methods of computing loads are covered by *NEC* Article 220. For residential occupancies, the loads are computed by specific type: lighting, small-appliance circuits, laundry circuit, special (usually heavy duty) fixed or stationary appliances, heating, central air conditioning, and similar loads.

Lighting, small-appliance, and laundry circuit loads   Compute the lighting loads on the basis of 3 volt-amperes (VA) per sq ft. This will include the receptacles for lamps, radios, TVs, vacuum cleaners, etc. It makes no difference whether the lighting circuits are 15- or 20-amp circuits—they total the same. Temporarily do not consider any demand factor; how to do so is explained later.

For small-appliance and laundry circuits, if each such 20-amp circuit can provide 2400 VA, allow only 1500 VA for each circuit in your computations because all these branch circuits will not be fully loaded at the same time. Since a minimum of two small-appliance circuits and one laundry circuit is required by the *NEC*, allow 4500 watts in your computations. However, if you do install more than two small-appliance branch circuits (commonly done), be sure to apply the 1500 VA allowance for each circuit actually installed. A special exception excludes from this calculation the individual branch circuit for refrigeration equipment, if you choose to install it.

Now add together the volt-amperes computed for lighting and the 4500 volt-amperes of the preceding paragraph. Count the first 3000 VA of the total at 100%, but the remainder above 3000 VA at 35%. Examples are shown later in this chapter.

Special appliance loads   As used here, "special appliances" include all appliances not supplied by a lighting circuit, a small-appliance circuit, or the laundry circuit. Examples are ranges, range tops, ovens, water heaters, clothes dryers, kitchen waste disposers (and sometimes, but not always, room air conditioners). Each such appliance must be supplied by its own circuit serving no other loads. In your computations the load in watts stamped on the nameplate of the appliance must be used at its full volt-ampere rating, with several exceptions.

If there are four or more *fixed* appliances not including, or in addition to, any ranges, clothes dryers, space heating equipment or air-conditioning equipment, *and* if all are supplied by the same service-entrance conductors or the same feeder (almost always the case in a residence), then a demand factor of 75% may be applied to the total load ratings of these items. This rule never includes clothes dryers. They are calculated under a different section at a minimum of 5000 VA (or more if indicated on the actual nameplate). If there are multiple household dryers, as in a central laundry room in a multifamily dwelling, the *NEC* has a demand factor table for this calculation. The table applies only in a household context; it doesn't apply to the commercial dryers in a laundromat.

A special rule applies to load calculations involving certain major single-phase loads, including electric clothes dryers and ranges, when they are connected to a three-phase, four-wire system. The load must be calculated based on twice the maximum number of appliances connected between any given pair of phase legs. Apartment buildings frequently require this type of calculation, and it is covered in depth in Chapter 25.

A range is actually a multi-unit load assembled as a single appliance. The watts rating of the range is the total of the oven load, plus the total load of all burners turned to their highest setting. These heating elements are unlikely to all be in use at any one time. Therefore a load of only 8000 watts (8 kW) may be used for any range rated at not over 12 000 watts (12 kW). If the range is rated at over 12 000 watts, start with 8000 watts; then for each kilowatt (or fraction thereof) above 12 000 watts, add 400 watts. For example, a range might have a rating of 16 kW, which exceeds 12 kW by 4 kW. Add 400 watts for each of the extra kilowatts, or 1600 watts altogether; add that to the first 8000 watts for a total of 9600 watts.

For separate ovens and range tops, if one range top and not more than two ovens are in the same room and supplied by one circuit, add the total of their separate ratings, and proceed as with self-contained ranges discussed in the preceding paragraph. In all other cases, the nameplate rating of each item must be used separately, and no demand factor may be used. Installing a separate circuit for each range top and each oven is advantageous because it simplifies maintenance and replacement, and voltage drop will be lower. Note that ordinary portable hot plates, toaster ovens, and small roasters are not range tops or ovens. (The *NEC* term for a range top is "counter-mounted cooking unit," and for a separate oven is "wall-mounted oven.")

Other loads   There are no demand factors for other residential loads; use their full rating in watts in computing the total for the building. If there are loads that would never be used at the same time (for example, space heating and air conditioning), count only the larger load. For space heating, use the total watts rating of all the elements; if the circuits are arranged so not all the elements can be on at the same time, count only the maximum load that can be used at one time. Many times the equipment will consist of a motor, or a motor plus other equipment. If the equipment is rated in watts, use that figure divided by volts. If it is rated only in amperes, use that figure. For motors as such with horsepower ratings, ignore the nameplate and use the full-load current as given in the horsepower tables at the end of *NEC* Article 430. For motor-operated appliances, on the other hand, such as garage door openers, dishwashers, and kitchen waste disposers, ignore any horsepower ratings and use the nameplate full-load current.

Computing the service load for a specific house   Sample computations for two houses follow. These computations determine the size of the *ungrounded* wires in the service equipment. Often the grounded neutral wire in the service may be smaller than the ungrounded wires; that is discussed later in the chapter. Two methods of computing the service load are given. After studying the samples, practice computing the service loads for other houses. Start with: (1) the house in which you now live as it is now wired, (2) the same house as you would like it after adding more appliances, air conditioning, etc., and (3) the "ideal" house in which you would like to live. Keep in mind that these *NEC* computation methods will give you only the minimum size service that the *NEC* considers necessary for safety. You may want to use wires and other equipment larger than the minimum to allow for future loads.

Common method For a house with an area of 2000 sq ft (determined as outlined earlier for calculating the number of circuits) and assuming no major electrical appliances such as a water heater and range, the computations are as follows:

	Gross computed VA	Demand factor	Net computed VA
Lighting, 2000 sq ft at 3 VA	6000		
Small-appliance circuits (minimum)	3000		
Special laundry circuit	1500		
Total gross computed VA	10 500		
First 3000 VA		100%	3000
Remaining 7500 VA		35%	2625
Total net computed VA			5625

The ampere load is $5625/240$, or 23.4 amps. It might appear that 10 AWG wire, having an ampacity of 30, would be large enough for the service. But the *NEC* requires a minimum 100-amp service for a one-family house; therefore select wires with an ampacity of 100 or more.

Now consider a larger house in a semi-rural area. The house has its own independent water system and an oil-burning furnace. Appliances include an electric range rated at 12 kW, a permanently installed bathroom heater, a water heater rated at 3500 watts, a clothes dryer rated at 5000 watts, two ¼-hp motors (one for the oil burner and one for the furnace fan), and one ½-hp motor for the water pump. If the house has an area of 3000 sq ft, a minimum of 9000 VA must be included for lighting circuits. The computation is as follows:

	Gross computed VA	Demand factor	Net computed VA
Lighting, 3000 sq ft at 3 VA	9000		
Small-appliances circuits (minimum)	3000		
Special laundry circuit	1500		
Total gross computed VA	13 500		
First 3000 VA		100%	3000
Remaining 10 500 VA		35%	3675
Range		Minimum	8000
Fixed appliances:			
Bathroom heater	1500	100%	1500
Water heater	3500	100%	3500
Clothes dryer	5000	100%	5000
Total net computed VA, less motors			24 675

The ampere load, not including motors, is 24 675 divided by 240, or 102.8 amps. The ampere load of motors must be added separately. Use the figures from *NEC*

Table 430.248, which shows that the ¼-hp motors must be counted at 5.8 amps on each ungrounded leg, and the ½-hp 230-volt motor will consume 4.9 amps, all at 240 volts. The *NEC* requires that the largest motor must be included at 125% of its rated amperage, so for the ½-hp use 125% of 4.9, or approximately 6.1 amps. The total for the motors is 5.8 plus 6.1, or 11.9 amps. If the 115-volt motors had been of different sizes, you would need to keep track of the loading on each ungrounded leg and make the sizing decision based on the worst case. For the present example, add 11.9 amps to the 102.8 amps already determined for other loads, making a total of 114.7 amps. You should install a service of at least 125-amp and preferably 150-amp capacity; if you use a fused switch it will have to be rated at 200 amps (although you would be free to select smaller fuses). In any case the service-entrance wires will have to have an ampacity of at least 115; larger ones would be desirable, and required if you decide to build in additional service capacity to permit adding future loads.

Optional method   Instead of following the procedures already outlined, *NEC* 220.82 permits an optional method that many prefer. Include lighting at the usual 3 VA per sq ft, the small-appliance circuits at 1500 watts each, and the laundry circuit at the usual 1500 watts, but do not apply a demand factor to the total of these three items. List the range (or oven and counter units) at their full nameplate ratings instead of the arbitrary 8000 VA. Add all other loads at their actual volt-ampere ratings. Add all these loads together and then apply a demand factor: the first 10 000 volt-amperes at 100%, and the remainder at 40%. The computation for the 3000-sq-ft house just considered would be:

	Gross computed VA	Net computed VA
Lighting, 3000 sq ft at 3 VA	9000	
Small-appliances circuits (minimum)	3000	
Special laundry circuit	1500	
Range, nameplate rating	12 000	
Fixed appliances:		
Bathroom heater	1500	
Water heater	3500	
Clothes dryer	5000	
¼-hp motor, oil burner	700	
¼-hp motor, furnace blower	700	
½-hp motor, water pump (at 125%)	1470	
Total gross computed VA	38 370	
First 10 000 VA at 100% demand factor		10 000
Remaining 28 370 VA at 40% demand factor		11 348
Total net computed VA		21 348

Add the motors at their volt-ampere equivalent: the ¼-hp at 5.8 A × 120 V or approximately 700 VA each, and the ½-hp at approximately 1175 VA plus 25% as in the previous example, or 1470 VA. The total computed by this optional method is 21 348 watts, which at 240 volts is 89 amps, compared with 115 amps as computed in the usual way in the previous example. By this optional method the house qualifies for the minimum 100-amp service, but a larger service will do a better job of meeting the present and future needs of the occupants.

If a house has electric heat or air conditioning or both, take the result from the optional calculation just outlined (89 amps in this case) and, after considering the following six possibilities, add to it only the largest result based on the actual nameplate ratings:

■ 100% of the air conditioning load.

■ 100% of heat pump compressors, unless the controller prevents simultaneous operation with supplemental heating if present.

■ 65% of central electric heating equipment, including supplemental heating in heat pumps. The compressor need not be counted if the controller prevents operation at the same time as the supplemental heat.

■ 100% of electric thermal storage and similar systems that will run continuously (defined as at least 3 hours) at full load. Some solar heating designs, for example, circulate solar heated air over various thermal masses, which then give that heat back to the building at night. During extended cloudy periods, many systems have an electric backup that adds heat to the thermal mass. Some utilities meter this backup at reduced rates during off-peak hours in exchange for keeping it off line during periods of higher demand. When the heating system does run, it may run steadily for a considerable time. Although these systems are a form of central heating (which is normally given a 65% demand factor), the *NEC* prohibits giving them that classification, assuring they will be calculated at 100%.

■ 65% of electric space heating using fewer than four separately controlled units.

■ 40% of electric space heating using four or more separately controlled units, such as baseboard convector units.

Determining size of the grounded neutral in the service   First of all, there is no neutral in a two-wire service or two-wire branch circuit, and both the grounded and the ungrounded wires must be the same size. But in a three-wire, 120/240-volt service there *is* a neutral, grounded as explained in the grounded neutral wire discussion on page 131. In most cases the grounded neutral service wire may be smaller than the two ungrounded wires.

If a load operates at 240 volts, the neutral wire of the service carries none of the current consumed at 240 volts. A range (or a separate oven plus a counter unit) is a combination appliance operating sometimes at 120 volts and sometimes at

240 volts depending on where the burners are set between the highest and the lowest heats.

To determine the size of the grounded neutral in your service according to *NEC* 220.22, calculate as follows:

a. Load for ungrounded wires as discussed in preceding
   paragraphs ...._____VA

b. Total of all 240-volt loads plus the range (or separate
   oven and cooking units) and clothes dryer ...._____VA

c. Total of 70% of the range (or separate oven and cooking units)
   and clothes dryer load ...._____VA

d. Subtract *b* from *a*, then add *c* ...._____VA

e. Divide volt-amperes of *d* by 240, which results in the
   ampacity required for the grounded neutral ...._____amps

If the total of *e* is more than 200 amps, count only 70% of the amount over 200 amps. For example, if the total is 280 amps, to the first 200 amps add 70% of the remaining 80 amps, which is 56 amps ($0.7 \times 80 = 56$), for a total minimum ampacity of 256 amps. For residential work, this calculation usually governs the neutral wire size. However, keep in mind that the grounded neutral also returns fault currents to the power source and must never be smaller than the basic size of the grounding electrode conductor of the installation, as covered in *NEC* Table 250.66. This minimum sizing (per Table 250.66) rule is actually somewhat more complicated, especially on industrial jobs, but the Table 250.66 result should suffice as the lower limit on residential applications. Given the amount of load connected line-to-neutral, the limiting factor in these applications is usually the load requirement and not the fault-current return capability.

The methods outlined above are also correct for calculating the size of the neutral wire of a feeder in a larger installation, with one important difference. Unless the feeder neutral also may return fault current, which is only possible in the case of a regrounded feeder neutral serving second buildings (refer to the page 335 discussion on grounding requirements for outbuildings), you only need consider the connected load in determining the neutral size.

SERVICE EQUIPMENT MUST BE SUITABLE FOR ITS FUNCTION

There are times when it is necessary to disconnect parts or all of a wiring system, as in the case of fire or when work must be done on parts of the installation. As covered in Chapter 4, a device to disconnect the entire building from its source of supply must be provided at or near the point where the service wires enter the building. In addition, devices that disconnect each separate branch circuit must be provided. In order to protect against short circuits, ground faults, and overloads, overcurrent protection is required for the installation as a whole, as well as for the individual branch circuits or feeders. The equipment to disconnect the entire building and the overcurrent device to protect the entire installation are what the *NEC* calls "service equipment." The service equipment may contain up to six main circuit breakers or fused switches—these main breakers or switches have the combined function of disconnecting and protecting the installation as a whole. The devices to control and protect individual branch circuits are always (in residential and farm installations) included in the same cabinet as the main breakers or switches. The equipment used must be listed and must be marked as suitable for service equipment.

Solid neutral, or SN, is a term frequently used to describe service equipment as well as much other equipment. It means that the grounded neutral conductor is not switched or protected by an overcurrent device. *NEC* 240.22 and 404.2(B), respectively, prohibit an overcurrent device or a switch or breaker in the grounded wire unless the device is one that opens the grounded wire at the same time that it opens the ungrounded wires. Such devices seldom if ever are used in residential or farm wiring. Yet *NEC* 230.70 requires that means be provided for disconnecting *all* conductors, including the grounded neutral, from the service-entrance wires. Contradictory as this sounds, the solution is simple. The incoming grounded neutral wire is connected to the grounding busbar in the service equipment using a solderless connector, which allows it to be disconnected by removing a nut or bolt or cap screw, and that satisfies the requirements (see also *NEC* 230.75). Thus three-wire, 120/240-volt service equipment will contain only two circuit breaker positions (or two fuses).

Neutral (grounded) busbar—when to bond and when to insulate   Service equipment as purchased includes a "grounding busbar," which is a piece of copper, brass, or plated aluminum containing usually all the following: a solderless connector to which the incoming neutral wire is connected; one or more connectors for the neutral wires of 120/240-volt, three-wire branch circuits of large ampere rating such as for a range; one connector for the ground wire; and numerous solderless connectors or terminal screws for the grounded wires of all 120-volt circuits (and the bare grounding wires of nonmetallic-sheathed cable, if that wiring method is used). In equipment as purchased, this busbar may be insulated from the cabinet, but if the equipment is used as service equipment, the busbar must be bonded (grounded) to the cabinet. The bonding is easily done if it did not come this way. The busbar in some brands includes a heavy screw that you must tighten securely to bond it to the cabinet. In other brands, a short flexible metal strap is already bonded to the cabinet and you must connect the free end to one of the connectors on the grounding busbar. But note that while the grounding busbar must be bonded to a cabinet that is used as service equipment, it must not be bonded but rather must be insulated from a cabinet that is used downstream from the service equipment.

Lighting and appliance branch-circuit panelboards   A lighting and appliance branch-circuit panelboard is one that has more than 10% of its overcurrent devices protecting lighting and appliance branch circuits, defined as circuits of 30 amps or less and having a wire connected to the supply neutral. A two-pole breaker counts as two devices in this calculation. The *NEC* requires these panel-boards with small overcurrent devices to have individual protection (defined as not more than two protective devices) ahead of them. In addition, these panels must not include provisions for more than 42 circuit positions, and they must be designed to physically reject more than their designed capacity.

In general, the *NEC* does not insist that panelboards have specific levels of protection ahead of them; it only requires that the busing be adequate to carry the computed load. In effect the *NEC* relies on appropriately computed (and inspected) load analyses to protect panelboards from overload. However, due to historic concerns about smaller equipment being misapplied through frequent modifications, and frequently inadequate inspectional supervision over this work, the *NEC* imposed the additional restrictions in these cases. This category clearly includes nearly every residential panelboard.

Selecting circuit-breaker service equipment An individual circuit breaker looks much like a toggle switch. On overload it opens itself; reset it as shown in Fig. 6–9 after you have corrected the trouble that caused it to trip. (In some brands it is not necessary to force the handle beyond the OFF position before resetting.) Breaker-type service equipment may contain up to six separate main breakers, making it necessary to turn off up to six breakers to disconnect the entire building. Equipment containing only one main breaker is preferable.

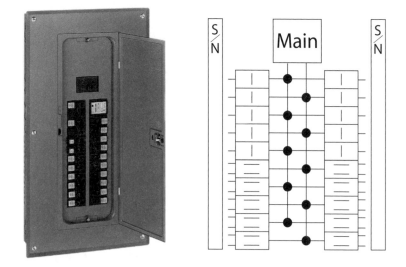

Fig. 13–3 A typical panelboard that could be used for service equipment. It has a main breaker and space for many branch-circuit breakers. The wiring diagram shows only single-pole breakers, but two-pole breakers may also be used. *(Square D Company)*

Service equipment as purchased usually includes the main breaker(s) plus an arrangement of busbars into which you plug as many breakers of the required ratings as needed to protect individual branch circuits. One single-pole breaker protects one 120-volt branch circuit; a two-pole (double-pole) breaker protects one 240-volt circuit (or a three-wire, 120/240-volt circuit such as for a range). A two-pole breaker has a single handle and occupies twice the space of a single-pole breaker. So-called "thin" or "piggyback" breakers are available that occupy only half the space of ordinary breakers; two of the small variety can be plugged into the space for one ordinary breaker. This is acceptable as long as the panelboard is designed for use with small breakers. The panelboard installation instructions will say if and where such breakers may be used within it. Do not attempt to override those instructions; you will end up trying to force the breaker past a rejection mechanism, which may result in physical damage to the panel as well as the circuit breaker. There are breakers available without the rejection feature, because sometimes you have to work in an old panel built before the *NEC* change. It is an *NEC* violation to use them in any other panel, and they are marked accordingly.

A typical breaker-type service equipment is shown in Fig. 13–3, together with its internal wiring diagram. Note that it contains one main breaker. Figure 13–4 shows breaker equipment of the "split bus" type (prohibited in a few localities).

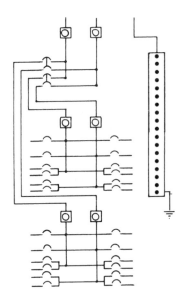

The top portion contains two main breakers, so that both breakers must be turned off to disconnect the entire installation. This panelboard qualifies for use as a lighting and appliance branch-circuit panelboard. (There are split-bus arrangements with six main breakers in the upper portion, one or two of which protect groups of breakers in the lower portion. These cannot be installed today, but a special exception allows those now in use to continue.)

In a few areas it is common practice to use an outdoor cabinet containing the meter socket and six 2-pole main breakers, as shown in Fig. 13–5. One of the breakers protects a panelboard inside the house containing all 15- and 20-amp branch-circuit breakers; the other five protect 240-volt circuits in the house. This panelboard must be wired so that it does not qualify as a lighting and appliance

Fig. 13–4 Diagram for typical "split bus" panelboard. Each of the two-pole breakers in the upper part protects a group of breakers in the lower part.

branch-circuit panelboard, or it will require individual protection ahead of it. Typical uses include having all of the branches rated above 30 amps, or restricting all breakers rated 30 amps and below to applications that don't require a neutral connection. In those cases (30 amps and below but no neutral) the *NEC* allows this type of arrangement without a main provided it constitutes the service equipment, recognizing that any modifications to service equipment are likely to be inspected.

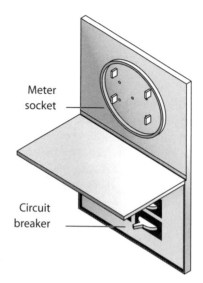

Selecting fused equipment   In very old houses you may find service equipment that consists of a main switch with two hinged blades and an external handle used to turn the entire load in the building on and off. The switch cabinet also contains two main fuses. Usually it also contains as many

Fig. 13–5 This combination of meter socket and six circuit breakers may be used as long as it does not qualify as a lighting and appliance branch-circuit panelboard.

smaller fuses as needed to protect all the individual branch circuits in the house; sometimes these fuses are in a separate cabinet. Today such switches are not used as service equipment.

Instead of a switch with an external handle, modern fused service equipment has two main fuses mounted on a pullout block as shown in Fig. 13–6. With the pullout in your hand, insert the fuses into their clips. One side of each fuse clip has long prongs; the entire pullout has four such prongs. The equipment in the cabinet has four narrow open slots, but no exposed live parts; the live parts are behind the insulation. Plug the pullout with its fuses into these slots; the four prongs on the pullout make contact with the live parts, completing the circuit. Plugging the pullout into its holder is the same as closing a switch with hinged blades; removing it is the same as opening such a switch. You can insert the pullout upside down to leave the power turned off.

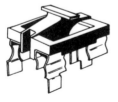

Fig. 13–6 In fused service equipment, cartridge fuses are installed in insulating blocks. When such a pullout block is removed from its holder, it has the function of a switch.

Besides the main pullout with large fuses protecting the entire load, the service equipment contains additional pullouts to protect 240-volt circuits with large loads such as ranges, plus it contains as many fuseholders for plug fuses as needed to protect individual branch circuits—one for each 120-volt circuit, two for each 240-volt circuit. A switch of this type is shown in Fig. 13–7. The wiring diagram of the switch is the same as shown in Fig. 13–3, except that fuses are used in place of breakers.

Fig. 13–7 Typical fused service equipment. It has a 100-amp main pullout, plus additional cartridge fuses on pullout blocks, and fuseholders for plug fuses to protect the branch circuits.

For fuses of the cartridge type, the 60-amp holders accept fuses rated from 35 to 60 amps. The 100-amp holders accept fuses rated from 70 to 100 amps. The 200-amp holders accept fuses rated from 110 to 200 amps. Holders for plug fuses accept fuses up to the 30-amp size, the largest made in the plug type. Thus you can always install the proper fuse to match the ampacity of the wires you are using.

The equipment shown in Fig. 13–7 is almost never installed in new installations today, having been essentially replaced in the market by circuit breakers (Fig. 6–9). However, you will often find it in existing installations and you need to be familiar with its parts.

CHAPTER 14

Residential Lighting

THE LIGHTING IN A HOME can help the occupants perform tasks more easily, feel safer and more comfortable, and enjoy the home to its full potential. This chapter offers lighting fundamentals and provides guidance in the selection, installation, and use of suitable lighting fixtures and the lamps that go with them. (A complete "light bulb" is properly called a lamp; only the glass part of the lamp is the bulb.) Note that the international term for lighting fixture is *luminaire,* as used in recent editions of the *NEC,* but this book will continue with the old terminology until the new wording is in more common use. This chapter focuses on residential applications. The more sophisticated lighting systems in commercial occupancies, including various types of high-intensity discharge (HID) lighting, are covered in Chapter 29.

MEASURING LIGHT—HOW MUCH DO YOU NEED?

Good lighting contributes to personal comfort, reduces fatigue, supports task efficiency, and promotes safety by preventing accidents caused by poor visibility. Becoming familiar with general standards for lighting levels and with types of lighting equipment available will help you design an optimum lighting system. The following lighting terms will be useful in understanding basic technology of lamps and fixtures.

Candelas (candlepower) measure luminous intensity   The candela is the unit of luminous intensity of light emitted by a light source in a given direction. Also called candlepower, it describes the amount of light (lumens) in a unit solid angle, assuming a point source of light. As the light rays travel away from the source, this solid angle, also called the steradian, covers an increasingly larger area, but the angle remains the same, as seen in Fig. 14–1.

Lumens measure total light output of a source   The time rate or flow of light, technically referred to as "luminous flux," is measured in lumens. One lumen is the amount of light that falls on a 1-sq-ft surface area that is 1 ft equidistant from a source with an intensity of 1 candela, as shown in the left side of Fig 14–2. Technically, this drawing should show segments of a sphere as in Fig.

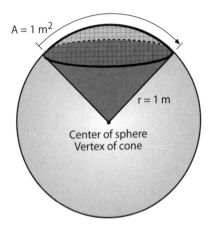

Fig. 14–1 A steradian illustrated. A steradian is, in effect, the flare angle of a cone that, when striking the surface of a sphere at any distance, cuts off an area (shown crosshatched in the figure) equal to the square of the distance. In the case of a projected beam of light, the number of steradians equals the projected area divided by the distance squared. A steradian covers about one-twelfth of the area of an entire sphere.

14–1, but a plane is easier to visualize. When you buy a lamp, you will often find the number of lumens it produces printed on the carton.

Footcandles measure degree of illumination for a unit surface
Perhaps the most practical of lighting measurements, the *illuminance* or degree of illumination at any point or surface is measured in *footcandles* (fc) in the English system, or *lux* (lumens per square meter) in the metric system.

To provide some starting point of known values in footcandles, bright sunlight on a clear day varies from 6000 to 10 000 fc. In the shade of a tree on the same day there might be roughly 1000 fc. On the same day the light coming in through a window on the shady side of a building will be around 100 fc, while 10 ft back it will drop to something between 7 and 15 fc. At a point 4 ft directly below a 100-watt lamp without a reflector, and backed by a black ceiling and walls, which have negligible reflecting power, there will be about 6 fc.

When 1 lumen of light falls on an area of exactly 1 square foot, that area is said to be lighted at 1 footcandle; 10 lumens falling on 1 square foot produces 10 footcandles. The illumination in footcandles does not vary with distance from the light source as long as the number of lumens of light falling on each square foot does not change. If all the light produced by a source delivering 1 lumen is focused on an area of 1 square foot, that area is illuminated to 1 footcandle no matter what the distance. In practice it is impossible to concentrate light to this degree because reflectors are not perfect; they absorb some light and allow some to spill in various directions. As a starting point, however, the relation can be considered correct: 1 lumen of light on 1 sq ft of area produces 1 fc.

To illustrate, assume that an area 12 ft × 12 ft is to be lighted to 15 fc. The total area is 144 sq ft, and that will require 144 lumens to provide 1 fc. Fifteen fc will require 144 × 15, or 2160 lumens. Common incandescent lamps run from 870 lumens in the 60 watt size to 1750 lumens in a 100 watt size, to 2880 lumens for 150 watt lamps, and 4010 lumens for a 200 watt lamp. Therefore you might expect that a 150 watt lamps would provide the required illumination (actually, about 20 fc). However, due to practical considerations including lack

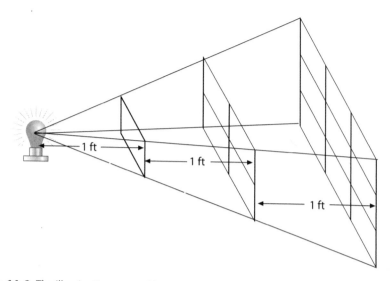

Fig. 14–2 The illumination on an object varies inversely as the square of the distance from the light source.

of reflectivity of walls and ceilings, the actual result would probably end up between 4 and 6 fc.

Luminance describes the brightness of what we see   Brightness and footcandles of illumination are not the same thing. A black sheet of paper illuminated to 10 fc will not seem as bright as a white one illuminated to 5 fc because the white paper reflects most of the light falling on it while the black absorbs most of it. Luminance is defined as intensity in a given direction divided by a projected area, as a person would see it. Commonly, the term is used to define how comfortable our vision is when looking directly at a luminaire, or fixture, or when looking at a surface. It is important to consider the direct luminance or brightness of a fixture when selecting and installing lighting equipment.

Law of inverse squares describes relationship between distance and illumination   The number of lumens of light produced by a light source is constant, other conditions remaining the same. A lamp produces the same number of lumens whether the observer is 1, 2, or 3 ft or any distance away. Nevertheless it is easier to read a newspaper, for example, when you are close to a light source rather than farther away.

Looking at Fig. 14–2, assume that the lamp produces enough lumens so that 10 lumens falls on an area of 1 sq ft when 1 ft away. If the illuminated area is moved to 2 ft away, those same 10 lumens now fall on an area of 4 sq ft, and at a distance of 3 ft they fall on 9 sq ft.

The area lighted was first 1 sq ft, then 4, then 9, yet the total amount of light remained the same: 10 lumens. Obviously it will be harder to read a label at the

3-ft distance than at the 1-ft distance because there is only one-ninth the illumination on the label. *The illumination of a surface varies inversely as the square of the distance from the light source.* This is the "law of inverse squares."

Due to ambient conditions, the inverse-square law seldom applies directly. The law is correct when all the light comes from one dimensionally small source, such as a small lamp, and further that none of the light from the source hits a surface and is then reflected back into the light meter. Such conditions rarely exist. Illuminance at any given location includes not only the light from some particular fixture under test but also light from other fixtures in the room and light that is reflected from the ceilings and walls. Generally the law of inverse squares can be confirmed by a footcandle meter only if the distance from light source to meter is at least five times the maximum dimension of the light source. Nevertheless the footcandle meter is a convenient device.

Lightmeters measure footcandles   Figure 14–3 shows a direct-reading footcandle meter, commonly known as a "lightmeter." Simply set the instrument at the point where the illumination is to be measured, and read the footcandles directly on the scale. The device consists of a photoelectric cell, which generates electricity when light falls on it. The indicating meter is simply a microammeter (a microampere is one-millionth of one ampere), which measures the current generated; the scale is calibrated to read in footcandles. The use of this instrument is invaluable, especially in commercial work, because it takes the guesswork out of many lighting problems.

Fig. 14–3 The lightmeter indicates footcandles (or lux) of light directly on its scale. *(Extech Instruments Corp.)*

Footcandle requirements for various jobs   There can be no absolute standard for lighting. The requirements for different individuals vary. Furthermore, what was considered appropriate yesterday may be considered unacceptable today, and what we accept today may not meet tomorrow's needs. All that can be given are today's commonly accepted standards. The more critical the task, the higher the level of illumination required. The more prolonged the task, the greater the amount of light needed. For example, in the relatively poor light of evening dusk you can quite easily read a paragraph of a newspaper, but it is almost impossible to read an entire page.

Continuing concerns about the availability and cost of energy have prompted reevaluations of just how much light is required in what locations. Over time, building code organizations and standards writing bodies have reduced general lighting requirements. For example, allow 10 fc in hallways, stairs, and spaces primarily devoted to conversation, relaxation, or entertainment; this is down from 20 fc considered standard several decades ago. However, be sure to provide adequate task illumination in specific areas. For example, in a living room provide table or floor lamps to increase the level to 30 to 50 fc when needed in areas where someone may be reading. For prolonged studying, 75 fc would be a wise minimum. Sewing requires from 50 to 200 fc; the finer the work and the darker the material, the more light is needed.

In the kitchen and laundry, consider 30 fc the minimum, with about 50 fc in areas where cooking, dishwashing, and ironing are done. Detailed food preparation areas may require as much as 150 fc. The dining room generally needs only about 15 fc if used exclusively for dining, but consider that many family dining rooms are also used for activities such as homework and hobbies. The bathroom should have 30 fc for general illumination and at least 75 fc at the mirror for makeup and shaving.

Photometry describes the light output of fixtures   This data includes candle-power distribution charts, isofootcandle charts and coefficient of utilization tables. Fixture manufacturers publish this information for lighting designers.

Candlepower (intensity) distribution curve   This curve, generally a polar graph, represents the variation of luminous intensity (candelas) of a lamp or fixture on a plane through the light center. If a fixture has a symmetrical distribution, such as an incandescent downlight unit, a single plane representation is sufficient.

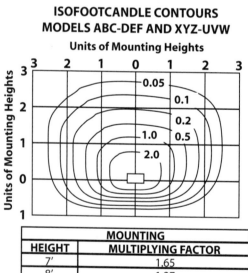

Isofootcandle chart (or diagram)   These charts are drawn with a series of concentric lines, where a single line represents a particular footcandle value of the light coming from the fixture. (See Fig. 14–4.) The light intensity is projected on a plane located at some angle to the source and the distance from the source is indicated. Appropriate multipliers can determine the footcandle values for other mounting dimensions.

ISOFOOTCANDLE CONTOURS
MODELS ABC-DEF AND XYZ-UVW
Units of Mounting Heights

MOUNTING	
HEIGHT	**MULTIPLYING FACTOR**
7'	1.65
8'	1.27
9'	1.00
10'	0.84
11'	0.74

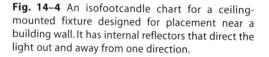

Fig. 14–4 An isofootcandle chart for a ceiling-mounted fixture designed for placement near a building wall. It has internal reflectors that direct the light out and away from one direction.

Coefficients of utilization   These numbers refer to the ratio of lumens that actually reach a work plane to the total lumens generated by the lamp/fixture combination, since some light is trapped inside a fixture, and is absorbed at other locations in the room.

WHAT QUALITY OF LIGHT DO YOU NEED?

The degree of illumination is not the only important factor in good lighting. The light must also be free from glare. Surface brightness must be controlled in order to prevent discomfort. Even the reflective qualities of ceilings and walls contribute to the quality of light in a room. And the color of light produced by different types of lamps can complement or detract from a room's furnishings. Understanding the qualities of light can help you make good choices in selection and placement of fixtures.

Glare—cause and prevention   Glare is caused by a relatively bright area within a relatively dark or poorly lighted area; glare is likely to cause annoyance, discomfort, eye fatigue, and interference with vision. A common cause of glare is exposed lamps placed to intrude on one's field of vision while one is trying to focus on objects or tasks. The glare of a single unshaded table lamp in an otherwise dim room would not be comfortable lighting for reading or other activities. Glare of an equally objectionable type can be caused by reflection, such as from polished metal surfaces or poorly placed mirrors.

Reading in direct sunlight is difficult. On the other hand, it is not difficult to read in the shade of a tree on a bright day, even though the footcandles are much lower than in direct sunlight. There is more to good lighting than footcandles of illumination. Direct sunlight comes from essentially a single point and causes glare. In the shade of the tree the light does not reach you from a single direction, but rather is diffused and therefore does not cause harsh shadows and glare. With careful attention to the design and placement of lighting fixtures in a room, you can avoid creating sources of glare.

Surface brightness—lower is better   When looking at an exposed 300-watt lamp of clear glass, you see a concentrated filament less than an inch in diameter. The diameter of the lamp itself is a little over 3 in., but because the filament is so bright, the glass bulb is barely visible. Most of the lamps we use are "frosted" or "soft white"—the filament is not seen, but rather the entire bulb. The same total amount of light, which in the case of the clear glass bulb was concentrated in an area of about 1 sq in., is now distributed over a much larger area of about 7 sq in. The total amount of light is the same, but the brightness of the larger area is greatly reduced—the "surface brightness" is lower. It is still uncomfortably bright, however, if looked at directly.

Put the lamp inside a globe of translucent (not transparent) glass about 8 in. in diameter; this has an apparent area of about 50 sq in. instead of the 7 sq in. of the frosted bulb. The total amount of light is still the same, but it is far more comfortable to the eye because the surface brightness of the light source has been reduced still more. Consider the most common fluorescent lamp: 32 watts and 48 in. long. The surface brightness is very low because the total surface area of the tube is very large for the amount of light emitted, and there are no bright spots. For comfortable lighting, use light sources that have low surface brightness.

Reflectance   Ceilings and walls in colors that reflect as much light as possible contribute to optimal lighting. A matte (not glossy) finish helps to avoid the mirror effect of glare-producing spots. Different surface colors reflect light in greatly varying degrees. The percentage of light reflected is called the "reflectance" of the surface. White, pale pink, pale yellow, or ivory-colored surfaces reflect 85 to 80%. Cream, buff, gray, or light blue reflect 70 to 50%. Darker colors reflect still less, down to 5 to 2% for black. Woods in natural finish rarely reflect as much as 50%, and sometimes as little as 5%.

Color of light   The color of an object appears different in natural light than it does in artificial light. Light produced by an incandescent lamp does not have the same proportion of colors that exists in natural light; it has more orange and red, less blue and green. There are special lamps that duplicate more closely the mixture and proportions of colors existing in natural light. Often described as "daylight" lamps, they approximate the light from a northern sky rather than direct sunlight. Skin tones are flattered under these lamps, but less light is produced per watt.

LAYERED LIGHTING AND SPECIAL EFFECTS

Today, residential lighting design calls for layered lighting. This concept recognizes that lighting serves three functions in addition to illuminating a general area: It can be used for calling attention to specific features, and for the obvious necessity of making it possible to focus on a specific task, and for some locations the fixtures themselves are chosen as much for their decorative value as for their practical function. A single fixture simply can't perform all these functions. A chandelier can be an artwork unto itself, and on a dimmer switch it may provide both dining and task illumination, but it would probably not spotlight an architectural detail. Under cabinet lighting in kitchens works well for task applications, but contributes nothing by way of decoration and provides little general lighting for the room. Furthermore, a single source of direct lighting can produce a sharp contrast between bright light and deep shadow; for some purposes this might lend an attractive sense of drama, but for reading and other household activities the contrast is annoying and tiring to the eyes. A minimum of two sources of light, for example a floor lamp and an overhead fixture, lead to greater comfort. Choose lamps or fixtures that provide diffused light. To apply the layering concept to a family room, for example, you might provide for ambient, track, and portable light to help facilitate the purposes of that room.

Direct and indirect lighting—include both kinds   Good lighting in homes is neither completely direct nor completely indirect. Most light sources should direct light on the areas below as well as reflecting some light off the ceiling. Direct lighting is produced by fixtures that aim all their light on the area to be lighted, with none reflected from ceiling or walls. Indirect light is produced by fixtures that throw all their light on the ceiling and walls, which reflect it to the areas to be lighted. Direct lighting can produce glare, sharp shadows, and uneven levels of light. Yet it can be quite effective in some situations, comparable to usage in stores where track or recessed fixtures in the ceiling might be used to light particular displays. This is an example of the principle of intensively lighting only that which must be lit.

Indirect lighting of all kinds is relatively inefficient, but it can be made quite effective with fixtures designed to distribute the light properly on a light-colored ceiling. The popular torchiere floor lamps illustrate the principle. Standing about 6 ft tall, they produce very little direct light. Most of the light goes up, preferably aimed at a white ceiling to reflect it back into the room.

Lighting design techniques   Lighting that creates special effects can raise the level of a lighting installation from ordinary to sophisticated. One well known technique, sometimes referred to as highlighting, uses a focused light source that will produce about five times the brightness of the surrounding surface in order to accent a painting or architectural detail. Another technique might be thought of as the inverse of highlighting: wall washing provides essentially even brightness on the vertical surface. Objects hanging in front of or attached to the wall tend to be seen as part of a whole. The lit wall attracts attention, and it provides quite a bit of ambient illumination in the process. There are several techniques for this.

- The first technique (sometimes referred to as point source) uses a row of recessed fixtures, usually incandescent, that are located closer together than their distance from the wall. The spacing between adjacent fixtures varies with the ceiling height and the desired surface light intensity.

- Instead of point sources, if the ceiling height is from 8 to 10 ft consider a diffused source using linear fluorescent lamps (or compact lamps in some cases).

- Instead of aiming the light at the wall, consider placing the fixtures close to the wall, either behind a valance or recessed, and aim them almost straight down so there are no hot spots on the wall, but rather a more uniform curtain of light as the beams graze the wall at a shallow angle. Either incandescent or fluorescent sources can be used in this case.

- If the wall is heavily textured, consider combining the point-source method with a placement at a grazing distance from the wall, that is, between 6 and 12 inches away. You can use recessed incandescent or even low-wattage HID fixtures (covered in Chapter 29) to achieve this effect.

When new construction or major renovations are planned, wall washing or accent fixtures can be structurally mounted within architectural features, such as cornices or box beams, that conceal the light source. Other approaches include backlighting and uplighting. Backlighting directs diffuse light through translucent panels made of a wide variety of materials including thin marble, acrylic, stained glass, and even fabric. The usual procedure involves mounting rows of linear fluorescent strips behind the translucent surface. Although we usually think of fixtures being mounted in ceilings for this work, you can also light a wall from below, using a floor or low surface placement, particularly if ceiling fixtures can't be used. The usual term for this is uplighting.

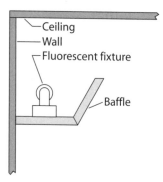

Fig. 14–5 Cove lighting does away with dark areas on the ceiling.

Most fixtures and combinations of fixtures that are used to light a room leave large parts of the ceiling relatively dark. Cove lighting is a technique for washing a ceiling in even light. Cove lighting consists of a series of single-tube fluorescent fixtures placed end to end within a cornice that extends around the room's perimeter near the ceiling. See Fig. 14–5. The cornice should be at least 12 in. below the ceiling, and in that case the center of the fluorescent tube should be about 2½ in. from the wall. As the distance below the ceiling increases, the distance from the wall should be increased somewhat, up to about 3½ in. for a distance of 24 in. below the ceiling. A reflector behind the tube, aimed at about 20 deg above the horizontal plane, produces a more even level of light on the ceiling. Note the baffle, which must be high enough to prevent the fluorescent tube from being seen from anywhere in the room, and low enough so there is "adequate space" to allow the tube, fixture, and ballast to be serviced as required.

Fixture styles and features Recessed lighting, using a variety of downlights, and track lighting are major elements in modern lighting design. Being aware of their many available features will help you use them effectively.

Downlights A straight downlight, or direct luminaire, uses an internal reflector to redirect, spread, or concentrate the light. Downlights with open reflectors are usually round and large enough to facilitate relamping and heat dissipation, with the diameter of the reflector bottom ranging from 6 to 9 inches. Downlights that use asymmetrical reflectors and lenses create a distinctive light beam pattern well suited for accent lighting as well as wall-washing applications.

The downlight's lens is used to widen or narrow the beam pattern. A baffle (either black grooves on the inner surface of the fixture, or a louver) will conceal a very bright light source from normal view. Choose a downlight equipped with a diffuser to reduce surface brightness.. A dropped diffuser, usually made of opalescent glass or comparable plastic, extends slightly below ceiling level to spread the light, reducing source brightness. In addition, a solid lens or diffuser can be used by a manufacturer to achieve a damp or a wet location listing for the fixture, allowing it to be used in shower stalls and in exterior porch ceilings. Solid-lens recessed fixtures also have a smaller required set-back from storage areas in clothes closets, making them more flexible for that application.

Track lighting   Use of track lighting began in commercial occupancies and has been used in dwellings for many years. A tremendous variety of fixtures can be positioned at any point along the track's length for washing walls with light and highlighting areas or objects of interest. The newer flexible tracks offer great design versatility. Track lighting is also available in smaller profile versions useful for such purposes as illuminating a kitchen counter from under the upper cabinets. The fittings that attach the fixtures to the track are designed by their manufacturer to mate precisely with their lighting track, and therefore cannot be used on a competing track. The *NEC* requires that track fittings be designed to preserve polarization and grounding in the process of suspension directly from the track.

Most track is rated 20 amps at 120 volts, although the smaller profile versions can be less. It is commonly available in two- or three-circuit multiwire versions as well, allowing independent control of different sets of fixtures over its length, or increased loading by representing multiple circuits over its length (covered in the discussion of application of three-wire circuits on page 375). It is available in ratings above 20 amps, but only as "heavy-duty lighting track" requiring every fixture to have supplementary overcurrent protection in its support fittings. Most track systems can be fed either in the middle or the end. Don't run it within 5 ft of the floor unless it has physical protection or operates at not over 30 volts. Track runs must be secured at each end and every 4 ft unless identified for support at different intervals.

HOW TO SELECT INCANDESCENT LAMPS

Incandescent lamps are efficient at only one thing: producing heat. About 12% (or less) of the input watts to this lamp is turned into visible light, the remainder being infrared light energy and nonradiative heat, making it a very inefficient light source. In addition, the waste heat increases the air-conditioning burden, which further decreases the overall energy efficiency of the total occupancy. There are some new designs that involve different gases within the bulb and slight differences in the filament, resulting in marginal improvements in efficiency. For example, you can find 67-watt bulbs that compare with conventional 75-watt bulbs. Tungsten-halogen lamps fall in the same category. They use incandescent filaments in pressurized quartz capsules containing certain elements that cause vaporized tungsten to redeposit on the filament instead of the capsule walls, allowing for higher temperature operation with whiter light quality and longer life, but very marginal gains in efficiency. The comparatively new TH HIR (halogen infrared-reflecting) lamp further improves efficiency by redirecting heat energy back to the filament by means of a coating on the quartz glass bulb surrounding the filament, boosting the light output. These are improvements, but the primary output of these lamps is still heat. They are not in the same league with different systems, such as fluorescent lighting, in terms of energy efficiency. That said, incandescent lamps are almost foolproof in their simplicity, and their light is aesthetically pleasing to many.

Incandescent lamps, however inefficient, are still widely used and need to be chosen correctly. Lamps differ in their life expectancy, light output, voltage ratings, and some even have specified base positions in use (such as the directive to "burn base down").

Voltage affects lamp life and light output   If an incandescent lamp is operated at lower than its designed voltage, its life is prolonged considerably but the watts, the lumens, and the lumens per watt drop off rapidly. If it is operated at a voltage above normal, its life is greatly reduced, although the watts, the lumens, and the lumens per watt increase. For lowest overall cost of illumination, operate lamps at the voltage for which they were designed.

As an example, an ordinary 100-watt lamp designed for use on a 120-volt circuit will, when burned at 120 volts, have a life of about 750 hours, produce 1750 lumens, and consume 100 watts, resulting in 17.5 lumens per watt. When burned at 108 volts (90% of 120 volts), its life will increase by about 400% or 3000 hours and produce 70% of 1750 or about 1190 lumens; it will consume 85% of normal or about 85 watts, resulting in 83% of normal efficiency or about 14.5 lumens per watt.

A lamp burned at its rated voltage represents the lowest total cost of light, considering the cost of the lamp itself and the cost of the power consumed during its normal life. A 100-watt lamp during its normal 750-hour life consumes 75 kWh of power, costing approximately eight times as much as the lamp itself.

Special locations   For some locations such as factories and commercial establishments it makes sense to use lamps with a life much longer than that of ordinary lamps. Some lamps are located in areas where changing them involves a significant maintenance cost, to the point of making lower efficiency in the context of greatly extended service life justifiable. Consider 130-volt lamps in such areas, particularly where some reduction in lumens is inconsequential. When burned at the usual 120 volts (92% of their rating) they last roughly 300% longer than a comparable 120-volt lamp, though at a considerable reduction in brightness. These lamps are readily available in electrical supply houses at prices little different from conventional lamps.

Lamp bases   There are various standardized kinds and sizes of lamp bases matched to the watts, the physical size, and the purpose of the lamp. In the screw-shell type base the largest is the mogul (1.555-in. diameter), used on 300-watt and larger lamps. Smaller bases are the medium (also called "Edison"), used

Candelabra
|0.465 in.|

Intermediate
|—0.651 in.—|

Medium
|——— 1.037 in. ———|

Mogul
|——————— 1.555 in.———————|

Fig. 14–6 The screw-shell bases used on lamps are standardized to the dimensions shown. The illustrations are actual size.

on ordinary household lamps, and the intermediate, and candelabra. All four are shown in actual size in Fig. 14–6.

Some lamps come with two separate filaments, and are designed to be used in a special base arranged so that either of the filaments or both filaments can be turned on. A lamp with a 100-watt filament and a 200-watt filament could produce 100, 200, or 300 watts as desired. Some consumer lamp packaging describes these as "three-way" lamps.

Lamp designations The mechanical size and shape of an incandescent lamp are designated by standardized abbreviations such as A-19, PS-35, F-15. The letter designates the shape in accordance with the outlines shown in Fig. 14–7. The numeral designates the diameter in eighths of an inch. Thus an A-19 lamp has a conventional bulb shape (the "A" stands for "arbitrary") and is $19/8$ or $2\frac{3}{8}$ in. in diameter.

Lumiline lamps—not for new installations These are tube-shaped incandescent lamps that have a special contact at each end. At one time they were in common use, but they are inefficient lamps, producing less than 10 lumens per watt. For that reason they are seldom used in new installations but are available for replacements in 30-, 40-, and 60-watt sizes.

Reflector lamps for floodlighting and spotlighting Special lamps called Type R or PAR have a silver or aluminum reflector deposited on the inside of the glass bulb, and are used for floodlighting or spotlighting. The most common uses for these lamps involve exterior locations such as walks and driveways. A thinner-walled interior version is frequently used in recessed or track lighting. The lamp's shape and interior coating drive most of the light out of the bulb in one direction. The tightness of the beam spread determines whether they are "spot" lights or "flood" lights.

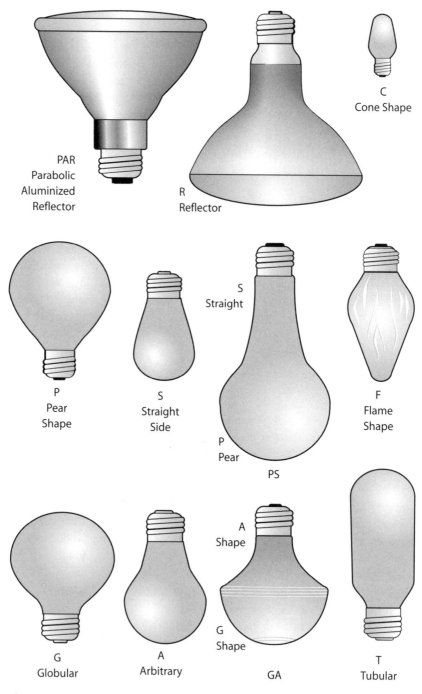

PAR
Parabolic
Aluminized
Reflector

R
Reflector

C
Cone Shape

P
Pear
Shape

S
Straight
Side

S
Straight

P
Pear
PS

F
Flame
Shape

G
Globular

A
Arbitrary

A
Shape

G
Shape
GA

T
Tubular

Fig. 14–7 The shape of lamps is well standardized.

Low-voltage incandescent lamps have specialized uses   Low-voltage (LV) incandescent lamps operate between 6V and 75V. With a more compact filament compared to a 120V lamp, the LV lamp provides a more precise beam control. Based on Ohm's law: If $E=IR$ (page 33), then as the voltage decreases the resistance can decrease as well, assuming comparable current to heat the filaments to the same degree. In practice, this means that LV filaments are shorter in length and thicker in cross section. This is why 12V vehicle headlights last as long as they do, even with constant vibration. The shorter length allows for a very focused beam.

There are three different lines of LV lamps operating at 12V: the MR, PAR, and AR lamps. Using a compact TH bulb and a two-pin base, the line of MR (miniature reflector) lamps, includes a variety of beam patterns for two different models, or sizes. The 2-in. diameter MR-16 lamp comes in 20, 42, 50 and 70W ratings. The 1 3/8-in. diameter MR-11 lamp comes in 20, 30 and 50W ratings. Both types have a variety of beam widths from narrow spot to narrow and wide flood.

All MR lamps have a heat-rejecting dichroic film coating on their reflectors, so that a portion of the infrared heat passes through the rear of the reflector. The MR-16 lamp is available with a front lens, which permits improved beam control while also providing a physical barrier for the tungsten-halogen capsule and the dichroic reflector. Both lamp types are widely used in track-mounted heads and recessed downlights to serve a variety of display and accent applications.

Using screw terminals, the PAR 36 and PAR 56 lamps have a filament shield that blocks direct view of the filament. This makes it possible to provide glare control for lamps that are exposed to view, such as track-mounted open heads. The metric-sized AR (aluminum reflector) lamp, which also has screw terminals, compares with the PAR 36 lamp in size and performance. Some of these lamps (the 37 and 56 mm sizes) have lenses available as an option, and some don't, including the 48, 70, and 111 mm sizes.

Low-voltage lighting systems have special *NEC* coverage in Article 411
When these systems operate at 30 volts or less, they often qualify for special treatment, including the option to use exposed bare conductors if run at least 7 ft above the floor (or lower if specifically listed for this duty). If the wiring is to be concealed, however, then Chapter 3 wiring methods must be used unless the power is so limited as to qualify as a Class 2 circuit (refer to the discussion beginning on page 364). They must be supplied from no more than a 20-amp branch circuit, and they must use a listed system. This means that you cannot assemble these systems piecemeal in the field; all the components must be listed by a qualified testing laboratory. Note that these systems are among the few in the *NEC* for which grounding one of the system conductors is expressly forbidden; they must be operated as ungrounded systems. This requirement dovetails with the requirement that these systems must only be supplied by an isolating transformer so there will be no connection (other than inductive coupling) with the 120-volt branch circuit on the primary side.

FLUORESCENT LIGHTING SAVES ENERGY

Fluorescent lighting was introduced commercially in 1938. Compared with an ordinary incandescent (filament-type) lamp, the operation of a fluorescent lamp is quite complicated and requires auxiliary equipment.

Tubular fluorescent lamps and ballasts The fluorescent lamp produces light by activating a coating of phosphors on the inner wall of the bulb by means of ultraviolet energy generated by a mercury arc. This tubular hot cathode fluorescent lamp, also called a low-pressure electrical discharge source, consists of a glass tube capped at either end with a coiled wire filament called an electrode, or cathode. The inside surface of the tube is coated with a phosphor powder coating, while the tungsten wire electrodes have a coating (called an "emission mix") that increases their ability to emit electrodes. Figure 14–8 shows the construction.

Light generation in a fluorescent lamp involves a series of operations. When the lamp is first energized, a starting voltage source establishes an electric arc—a high velocity electron flow—across a gap that has a gas atmosphere containing argon and mercury. This stream of electrons constitutes an electric current, which heats the argon gas, which in turn heats the mercury changing it to a vapor. The collision between an electron in the arc and an atom of mercury imparts energy to an electron belonging to a mercury atom. This atom, being unstable in a high-energy state, quickly falls back to its original (lower) energy state. As it does so, a very precise amount of energy is released from the atom. This released energy takes the form of a photon of ultraviolet light.

Then, the phosphor coating on the tube wall absorbs the ultraviolet energy, and in turn emits visible light. Both the chemical and physical characteristics of the phosphor material determines the color of the light produced.

Once a lamp is started, the gas atmosphere offers little or no electrical resistance, so an auxiliary device called a ballast serves to limit the amount of current passing through the lamp's arc. The magnetic type ballast is essentially a choke coil with some other features. In the 1980s, improved ballasts were introduced for fluorescent lamps using solid-state electronic components in

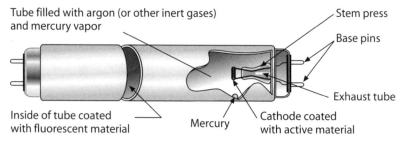

Fig. 14–8 Construction of a fluorescent lamp.

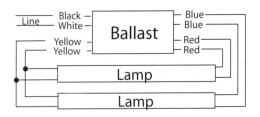

Fig. 14–9 Internal fixture wiring for a two-lamp fixture using a rapid-start (or programmed-start) electronic ballast.

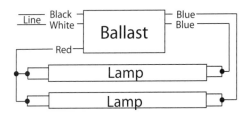

Fig. 14–10 Internal fixture wiring for a two-lamp fixture using an instant-start electronic ballast.

place of the traditional core and coil transformer operating at 60Hz, which greatly boosts lighting system efficiencies. An electronic ballast optimizes the shape of the input voltage waveform sent to the lamp and it uses frequencies in the 20kHz to 45 kHz range, which excites the phosphors to a higher degree. These systems also have the ability to interface with sophisticated occupancy and other controls, such as daylighting dimming response to sunlight entering a window.

The antiquated "pre-heat" method of fluorescent lighting required an external switch that would interrupt the flow of current to the ballast, resulting in a high-voltage "kick" as the magnetic field collapsed. This started the arc, and the current that followed was moderated by the magnetic choke supplied by the ballast windings. This style of fixture is seldom seen today because of a very poor power factor and substantial energy inefficiencies.

There are two basic approaches to starting the initial arc using today's ballasts. A rapid-start fixture uses a small amount of current to heat the filament so it will produce the required initial flow of electrons. The instant-start fixture does not preheat the filament. Instead, a much higher voltage is applied directly across the tube when starting, sufficient to ignite the lamp without preheating the filaments. There is an important trade-off to consider when making this choice. The low-power heating current consumes a small but noticeable amount of energy that the instant-start fixture does not waste. On the other hand, an instant-start voltage pulse is much tougher on the filament, tending to wear it out faster than a rapid-start design. Therefore, if the fixture will be subject to repeated on-off cycles each day, choose a rapid-start design, or its modern electronic equivalent, the "programmed start" ballast. If, when started, it will generally be left on for the day, then choose an instant-start design.

The wiring diagrams in Figs. 14–9 and 14–10 illustrate the differences. The rapid-start model has current flowing through the filaments, and the instant-start model has both filament leads brought together, since current will not be used to pass from one end of the filament to the other. These designs allow for a single, traditional, bi-pin-end fluorescent tube to be used in either fixture. Note that some instant-start fixtures, especially older T12 systems (including the older 8-ft instant-start models), use only a single-pin contact on each the end of the tube. These tubes can only be used in instant-start fixtures. The two-lamp ballast shown in Fig. 14–10 also shows that the two tubes are wired in parallel, so if one lamp fails, the other lamp can remain lit. Three- and four-lamp ballasts are also available.

Compact fluorescent lighting   The fastest growing area in fluorescent technology is the compact fluorescent lamp (CFL), which uses a narrow tube ($\frac{1}{2}$ to $\frac{5}{8}$-in. diameter) that is looped or folded back on itself and terminated in a base. CFL sources are small enough to replace incandescent lamps in diffuse-source applications. The initial versions of this lamp use an integral starter, operate through a simple choke ballast, and have a plug-in two-pin base. Called a preheat lamp, it has a rated life of up to 10 000 hrs and even when ballast losses are included it offers four times the efficacy of an incandescent lamp. Other versions of the CFL use a four-pin base, operate on a rapid-start ballast (either magnetic or electronic) and are dimmable. The operating characteristics are generally the same as the two-pin base lamp.

Advantages of fluorescent lighting   The greatest single advantage of a fluorescent lamp is its efficacy (luminous efficiency). Per watt of power used, it produces two to four times as much light as an ordinary incandescent lamp. Its life is very much longer than that of an incandescent lamp. Being more efficient, it produces much less heat, which is important when a large amount of power is used for lighting, especially if the lighted area is air-conditioned.

Another major advantage is that the fluorescent lamp, being of relatively large size (in terms of square inches of surface compared with total light output), has relatively low surface brightness, which in turn leads to less reflected glare and less shadow. Exposed fluorescent lamps, however, are bright enough to produce some glare discomfort, and enclosures are usually desirable.

Disadvantages of fluorescent lighting   Ordinary incandescent lamps operate in any temperature. Fluorescent lamps are somewhat sensitive to temperature. The ordinary type used in homes will operate in temperatures down to about 50°F. For industrial and commercial areas other types are available that will operate at lower temperatures, some as low as –20°F.

Due to the mercury content of fluorescent lamps, disposal of the lamps is subject to varying degrees of regulation depending on locale. Check with local hazardous waste regulations for information on proper disposal.

Issues in fluorescent lamp design choices   The most popular fluorescent in general use for years, and the reference when comparing improvements, the 4-ft long, rapid-start T12 F40 lamp ($1\frac{1}{2}$-in. diameter) is now considered inefficient and thus obsolete. Two newer lamps with smaller diameters, the T8 (1-in.

diameter) and the T5 (⅝-in. diameter), in a variety of lengths replace the T12 lamp in most applications. A complete family of T8 lamps is available in four straight-tube lengths and three bent-tube lengths. The 4-ft 32W lamp rated at 265 mA can be a direct substitute for the F40T12 lamp.

Technology upgrades in T8 linear fluorescent lamps include new lamp coatings and a reduction in the quantity of mercury in the lamp, while also allowing a thinner phosphor coat. In addition to color rendering improvements, advanced phosphor coatings permit less watts/same light or same watts/more light in a lamp/ballast system. Both the T8 improvements and developments in ballast products greatly expand the selection. For example, some 4-ft F32 T8 lamps (with matching ballasts) are rated for 24 000 hours with 92% lumen maintenance.

Made to metric measurements, the T5 fluorescent lamp family is available in nominal lengths of 22-, 34-, 46-, and 56-in. One ballast can serve either one or two standard T5 lamps of any wattage (14, 21, 28, and 35W), since all lamps in this line operate at the same current (170 milliamps). A companion line of T5 HO—meaning high output—lamps (24, 39, 54, and 80W) is used when twice the light output from a single lamp is needed. Since they operate on different currents, each HO lamp wattage requires its own ballast. At 83 to 94 lm/W, these lamps are about 10 to 15% less efficacious than standard T5 lamps and up to 8% less efficacious than T8 systems.

Table 14–1 LAMP EFFICACY COMPARISONS

Lamp type	Rated lamp wattage	Light output in lumens	Lumens per watt
Incandescent			
PAR 38 halogen flood	90	1270	14
IR PAR 38 halogen flood	60	1150	19
Low voltage MR16 BAB	20	280	14
A-19	60	880	15
A-19 halogen	60	960	16
Tubular halogen	300	6000	20
Fluorescent			
CFL self-ballasted	26	1550	43
T12 lamp w/magnetic ballast	46	2800	61
T8 lamp w/electronic ballast	37	2850	77

The efficacy, average rated life, and the lumen depreciation characteristics of the lamps are important concerns in selecting and operating fluorescent systems. Table 14–1 shows typical efficacies of the lamps discussed in this chapter. Assuming normal conditions, the efficacy of a fluorescent lamp depends on the operating current and the lamp's phosphor composition. Ambient temperature also significantly affects efficacy, because maximum lumen output occurs when the coldest spot on the bulb wall is about 77°F for the T12 and T8 fluorescent lamps. The newer, even thinner T5 and T5 HO lamps provide maximum lumen output at a 95°F bulb wall temperature. The reason for the differences in optimal wall temperature have to do with the fact that the smaller the tube, the closer the phosphors are to the arc path, resulting in a more intense excitation, but also making the lamp more temperamental with respect to ambient conditions.

Since reducing energy consumption is an important consideration today, automatic occupancy sensing devices are used to turn off fluorescent fixtures in an area that is unoccupied. Remember however, that fluorescent lamp life is significantly dependent on the number of hours the lamp operates per start, since some electrode emission coating is depleted every time it is turned on. This is a good example of where a choice between rapid-start and instant-start designs depends on the expected frequency of cycling.

Ballast designs respond to occupancy requirements   An understanding of pertinent U.S. Department of Energy (DOE) rules is important before selecting linear fluorescent ballasts. Effective April 1, 2005, for the most popular lamps, DOE raised the Ballast Efficiency Factor from the energy-efficient (magnetic) T12 ballast to levels met only by the electronic T12 ballast. However, weighing many economic factors, consider T8 electronic ballasts systems as the minimum standard of the future, rather than even considering T12 electronic ballast systems. These newer ballasts can include features such as end-of-life sensing and automatic shut-off, and the ability to "soft start" a fluorescent lamp as a way of extending lamp life. The programmed-start ballast is a more sophisticated version of the rapid-start type discussed previously. It provides lamp electrode heating during lamp starting, and it is recommended for applications where lamps are frequently switched on and off. This ballast design precisely controls the timing of the application of starting and warming cathode voltage to the lamp.

A second important feature of electronic ballasts is the ability to select the amount of power you want to deliver to the lamps. This is achieved by specifying what is called the Ballast Factor (BF). Low power (74 to 78 BF) is for retrofit installations where lighting levels can be reduced. Standard power (87 to 88 BF) is for areas where light levels are to be maintained. Reference and high power (1.00 to 1.20 BF) is useful when reducing the number of lamps within a fixture during a retrofit project.

Life of fluorescent lamps   The life of an ordinary incandescent lamp varies from 750 to 2500 hours depending on the size, type, and purpose of the lamp. It does not make any difference how many times the lamp is turned on and off. A fluorescent lamp that is turned on and never turned off will probably last well over 30 000 hours, about four years. If it is turned on and off every few minutes, it may not last 500 hours. The reason lies in the limited amount of electron-emitting material on the cathodes; a specific amount of it is used up every time the lamp is turned on; when it is all gone, the lamp is inoperative. It is not possible to predict the exact number of starts the lamp will survive, and ordinary operation of the lamp also consumes some of the material, but the fact remains that the more often a fluorescent lamp is turned on, the shorter its life will be.

The published figures for approximate life are based on the assumption that the lamp will be lighted for 3 hours every time it is turned on; for the older 40-watt lamp the average was 18 000 hours, and with today's electronic ballasts the life is even greater. Under ordinary operation, fluorescent lamps will last 5 to 15 times longer than ordinary incandescent lamps.

Rating of fluorescent lamps   An ordinary incandescent lamp marked 40 watts will consume 40 watts when connected to a circuit of the proper voltage. A fluorescent lamp rated at 40 watts will also consume 40 watts within the lamp, but additional power is also consumed in the ballast; this additional power is from 10 to 20% of the watts consumed by the lamp itself, and must be added to the watts consumed by the lamp itself to obtain the total load in watts of the fixture. Here again, modern electronic ballasts greatly reduce the energy losses in the ballast.

Power factor of fluorescent lamps   Ordinary incandescent lamps have a power factor (pf) of 100%. A single fluorescent lamp connected to a circuit using an old-fashioned simple magnetic ballast has a power factor between 50 and 60%, resulting in much higher current than would be assumed from the power of the lamps. Fortunately this is not as serious as it sounds. The common method is to have either two or four lamps per fixture and, in addition to the usual ballast, to use power-factor-correction devices built into the same case with the ballast, which brings the power factor up to about 90% or higher.

Most two- and four-lamp fixtures on the market today are the high-power-factor type, and especially so in the case of fixtures for commercial and industrial use. But if you buy "bargain" fixtures for the home, be sure they are of the high-power-factor type.

Color of light from fluorescent lamps   The colors of flowers, clothing, and so on look different under ordinary incandescent lamps than in natural sunlight because incandescent lamps produce light rich in red. The color of the light produced by fluorescent lamps is determined entirely by the chemical and physical characteristics of the phosphor or powder deposited on the inside of the tube. With the old T12 lamps, the most ordinary color is "white," but there are many varieties including deluxe warm white, warm white, white, cool white, and deluxe cool white.

The differences lie in the proportion of red and blue in the light produced by the particular lamp. The "warm" varieties emphasize red and yellow (like light from incandescent lamps), whereas the "cool" varieties emphasize blue (more like natural sunlight). The kind of white to use depends on the effect desired.

If you look at a black-and-white printed page under each of the "white" colors in turn, you will see little difference. But if you look at a colored page, colored fabrics, or food under each kind, you will see much difference. Under deluxe cool white, people's complexions appear much as in natural light, but the deluxe warm white will flatter them a bit, adding a ruddy or tanned appearance similar to the effect of ordinary incandescent lamps. The cool white provides quite a good appearance also, but with a tendency toward paleness. Use deluxe warm white if your preference in home furnishings leans toward warm colors (red, orange, brown, tan); use deluxe cool white if you prefer the cool colors (blue, green, violet).

These "colors" are often hard to find today, and they frequently don't apply to the current workhorse of the industry, the T8 lamp systems. These (T8) lamps often have numerical designations such as "850" or "750" to describe the phosphors being used. Consult the lamp manufacturer's catalog information to make the best decision. Be aware that as you move away from the cool white type of colors into the warmer ranges, the efficacy of the lamps tends to decline.

Some sizes of fluorescent lamps are also available in blue, green, pink, gold, and red. Their efficiency in colors is extraordinarily high compared to colored incandescent lamps. For example, a green fluorescent lamp produces about 100 times as much green light per watt as a green incandescent lamp. Colored lamps are used where spectacular color effects are needed, as in theater lobbies, lounges, stage lighting, and advertising.

CHAPTER 15

Residential and Farm Motors

THIS CHAPTER DISCUSSES some properties and characteristics of motors in general, then various kinds of motors, and finally the wiring of the kinds of motors generally used in homes and farms. Chapter 30 covers the wiring of motors used in industrial or commercial projects.

HOW ELECTRIC MOTORS ARE RATED

Electric motor ratings fall into two general categories, mechanical and electrical. This portion of the chapter covers the mechanical ratings, which allow you to decide if the motor will do its job properly and without failing prematurely. How much power does it have, and how fast does it turn? Will its starting torque be enough to accelerate its load? Subsequent portions of this chapter cover the electrical ratings that you need to know in order to safely connect the motor to a circuit.

Horsepower One horsepower is defined as the work required to lift 33 000 pounds one foot (33 000 foot-pounds) in one minute, which is equivalent to lifting 550 pounds one foot in one second. One horsepower is equal to 746 watts. The horsepower rating of a motor is stamped on its nameplate. Rated horsepower indicates the amount of power the motor will develop—the greater the horsepower, the more mechanical energy the motor can bring to bear on a given task per unit time.

Starting capacity Motors deliver far more power while starting, particularly under load, than after accelerating the load to full speed. The proportion varies with the type of motor; some types have starting torques four or five times greater than at full speed. Naturally the current consumed while starting is much greater than while the motor is delivering its rated horsepower at full speed. The motor will overheat quickly if too large a start-up load prevents it from reaching full speed. The size of a machine's start-up load is an important factor in selecting a motor to run the machine.

Overload capacity Almost any good motor, after reaching full speed, will develop $1\frac{1}{2}$ to 2 times, or more, its rated horsepower for a short time. But no motor should be overloaded continuously, because overloading leads to overheating, and that in turn leads to greatly reduced life of the windings. Rewinding a burned-out motor is costly; for smaller size motors, replacement is usually more economical than rewinding. Nevertheless, this ability of a motor to deliver more than its rated horsepower is very convenient. For example, ½ hp may be just right for sawing lumber, but when a tough knot is fed to the saw blade, the motor can instantly deliver 1½ hp and then drop back to its normal ½ hp after the knot has been sawed. If the motor is undersized, and if an "overload protection device" has been properly installed, the device will stop the motor before the motor windings burn out from the heat developed when windings carry far more current than intended by their design. An overload protection device is not the same as an overcurrent device, as explained later.

Replacing gasoline engines with electric motors   A gasoline engine is rated at the maximum horsepower it can deliver continuously at a given speed; it has no overload capacity. Thus an engine rated at 5 hp can deliver 5 hp continuously when new, gradually diminishing with age, but unlike an electric motor it cannot deliver more than 5 hp at any time. Thus it is sometimes possible to replace an engine with an electric motor of slightly lower horsepower rating. If a gasoline engine always runs smoothly and seldom labors and slows down, it can be replaced by an electric motor of a lower horsepower. But if the engine is always laboring at its maximum power, the motor that replaces it should be of the same horsepower as the engine because no motor will last long if continuously overloaded.

Power consumed by a motor   A motor delivering 1 hp—746 watts—is actually consuming about 1000 watts from the power line, the difference of 254 watts being lost as heat in the motor, friction in the bearings, the power that it takes to run the motor even while idling, and similar factors. The watts that a motor draws from the power line are in proportion to the horsepower the motor is delivering. The approximate power consumption for a 2-hp motor is as follows:

While starting	4000 watts
While idling	400 watts
While delivering ½ hp	750 watts
While delivering 1 hp	1150 watts
While delivering 1½ hp	1500 watts
While delivering 2 hp	2000 watts
While delivering 2½ hp	2600 watts
While delivering 3 hp	3300 watts

These values represent true power only; the actual current taken exceeds the amount required to do the work. This additional current represents current required to magnetize the motor windings and lags the current doing the useful work of rotating the load, and the current required to overcome frictional resistance. The total current multiplied by the voltage is expressed as "volt-amperes" (or "VA"), as covered in Chapter 3. For example, a 1-hp, 120 V motor may draw over 1900 VA under full load. Of this, 746 watts is work delivered to the load, 254 watts overcomes mechanical resistance, and 900 VA is magnetizing current taken from and subsequently returned to the line as the fields build and collapse every cycle. A conventional power meter reading true power records this as 1 kW, but the wiring must carry 1900 VA. The ratio of true power (1 kW) to apparent power (1900 VA) is the power factor of the motor circuit.

Motors are designed to operate at greatest efficiency (meaning with the least power consumption per hp delivered) when operating at their rated horsepower. Efficiency usually falls off when delivering more or less than rated horsepower. Thus it costs a little more per hour (due to increased power consumption) to run a 1-hp motor at half load than to run a ½-hp motor at full load. It is more economical to use a motor of the proper size for a machine than to use a larger motor you might have on hand. On the other hand, a motor that is continuously overloaded will not last long. Use the right size and right kind of motor for the load.

Speed of ac motors   The most common 60-Hz ac motor runs at a theoretical 1800 r/min, but at an actual 1725 to 1750 r/min delivering its rated horsepower. When the motor is overloaded, its speed drops. If the overload is increased too much, the motor will stall and burn out if not properly protected by an overload device. Low voltage also reduces a motor's speed as well as its horsepower output. A motor operating at a voltage 10% lower than normal will deliver only 81% of its normal power.

Motors of other speeds are available, running at theoretical speeds of 900, 1200, and 3600 r/min, with actual speeds a little lower. The slower-speed motors cost more, are larger, and it usually takes longer to find replacements for them, although they do run significantly quieter. The more readily available 1800 r/min motor will serve most purposes. If a very low speed is necessary, use one of the 1800 r/min type with built-in gears that reduce the shaft speed to a much lower figure, or use a separate reduction gear. In terms of energy conservation, however, gears inevitably involve additional friction losses, compromising efficiency.

The speed of ordinary ac motors cannot be controlled by rheostats, switches, or similar devices because it is directly related to the number of poles in the motor's windings and the frequency of the supply system. Neither of these factors is subject to change by a simple external controller. Some small motors, including universal motors and shaded-pole motors, can be controlled by voltage regulation, but that is an insignificant portion of the entire ac motor population.

Varying the speed of larger, conventional ac motors has become a major area of interest, particularly in commercial and industrial occupancies, largely for reasons of energy conservation. For example, an economical approach to meeting ventilation needs in a building would be to have remote sensors trigger variable speed motor drives to alter the rate at which exhaust fan blades turn according to the amount of air circulation required. Since the horsepower required to move a fan varies with the cube of the impeller speed, even a small reduction in required speed makes a significant difference in energy consumption. These drives involve varying the frequency of the electric power delivered to the motor through the use of devices that rectify 60-Hz power and then send it to the motor(s) at whatever frequency will produce the required motor result. Variable speed drives are notorious sources of harmonics in power systems; for a more complete explanation refer to the nonlinear load discussion on page 47.

Service factor relates motor output to temperature   At one time motors were rated only on the basis of a temperature rise above the ambient temperature (the air temperature at the motor location with the motor not running) while delivering their rated horsepower. Ordinary motors were based on a temperature rise of 40°C (72°F).[1] The actual temperature of a motor installed in a hot location, such as a pump house on a farm where the temperature might be 115°F, would then increase by 72°F to a final temperature of 187°F, not far below the boiling point of water. This would feel exceedingly hot to the hand but would not harm the motor.

Over the years, the heat-resisting properties of insulations on wires used to wind the motor, and the materials used to insulate the wires from the steel in the motor, have been vastly improved. This has made it possible to reduce the physical size of motors significantly—today's 10-hp, three-phase motor isn't much bigger than a 3-hp motor made in 1945. Although these smaller motors run much hotter than the old larger size, they will not be burned out by temperatures that would have destroyed the 1945 motor. They often exceed 212°F (the boiling point of water) but are not damaged.

Most motor ratings now include a "service factor" stamped on the nameplates, ranging from 1.00 to 1.40. A service factor of 1.00 indicates that if the motor is installed in a location where the ambient temperature is not over 40°C (104°F), the motor can deliver its rated horsepower continuously without harm. A service factor of 1.15 means that a motor can be used at up to 1.15 times its rated horsepower under the same conditions. Multiply the rated horsepower by the service factor; a 5-hp motor, if its service factor is 1.15, can be used continuously as a 5.75-hp motor. But remember that a motor will operate most efficiently and last longer if used at not over its rated horsepower. The service factor provides temporary extra power if needed.

---

1. Do not confuse a change in the readings of two different thermometers with their actual readings. When a Celsius (formerly called centigrade) thermometer reads 40°, a Fahrenheit thermometer reads 104°. When a Celsius thermometer changes by 40°, the Fahrenheit changes by 72°. Thus if the Celsius changes from 40° to 80° (a 40° rise), the Fahrenheit changes from 104° to 176° (a 72° rise). One degree on the Celsius scale is the equivalent of 1.8° on the Fahrenheit scale.

Most open motors larger than 1 hp have a service factor of 1.15. Fractional-horsepower motors often have a service factor of 1.25, some as high as 1.40. If the ambient temperature is above 40°C (104°F), the motor should not be operated at its full rated horsepower without a fan or blower in the area, and certainly not *over* its rated horsepower even with a fan or blower. Regardless of temperature, install motors where plenty of air is available for cooling.

Note that some motors now being made are still rated on the basis of a 40°C temperature rise and do not have a service factor.

TYPES OF MOTORS COMMONLY USED

Since most residential and farm occupancies are served by single-phase electrical systems, the most common motors are single-phase motors. Alternating current motors get their power from a magnetic "kick" that a 60-Hz electrical supply can deliver 120 times each second, one push from like magnetic poles opposing each other at each current maximum. However, when the motor is at a standstill, those kicks would immobilize the rotor, tending merely to jerk it back and forth 120 times each second. In the case of a three-phase motor, those magnetic kicks come in a rotating sequence, first phase A, then B, then C, then A again, etc. This creates a perfectly timed rotating field within the motor, and the rotor accelerates to match it. Single-phase motors don't have that advantage. They need an additional winding for starting that creates, during the starting sequence, kicks that occur at a different but consistent moment in time from the main winding.

Split-phase motors   For example, a split-phase ac motor has a starting winding that uses thinner wire than the main winding. As such, it has a higher resistance than the main winding. Since it has a higher resistance, it has a different, higher power factor; its current maximums occur closer in time to the applied voltage maximums (review power factor topic beginning on page 45). This means that its current maximums occur at different moments in time from the main winding, which has very little resistance and is an almost purely inductive load at standstill. Pure inductive loads have current maximums that lag the voltage by 90 electrical degrees. As an example, such an arrangement might produce a push at 85 degrees lagging and 15 degrees lagging.

Magnetic fields develop in response to current flow, so the magnetic forces from the two windings then act on the rotor at those different instants in time. This creates a rotating field that will accelerate the motor, although not with the same initial torque as a three-phase motor which always has three perfectly distributed magnetic impulses. Being thinner wire, if such a starting winding stays in the circuit it will burn out, so these motors have automatic means to cut out the starting windings as the motors approach full speed. At full speed the momentum of the rotor always carries it to a point where the next push tends to continue its direction.

A split-phase motor operates only on single-phase alternating current. It is the simplest type of single-phase motor made, which makes it relatively trouble-free; there are no brushes, no commutator. It is available only in sizes of ⅓ hp and smaller. It draws a very high current while starting. Once up to full speed this type of motor delivers just as much power as any other type of motor, but it is not capable of starting heavy loads. Therefore do not use it to drive any machine that is hard to start, such as a water pump, or air compressor that has to start against compression. Use it on any machine that is easy to start, or one where the load is thrown on after the motor is up to full speed. It is suitable for grinders, saws, lathes, and similar equipment of the kind used in home workshops.

Capacitor motors   A capacitor motor also operates only on ordinary single-phase alternating current. It is similar to the split-phase type with the addition of a capacitor which enables it to start much-harder-starting loads. The capacitor further shifts the power factor of the starting winding, causing the internal magnetic pushes to be more evenly distributed in time, increasing the effectiveness of the starting winding. There are really two types of capacitor motors: (1) the capacitor-start type, in which the capacitor is cut out of the circuit after the motor has reached full speed, and (2) the capacitor-start, capacitor-run type, in which the capacitor remains in the circuit at all times. This type has a more robust starting winding capable of remaining in the circuit. The capacitor remains connected to the starting winding, keeping the magnetic pushes in the motor from the two windings 90 electrical degrees apart. In effect, such a motor approximates a two-phase motor.

There are several grades of such motors available, ranging from the home-workshop types which will start loads from 1½ to 2 times as heavy as the split-phase type will start, to the heavy-duty type which will start almost any type of load. Capacitor motors are usually more efficient than split-phase motors, using fewer watts per horsepower. The current consumed while starting is very much less than that of the split-phase type. Capacitor motors are made in any size but are commonly used only in sizes up to 7½ hp.

Repulsion-start induction-run motors   Often called an "RI" motor, this type again operates only on single-phase alternating current. It has very high starting ability suitable for the heavier jobs; it will "break loose" almost any kind of hard-starting machine. The starting current is the lowest of all types of single-phase motors. It is available in sizes up to 10 hp. These motors have a commutator connected to the windings on the rotor that allows for precise positioning of the rotor poles (by moving the brush positions) in relation to the field poles to produce maximum starting torque. Commutators and brushes require skilled maintenance. With the increased availability of three-phase power, use of RI motors is decreasing.

Universal motors   A universal motor operates either on direct current or single-phase alternating current. However, it does not run at a constant speed but varies over an extremely wide range. While idling, a universal motor may run as fast as 15 000 r/min, but under a heavy load may slow down to 500 r/min. This of course makes the motor totally unsuitable for general-purpose work. It is used only where built into a piece of machinery whose load is substantially constant and predetermined. For example, universal motors are found on vacuum cleaners, sewing machines, and drills. Their speed can be controlled within limits by a rheostat, as on a sewing machine.

Dual-voltage motors   Most motors rated at ½ hp or more are constructed to be operated on either of two different voltages, such as 120 or 240 volts. Single-phase motors of this type have four leads. Connected one way, with the windings in parallel, the motor will operate as a 115-volt motor. Connected the other way, with the windings in series, the motor operates as a 230-volt motor. See Fig. 15–1.

Dual-voltage motors have two different voltage ratings and two different current ratings on the nameplate. For example, it may be marked VOLTS 115/230, AMPS 24/12, which means that while delivering its rated horsepower it will consume 24 amps if operated as a 115-volt motor, but only 12 amps if operated as a 230-volt motor.

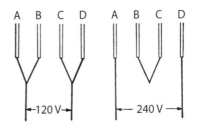

Three-phase motors   This type of motor, as the name implies, operates only on three-phase alternating current. Three-phase motors are the simplest of all kinds of motors and are available in any size. They cost less than any other type, so use them if you have three-phase current available. But do not assume that

Fig. 15–1 Larger motors are made to be operated on either 120- or 240-volt circuits, depending on the connections of the four leads from the motor, as shown in the diagram above. It is always wise to operate the motor at 240 volts if possible.

because you have a three-wire service that you have three-phase current. In all likelihood you have three-wire, 120/240-volt, single-phase current. If in doubt, consult your electric utility. A three-phase, dual-voltage motor has nine leads, and great care must be used to connect them properly for the voltage selected.

Direct-current motors   Direct current (dc) is still found in the downtown sections of a few large cities and perhaps in a few small towns. All battery-type small generating plants, as once used on many farms and now used mostly in marine applications, are direct current. There are many types of dc motors, among them the series, the shunt, the compound types, and others. However, too few dc motors are in general use to warrant a full discussion here.

One important difference in operating characteristics between ac and dc motors is that the ac motor will run at a specific speed even if there are considerable changes in loads and voltages, and that speed cannot be controlled by a rheostat or similar device. A dc motor, on the other hand, even if designed for 1800 r/min, will run at that speed only with a specific combination of voltage and load. If either voltage or load changes, the speed changes. The speed can be controlled by a rheostat, although once adjusted it will change with any change in voltage or load. This feature is a benefit in sensitive industrial drive applications, where it allows for constant torque refinement in response to sensor input. For example, a paper coating application may require a certain tension on a reel of paper moving through a coating head. The dc drive controls can be arranged to sense the tension and automatically adjust the motor performance accordingly. These applications are beyond the scope of this book.

Reversing motors   The direction of rotation of an RI motor can be reversed by changing the position of the brushes. On other types of single-phase motors, reverse the two leads of the starting winding in relation to the two leads from the running winding. If a motor must be reversed often, a special switch can be installed for the purpose. To reverse the rotation of a three-phase motor, simply reverse any two of the three wires running between the controller and the motor; do this in the controller. For frequent applications there are reversing controllers designed for this purpose. If one of the wires is grounded, reverse the other two.

Problems with large motors   The size of farms is constantly increasing, leading to the use of larger sizes of farm machinery, requiring bigger motors: 10 hp, 25 hp, and even larger. But most farms have only a single-phase, three-wire, 120/240-volt service. That requires only two high-voltage lines to the farm (actually only one if the utility uses a wye primary distribution; review Fig. 9–2) and only one transformer. Single-phase motors are not usually available in sizes larger than 7½ hp, although a few larger ones are made. But before buying even a 5-hp, single-phase motor, check with your power supplier to see whether the line and the transformer serving your farm are big enough to operate such a motor.

Single-phase motors 5 hp and larger require an unusually high number of amperes while starting, and the line and transformer often are too small to start such a motor. If you operate the motor only a comparatively few hours per year, your utility will object to installing a heavier line and transformer, just as the farmer would not buy a 10-ton truck to haul 10 tons a few times per year, while using it for much smaller loads most of the time.

In a few localities, at least some of the farms are served by a three-phase line, requiring three (usually four) wires to the farm, and three transformers. If you are fortunate enough to have three-phase service, your problems are solved. Simply use three-phase motors, which cost considerably less than single-phase; being simpler in construction, there is rarely a service problem. (Note: If you have three-phase service it will be either a delta connection with three-phase power at 240 volts, plus the usual 120/240-volt, single-phase for lighting, appliances, and other small loads, or it will be a wye connection, resulting in 208-volt, three-phase for power loads and 120 volts for lighting and smaller loads. The trend in utility practice is from the former toward the latter, so be sure to ask.)

Phase converters allow use of three-phase motors on single-phase lines

If you have only the usual single-phase, 120/240-volt service, and still need larger motors, what can you do? One solution is to use smaller machinery requiring motors not over 3 hp, but that is a step backward because it increases the cost of labor. There is another solution. Phase converters permit three-phase motors to be operated on single-phase lines. The phase converter changes the single-phase power into a sort of modified three-phase power that will operate ordinary three-phase motors, and at the same time greatly reduces the number of amperes required while starting. In other words when operating a three-phase motor with the help of a phase converter, the same single-phase line and transformer that would barely start a 5-hp, single-phase motor will start a 7½-hp or possibly even 10-hp, three-phase motor; a line and transformer that would handle a 10-hp, single-phase motor (if such a motor could be found) would probably handle a 15- or 20-hp, three-phase motor. Phase converters are expensive, but their cost is partially offset by the lower cost of three-phase motors, and by the economy of operating larger motors than are possible without the converter.

There are two types of phase converters: the static type with no moving parts except relays, and the rotating type. The static type must be matched in size and type with the one particular motor to be used with it; generally there must be one converter for each motor. The rotating type of converter looks like a motor but can't be used as a motor. Two single-phase wires run into the converter; three 3-phase wires run out of it, one of which (the "manufactured" or "derived" phase) originates at the converter and has no direct connection with the incoming single-phase supply. Usually several motors can be used at the same time with a rotating converter; the total horsepower of all the motors in operation at the same time might be at least double the horsepower rating of the converter. Thus if you buy a converter rated at 15 hp, you can use any number of three-phase motors totaling not over 30 to 40 hp, but the largest may not be more than 15 hp, the rating of the converter. The converter must be started first, then start the motors, first the largest, then the smaller ones. The *NEC* requires the manufacturer to mark the nameplate with both the maximum total load in kVA or horsepower, and the maximum (and minimum) single load in kVA or horsepower, so you'll know what loads you can connect.

But some words of caution are in order. A three-phase motor of any given horsepower rating will not start as heavy a load when operated from a converter as when operated from a true three-phase line. It will often be necessary to use a motor one size larger than is necessary for the running load. This will not significantly increase the power required to run the motor once it is started. The converter must have a horsepower rating at least as large as that of the largest motor.

The voltage delivered by the converter varies with the load on it. If no motor is connected to the converter, the three-phase voltage supplied by it is very considerably higher than the input voltage of 240 volts. Do not run the converter for significant periods without operating motors at the same time or it will be damaged by its own high voltage. Do not operate only a small motor from a converter rated at a much higher horsepower because the high voltage will damage the motor or reduce its life. This is why the nameplate includes a minimum single load.

Before purchasing a power converter, check with your power supplier; some do not favor or permit converters. If they do permit converters, the line and the transformer serving your farm must be big enough to handle all the motors you propose to use. In addition, use caution when connecting single-phase loads on the load side of a phase converter; such loads must never be connected to the manufactured phase. The *NEC* requires that this conductor have a distinctive marking anywhere it is accessible, so there shouldn't be any confusion. The usual procedure is to always use yellow insulation for this wire.

FIVE REQUIREMENTS FOR EVERY MOTOR INSTALLATION

The *NEC* sections that govern the installation of motors are extremely detailed because they cover all motors from the tiniest to those developing hundreds or thousands of horsepower and operating at thousands of volts. Such large motors are discussed in Chapter 30, but the wiring of motors for homes and farms can be covered by a few simple rules covering five basic components of motor circuits, as follows:

Determine minimum wire size   All wires in a motor branch circuit all the way up to the motor must be large enough to carry the starting current. Voltage drop should also be taken into account in regard to length of the circuit. The *NEC* requires the wires to have an ampacity of at least 125% of the Code-table ampere rating of the motor. Multiply the motor ampere rating from the tables at the end of *NEC* Article 430—not the nameplate rating—by 1.25, which is 125%. Then use wire with an ampacity at least as great as the figure determined, using the 75°C column of *NEC* Table 310.16, assuming the other devices in the motor circuit have 75°C-rated terminations (refer to termination temperature discussion on page 106). A special rule allows the terminations at the motor to automatically qualify for 75°C status, even on sizes under 100 amps that would normally require 60°C consideration. As always, though, remember that the wire itself together with every connected device must carry the same recognition (likely but not certain in this case). If any part doesn't, then you need to drop back to the 60°C column and probably use a larger wire.

Next consider the one-way distance to the motor to see whether the wire size you have selected will limit the voltage drop to a reasonable value. Because the motor will consume many more amperes while starting, the drop will be much greater on starting than while running. If the motor must start loads that are excessively hard-starting, the voltage drop might be too much to allow the motor to reach full speed, and that might lead to motor burnouts. To simplify the whole problem of correct wire sizes, use Table 15–1. If the machine that the motor drives is of the very-hard-starting type, use wire one size larger than shown.

Table 15–1 MAXIMUM ADVISABLE DISTANCES (IN FEET) FROM SERVICE EQUIPMENT TO SINGLE-PHASE MOTOR FOR DIFFERENT WIRE SIZES

Motor			Wire sizes								
Horse-power	Volts	Am-peres	14	12	10	8	6	4	2	1/0	2/0
1/4	115	5.8	55	90	140	225	360	575	900	1500	1800
1/3	115	7.2	45	75	115	180	300	450	725	1200	1500
1/2	115	9.8	35	55	85	140	220	350	550	850	1100
3/4	115	13.8	25*	40	60	100	150	250	400	600	800
1	115	16.0	—	35	50	85	130	200	325	525	650
1 1/2	115	20.0	—	25*	40	65	100	170	275	425	550
2	115	24.0	—	—	35	55	85	140	225	350	450
3	115	34.0	—	—	—	40*	60	90	160	250	325
1/4	230	2.9	220	360	560	900	1450	2300	3600		
1/3	230	3.6	180	300	460	720	1200	1600	2900		
1/2	230	4.9	140	220	340	560	875	1400	2200		
3/4	230	6.9	100	160	240	400	600	1000	1600	2400	
1	230	8.0	85	140	200	340	525	800	1300	2100	
1 1/2	230	10.0	70	110	160	280	400	675	1100	1700	2200
2	230	12.0	60	90	140	230	350	550	900	1400	1800
3	230	17.0	—	65*	100	160	250	400	650	1000	1300
5	230	28.0	—	—	60*	100	160	250	400	650	800
7 1/2	230	40.0	—	—	—	70*	110	175	275	450	550
10	230	50.0	—	—	—	—	90*	140	225	350	450

Figures below the wire sizes indicate the one-way distance in feet (not the number of feet of wire in the circuit) that each size wire will carry the amperage for the size motor indicated in the left-hand column, with 1 1/2% voltage drop. A dash indicates that the wire size in question is smaller than the minimum required by the *NEC* for the horsepower involved, regardless of the circumstances. Figures are based on single-phase AC motors.

* If you are using Type T or TW wire, the size of wire at the top of the column is too small for the amperage; a different type of wire may be suitable. See *NEC* Table 310.16 (Table A-1 in the Appendix of this book).

Extension cords made of the usual 16 AWG wire should never be used even with fractional-horsepower motors. A cord on a portable motor is necessary but if a longer extension cord is added, the voltage drop in the cord during the starting period might be so great that the motor will never reach full speed and thus will not switch off its starting winding; a damaged motor might easily result. If you do use an extension cord, use one with large enough wires to reduce the voltage drop to a minimum.

Provide required disconnecting means   Every motor must be provided with a method of disconnecting it from the circuit. For a portable motor, the plug and receptacle can serve as the disconnecting means provided the receptacle horse-power rating is large enough for the motor. To determine this, review the horse-power rating table in the UL *"White Book"* (see "Guide card information" on page 8), under the category "Receptacles for Attachment Plugs and Plugs." Here are some single-phase examples: A 125-volt, 15-amp receptacle can be used with up to a ½-hp motor, and a 125-volt, 20-amp receptacle is good for 1 hp. A 250-volt, 15-amp receptacle works for up to 1½ hp, and the 20-amp variety works for up to 2 hp. If the motor is ⅛-hp or smaller, no separate disconnecting means is required; the branch-circuit breaker or fuse will serve the purpose. All other motors must have a switch or circuit breaker as a disconnecting means, and it must open all ungrounded wires to the motor and its controller (if it has a separate controller). The *NEC* also permits a "molded case switch" for this purpose. This turns on and off like a circuit breaker, and can do so safely even while a fault is in progress. However, it has no provision for an automatic trip, thus it is equivalent to a switch. In general, every motor-circuit disconnect switch must carry a rating at least equal to 115% of the full-load current rating for that motor as listed in the tables at the end of *NEC* Article 430, or, if rated in horse-power, must at least equal the motor horsepower rating.

Fig. 15–2 A switch of this type, with plug or cartridge fuses, may be used with small motors. If the switch has two fuses, it is for a motor operated at 240 volts; if it has only one fuse, it is for a motor operated at 120 volts. *(Square D Company)*

A switch of the general type shown in Fig. 15–2 (or a larger one with cartridge fuses for a larger motor), or a circuit breaker in an enclosure may be used as the discon-necting means. If the motor is larger than 2 hp, the switch must be rated in horsepower not less than the horsepower rating of the motor; many switches are rated both in amperes and in horsepower. If the motor is 2 hp or smaller, you may use the same kind of switch rated only in amperes if the ampere rating of the switch is at least double the ampere rating of the motor. Note that all circuit breakers are equivalent to switches rated in horsepower. You may also use a general-use-ac-type snap switch of the kind you use to control lights if its ampere rating is at least 125% of the motor ampere rating. In any case use a single-pole switch for motors on 120-volt circuits, and a two-pole switch for motors on 240-volt circuits. Always be sure that the disconnecting means is open (off) before working on a motor, a motor circuit, or the driven machinery.

Provide short-circuit and ground-fault protection   A motor and its circuit wires must be protected against short-circuit or ground-fault currents by a circuit breaker or fuse in each ungrounded wire. Because the starting current for a short time is much greater than its full-load running current, the circuit breaker or fuse must have a high enough ampere rating that it will not trip or blow during the starting cycle. Its ampere rating must be at least 125% of the ampere rating of the motor, but it may be increased to as much as 175% if time-delay fuses are used and 250% if conventional circuit breakers are used. Here again, base your calculations on the table values at the end of *NEC* Article 430, and not on the nameplate full-load current. Detailed coverage of motor branch-circuit and ground-fault protection begins on page 589.

If those table values seem to contradict everything said up to this point, that the branch-circuit overcurrent devices may not have an ampere rating greater than the ampacity of the wire in the circuit, keep in mind that each motor must be provided with an overload (not overcurrent) device that protects the motor against overloads or failure to start, as explained later in this chapter. The branch-circuit overcurrent device therefore protects only against short circuits or ground faults. In fact, for this very reason *NEC* Article 430 refers to these devices only as "short-circuit and ground-fault protective devices" and not overcurrent devices.

Provide controller to start and stop motor   Every motor must have a means of starting and stopping it. This is called a "controller." The controller may be an automatic device, part of a refrigerator, water pump, or any other equipment that starts and stops automatically. In that case you can safely assume that the proper controller comes with the equipment. The following discussion is about manually operated controllers.

If the motor is portable and rated at ⅓ hp or less, the plug and receptacle serve as the controller. If, however, it is larger than ⅓ hp, proceed as for stationary or fixed motors as discussed in following paragraphs.

If the motor is 2 hp or smaller, and used at not over 300 volts, you may use the disconnecting means (as already discussed under that heading) as the controller. If the motor is larger than 2 hp, a separate controller must be installed, rated in horsepower not less than the horsepower of the motor.

Manual motor controllers   Regardless of the size of the motor, instead of using switches of the kind described, special motor starters of the type shown in Fig. 15–3 are generally used as controllers. They are rated in horsepower; use one rated at not less than the horsepower of the motor. They have built-in motor overload protection (discussed in a later paragraph). For smaller (fractional-horsepower) motors use the kind shown in the left-hand part of the illustration; it isn't much larger than an ordinary toggle switch and controls the motor just as an ordinary switch would. For larger motors, use the kind shown in the right-hand part of the illustration. It has push buttons in the cover for starting and stopping the motor. It is also available without the push buttons, which are then installed separately at some convenient distant location. Motor starters and their method of operation are discussed in more detail on page 593.

Fig. 15–3 Controls of these types are "manual motor controllers." They are used to start and stop a motor; they also contain motor overload devices. If marked SUITABLE AS MOTOR DISCONNECT they can additionally function as the required disconnecting means. *(Square D Company)*

If you intend to use one of these devices as a motor disconnect instead of providing an additional switch similar to the one in Fig. 15–2, be certain that they have the additional marking SUITABLE AS MOTOR DISCONNECT. In addition, for motors larger than 2 hp, these manual motor controllers must be located on the load side of the final motor branch-circuit short-circuit and ground-fault protective device.

"In sight from" requirements A motor disconnecting means must be in sight from the motor controller; there are no exceptions that normally apply to this rule. In addition there must be a disconnect in sight from the motor and its driven machinery. The *NEC* defines "in sight from" as being visible and not over 50 ft from the specified location. If you can see one component from the other but they are more than 50 ft apart, they are not "in sight from" each other. This concept, which originated in the *NEC* motor article, is now formally defined in *NEC* Article 100 for use throughout the *NEC*.

If the controller is not in sight from the motor and its driven machinery, you must install an additional disconnecting means that is in sight from the motor and its driven machinery. This additional disconnecting means must meet the requirements for the disconnecting means already discussed, but if it is a switch it need not have fuses. There is an exception that allows the in-sight disconnect to be omitted if the disconnecting means for the controller can be individually locked in the open position; however, effective with the 2002 *NEC* that exception is now limited to 1) industrial occupancies with written safety procedures that ensure that only qualified persons will service the motor, or 2) other installations if the additional disconnect would introduce additional hazards or would be impracticable. For example, it would be plainly impracticable to place a disconnect 50 ft down a well shaft to be "in sight" (not over 50 ft distant) from a submersible pump motor 100 ft down the same shaft. Variable frequency drives should not have the motor disconnected unless the drive itself is disconnected, and multi-motor equipment may cause hazards unless a coordinated stop is arranged. Large motors (over 100 hp) only need isolation switches anyway, so their disconnects can be remote, and additional disconnects in hazardous (classified) locations only exacerbate the explosion hazards (refer to the end of Chapter 31 for more information). For many common residential, commercial, and farm applications, however, there must now be an additional disconnecting means located within sight of the motor regardless of the lockability of the controller disconnect. This was a major change in the Code. The unrestricted remote locking rule had been in the *NEC* more than 60 years.

Provide motor overload protection   The branch-circuit breaker or fuses, if of an ampere rating high enough to carry the starting current of the motor, will rarely protect the motor and its supply wires against damage caused by an overload that continues for some time, or by failure to start. Therefore a separate overload protective device must be installed to protect the motor and its branch-circuit wires against such damage.

The *NEC* does permit manually controlled portable motors rated 1 hp or less to be plugged into an ordinary 15- or 20-amp 120-volt circuit. All others must be provided with separate overload protection.

Often, and especially in the case of smaller motors, the motor-overload device is built into the motor and is called a "thermal protector." A thermal protector has one or more heat-sensing elements that protect the motor from dangerous overheating due to overloads, or failure of the motor to start, by opening the circuit when the motor temperature exceeds a safe value. Usually a thermal protector must be reset by hand when it trips; in some cases it resets automatically after a short interval. Of course the overload condition should be corrected before a motor is put back into operation. Motors equipped with thermal protectors must be marked THERMALLY PROTECTED on their nameplates. Never use a motor with an overload device capable of automatic restarting after an overload trip if such a restart could result in an injury. For example, if a saw motor were to restart after it cooled down following an overload, service personnel trying to clear the saw could be severely injured.

If a motor is permanently installed and not provided with built-in protection, separate overload protection must be provided. For the kinds of motors used in homes and on farms, the overload device must have an ampere rating generally not more than 25% above the ampere rating of the motor. This additional 25% limit is what makes the overload protection suitable for protecting the wires as well as the motor from overload, since the minimum wire size is also based on being 125% the motor full-load current. You may have noticed that throughout this chapter wire sizes, disconnect sizes, and short-circuit and ground-fault protective devices have been related to the tables at the end of *NEC* Article 430. Running overload protection is different. This is the one motor application where you must use the nameplate full-load current rating, and not the table values.

For running overload protection, you will most likely use "heater coils" that are installed inside your motor controller if it is the type shown in Fig. 15–3. The heater coils are rated in amperes and must be selected to match the ampere rating of the motor. If properly selected, the overload device(s) will carry the normal running current of the motor indefinitely, will carry a small overload for some time, but will trip the controller and shut off the motor quickly in case of heavy overload or failure to start. Such a starter will then serve as the controller as well as the overload protection device. It must never serve as the disconnecting means unless separately marked SUITABLE AS MOTOR DISCONNECT. A more detailed discussion of motor and branch-circuit running overload protection begins on page 599.

Note that motor overload devices are quite capable of handling the normal motor current as well as the higher current that might result from a nominal overload or from failure to start. But they are entirely incapable of handling the very much higher currents that would develop in case of a short circuit or ground fault; such currents are interrupted by the motor branch-circuit short-circuit and ground-fault protection. Unlike other branch circuits, motor circuits usually have their protective functions split between two devices, one for overloads and one for shorts; taking away either protective device from a motor circuit creates a hazard.

MAINTAINING RESIDENTIAL AND FARM MOTORS

Motors require very little care. The most important is proper oiling of the bearings. Use a very light machine oil, such as SAE No. 10, and use it sparingly— many people oil motors too much. Never oil any part of the motor except the bearings; under no circumstances put oil on the brushes, if your motor has brushes.

In most commonly used motors, there is a choice of sleeve bearings or ball bearings. If the motor is to be operated with the shaft in the ordinary horizontal position, there is no need for ball bearings. If the motor is to be operated with the shaft in an up-and-down position, ball bearings should be used because the usual sleeve-bearing construction lets the oil run out. Ball bearings are also better able than sleeve bearings to absorb the weight of the rotor. Ball bearings of some types are filled with grease and permanently sealed, eliminating the nuisance of greasing. If this type is purchased, be sure the bearings are double-sealed, that is, with a seal on each side of the balls. Some bearings are sealed on only one side so that the grease can still get out on the other side.

CHAPTER 16
Installing Service Entrances and Grounds

AN OVERVIEW OF THE WIRING SYSTEM, positioning the service in that overall picture, is presented in Chapter 4. Selection of the service-entrance wires, rating of the service equipment, number of branch circuits, and similar essentials are examined in Chapter 13. The actual installations of the materials selected are explained in the present chapter. Many variations are possible in the selection and arrangements of the different parts. The service-entrance wires may come in through conduit or they may be in the form of service-entrance cable; the supply may be overhead or underground; the meter may be inside or outside; the overcurrent devices may be circuit breakers or fuses, etc. Be sure to check with your power supplier for required location of the meter; usually the meter must be installed outdoors.

Starting with a look at how overhead and underground service conductors arrive at a building, the chapter then covers two methods of installing service conductors on the side of a building, first using conduit and then service-entrance cable, including selection and routing, and how to bring the service conductors into the building. Metering and making connections in service conductors are discussed. The last part of the chapter covers how to install an appropriate grounding system, concluding with what other systems need to be bonded to it. It will be helpful for you at this point to review the Chapter 9 discussion defining the differences among the terms ground, grounded, and grounding. Also review the general topics related to services in Chapter 4.

FROM THE STREET (OR UTILITY RIGHT-OF-WAY) TO THE BUILDING

NEC Article 100 defines service-drop wires as "the overhead service conductors from the last pole or other aerial support to and including the splices, if any, connecting the service-entrance conductors at the building or other structure." If the wires are underground rather than overhead, they are called service laterals instead of service-drop wires.

Overhead services begin with the service drop Service-drop wires are usually, but not necessarily, furnished and installed by the power supplier. The insulators that support the wires on the building are sometimes furnished by the owner or contractor, depending on the location of the service point—the point where the owner's responsibility for the premises wiring begins as defined by law and regulations. If the service point is, for example, the point of attachment to the building (frequently the case), the power supplier determines the size of the service-drop wires, which are often smaller than the service-entrance wires. (The distribution practices of electric utilities are controlled by a different code.) If this leads you to wonder about too much voltage drop, remember the drop is ahead of the meter, thus the occupant of the building does not pay for wasted power. Moreover, the power supplier can adjust the voltage so it is correct at the service equipment. If the service point is at the street, the usual *NEC* rules apply to the size of the drop conductors.

Service insulators provide support   Insulators must be provided for supporting the service-drop wires where they reach the building (point *A* in Fig. 16–10). These may be the simple screw-point insulators shown in Fig. 16–1; according to the *NEC* they must be kept a minimum of 6 in. apart. More frequently used is the type shown in Fig. 16–2, known as service brackets or secondary racks. Triplex cable is being used in service drops with increasing frequency and requires only one insulator; the cable consists of two insulated wires (or three—"quadruplex"—for applications needing another insulated wire) spirally wrapped around a bare neutral, as shown in Fig. 16–3.

In ordinary residential construction, the insulators should be installed as high above the ground as the shape and structure of the building permits, but below the level of the service head. If a building is very high, just observe the clearances required by the *NEC* as discussed under the next heading. *NEC* 230.54 requires the point of connection of the service-entrance wires to the service-drop wires (point *C* in Fig. 16–10) to be lower than

Fig. 16–1 Screw-point insulators of this type are used to support outdoor wires.

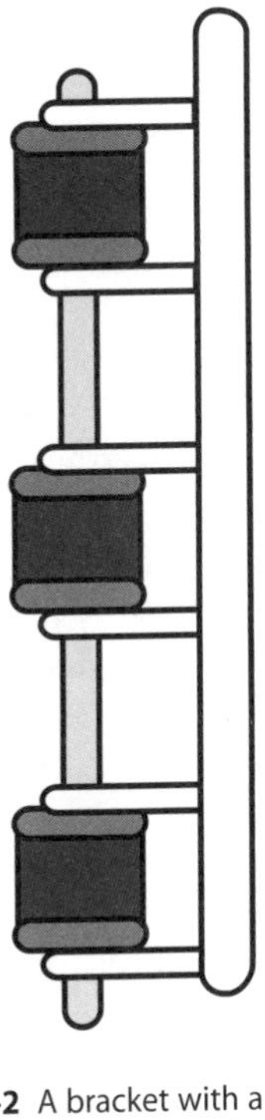

Fig. 16–2 A bracket with a number of insulators may be used instead of separate insulators.

Fig. 16–3 Triplex (or quadruplex for three-phase, four-wire) cable of this kind is increasingly used by power suppliers in the service drops to their customers' buildings.

the service head (point *B* in Fig. 16–10). A difference of a foot or so is sufficient. If the structure of the building makes this impossible, the same *NEC* section permits locating the insulators and the service head at approximately the same level but not more than 24 in. apart. In that case, drip loops as shown in Fig. 16–4 must be installed, and the splice between the drop and the entrance wires must be made at the lowest part of the loop. The goal is to prevent water from following the wires into the conduit or cable. Such splices must be taped carefully. New styles of splicing arrangements are available that include integral insulation over the splicing device.

Clearance of service-drop wires   *NEC*

230.24 specifies the clearances that service-drop wires must maintain from other objects:

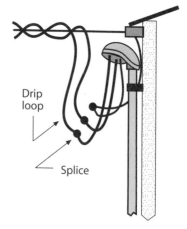

Drip loop

Splice

Fig. 16–4 If it is impossible to locate the service head higher than the insulators, install a drip loop in each wire. Be sure each splice is insulated with tape, or use listed assemblies that include an insulating covering.

- Minimum of 8 ft above a roof. If there will be pedestrian or vehicular traffic on the roof, such as on many parking garages, then the normal grade clearances apply. If the roof has a 4-12 pitch (at least 4 in. of rise in every 12 in. of run) or steeper, the clearance drops to 3 ft if the voltage doesn't exceed 300 volts between conductors. The clearance drops to 18 in., with the same voltage limitation, if the drop conductors connect to a service mast extending through a roof overhang and the path of the drop conductors limits the length over the roof to 6 ft on a total horizontal span of 4 ft. This occurs on roofs with wide overhangs when the drop approaches the building from an angle instead of being perpendicular to the building wall. In measuring roof clearances, mentally extend the roof plane 3 ft in all horizontal directions and continue the clearances above that extension, with one exception: The final span, where the drop attaches to the building, is exempt from this rule. That exemption prevents a porch on the gable front of a building from effectively obstructing access to the side of the building by a drop run from the street in front because the highest point on the side wall doesn't meet the 8-ft criterion of the main rule. See Fig. 16–5.

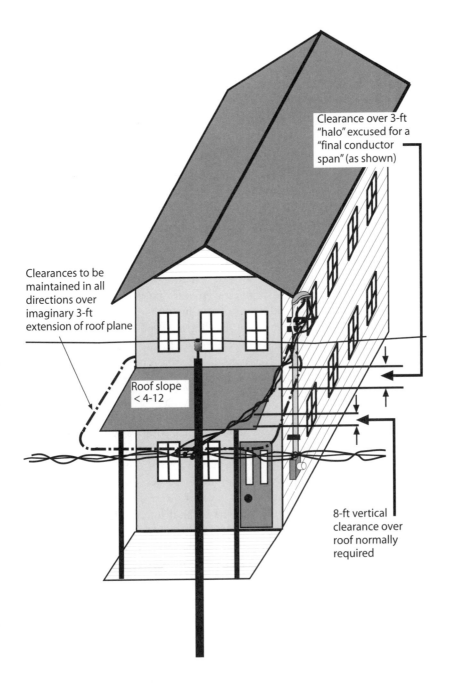

Clearance over 3-ft "halo" excused for a "final conductor span" (as shown)

Clearances to be maintained in all directions over imaginary 3-ft extension of roof plane

Roof slope < 4-12

8-ft vertical clearance over roof normally required

Fig. 16–5 The point of attachment for the final span of this service drop need not be 8 ft above the imaginary 3-ft extension of the roof plane, but any open runs of service conductors, and the service drop itself where above the actual roof surface, must meet the 8-ft clearance requirement.

▧ Minimum of 10 ft above finished grade, sidewalks, or from any platform or projection from which they might be reached. This clearance applies to the bottom of any drip loops, so the actual span clearance may be somewhat higher. This dimension applies only to systems using triplex or quadruplex running not over 150 volts to ground.

▧ Minimum of 12 ft above residential driveways and commercial areas such as parking lots and drive-in establishments not subject to truck traffic, provided in each case the voltage isn't over 300 volts to ground.

▧ 15 ft above commercial areas, parking lots, agricultural or other areas not subject to truck traffic, as in the 12-ft classification, except in this case the voltage to ground could exceed 300 volts.

▧ 18 ft above public streets, alleys, roads, and driveways on other than residential areas.

For any open service conductors (not Type SE cable, for example), *NEC* 230.9 requires a minimum clearance of at least 3 ft from windows (unless installed above windows), doors, porches, fire escapes, and similar locations from which the wires could be reached, and points of attachment for final spans must allow for the same clearance above the local standing level (porch, fire escape, etc.) as would be required from grade. For a glass surface to qualify as a window in this case, it must be able to open. Never install a service drop in a way that could obstruct access to a building opening through which materials may be moved, such as a hayloft door. See Fig. 16–6.

Underground services use service laterals   When the power supplier's wires to a building are installed underground, they constitute a service lateral. The lateral ends where the wires enter the building and connect to the service-entrance wires at any enclosure. In the case of outdoor metering, that enclosure would usually be the outdoor meter socket. Figure 16–9 shows an example of a service lateral ending at a pull box just inside the building wall. If the wires enter the building and continue without interruption directly to the service equipment, the continuous wires at the point where they enter the building become service-entrance wires, and the entire interior portion of the lateral must comply with *NEC* requirements for the service-entrance wires.

Selecting underground service wires   Where service wires are to run underground, either of two methods may be used. Any kind of wire suitable for use in a wet location (any wire with a "W" in its type designation) may be installed in an underground raceway. But in most installations of the kind covered in this part of the book, a special cable known as Type USE (for Underground Service Entrance) is used. It may be buried directly in the ground. The single-conductor type is shown in Fig. 16–7, but it is also available in multiconductor type. The individual conductors are insulated with an especially moisture-resistant compound, with another final layer of insulation that is mechanically very tough besides being moisture-resistant.

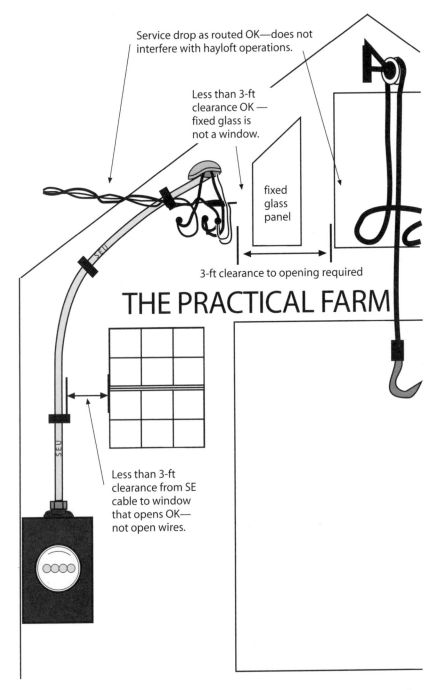

Service drop as routed OK—does not interfere with hayloft operations.

Less than 3-ft clearance OK — fixed glass is not a window.

fixed glass panel

3-ft clearance to opening required

THE PRACTICAL FARM

Less than 3-ft clearance from SE cable to window that opens OK— not open wires.

Fig. 16–6 Service drops must not be run where they are likely to be touched or where building operations would likely interfere with them.

Fig. 16–7 Type USE cable. It is usually used in the multiconductor variety.

In *NEC* 300.5 you will find the requirements that must be observed in installing any underground wires, including service-entrance wires. The burial depths are measured from grade down to the top of the underground installation, which means that the depth of the trench must equal the *NEC* requirement, as follows, plus the thickness of the conduit or cable involved:

- If installed under a building, all conductors (including those normally acceptable for direct burial) must be installed in a raceway extending beyond the outside wall of the building.

- If installed underground but not under a building, cable buried directly in the ground must be buried at least 24 in., with an allowance to bring that up to 18 in. if there is a 2-in.-thick concrete warning slab set in the trench above the cable. The same allowance applies to a 4-in. slab at the surface, such as a sidewalk that extends beyond the cable location by at least 6 in., or a driveway or parking area similarly constructed and used only for dwelling purposes. Rigid nonmetallic conduit also uses the 18-in. depth, with an allowance to bring that up the same 6 in. to a 12-in. burial depth with 2-in. slab of concrete in the trench. Although this last point doesn't apply to service work, you should be aware that only in residential occupancies, if the cable serves as a branch circuit and not as a service, and if the circuit has GFCI protection and isn't used on more than a 20-amp circuit, the burial depth comes up to 12 in.

- If you use steel conduit, the required depth is only 6 in.

- If you run wire under public streets, alleys, driveways, and parking lots, the minimum burial depth is 24 in. regardless of the wiring method or degree of concrete encasement.

There are other points to consider as well, regardless of burial depth:

- When using direct-burial cable, the cable must be protected against possible physical damage. Use only clean earth. If the backfill contains rocks, slate, or similar materials, the cable must first be covered by sand or clean soil.

- Where a direct-burial cable enters a building, it must be protected by a raceway extending from underground (equal to 18 in. or the required wiring depth, whichever is less) to the point of entrance; where it rises on a pole (or high on the side of a building), similar protection must be provided to a height of at least 8 ft above ground level.

Protecting underground wires If there is any likelihood of future disturbance by digging, or if the cable is located where there might be deep ruts in driveways such as on farms, it is wise to place a board or similar obstruction over the cable to serve as a warning. In the case of service conductors buried at least 18 in., the

NEC now requires a warning ribbon to be placed in the trench at least 12 in. above the underground installation unless the wiring is encased in concrete. The goal is to alert the excavator in time to avoid hooking the cable. If you use single-conductor cables, bunch them instead of having them spread apart. Do not pull any kind of cable tight from one end to the other; "snake" it a bit to permit expansion and contraction under the action of temperature or frost. In addition, if your underground installation rises to a meter socket or other enclosure, you need to be sure frost heaves don't turn the riser conduit into a spear that will damage the electrical equipment to which it is connected. Also beware of construction on fill, because earth settlement can pull raceways apart over time. Using an expansion fitting of the type shown in Fig. 11–23 in a vertical riser would solve these problems. Figure 16–8 shows a terminal adapter and an expansion fitting below it at a meter socket.

If the power supplier's area distribution is underground, the conductors will begin at a street main or at a pad-mounted transformer, in either case at a location likely under the exclusive control of the utility. If the area distribution is overhead, the underground service conductors will begin at the power supplier's pole. Poles are generally considered locations subject to physical damage, and that means using the heavier-wall Schedule 80 product in the case of rigid nonmetallic conduit. Even if you have used Type USE cable, suitable for direct burial, the *NEC* requires mechanical protection such as conduit for the cable in the aboveground portions. Although the *NEC* requires conduit to extend only 8 ft up the pole, it is usually easier to run it all the way to the top of the pole and install a service head at the top of the conduit. At that point all the wires of the cable can be spliced to the power supplier's wires. Some power suppliers require that the metal conduit stop at about 8 or 10 ft above ground level, in which case continue the cable to the top of the pole but make a strong watertight seal at the top end of the conduit where the wire conductors (individual wires or multiconductor cable) emerge from the conduit.

If you use steel conduit, don't forget that metal service raceways must always be bonded, even where physically separated from the service equipment. Because this is the line side of the service equipment, this isn't difficult to arrange. Just install a bonding bushing on the upper end of the steel riser and connect a bonding jumper, generally the same size as the neutral, from the bushing to the grounded circuit conductor continuing up the pole. You can also make this connection in the earth at the lower end, but you may have to do some searching for suitable grounding hardware. Grounding hardware suitable for use in the earth (or concrete) is marked DIRECT BURIAL or DB. If the steel conduit directly transitions to nonmetallic conduit, the bonding wire will have to run along the outside of the nonmetallic conduit all the way to its end. Although *NEC* 250.102(E) normally restricts the length of such bonding conductors to 6 ft, an exception allows a longer length on poles to address this practical problem.

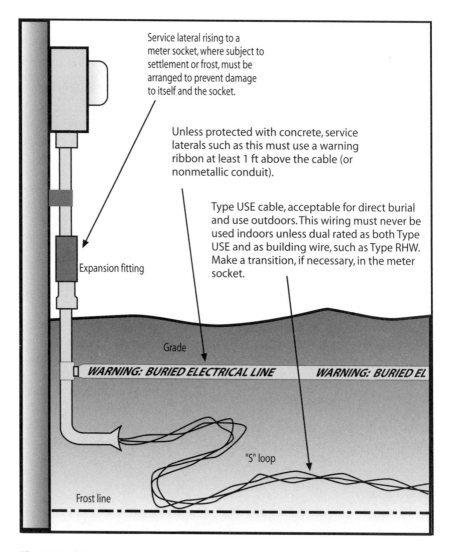

Service lateral rising to a meter socket, where subject to settlement or frost, must be arranged to prevent damage to itself and the socket.

Unless protected with concrete, service laterals such as this must use a warning ribbon at least 1 ft above the cable (or nonmetallic conduit).

Type USE cable, acceptable for direct burial and use outdoors. This wiring must never be used indoors unless dual rated as both Type USE and as building wire, such as Type RHW. Make a transition, if necessary, in the meter socket.

Expansion fitting

Grade

WARNING: BURIED ELECTRICAL LINE   WARNING: BURIED EL

"S" loop

Frost line

Fig. 16–8 This expansion fitting in this installation protects the meter socket from frost-heaving and earth settlement. Note the restriction on indoor use of conventional Type USE cable, unless it is dual-rated.

Be aware that simply driving a ground rod at the pole and making a grounding connection accomplishes nothing as far as solving your electrical problem because it leaves the relevant parts yet unbonded. There is a variation on this problem that arises when using nonmetallic conduit with steel 90-deg sweeps. This is done when difficult pulling is anticipated, and the installer doesn't want the pulling line under tension to saw through the inner radius of a nonmetallic sweep, which can easily happen. The *NEC* now allows such sweeps to be unbonded, provided they are buried so their highest point is still at least 18 in. below grade.

The bottom end of the protective conduit must extend at least 18 in. below grade (or minimum wiring depth if less) and it should be arranged so the cable will not be damaged where it exits. Either point the conduit straight down into an even deeper trench, or install a 90-deg sweep on the end at the bottom of the trench. Install a bushing on the end of the conduit even if it is nonmetallic conduit. Be sure, especially in northern climates, to provide an S-shaped loop in the cable to help prevent damage to the cable caused by movement of the earth (frost-heaving) during the transition from winter to spring.

At the house end, if the meter is the outdoor type, follow the same procedure as at the pole, using a short length of conduit from the meter socket down into the trench. But if the cable is to run through the foundation to an indoor meter, follow the procedure shown in Fig. 16–9. After the cable has been installed, fill the openings around the conduit and inside the conduit with commercially available "duct seal" compound to prevent rainwater, melted snow, or leaking gas from underground sources from following the cable into the building.

SERVICE-ENTRANCE WIRING ON A BUILDING

The wires from the point where the service drop (or lateral) ends, up to the service equipment, are service-entrance wires. They may be Type TW, RHW, THW, or any other type suitable for outdoor (wet) locations—in other words any type that has a "W" in its designation. They may be separate wires brought in through conduit, or wires made up into service-entrance cable approved for the purpose.

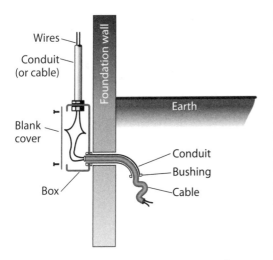

Conduit can be used to meet an overhead supply Figure 16–10 shows the essential elements of this type of service. At the top of the service conduit (point B in Fig. 16–10), the NEC requires a fitting that will prevent rain from entering the conduit. A fitting of this type is shown in Fig. 16–11; it is known by various names such as service head, entrance cap, or weather head. It consists of three parts: the body, which is attached to

Fig. 16–9 If underground conductors cannot be run directly to the service equipment, use the construction shown in changing from underground to interior-type wiring.

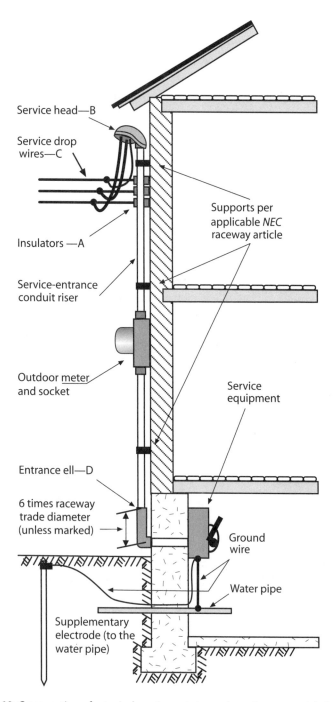

Service head—B

Service drop
wires—C

Insulators —A

Service-entrance
conduit riser

Outdoor meter
and socket

Entrance ell—D

6 times raceway
trade diameter
(unless marked)

Supplementary
electrode (to the
water pipe)

Supports per
applicable *NEC*
raceway article

Service
equipment

Ground
wire

Water pipe

Fig. 16–10 Cross section of a typical service entrance using wires to conduit. Locate the service head *B* higher than the insulators *A*.

the conduit; an insulating block to keep the wires apart where they emerge; the cover that keeps the rain out and holds the parts together.

Entrance using conduit A typical installation is shown in Fig. 16–10. The size of the conduit and fittings will depend on the size of the service wires as discussed in Chapter 13.

Conduit bodies used for service entrance work (and other large conductors) At the

Fig. 16–11 A service head is installed at the top of the conduit through which wires enter the building. *(Hubbell Electrical Products)*

point where the conduit enters the building (point *D* in Fig. 16–10), it is customary to use an entrance conduit body of the type shown in Fig. 16–12. With the cover removed, it is a simple matter to pull wires around the right-angle corner. Be careful to observe the sizing rules in *NEC* 314.28(A)(2) and (3) in choosing this fitting. The distance between the centers of the conduit stops (roughly equal to the long dimension of the cover) must not be less than 6 times the trade diameter of the entering conduits, or a full 12 in. for a 2-in. conduit. The distance between the conduit stop in the short dimension and the inside of the cover must at least equal the minimum bending radius for the size of conductors used, based on *NEC* 312.6(A). For 3/0 AWG copper conductors, rated 200 amps, that dimension is 4 in. Conduit bodies (which generally look like the letter "L"; those with the opening in the back are called "LB") that meet these spacings are generally referred to as "mogul" fittings; conventional conduit bodies won't meet *NEC* requirements.

The only other option is to see if a conduit body of smaller dimension has been listed for a conductor fill that meets your needs. For example, there are no

Fig. 16–12 A conduit body is used at the bottom of the conduit where it enters the building. The one shown doesn't meet the normal spacing rules, so it has an allowable conductor fill as marked.

"mogul" rigid nonmetallic conduit bodies, so they are all (if listed to U.S. standards) routinely investigated for allowable wire fill. Look for a "3 4/0 XHHW MAX" or similar marking. If you aren't using XHHW wires, or if you're using combinations of sizes, just compare the actual cross-sectional area of the conductors with the area represented in the listing instructions. This fitting also must be raintight and arranged to drain; drill a small weep hole in its bottom end before you pull the wire.

After completing the pull, and assuming the raceway passes from the outdoors to an indoor location with air that can be significantly warmer or cooler than outdoor air, stuff some sealing putty around the wires where they enter the raceway. This will obstruct the circulation of humid air to a point where condensation could occur at the colder end. The *NEC* has routinely enforced this principle on raceways passing from cold storage areas into warm rooms; it also applies to raceways (or cable sleeves) "passing from the interior to the exterior of a building."

Installing service conduit and pulling wires For now, assume that you have made an opening into the building for the service conduit to enter, and have marked the location of the meter, usually about 5 ft above ground level. (Some power suppliers have specified heights for meters, so check before deciding where to locate it.) In some cases, you can preassemble the service conduit, including the meter socket, before installing it on the wall. Follow the installation requirements in Chapter 11 for the wiring method you are using. Inside the building, the conduit must be anchored to the service equipment using the locknut-and-bushing procedure shown in Fig. 10–13, but instead of an ordinary bushing, use a grounding bushing as described later in this chapter.

The *NEC* requires the conduit system to be in place between points of access before pulling wires, so this step occurs after the conduit riser is secure. The conduit for a typical single-phase service will contain three wires. The neutral if insulated must be white if available. The other two wires should be red and black (or both black if the two wires need not be distinguished); larger sizes usually are available only in black, however, and if this is the case the neutral (which must always be distinguishable) must be reidentified with white tape or paint at all terminations. Cut three lengths to reach from the meter socket to the top, plus about 3 ft extra. Pull these three wires into the conduit; connect to the meter socket as shown in Fig. 16–13; let the extra length project out of the top through the entrance cap. Since the conduit is very short you will probably be able to push the wires through it without any trouble; if you wish, use fish tape. Cut three more lengths to reach from the meter socket to the service equipment inside the house, allowing sufficient length so they can be neatly placed along the walls of the cabinet to reach the most distant connector instead of being run loosely to their connection points; this applies especially to the neutral.

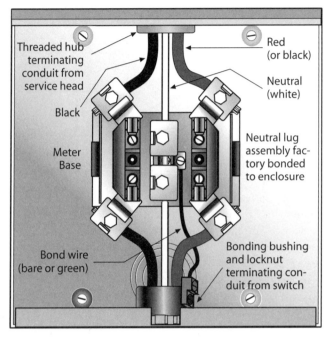

Threaded hub terminating conduit from service head

Black

Meter Base

Bond wire (bare or green)

Red (or black)

Neutral (white)

Neutral lug assembly factory bonded to enclosure

Bonding bushing and locknut terminating conduit from switch

Fig. 16–13 How wires are connected to the meter socket. The neutral wire is always connected to the center terminal. The threaded hub is presumed to have sufficient continuity, but the conduit connection at the bottom requires additional bonding, as shown.

Electrical metallic tubing (EMT)—alternative to rigid conduit   Instead of rigid conduit, EMT may be used in the service entrance. The procedure is the same as outlined for rigid conduit except that threadless couplings and connectors are used at the ends instead of the threaded fittings used with rigid conduit. Due to a product standard change, relatively few compression-style EMT fittings are listed as suitable for use in wet locations, so pay close attention to the installation instructions on this point for the outdoor portion of the job. Entrance caps and entrance ells of a slightly different design that clamp to the EMT are used. However, caps and ells designed for rigid conduit may also be used, because the external threads on the EMT connectors will fit threaded fittings designed for use with rigid conduit. Most mogul conduit bodies are made only with conduit threads anyway.

Masts help to maintain clearance   In wiring the rambler or ranch-house type of residence, it is often impossible to maintain the prescribed minimum clearances. In such cases, use a mast as shown in Fig. 16–14. When using masts, the minimum clearance above the roof overhang is only 18 in. if the overhang does not exceed 4 ft. Such conduit masts must be able to support the service-drop wires and should not be smaller than 2-in. rigid metal conduit unless braced or guyed for extra strain support. Some of the conduit fittings in mast kits are designed to fit either 2- or 2½-in. conduit without threading. Flashing is usually included in each kit to make a watertight opening through the roof overhang.

Service-entrance cable can also be used to meet an overhead supply Instead of separate wires inside the conduit, service-entrance cable is also commonly used. The most usual kind is Type SE shown in Fig. 16–15. It contains two insulated wires, black and red. The bare uninsulated wire consists of many strands of small aluminum or tinned copper wires wrapped spirally around the insulated wires. Over the neutral, there is a layer of plastic shielding tape for protection against moisture, and a final nonmetallic jacket, usually gray. If you wish, paint it to match the color of the house.

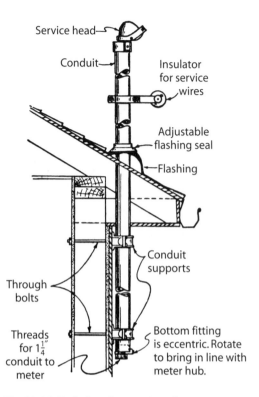

Fig. 16–14 Typical service mast installation.

All the small wires in the bare neutral collectively are sometimes equal to the same size as each insulated wire, but often are one size smaller as usually permitted for neutrals of services. All the small wires of the neutral must be bunched together and twisted at the terminal to become a single wire, as shown in Fig. 16–16. It is then handled as if it were an ordinary stranded wire.

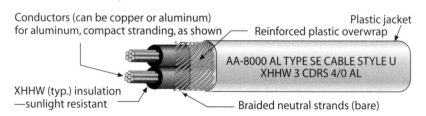

Fig. 16–15 Service-entrance cable, Type SE. The neutral consists of many strands of bare wire wrapped spirally over the insulated wires.

Secure the cable to the building within 12 in. of the meter socket, entrance cap, entrance ell, service equipment, cabinet, etc., and at additional intervals not over 2½ ft. The service head is slightly different from the type used with conduit in that it is supported on the building instead of being supported by the conduit; Fig. 16–17 shows one style.

Fig. 16–16 The small wires of the neutral are bunched together into a single wire for connection to a terminal. ▶

Strands of bare——
wire, twisted

◀ **Fig. 16–17** A typical service head for service-entrance cable. *(The Halex Company)*

Cable is anchored to the meter socket by means of waterproof connectors; two types are shown in Fig. 16–18. Such connectors incorporate soft rubber glands; tightening the locking nut or screws compresses the rubber, making a watertight seal around the cable. In use, the connectors are screwed into the threaded openings of the meter socket (the threads are usually already treated with a water-proofing compound). Next the cable is slipped through the rubber gland, and the locking nut or screws are tightened. Indoors, and on the bottom of the meter socket, waterproof connectors often are not required; ordinary connectors of appropriate size can be used if the inspector agrees. Don't assume that a connector is watertight just because it has the neoprene gland construction. Watch for a separate RAINTIGHT marking on the carton.

At the point where the cable enters the building, the *NEC* requires that water be prevented from following the cable into the building. You can do this by arranging the geometry of your lower cable run to shed all water prior to arriving at the service equipment, but the simplest and most workmanlike way to seal this point is to use a sill plate such as shown in Fig. 16–19. Soft waterproofing compound that comes with the plate is used to seal any opening that may exist.

Special considerations for routing and selecting service conductors   In most cases it is simple to let the service-entrance conduit or cable run straight down from the weatherhead, through the meter socket, and then continue straight down to the point where it enters the building, as shown in Fig. 16–10.

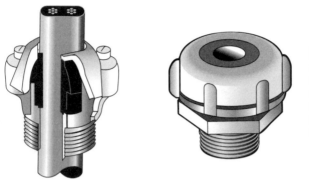

Fig. 16–18 Watertight connectors for service-entrance cable. The rubber gland in the connector squeezes against the cable, making the installation watertight.

Sometimes that is impossible, in which case do a neat job. Never run conduit or cable at an angle to the sides of the building unless you are against the overhang or rake on the gable end of a building, in which case follow its angle exactly. Otherwise let the runs be either straight up and down or horizontal.

Sunlight exposure The *NEC* now prohibits conductors being run exposed to direct sunlight unless they are listed SUNLIGHT RESISTANT. UL has amended the guide card provisions on Type SE cable to say that its insulated conductors and its outer jacket "are suitable for use where exposed to sun." That may solve the problem with SE cables (the product standards require a more severe test for the actual SUNLIGHT RESISTANT marking than individual SE cable conductors actually undergo), but the issue for pipe services is more difficult. If you are planning a conduit (or tubing) service, and it isn't on the north side of the building, be sure to ask the local inspector what the policy is in the local jurisdiction, and check with local supply houses to see if building wire with the appropriate listing recognition is available. The 2005 *NEC* also recognizes appropriately listed sleeving or tape applied over conventional insulation.

Fig. 16–19 Install a sill plate where cable enters the building. Soft sealing compound seals the opening to keep out water.

Ampacity allowances for residential uses The ampacity rating of the cable generally follows the normal rules as presented in the page 105 discussion on using *NEC* Table 310.16 for selecting conductors. However, for single-phase, 120/240-volt applications involving a single dwelling unit, the *NEC* gives extra leeway in sizing service conductors, as well as feeder conductors that stand as the main supply to the lighting and appliance branch-circuit panelboard for that dwelling. In these applications *NEC* Table 310.15(B)(6) allows a general single-wire size reduction from what you would normally need to provide. For example, a 100-amp service gets by with 4 AWG copper (or 2 AWG aluminum) instead of the normal 3 AWG copper (1 AWG aluminum), and a 200-amp service uses 2/0 AWG copper (4/0 AWG aluminum) instead of 3/0 AWG copper (250 kcmil aluminum). This occurs because on a true single-phase distribution, the total heating from three wires, one of which carries only unbalanced current, must always be less than that produced by three wires carrying full current such as to a three-phase motor. In effect, this special table is an ampacity bonus instead of derating. The other uniquely motivating factor involves the inherent diversity in a dwelling load profile; if a dwelling carries a high current at some time, that magnitude isn't likely to continue very long before something shuts off. The same allowances apply to individual wires run in a raceway.

Size of neutral wire   The neutral of the service entrance may in many cases be smaller than the insulated wires. This subject is covered in the grounded neutral topic beginning on page 235. Many people use a rule of thumb that the neutral can be one size smaller than the hot wires if the hot wires are 6 AWG or larger. This rule of thumb is only an approximation, and while usually acceptable, it is a good idea to actually calculate the minimum size permitted for the neutral.

Uninsulated wire in service entrance   In the wiring of houses and farms, the neutral wire in the service entrance is grounded. *NEC* 230.41 permits (but does not require) an uninsulated wire for the grounded neutral wire of the service entrance. If you use Type SE service-entrance cable, the neutral is uninsulated but covered with an outer cable jacket. If the underground service is made using individual wires in conduit, the neutral may be bare if copper, but not if aluminum. If individual Type USE conductors are used, buried directly in the earth, the neutral if copper may be bare "where bare copper is judged to be suitable for the ground conditions," which means "if the inspector agrees." It is the writer's opinion that there are few locations where soil conditions are not suitable for bare copper. If all the wires are in the form of a multiconductor Type USE cable, which has a moisture- and fungus-resistant outer cover, the neutral may be bare whether copper or aluminum.

SERVICE CONDUCTORS ENTERING BUILDINGS

In frame construction, it is a simple matter to bore a hole through the wall into the building to provide entrance for service conductors. Hard labor is involved if the building is of brick or concrete construction. Use one of the methods shown in Fig. 21–20, depending on circumstances. For really large holes, especially where reinforcing steel must be cut in the process, use hole saws with diamond-impregnated teeth designed for core boring in concrete. Carbide-tipped drills are quickly ruined by hitting reinforcing steel.

Passing through thermal insulation presents quite a different problem due to its potentially devastating impact on ampacity. The ampacity tables assume at least some circulation of air and an absence of barriers to the radiational ability of the wiring method. Wiring embedded in thermal insulation has neither characteristic. Review the discussion in Chapter 11 on this topic, which applies with particular urgency to the larger service and main feeder wires. The wires on either side of a wall penetration will work as a heat sink for a short pass straight across, but running along the long dimension of an insulated stud cavity is asking for trouble.

LOCATION AND WIRING OF SERVICE EQUIPMENT

Review the Chapter 4 discussion on service locations in terms of "nearest the point of entrance." The point of entrance is the point of penetration of the outer skin of the building envelope. For example, entering a stud cavity on the living room wall and descending into the basement to the service equipment is a violation, because the readily accessible point nearest the point of entrance is in the living room. On the other hand, running the service conductors on the outside of the building and then passing directly (or with insignificant offset) through the same stud cavity from the outdoors into the basement is perfectly acceptable. If that option isn't available, either encase the service run in at least 2 in. of concrete (or other option in *NEC* 230.6), or use a combination meter and service disconnect outside the building. Make the grounding connections at that point, and come into the building with a feeder. Since the feeder has full overcurrent protection at the service equipment, you may run it anywhere in the building.

Service equipment   Most of the details of the service equipment are explained in Chapter 13, which you should review in conjunction with this discussion. Run the hot incoming wires to the terminals marked MAIN and the neutral to the grounding busbar. Spend a few extra minutes to run the wires neatly within the cabinet, bending them to run parallel to the sides or ends of the cabinet. You can then take pride in your workmanship.

Provide adequate workspace for service equipment   Access must be provided around all electrical equipment, even receptacles and snap switches, to allow for safe operation and maintenance. Electrical equipment likely to require testing or maintenance while energized, a category universally deemed by the inspection community to include service equipment, must meet specific workspace dimensions. It is a safety hazard to force electrical maintenance personnel into contortions while they work on equipment. They are more likely to drop tools onto energized parts and their escape access would be compromised.

Provide a clear workspace from the floor (or platform) to the height of the equipment or 6½ ft, whichever is higher, and equal in width to the equipment but never less than 2½ ft wide. That area must be entirely clear of obstructions, with the point of reference being the plane of the front of the equipment. However, related electrical equipment can intrude, but not more than 6 in. For example, you could put a 12-in. deep surface-mounted wiring trough under a 6-in. deep panelboard mounted on the same surface. In any case, equipment doors must be able to open a full 90 degrees. Make sure the height of any switch or circuit breaker handles does not exceed 6 ft 7 in. above the floor (a topic more fully addressed in Chapter 30 on page 588).

The minimum depth of the workspace for most residential work is 3 ft, also measured from the plane defined by the front of the equipment. However, if the voltage exceeds 150 volts to ground, which does not include conventional 120/240V or 208Y/120V systems, the workspace depth increases to 3½ ft if the opposite wall is a grounded surface (defined to include concrete, masonry, or tiled surfaces). If, with voltage exceeding 150 volts to ground, the opposite wall consists of similar equipment, potentially involving work on live parts on both sides of the workspace at the same time, the minimum distance increases to a full 4 ft.

In addition to providing workspace for maintenance personnel, you must also assure that no foreign systems pose a dripping or access hazard to the electrical equipment. The zone equal to the width and depth of the equipment and extending from the floor to 6 ft above the equipment, or to the structural ceiling (not a suspended ceiling) if lower, must normally be kept clear of foreign systems. No matter how high such foreign piping may lie, you must provide a drip shield if a break or condensation could occur above the equipment. This doesn't prevent you from providing sprinkler protection for an electrical room, but the sprinkler piping must comply with these rules. Be aware of coordination issues with other trades. For example, if the plumber gets there first and runs a 3-in. drain on the wall below where the service panel will go, then the panel has to be supported far enough off the wall so that its projected footprint, from floor to 6 ft above the panel, does not intersect any foreign system.

Soldered connections are prohibited   *NEC* 230.81, 250.8, and 250.70 prohibit the use of soldered connections in service wires, in the ground wire, or in a grounding wire. The reasons are not difficult to understand. Service wires are ahead of ("upstream" from) the main overcurrent devices. A high fault current might melt the solder, resulting in a loose connection and heating that could lead to a serious fault current and possibly even an explosion in the service equipment. In the ground wire or in a grounding wire, a melted soldered joint would result in a poor ground connection or no ground at all. Soldered joints are a thing of the past. Use solderless (pressure) connectors in all cases.

Meter and socket   The power supplier decides whether the meter is to be located indoors or outdoors; in most cases it will be the outdoor type shown in Fig. 16–20 with its socket installed exposed to the weather. The power supplier furnishes the meter. The socket is sometimes furnished by the power supplier, sometimes by the owner; in either case it is installed by the contractor. Some utilities require meter sockets with bypass switches that allow their personnel to remove the meter without disconnecting power to the occupancy, particularly in the case of commercial operations. Be sure to check with the local utility metering department to verify which socket they will accept. Remember that the utility need not energize your installation just because it meets *NEC* requirements; it must meet their service requirements as well.

PROVIDE A SAFE GROUNDING SYSTEM FOR THE SERVICE

As discussed in Chapter 9, the *NEC* differentiates between grounded conductors (which carry current in normal operation) and equipment grounding conductors (which carry current only under abnormal fault conditions). If insulated, the grounded conductor of a circuit is almost always white. The grounding conductor may be bare, or green if insulated; if armored cable is used, or a metal raceway (other than flexible metal conduit), no separate grounding conductor is required. At the service equipment a single ground wire (which the *NEC* calls the "grounding electrode conductor") is used to connect both the grounded and the grounding wires to the earth, a function achieved by the main bonding jumper that bonds the neutral busbar of the service equipment to its cabinet by means of a screw or flexible strap. Refer to the page 237 discussion on bonding the neutral (grounded) busbar.

Grounding electrode system

Most qualified grounding electrodes at a given building or structure must be bonded together as part of a common grounding electrode system. The *NEC* identifies seven electrodes, any or all of which (other than certain "local metal underground systems or structures"), if present, must be employed as electrodes.

In addition to being present, qualified electrodes must be at the building or structure served. A grounding electrode conductor (see the next topic) that extends great distances loses some of its effectiveness to inductive reactance.

Fig. 16–20 A detachable outdoor meter and the socket in which it is installed. Meters of this type are exposed to the weather.

The *NEC* does not insist that any particular one of these electrodes be installed. However, with one exception, any of the identified electrodes that meet both of these conditions must be used: Electrode is (1) present, and (2) located at the building or structure served. The exception is the last item on the following list, the local metal underground systems or structures—these items are qualified electrodes, but including them in the grounding electrode system is optional. The seven eligible electrodes are:

▪ A metal underground water pipe, in contact with earth for at least 10 ft. If the metal water pipe originates as the drop pipe within a well casing, but has at least 10 ft of direct earth contact prior to entering the building, it qualifies as an electrode, but it must also be bonded to the well casing. For *NEC* purposes the electrode portion of a water system effectively ends within 5 ft of the building entrance, and that is the only portion eligible for electrical connections. However, industrial and commercial occupancies with qualified maintenance staff are still permitted to make remote connections provided the entire length of the water pipe running back to the building entrance is exposed (could be above a suspended ceiling).

▪ An effectively grounded steel building frame.

▪ A concrete encased electrode consisting of not less than 20 ft of bonded reinforcing metal or 4 AWG minimum bare copper wire, all encased within a concrete footing.

▪ A ground ring encircling the building or structure, in direct contact with earth at least 2½ feet down, and consisting of bare copper not smaller than 2 AWG.

▪ A rod or pipe electrode, installed as covered below.

▪ A plate electrode, installed as covered below.

▪ Some other local metal underground system or structure, such as underground tanks or piping systems. This category generally includes metal well casings.

Underground water pipes don't have the reliability they once had. In a large city with cast iron water mains and everyone connected through copper water laterals, the total resistance to ground from any given connection may be less than 1 ohm, about as good as possible. On the other hand, today's power systems inject into municipal water systems sporadic dc currents and ac waveforms with multiple harmonics, emanating from such mundane sources as half-wave rectifiers in hair dryer speed controls. Many water utilities are trying to separate themselves from the power system for that reason. When they succeed, by installing such devices as dielectric unions, what was an excellent ground becomes no ground at all, and the property owner may not recognize the technical significance of what has taken place. For this reason the *NEC* requires that a water pipe electrode be supplemented by another qualified electrode. This is not because there is anything wrong with a water pipe electrode. As long as it is on the job, it may well be the best electrode on site. The supplementation rule has to do with the fact that the electrode, in the electrical sense, may quietly disappear without warning. The supplementary electrode must be viewed as a reserve electrode, one that will fully qualify with the water pipe removed from the picture. For example, if you are supplementing with a ground rod, and soil resistance is high, you must supplement with at least two ground rods a minimum of 6 ft apart (see below), because that is what you would have to do if no water pipe were present to begin with.

Metallic building framework is an excellent electrode, but seldom exists in houses. This concept applies to structural metal elements in direct contact with soil, or encased in similarly located concrete, or bonded to other qualified grounding electrodes. You might encounter it in prefab agricultural building construction, multifamily dwellings, commercial occupancies, and it is present in most industrial locations.

For new construction in rural areas, *concrete-encased electrodes* are excellent electrodes, but they are widely unappreciated. Suppose a new building is going up, with nonmetallic water service. You can go to the site after the basement walls are poured and drive some ground rods (discussed later in this chapter), which cost money in both materials and labor to drive properly. Or, knowing approximately where the service will end up, go to the job before the footings have been poured, taking along about 30 ft of 4 AWG bare copper. Lay at least 20 ft in the footing, perched up on small rocks so the concrete will fully envelop the copper. Leave 10 ft hanging out of the footing near the future service location. If you really want to be elegant (and add some corrosion protection to the copper), leave a short length of rigid nonmetallic conduit as a sleeve, to be placed at the eventual point of emergence from the floor slab when the concrete truck arrives. For little time and less money (not even a ground clamp of some form) you have an electrode that is at least one order of magnitude better than a ground rod in terms of lower impedance.

For generations, the *NEC* imposed a third condition on the electrodes required in the grounding electrode system, namely, that they be "available." Since once the pour is complete a concrete-encased electrode is no longer available, and since many building foundations are completed before the electrical contract is awarded, many qualified systems of reinforcing steel were not being included in the system as long as the availability criterion continued. Given the comparative quality of these electrodes, the grounding electrode system in these buildings suffered accordingly. One of the most important changes in the 2005 *NEC* is the removal of this qualification. Now (with an exception for existing buildings), a concrete-encased electrode must be used if it is present, whether or not someone had the foresight to make it available. This change will have enormous effects on how and when the construction industry schedules work and awards bids.

Ground rings are the principal electrode chosen on many small utility substations. The requirement to "encircle" makes them more difficult on larger projects.

If none of the principal electrodes is present or qualifies, you can always look to "*other local metal underground systems or structures*" or to rod, pipe, or plate electrodes. The first choice in a rural area may well be a metal well casing. Underground gas piping systems, however, must not be used as grounding electrodes. If existing "other electrodes" aren't feasible, or you choose not to use them for whatever reason, then you must supply an electrode, typically a ground rod or pipe. A ground plate at least 1 sq ft in size whether of ¼-in. (minimum) ferrous or 60 mil copper also qualifies if buried at least 2½ ft.

A single rod, pipe, or plate electrode must not exceed 25 ohms of resistance to earth. In areas where ground resistance is typically poor, the common practice is to immediately install two such electrodes spaced at least 6 ft apart (*NEC* minimum). That avoids the necessity for testing ground resistance, because the *NEC* takes the position that two tries is enough as a minimum, and doesn't ask for a third attempt. These electrodes vary dramatically in grounding effectiveness, based on soil moisture and electrolytes, and some measure of luck. In some areas a single rod routinely falls under the 25-ohm threshold; in other areas, even with moist earth, the resistance of a single rod routinely approaches or exceeds 1000 ohms.

Ground rods must be at least ⅝ in. in diameter unless of stainless steel or nonferrous rods or the equivalent, which can be as small as ½ in. if listed. The most common type is steel enclosed within a layer of copper. Driven pipe or conduit must be galvanized and at least ¾ in. in trade size. Either rod or pipe electrodes must be driven so at least 8 ft is in contact with earth. For the usual 8-ft ground rod, that means none of it should be visible after you are done. If you can't drive it vertically due to rock, drive it at an angle no more than 45 deg from vertical. If, and only if, that isn't possible, lay it in a trench so its entire length is buried at least 2½ ft.

Always ask qualified local people what to expect. There are special electrode constructions that condition the earth around them to reduce resistance in poorly conductive areas. Do not, however, ever use an air terminal (lightning rod) ground as an electrode for a power system. First establish an *NEC*-acceptable grounding electrode system, and then bond that system to the lightning protection grounding system.

Ground wire   The *NEC* calls this the "grounding electrode conductor" and explains the method of installing it in 250.64, and the minimum size in 250.66. The ground wire is usually bare but there is no objection to using insulated wire. It must never be smaller than 8 AWG but there is no objection to using a size larger than the minimum required. When using a made electrode (ground rod), the ground wire never needs to be larger than 6 AWG regardless of the size of the service wires. The ground wire, whatever its size, must be securely stapled or otherwise fastened to the surface over which it runs.

If the service wires are 2 AWG or smaller, the *NEC* permits an 8 AWG ground wire. Don't use it; use 6 AWG. Number 8 AWG must be enclosed in conduit or armor regardless of circumstances, but for 6 AWG this is not required. The cost of material when using 6 AWG is lower than when using 8 AWG and it takes less time to install the 6 AWG.

If the service wires are 1 or 1/0 AWG, the ground wire may not be smaller than 6 AWG. If it is installed free from exposure to physical damage and if it is stapled to the surface over which it runs, neither conduit nor armor is required.

If the service wires are 2/0 or 3/0 AWG, use 4 AWG ground wire. The ground wire must be fastened to the surface over which it runs. Conduit or armor is never required, but if installed where physical damage is likely, it must be otherwise protected. For example, if installed outdoors in an alley or driveway where it might be damaged by vehicles, protect it by installing a sturdy post or steel beam near the wire. If the service wires are 4/0 AWG or larger, you are involved with installations more typical of large commercial and industrial applications. Use the size given in *NEC* Table 250.66, and for more information refer to the discussion on grounding nonresidential systems beginning on page 541.

Ground wires preferably run without splice to their first principal electrode connection, although they can be spliced with a hydraulic crimping tool that makes an irreversible compression-type connection. They can also be welded using the thermite process (see Figs. 8–23 and 8–24).

Ground clamps   Select the type of clamp carefully. The ordinary sheet-metal strap type is prohibited by *NEC* 250.70. Be sure the ground clamp is listed for the particular electrode involved, including its metallurgical composition, and whether or not it will be buried in the earth. Clamps suitable for direct earth burial (or concrete encasement) are marked for that exposure (can be abbreviated DB). Ground clamps suitable for use on copper water tubing and for steel reinforcing bar connections have distinctive markings as applicable.

Ground-wire conduit or armor   It is best to install the ground wire without protective armor or conduit using the ground clamp shown in Fig. 16–21. But if for any reason you decide to install a ground wire protected by armor or conduit, observe carefully the details outlined in the following paragraphs.

If you use ground wire with armor similar to that of armored cable, use the clamp shown in Fig. 16–22. This clamps the armor of the wire, and the wire itself is connected to a terminal on the clamp. At the service equipment cabinet, use a connector of appropriate size to bond the armor securely to the cabinet; the wire itself is connected to the grounding busbar.

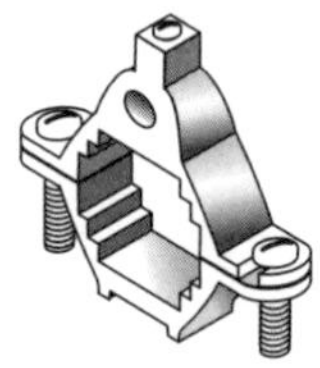

Fig. 16–21 Ground clamp for bare ground wire. If connected to copper water tubing, it must be marked to indicate suitability for this purpose.

If you use conduit, use a clamp of the type shown in Fig. 16–23. Thread the conduit into the clamp. At the other end anchor it securely to the service equipment cabinet with the usual locknut and a grounding bushing.

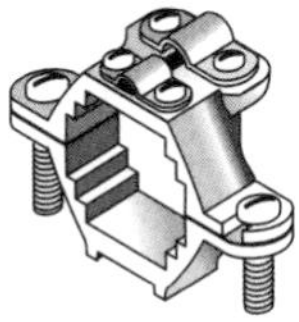

◀ **Fig. 16–22** Ground clamp for armored ground wire.

Fig. 16–23 Ground clamp for ground wire run through rigid conduit. ▶

Caution: If the conduit or the armor is poorly clamped, or clamped only at one end, or not clamped at all, the resultant ground is very much less effective than it would be if a wire without conduit or armor had been used. This is because the metal armor or conduit forms a "reactive choke" that causes high impedance and heat. To understand this, review the return path topic beginning on page 141. Another effect of the magnetic fields described in that chapter is to cause ac current to flow disproportionately over the outer margin of any wire. This "skin effect," as it is called, increases the resistance of very large wires beyond their dc resistance, because the current isn't using the entire cross section of the wire. In the case of grounding electrode conductors, the same principle makes the preferred current path the enclosing raceway instead of the wire. In fact tests have shown that if a 6 AWG ground wire is installed in ¾-in. rigid conduit properly bonded at both ends to the wire, and 100 amps flows to an electrode, 97 amps will flow over the conduit, and only 3 amps over the wire. If you frustrate this natural process on an ac system by leaving one or both ends of the conduit unbonded, and force the entire 100 amps to flow over the wire, the impedance will roughly double because of the increased reactance.

Water meters—potential grounding hazard   It is not unusual for a water meter to be removed from a building, at least temporarily, for testing. If the ground connection is not made on the street side of the meter, removing the meter leaves the system ungrounded, resulting in a temporary hazardous condition. In some cases the joints between the water pipes and the water meter are so poor, electrically speaking, that they are practically insulated. Therefore *NEC* 250.53(D)(1) requires that unless the ground wire is connected on the street side of the meter, a bonding jumper must be installed around the meter as shown in Fig. 16–24. The bonding jumper is also required around any other type of fitting, such as a union, that is likely to be disconnected at one time or another. Use two ground clamps and a jumper of the same size as the ground wire.

A similar jumper is required around any equipment in the incoming water pipe if there is likelihood that the electrical continuity of the pipe is or could be disturbed. A water softener could be temporarily disconnected for servicing. A

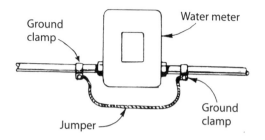

Ground clamp

Water meter

Ground clamp

Jumper

Fig. 16–24 A bonding jumper must be installed around the water meter and around other equipment, such as water filters, that could potentially cause a lack of continuity.

jumper is required around the softener if the ground wire is connected to a pipe running to the softener rather than to the pipe on the street side of the softener.

Grounding busbar   The service equipment cabinet contains a grounding busbar, sometimes called the "neutral busbar," to which all grounded wires must be connected; review the Chapter 13 discussion of this procedure. In the finished installation, the service conduit (and all conduit or cable armor in the branch circuits) is bonded to the service equipment cabinet. Therefore the main bonding jumper (screw or flexible strap) effectively bonds together the neutral busbar in the cabinet and all conduit and cable armor. However, *NEC* 250.92(B) requires a lower resistance between the service conduit and the service equipment than is provided by ordinary locknuts and bushings. The use of grounding bushings and bonding jumpers, or the equivalent, is required for service raceways (armored cable is not permitted in services).

Grounding bushings   Where the service conduit enters the service equipment cabinet, use a grounding bushing of the general type shown in Fig. 16–25 instead of an ordinary bushing. The bushing shown has a setscrew in its side that bites into the conduit. This prevents the properly installed bushing from turning, and the setscrew bites into the metal for the good ground continuity that is so important for safety. The bushing shown has a connector on the side for a bonding jumper to the grounding busbar. The bushing shown has an insulated throat, but all-metal bushings may be used if the service wires are 6 AWG or smaller. If the wires are 4 AWG or larger, bushings with an insulated throat must be used. Similar bushings made entirely of insulating material may be used if a locknut is installed between the bushing and the service equipment cabinet. Some other bonding method must be used in such a case, such as the grounding locknut in Fig. 16–26 applied at a clean knockout.

If you have used EMT, use a grounding-type locknut such as shown in Fig. 16–26 instead of the regular locknut on the EMT connector. You may also use a grounding wedge shown

Fig. 16–25 A grounding bushing.

in Fig. 16–27; install it on the inside of the cabinet between the wall of the cabinet and the locknut of the connector, or use a grounding bushing screwed over the male threads on the end of the EMT connector within the cabinet.

◀ **Fig. 16–26** A grounding locknut.

Fig. 16–27 A grounding wedge. ▶

Bonding jumpers   Your service equipment cabinet probably has concentric knockouts, described at the beginning of Chapter 10. If the conduit is large enough so that all the parts of the knockout have been removed, a grounding locknut or wedge as shown in Figs. 16–26 and 16–27 is all that is required. For a smaller conduit, some part of the knockout remains in the cabinet, and in that case you must install a grounding bushing and connect a bonding jumper from the grounding bushing to the grounding busbar, as shown in A of Fig. 16–28, which shows only the neutral wire of the service and the ground wire. The jumper wire must be the same size as the ground wire.

If you have used service-entrance cable with a bare neutral, the grounding bushing and the jumper are not required where the cable enters the cabinet, as shown in B of Fig. 16–28.

It is not necessary to use grounding bushings, grounding locknuts, or wedges where conduit or cables supplying feeders or branch circuits enter the cabinet, unless the grounding continuity is otherwise suspect.

BONDING FOR WATER AND OTHER SYSTEMS IS REQUIRED FOR SAFETY

Don't assume from the prior discussion that the only important connection to the water system is the one that leads to the earth.

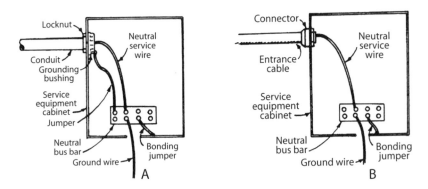

Fig. 16–28 Typical ground connections: at left, where conduit is used for the service entrance; at right, where Type SE or Type USE service-entrance cable is used.

Piping systems must be bonded   An interior metal water piping system must be bonded whether or not its below-grade extension qualifies as an electrode. That bonding connection can't depend on water meters, etc., and it must be accessible after you make it. The size of the bonding conductor is the same as the ground wire, which leads to the pleasant conclusion that one wire can do both functions, if you're able to run it so it does both. Simply run the wire to the water meter (and any filters and other continuity impediments), and run it through the ground clamp, but don't cut the end at the clamp. Let the unbroken end extend to the other side of any equipment requiring a bonding jumper, and then make the final connection. Other piping systems must be bonded if they may become energized, but usually no special arrangement needs to be made because the equipment grounding conductor supplying the equipment in effect also serves as the bonding conductor.

Communications systems have similar grounding requirements   For telephone systems, and the new network-powered broadband communications systems, a "primary protector" must be included, which requires a grounding conductor not smaller than 14 AWG run in as straight a line as possible to a qualified electrode. Qualified electrodes include the power system electrodes just described, or bonded interior metal water piping, or a rod or pipe electrode, which could even be a 5-ft ground rod. Coaxial cable systems as part of CATV systems must have their shields grounded at the building entrance, with a 14 AWG (minimum) grounding conductor run to one of the types of electrodes acceptable for power systems. Radio and television masts or metal supports must be grounded as well, also using one of the types of electrodes acceptable for power systems, connected with a minimum 10 AWG grounding wire.

Intersystem bonding   The above systems are governed by the same principles that govern power systems with respect to electrodes. In the event of an external electrical system insult, all electrical systems in a building should respond in relation to a common ground reference in order to minimize potentially catastrophic potential differences between wiring systems. To meet this safety objective, all building electrodes need to be made as one through bonding. In fact, the *NEC* requires that if there are no exposed nonflexible metal service raceways, and if the power system grounding electrode conductor is hidden as well, then you must provide a method of making these intersystem bonding connections. The first of two fine print notes attached to *NEC* 250.94 describes a bonded 6 AWG copper conductor extended to an accessible point as an acceptable means.

CHAPTER 17
Installing Specific Devices

IN PREVIOUS CHAPTERS, installations of electrical devices are considered in a general fashion. In this chapter the exact method of installing a variety of outlets using assorted materials is discussed in detail. Only methods used in *new work* (buildings wired while under construction) are explained. *Old work* (the wiring of buildings after their completion) is discussed in Chapter 21.

When a building is wired with any kind of conduit, the conduit is installed and the switch or outlet boxes are mounted while the construction is in its early stages. This is called "roughing in." Usually the wires are not pulled into the conduit until after installation of the lath and plaster or the wallboard; the switches, receptacles, fixtures, and so on can't be installed until the wall finishes have been completed. But to avoid repeating later in the book, the pulling in of wires is included in this chapter.

WIRING SIMPLE DEVICE AND FIXTURE OPENINGS

Each type of outlet is treated separately. In the illustrations in this chapter, each outlet is shown four ways, as it would appear: (1) on a blueprint, (2) in diagrammatic or schematic fashion, (3) in pictorial fashion using rigid metal conduit or EMT, and (4) in pictorial fashion using nonmetallic-sheathed cable. The primary purpose of this chapter is to show you how to connect the circuit wires (the wires carrying current in normal use) so that lighting fixtures, switches, receptacles, etc. will work properly. The equipment grounding wire of nonmetallic-sheathed cable and the bonding jumper from the green terminal of a receptacle to its box are omitted in the illustrations to keep them reasonably simple. This means that receptacles without the green grounding terminal screws are shown. The diagrams do not show the grounding wire of nonmetallic-sheathed cable. Figures 18–5 and 18–6 show how to connect the green grounding terminals and the grounding wire of the cable.

The basic details shown in Fig. 17–1 are also used in smaller illustrations that do not name the individual parts. Because the parts are named in the first figure they are easy to recognize in subsequent figures.

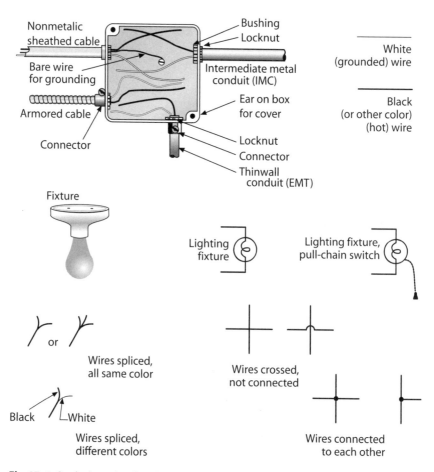

Fig. 17–1 Study these details well so that other diagrams in this chapter will be clear to you.

In each case the point from which the current comes is labeled SOURCE. The white grounded wire always runs from SOURCE without interruption by a switch, fuse, or breaker to each point where current is to be consumed at 120 volts. Splices are permitted if made in boxes, as in the case of the hot wires. When wiring with conduit, a black wire should be connected only to another black; a third wire should be red and connected only to red—this is not required but is common practice. In a three-wire circuit (covered at the beginning of Chapter 20), it is best to use red and black for the two hot wires. Green wire must never be connected to a wire of a different color.

Switches consume no power, so a white wire does not normally run to a switch; however, when using cable there is an exception explained later in this chapter. Grounded circuit wires are shown in the diagrams as light lines, ungrounded or hot wires are shown as heavy lines. This does not mean that one wire is larger than the other. Light and heavy lines are used only to make it easier for you to trace grounded and ungrounded wires.

Ceiling outlet, pull-chain control   This type of lighting outlet, with the wires ending at the outlet, is sometimes used in closets, basements, attics, and similar locations. The *NEC* requires wall-switch control of at least one lighting outlet in residential basements and crawl spaces, attics, and utility rooms used for storage or serviceable equipment, and similar control for commercial attics and crawl spaces with serviceable equipment, but pull-chain lampholders are still widely used for illumination in specific areas. It is the simplest possible outlet to wire; Fig. 17–2 shows it. The designation for the outlet as found on blueprint layouts is shown at *A;* the wires running up to it are not shown because blueprints show only which switch controls which outlet(s). At *B* is a wiring diagram for the same outlet. Assume you have already installed the outlet box for the fixture either directly to the structure, or by using one of the methods in the outlet box discussion beginning on page 399. Installing this outlet using conduit as at *C* in Fig. 17–2, assume you have properly reamed the cut end of the conduit, that you have properly anchored it to the box using a locknut and bushing, and that you have pulled two wires, white and black, into the conduit with about 10 in. protruding from the box. To complete the outlet all you have to do is connect the wires to the fixture and mount the fixture in one of the ways discussed in the next chapter.

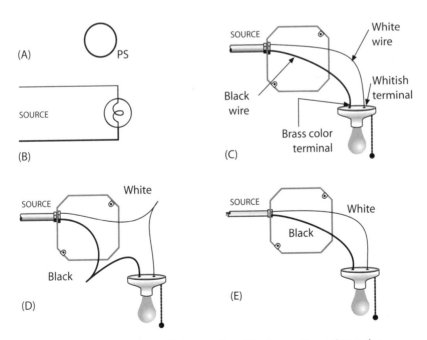

Fig. 17–2 A simple hookup of a pull-chain outlet, with wires ending at that outlet.

If the fixture is one with two leads (two wires instead of two terminals that the wires from the box can be connected to), the wiring is the same except that a couple of splices must be made in the box, as shown at D in Fig. 17–2. If you are using EMT instead of rigid metal conduit, follow the same procedure as with rigid conduit, except use the threadless fittings such as shown in Fig. 11–18 instead of the locknut and bushing used with rigid conduit.

Installation of this outlet using nonmetallic-sheathed cable is shown at E in Fig. 17–2. Connect the grounding wire of the cable to the box as covered in Fig. 18–5 and related text. Do the same for any wiring method with a separate equipment grounding wire, including interlocking-armor-style metal-clad cable. Review Chapter 11 for installation requirements for the various common wiring methods.

Lighting outlet, wires continued to next outlet   From the first outlet, the wires continue to the next; it makes no difference what is used at that next outlet. The only problem is how to connect at the first outlet the wires running from there to the next outlet.

This combination is pictured in Fig. 17–3. Comparing the outlet's blueprint designation at A with A of Fig. 17–2, you will find no difference. This might be confusing, but on blueprints the wires between different outlets are not indicated except that, as already mentioned, lines are shown from switches to the outlet(s) they control. Blueprints leave it to the installer to decide exactly how to hook together the various outlets in a proper manner. The wiring diagram for the combination is at B.

The outlet using conduit is shown at C. Notice that the new outlet shown at C in Fig. 17–3 is the same as at D in Fig. 17–2 except for the addition of two wires

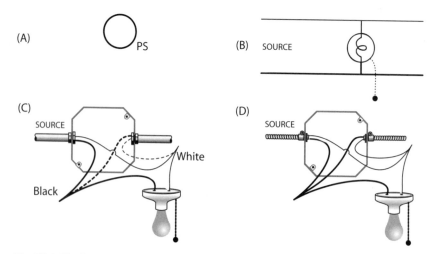

Fig. 17–3 The hookup shown in Fig. 17–2, but with the wires running on to another outlet.

running on to the second outlet (to the right); these two wires are shown in dashed lines to distinguish them from the first wires. Simply splice all the white wires together and all the black wires together with connectors such as Wire Nuts.

Receptacle outlets   In Fig. 17–4, *A* shows the blueprint designation for the outlet, and *B* shows the wiring diagram. To wire this outlet using conduit as at *C,* connect the wires to the receptacle, and connect a bonding jumper from the green terminal of the receptacle to the box, unless the receptacle is of the type that does not require the jumper. Refer to complete coverage in the grounding receptacle topic beginning on page 344.

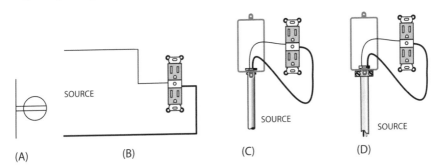

(A)   (B)   (C)   (D)

Fig. 17–4 Installation of a baseboard receptacle outlet.

At *D* is the same receptacle wired using nonmetallic-sheathed cable. The grounding wire is not shown but must be used. If you are using armored cable, proceed as with nonmetallic-sheathed cable. A bonding jumper must be connected from the green terminal to the box.

Receptacle outlet, wires continued to next outlet   In Fig. 17–5, *A* shows the wiring diagram for this circuit. The actual installations using conduit and cable are shown at *B* and *C*. Run the incoming white wire to the white terminal of the receptacle, and continue with another white to the next outlet. Do the same with black wire using the brass terminals of the receptacle. Because it is frequently necessary to do this, receptacles are provided with double terminal screws to allow two wires to be connected to the same terminal strap, as shown at *A* in Fig. 17–6. If conduit is used, you might find it more convenient to pull a continuous wire from SOURCE to box 1, and from there to box 2, then on to box 3, rather than pulling one wire from SOURCE to box 1, another from box 1 to box 2, another from box 2 to box 3. If you use a continuous length, allow a loop of wire to protrude about 6 in. at each box, as shown at *B* in Fig. 17–6. Remove the insulation as shown at *B,* and connect the wire under one of the terminal screws as at *C*.

For one type of circuit the *NEC* prohibits using the two-screw method shown at *A* in Fig. 17–6. On a multiwire branch circuit, the continuity of a neutral

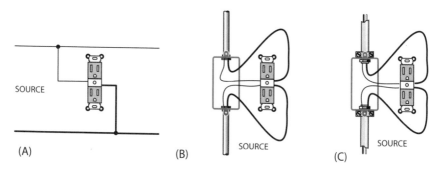

Fig. 17–5 Receptacle outlet, with wires running on to another outlet.

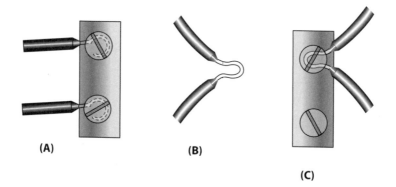

Fig. 17–6 Duplex receptacles have double terminal screws on each side for convenience in running wires on to another outlet.

cannot depend on a device. That means, for at least the white wire, you have no choice but to pigtail one side of the receptacle, as shown at *C* in Fig 17–8. That drawing does not show a multiwire circuit, but the pigtailed neutral is the same. If you are using conduit or tubing, you can also use a looped conductor, as at *C* in Fig. 17–6.

Light controlled by wall switch   This combination is very simple to wire. In Fig. 17–7, *A* shows the blueprint symbol and *B* shows the wiring diagram. The outlet wired using conduit is shown at *C*. Run the white wire directly to the fixture. Run two black wires from the outlet box to the switch, and connect them to the switch. Connect the upper ends, one to the black wire from SOURCE, the other to the fixture. In the optional method shown at *D*, a continuous black wire is brought all the way through to the switch box instead of having a splice at point X in *C*.

White wire exception when using cable   When the outlet described in the previous paragraph is wired using cable, either armored or nonmetallic-sheathed, as at *E* in Fig. 17–7, a difficulty becomes apparent. According to everything you have learned up to this point, both the wire from the outlet box to the switch and the wire from the switch to the fixture should be black wires because neither one is grounded. But all two-wire cable contains one black wire and one white. Should manufacturers, distributors, and contractors then be forced to stock a special cable containing two black wires? That would be impractical. *NEC* 200.7(C)(2) permits an exception for this. In the case of a switch loop (the wiring between an outlet and the switch that controls it), this *NEC* exception permits the use of a cable containing one white and one black wire, even though this does not comply with the general rule that reserves the white wire for grounded wires only. Under this exception the cable may be used provided that the white wire from SOURCE and the black wire from the switch loop are run directly to the fixture. That leaves only one place to connect the *white* wire of the switch loop, and that

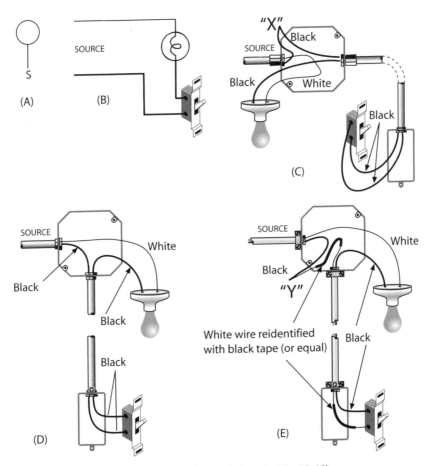

Fig. 17–7 Fixture outlet controlled by wall switch (see also Fig. 17–10).

is to the *black* wire from SOURCE in the outlet box. This is the only case where it is permissible to connect a white wire to a black (or color other than green) wire, and applies only if cable is used.

Study this rule well. When cable is used for a switch loop, the wiring up to the box on which the fixture is to be installed is standard. The fixture must have one white wire and one black (or color other than green) wire connected to it. Therefore the white wire on the fixture is the one that comes directly from SOURCE. The black wire on the fixture must be the black wire in the cable of the switch loop. That leaves two ends of wire: the black from SOURCE, and the white wire in the cable of the switch loop. Connect them together in the outlet box for the fixture. The same procedure is acceptable when using three-wire cable for three-way or four-way switches.

The 1999 *NEC* made a key modification in this traditional allowance for the use of white wire as an ungrounded conductor in a cabled switch loop, a procedure that had been in the *NEC* without change for more than 60 years. Due to concerns about unqualified persons being fooled by the ungrounded white wire running to the switch, now at all access points you must reidentify that white wire as ungrounded; use black paint, tape, or similar means. For the same practical reason, you can use cable assemblies with their white wires for circuits with loads that don't use grounded circuit conductors, such as 240-volt motors; just as with switch loops, you must reidentify such white wires.

Combining three outlets   The wiring of three different outlets having been explained, you can arrange one combination of three outlets as shown in Fig. 17–8. At *A* are the blueprint symbols, at *B* the wiring diagram, and at *C* the three outlets wired using conduit.

Taps   At times an outlet box is used simply to house a splice or tap where one wire branches off from another. Boxes installed only for splices are always closed with blank covers. *NEC* 314.29 requires that such boxes must be installed in locations permanently accessible without removing any part of the structure of the building.

In Fig. 17–9, *A* shows how taps or splices appear on blueprints, and *B* shows them in wiring diagrams. The procedure for making a splice or tap using conduit is shown at *C*. Simply connect all the black wires together, and all the white wires together. In the optional method shown at *D*, instead of three ends of each color, one loop of each color is used, formed by pulling a continuous wire through the box, the ends of the loop being skinned and joined to the remaining ends of the wires to the next outlet. With the now widespread use of twist-on wire connectors that require wire ends only (see Fig. 8–7), this method is rarely used today. At *E* of Fig. 17–9 is the same outlet using cable.

Outlet with switch, feed-through switch   Switches are connected in different ways to the outlets they control. The SOURCE wires do not always run, as they do in Fig. 17–7, through the ceiling to the outlet box on which the fixture is installed and from there on to the switch. Sometimes the SOURCE wires come in from

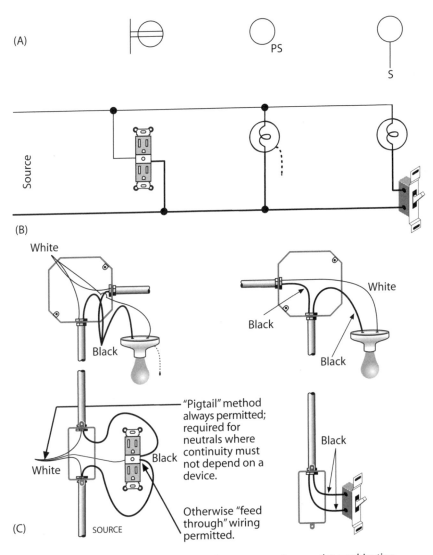

Fig. 17–8 The outlets of Figs. 17–2, 17–4, and 17–7 as one three-outlet combination.

below, run through the switch box and then on to the ceiling outlet where they end, as in Fig. 17–10. At A is the blueprint symbol, at B the wiring diagram, and at C the outlet using the conduit system. At D is an optional method of handling the wires at the switch box by pulling the white wire straight through. The black wire is also pulled straight through but with a loop that is later cut and the two ends are then connected to the switch.

The diagram for this outlet when wired with cable is not shown because there is no new problem. When the cable feeds through the switch box there is no diffi-

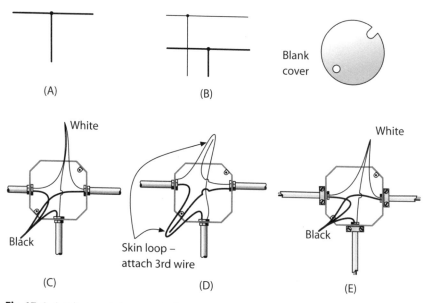

Fig. 17–9 A splice made in an outlet box.

culty with the colors of the wires in the cable, so there is no need to use the *NEC*'s white wire exception described earlier. The wires can be run through to completion of the outlet exactly as when using conduit as *C* in Fig. 17–10 shows.

Outlet with switch, with another outlet   This is simply a combination of two outlets previously discussed. The wiring of an outlet with wall-switch control is covered in the discussion relating to Fig. 17–7, and the wiring of an outlet with pull-chain control is covered in the discussion relating to Fig. 17–2. The two are combined in Fig. 17–11, the left-hand portion of which is identical with Fig. 17–7; to it have been added in dashed lines the wires of Fig. 17–2, making the new combination as shown. At *D* in Fig. 17–11 is an optional method of handling the wires through the outlet box, the wires from the fixture being connected at the points marked X. No new problems are involved in the cable methods, so they are not shown.

If the wires from SOURCE come in from below through the switch box instead of coming in through the ceiling outlet, the problem is different in that three wires instead of two must run from the switch box to the first outlet box, as at *B* in Fig. 17–12. A good way to analyze this combination is to consider first the right-hand outlet with the pull chain. Both a white and a black wire must run to this from SOURCE, uninterrupted by any switch, so that the light can always be controlled by the pull chain. Run the third wire from the switch to the left-hand outlet to control the outlet.

Analysis of *C* in Fig. 17–12 shows that it consists of a combination of *C* in Fig. 17–10 (wires in solid lines) and Fig. 17–2 (wires in dashed lines).

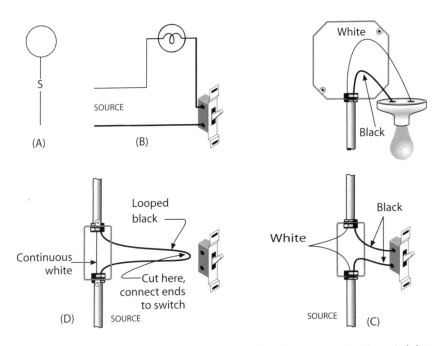

Fig. 17-10 The same as Fig. 17-7 except that the wires from SOURCE enter the switch box instead of through the outlet box on which the fixture is mounted.

If instead of conduit you use cable, as at D in Fig. 17–12, the problem is equally simple. If the wires are fed through the switch box, the colors of the wires present no problem. Simply connect white to white in the switch box, and then continue the white to each of the two fixtures. The remaining colors are as shown in the drawing.

Switch controlling two outlets   For a switch that is to control two outlets at the same time, wire the switch to control one fixture, then continue the white wire from the first fixture to the second, and do the same with the black.

The blueprint symbols, wiring diagram, and combination using conduit are shown at A, B, and C of Fig. 17–13. Compare C in this figure with C of Fig. 17–7; there is no difference except that these wires have been continued as shown in the dashed lines. To avoid all the splices shown in the outlet box for the first fixture, several of the wires may be pulled through as continuous wires to make a neater job.

If you use cable, there is again the problem of having to use in the switch loop a cable that has one black wire and one white wire, instead of the two black wires that should be used. Handle it as at D in Fig. 17–7. The white wire from SOURCE goes to each of the two fixtures. The black wire in the switch loop (on the load side of the switch) also goes to each of the two fixtures. That leaves only two unconnected wires, the black wire from SOURCE and the white wire in the switch loop; connect them together as permitted by the *NEC*'s exception explained

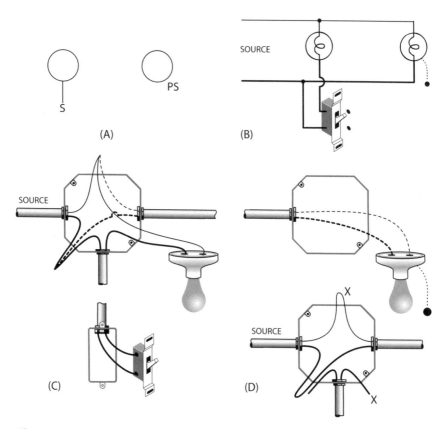

Fig. 17–11 The switch controls the first fixture; wires run on to a second fixture that is controlled by a pull chain.

earlier under the heading "White wire exception when using cable." Don't forget to reidentify the ungrounded white wire with black tape or paint.

Three-way switches   When a pair of three-way switches controls an outlet, there are many possible combinations or sequences in which the SOURCE, the two switches, and the outlet may be arranged; much depends on where the SOURCE wires come in. The three most common arrangements are:

> SOURCE → switch → switch → outlet
>
> SOURCE → switch → outlet → switch
>
> SOURCE → outlet → switch → switch

In another sequence the SOURCE comes into the outlet box, from which two runs are made, one to each switch. All these sequences are shown in Fig. 17–14a, A1 through A4; the wiring diagrams are shown in B1 through B4. Comparing B1, B2, B3, and B4, you will see little difference except for the location of the light that the switches control.

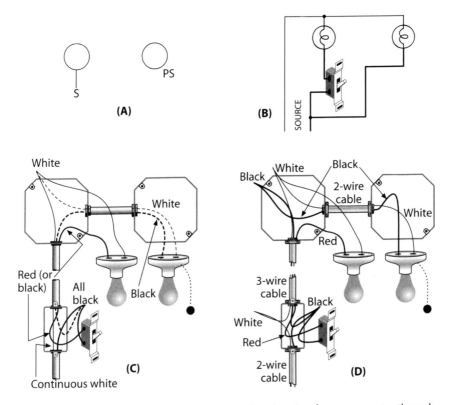

Fig. 17–12 This is the same as Fig. 17–11, except that the wires from SOURCE enter through the switch box.

Reviewing the subject of three-way switches as covered in Chapter 5, remember that one of the three terminals on such a switch is a "common" terminal that, when energized, always energizes one of the other terminals; the handle position merely decides which of the two wires is energized. This common terminal is usually identified by a different color from the other two. The exact location of this common terminal with relation to the other two varies with different brands; in drawings of three-way switches in this chapter the common terminal is the terminal that is alone on one side of the switch. To wire a three-way switch, run the incoming black wire from SOURCE directly to the common or marked terminal of either switch. From the corresponding common or marked terminal of the other switch, run a black wire directly to the proper terminal on the fixture. From the remaining two terminals on one switch, run wires to the corresponding terminals of the other switch, which are the only two terminals on that switch that don't already have wires connected to them. To complete the circuit, connect the incoming white wire from SOURCE to the fixture.

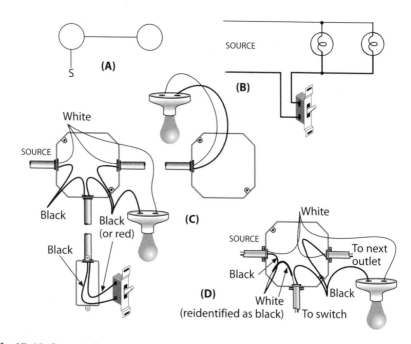

Fig. 17–13 One switch controlling two separate fixtures.

With these facts in mind, the wiring of any combination of three-way switches with conduit becomes quite simple. Assuming that the boxes and conduit have been properly installed, as shown in of Fig. 17–14b, *C1*, which covers the sequence of *A1*, pull the white wire from SOURCE through the switch boxes up to the outlet where the fixture is to be used; run the black wire from SOURCE to the common terminal of the nearest three-way switch. From the two remaining terminals of this switch, run two black wires to the corresponding two terminals of the second switch. From the common terminal of this switch, run a black wire to the fixture. All wires are black, but sometimes one red wire is used for identification purposes; any color may be used except white or green.

The wiring of the other combinations or sequences with conduit is equally simple if you remember the points of the two previous paragraphs; no diagrams are shown. It is good practice to draw the circuits in a fashion similar to *C1* of Fig. 17–14b.

This same combination wired with cable is shown in *C1 ½* of Fig. 17–14b. Note how the white wire from SOURCE is continued from box to box until it reaches the fixture. The black wire from SOURCE, as in the case of conduit, goes to the common terminal of the nearest switch. The three-wire cable between the switches contains wires of three different colors, of which the white has already been used, leaving the black and the red. Therefore run these two wires from the two remaining terminals of the first switch to the corresponding terminals of the second; the red and the black may be reversed at either end, being completely interchangeable. This leaves only one connection to make, and that is the black wire from the common terminal of the second switch to the fixture.

When some of the other sequences, such as *A2, A3,* or *A4* in Fig. 17–14a, are wired with cable, there is the usual difficulty relating to the colors of the wires in standard two- and three-wire cables. The red wire of three-wire cable is inter-changeable with the black. Many times, however, you must take advantage of the *NEC*'s exception permitting, in switch loops, white wire to be connected to black. Don't forget to reidentify the ungrounded white wire as other than white.

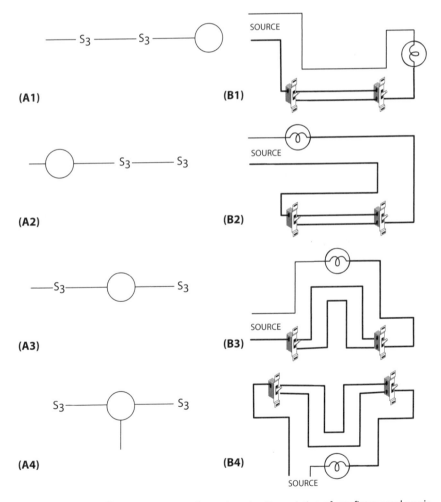

Fig. 17–14a Four different sequences of parts in a circuit consisting of one fixture and a pair of three-way switches.

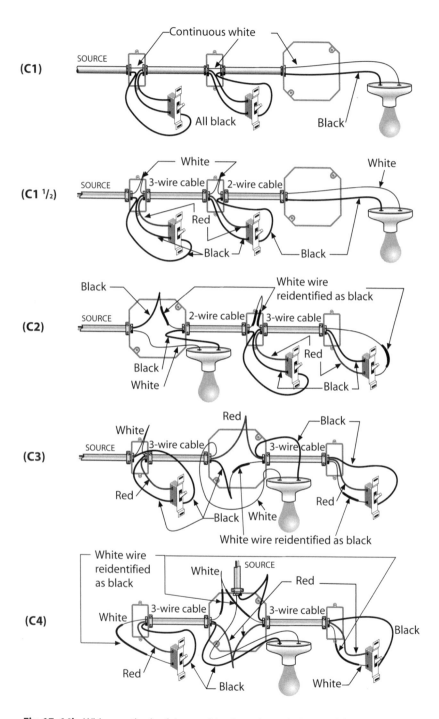

Fig. 17–14b Wiring methods of the combinations shown in Part 1 of this figure.

C2 shows the sequence *A2* and *B2* of Fig. 17–14a, wired with cable. The incoming cable from SOURCE contains one black and one white wire; run the white directly to the fixture. The *NEC* requires that the other wire on the fixture must not be white; consequently the black wire of the two-wire cable that runs to the first three-way switch is connected to the fixture; connect the opposite end of it to the common terminal of the first three-way switch. That leaves, in the outlet box on which the fixture is mounted, only two unconnected wires: the black wire from SOURCE and the white wire of the next run of cable. Connect them together, contrary to the general rule but permitted by the *NEC* for switch loops. From the first three-way switch to the second one, three-wire cable is used. Since one of these three wires is white, connect it to the white wire of the two-wire cable and continue it on to the common terminal of the second switch. That leaves one black and one red wire between the two switches; connect them to the remaining terminals of each switch, completing the installation.

In wiring the sequence of *A3* and *B3* of Fig. 17–14a, as pictured in *C3*, similar problems arise. There is a two-wire cable from SOURCE entering the first switch box; from that point a length of three-wire cable runs to the outlet box in the center, and from there another length of three-wire cable runs on to the second switch box. Continue the white wire from SOURCE from the first switch box directly to the fixture, as the *NEC* requires. Run the black wire from SOURCE directly to the common terminal of the first switch. Going on to the fixture, the second wire on the fixture must not be white; therefore make it black and continue it onward to the common terminal of the second switch. That leaves a pair of terminals on each of the two switches still unconnected, and your problem is to connect the two terminals on one switch to the two terminals on the other. There are yet two unused wires in each run of three-wire cable, a black and a red in the one, and a white and a red in the other. Therefore make one continuous red wire out of the two reds and connect it to one terminal on each switch; make a continuous white-black out of the other two wires and connect it to the remaining terminal on each switch. This completes the connections.

In the case of the sequence of *A4* and *B4* of Fig. 17–14a, as pictured in *C4*, the problems are similar, and you should have no difficulty in determining for yourself why each wire is the color indicated. The fundamental rule is that the white wire from SOURCE must go to the fixture, and the second wire on the fixture must not be white or green.

Pilot lights   A frequently used combination toggle switch and pilot light is shown in Fig. 17–15, together with its internal wiring diagram. This is simply a separate switch and a separate pilot light in a single housing. The arrangement of the terminals on another brand may be totally different from that shown. A pilot light is used at a switch that controls a light not visible from the switch, such as at a switch in the house controlling a light in the garage; it is used simply as a reminder that another light is on. Frequently today's practice is to use a "pilot handle" switch, which has a translucent backlit toggle that is on whenever the load is on. Don't confuse these with "lighted handle" switches that light whenever the load is off, making them easy to find in a dark room.

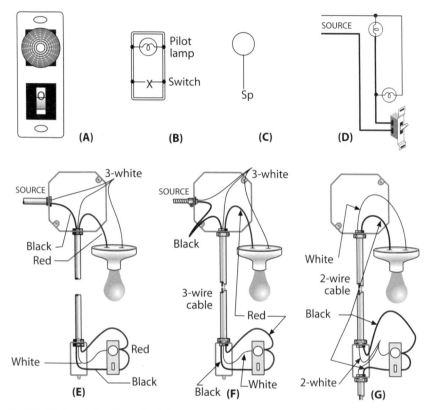

Fig. 17–15 A switch with a pilot light.

In the same figure at *C* is the blueprint symbol for this combination and at *D* the wiring diagram. Compare this with *B* of Fig. 17–13; it is simply a case of one switch controlling two different lights, one of which happens to be located at the same point as the switch.

Using the conduit system, the connections are shown at *E*. Run the incoming white wire from SOURCE directly to the first fixture and extend it from there to the second fixture, which in this case happens to be the lamp in the combination device. A pilot handle switch has this connection made internally. In the outlet box on which the fixture is mounted, connect a length of black wire to the black wire from SOURCE (or pull it long enough initially to pass through the outlet box without splice), and run it to the switch. From the other side of the switch, run a red wire to the first lamp (in the combination device, if necessary) and then on to the fixture, finishing the job. The *NEC* does not require any particular color code for these ungrounded conductors, so the return wire from the switch to the outlet could be black or any other color (not white or green). Many electricians prefer using black for a wire that is always on, and a color for one that is controlled by a switch, but that is entirely up to you.

Wiring the combination with cable, as shown at *F* in Fig. 17–15, presents no problem. If the wires from SOURCE come in through the switch box, the problem is still simpler, as *G* shows.

Switched receptacles   When entering a dark room it can be inconvenient and even dangerous to grope for a table or floor lamp to turn on. One solution is to have some of the receptacles controlled by a wall switch that turns them on and off together. Other receptacles can be left unswitched.

As discussed in Chapter 12, the *NEC* in 210.70(A) requires that every room of a house must have some lighting controlled by a wall switch; review that part of Chapter 12. The switch must control permanently installed lighting fixtures in kitchens and bathrooms, but in other rooms it may control receptacles into which floor or table lamps can be plugged.

If the receptacles are to be controlled by one switch, no particular wiring problems are encountered. But in some rooms they should be controlled by a pair of three-way switches. Then the wiring becomes a bit more complicated. There can be a great many different sequences of outlets and switches; the arrangement shown at *A* and *B* in Fig. 17–16 is perhaps as common as any. When conduit is used, as shown at *C* in that figure, the problem is simple. Run the white wire from SOURCE to each of the receptacles, connect the remaining terminal of each of the three receptacles together with black wire, and continue to the common terminal of one of the three-way switches. From there run two wires over to the other three-way switch. To the common terminal of this second switch, connect the black wire from SOURCE, and the job is finished. All the necessary wires in the boxes housing the receptacles will badly crowd ordinary switch boxes, so the preferable method is to use 4-in.-square boxes with raised covers designed to take a duplex receptacle; see Fig. 10–7.

To wire this same combination using cable might be difficult with the particular sequence shown because it requires four wires at some points and 4-wire cable is not always stocked by dealers. Therefore when using cable, if you can't get 4-wire cable you may need to modify the sequence to that shown at *D* in Fig. 17–16, which requires nothing more than 3-wire cable. The other option is to use two 2-wire Type NM cables. *NEC* 300.3(B)(3) allows this *for nonmetallic wiring methods only,* provided you don't create a condition in which current passes in only one direction through a magnetic enclosure, which is restricted by *NEC* 300.20(B). For example, suppose the two 2-wire cables enter a steel box through two different cable knockouts, and one of the cables consists of the two travelers. If the switched receptacles are supplying 10 amps of load, then 10 amps flows on one of the travelers through one of the two knockouts, and returns over a conductor passing through

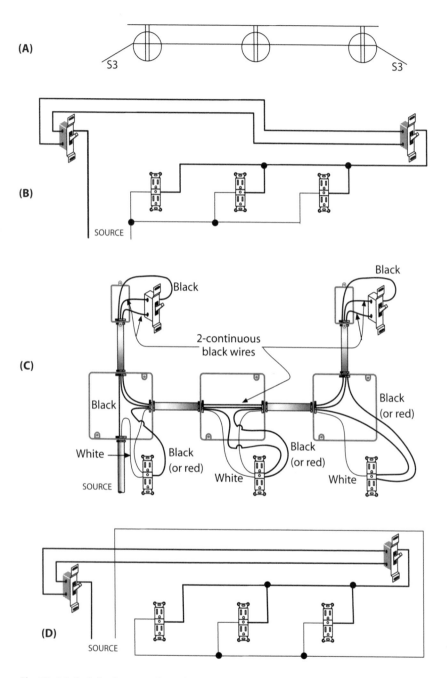

Fig. 17–16 Switched receptacle outlets.

the other knockout. The result would be inductive heating between the two cable entries, and the arrangement would be an *NEC* violation.

The solution is either to bring both cables in through the same knockout, or bring both cables in through adjacent cable knockouts in the end of a switch box (see Fig. 10–16 for an example), after having snipped the web between the two openings with your diagonal-cutting pliers ("dikes"). That makes the slot called for in *NEC* 300.20(B), effectively making the two holes into one. In general, cable clamps within boxes aren't listed for two cables; however, if you use a device box with a ½-in. conduit knockout on the end, there are a number of Type NM cable connectors that are listed for two 2-wire NM cables, solving the problem. Another solution is to use a nonmetallic device box. The best solution, however, is to find the 4-wire cable. Don't forget to reidentify the white wire in any case where it is no longer used as a grounded conductor.

Two-circuit duplex receptacles   Instead of having some receptacles switched for lamps and others permanently live, a better solution is to split conventional duplex receptacles so each half can be controlled independently. This allows half of each receptacle to be permanently connected for radio, TV, and so on, and the other half switched to control lamps plugged into them. Most receptacles of better quality have this feature, accomplished through a removable link between the two brass-colored terminal screws for the hot wires; the link can be pried out at the time of installation, thus splitting the ordinary receptacle. The procedure is shown in Fig. 17–17. You need only to remove the link on each side as needed. For example, if the grounded conductor side of the receptacle is used for both the switched load and the portion that will always be energized, there isn't any point to removing the link on that side. On the other hand, there are cases where the

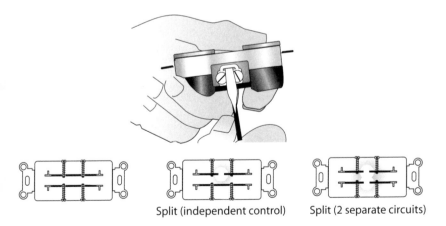

Split (independent control)   Split (2 separate circuits)

Fig. 17–17 Two-circuit receptacles have many advantages. In receptacles of better quality, the ordinary receptacle can be changed to the two-circuit type by breaking out a small metal strip as shown.

switched half of the receptacle is on an entirely different circuit than the unswitched half, and in such cases the links on both sides must be removed. As an example, the *NEC* allows receptacles in a dining room to be switched, but such switched receptacles must not be on one of the small-appliance branch circuits that normally must supply all receptacles in a dining room, and the switched receptacle must be in addition to receptacles meeting the normal perimeter spacing rules. In a case where only one duplex receptacle is supplied at such an outlet, splitting both sides of the duplex receptacle would be the only possible solution. Install such a two-circuit receptacle so the switched half is at the bottom, leaving the unswitched half at the top, more readily available for a vacuum cleaner and other things typically plugged in for a limited time. Take care, however, to read carefully about the 2002 *NEC* change that affects split receptacles, covered at the end of this chapter segment. The blueprint symbol for a two-circuit or split-wired receptacle is shown at *A* in Fig. 17–18.

The wiring for a two-circuit duplex receptacle presents no great problem. Run the white grounded wire from SOURCE to one of the terminal screws on the unbroken metal strap on the receptacle. Run the black wire from SOURCE to one of the terminal screws on the opposite side of the receptacle, and continue it to the switch. From the switch run another black wire back to the remaining terminal screw on the receptacle, on the side where the center of the strap is broken out. See *B* in Fig. 17–18, and for more detail see Fig. 17–19.

If you use cable to the switch, again you will have the problem that one of the wires is white where it should be black; take advantage of the *NEC*'s special dispensation and connect the white wire in the switch loop to the incoming black wire from SOURCE, as shown at *C* in Fig. 17–18. If you use three-way switches, the problem of wire colors is more complicated but may be solved exactly as in the case of the lighting outlets shown in Fig. 17–14.

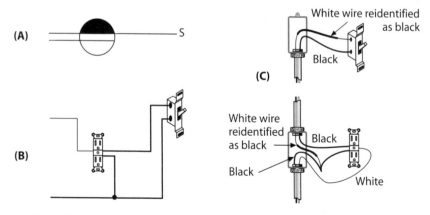

Fig. 17–18 Wiring a two-circuit receptacle using cable.

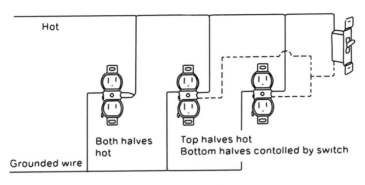

Fig. 17–19 More detail for wiring two-circuit receptacles.

Another common application for the two-circuit receptacle is in the two 20-amp small-appliance circuits in the kitchen. If these two circuits are merged into a single three-wire circuit (explained in Chapter 20), run the white grounded wire to the common terminal, the red and black wires to the other two terminals. When two different appliances are plugged into the same duplex receptacle, they will automatically be on opposite legs of the three-wire circuit. Don't forget that the continuity of the white wire on a three-wire or multiwire branch circuit must never depend on a device, and therefore the grounded side of the receptacle must be pigtailed

A recent change in the NEC places a very significant condition on the use of multiple devices on one yoke connected to more than one branch circuit. A means to simultaneously disconnect all ungrounded conductors arriving at a common yoke must be provided at the panelboard. Theoretically, this could be a multipole circuit breaker or a multipole switch installed adjacent to the panelboard. However, test lab guide card restrictions forbid multipole switches from being used for multiple circuits unless they are marked "2-circuit" or "3-circuit" as applicable. To this writer's knowledge no such switches are now on the market. This means that fuses could not be used to supply such multiple receptacles; the only practical method would be a multipole circuit breaker. In the case of a multiwire branch circuit, this is easy but it cancels one of the functional advantages of the traditional multiwire branch circuit, namely, if one circuit trips the other is still connected.

In addition, if two 2-wire circuits arrive at the device yoke, these circuits must still now originate from a multipole circuit breaker. In the case of the dining room split receptacle described previously, the multipole circuit breaker required here means the small-appliance branch circuit connected to this outlet could not be multiwired with a second small-appliance branch circuit, contrary to the usual configuration. Think carefully about the implications of this rule in planning your installation. It now applies in all occupancies, and any time more than one branch circuit supplies more than one device, even a switch-and-receptacle combination, on a common yoke, whether or not in a multiwire configuration. The concern is that an untrained person will assume all circuits are off when investigating a nonfunctioning load connected to one of the receptacles.

Combining outlets Earlier in this chapter, in Fig. 17–8, the outlets of Figs. 17–2, 17–4, and 17–7 were arranged in a single three-outlet combination. In similar fashion, outlets may be arranged in any desired combination with any desired total number of outlets, which then forms a circuit running back to the service equipment. To add an outlet at any point, connect the white wire of that new outlet to the white wire of the previous wiring (provided it is not a white wire in a switch loop, when using cable), and the black wire of the new outlet to any previous wiring where the black wire can be traced back to the original SOURCE without being interrupted by a switch. Simply connect the new outlet at any point on the previous wiring where a lamp connected to the two points in question would be permanently lighted. Review Chapter 5 if this point is not entirely clear.

WIRING TO RECESSED FIXTURES

The temperature inside recessed (flush) fixtures is higher than in fixtures in free air. The terminals will be the hottest part, and often their temperature will be higher than the temperature limits of the kinds of wire normally used in ordinary wiring. Assuming the fixture is UL listed (and in recessed fixtures it is especially important to use only UL-listed fixtures), it will be marked with the minimum temperature rating of the wire supplying the fixture.

Recessed fixtures usually involve high temperatures If a UL-listed fixture is unmarked or marked "60°C," it may be used with any kind of wire normally used in wiring. Most recessed fixtures are marked for a temperature considerably above 60°C. In that case you have a choice of two methods. The first is to wire the entire circuit up to the fixture with a wire having a temperature limit at least as high as the temperature shown on the fixture. Such wire is likely to be expensive but its use may be feasible if the circuit to the fixture is very short. (Fixture wire must not be used for the branch circuit supplying a fixture, nor in *any* branch circuit.)

The second method is to install a junction box, which may be an ordinary outlet box, not less than 12 in. from the fixture, and run a length of raceway from that junction box to the fixture. In that length of raceway install fixture wire of the proper temperature rating. In this case the outlet box, which need not be readily accessible, constitutes the outlet. Therefore the run from that box to the fixture (usually referred to as a "cold lead") is on the load side of the branch circuit, and fixture wire is perfectly acceptable. Using field-installed cold leads is seldom done today because the *NEC* requires most recessed fixtures to have thermal protection, by which a built-in thermal sensor shuts off the light if it overheats. The most common arrangements have the thermal sensors set near the recessed housing and then tested by UL or another test lab for effectiveness. A test lab can't test the effectiveness of thermal sensors if they are located in a field-installable outlet box that will be placed in the field at an unknown distance from the fixture. There are ways around this problem; obey any listing instructions that come with the fixture.

Prewired recessed fixtures are the norm   With the possible exception of some fixtures designed for use in poured concrete, almost any recessed fixtures you get today are prewired. In such fixtures the high-temperature fixture wire is factory installed and extended into a junction box factory-mounted on the side, together with any thermal protection, as shown in Fig. 17–20. Prewired fixtures have markings in the prewired junction box indicating how many conductors of what temperature rating can come into the box, and whether or not it is suitable for through wiring.

The concept of "through wiring" requires explanation because the way the manufacturers use the term could be considered counterintuitive. The *NEC* now allows either two-wire or multiwire circuit conductors supplying a fixture to pass through the fixture's wiring compartment by right, without necessitating a separate listing for through wiring. Identification for through wiring sounds like something that would apply to a branch circuit passing in and then out of the wiring compartment, whether or not the fixture is connected to it. But that is not the case; see Fig. 17–21. Ratifying the way the testing laboratories have been listing these

Fig. 17–20 This prewired recessed fixture is suitable for insulation contact. The thermal sensor is mounted on top of the housing.

fixtures for years, the *NEC* now only applies the concept of "through wiring" (in the sense that requires separate listing recognition) to additional circuits that aren't part of either the fixture supply circuit or one of its companions in a multiwire circuit.

Recessed fixtures frequently require clearance from insulation and combustible surfaces   *NEC* 410.66 requires a clearance of at least ½ in. between a recessed fixture and all combustible material such as wooden joists. Thermal insulation must be kept from above recessed fixtures, and not less than 3 in. from the sides, unless the fixture is listed as suitable for contact with insulation. Such fixtures bear the marking TYPE IC and may also be mounted in contact with combustibles. Read the manufacturer's directions carefully because some recessed fixtures can be either Type IC or not depending on the final trim

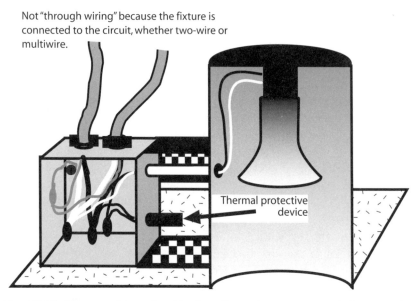

Not "through wiring" because the fixture is
connected to the circuit, whether two-wire or
multiwire.

Thermal protective
device

Fig. 17–21 The term "through wiring" applies only to a recessed fixture wiring
compartment when the wiring is not part of any portion of the two-wire or multiwire circuit
supplying the fixture.

that is installed. Fixtures supported by suspended ceilings must not exceed the
weight the ceiling is classified to support. Note that the insulation contact and
spacing rules now apply equally to fluorescent as well as incandescent fixtures.

WIRING GARAGES AND OUTBUILDINGS

If a garage is attached to the house, treat it as another room of the house.
Three-way switches are preferred for convenience, one at the outer door of the
garage and another at the door between garage and house. The *NEC* requires at
least one GFCI-protected general purpose receptacle outlet in every dwelling unit
garage that is attached to the building, and in every similar detached garage if it
is provided with electric power. If the garage is an outbuilding, there are many
considerations that must be addressed in order to make a safe installation.

Disconnecting means for outbuildings As noted in Chapter 4, you must
supply a local disconnecting means for all separate buildings and structures fed
from another building on the property, even an accessory building to a single-
family dwelling. The *NEC* does relax the rules a little for residential applications. For example, you can use snap switches as disconnects instead of
equipment suitable for use as service equipment (the normal rule). However,
each disconnecting means still must be grouped and marked to indicate its
function and the load served. Another residential compromise involves a waiver

of the reciprocal labeling rule (review Chapter 4) when multiple circuits supply a dwelling. The Code-making panel decided that reciprocal signage in such a building to the effect of "This is disconnect 1 of 2, controlling the overhead light; disconnect 2 of 2 located on the west side of the garage door controls the GFCI receptacle" would be excessive. The reciprocal labeling waiver does not, however, waive the identification rule on each switch, nor does it waive any other requirements covered in Chapter 4.

The remainder of this chapter covers comparatively simple residential outbuilding applications. The disconnecting rules for these cases boil down to the following:

- Ungrounded wires supplying a load intended to stay energized, such as a receptacle, must pass through a disconnecting means located at a readily accessible point nearest the point of entrance.

- A snap switch is a permissible disconnecting means, including a three-way switch with no identifiable off position.

- The switches associated with a single source of supply, such as a single branch circuit, must be grouped, although they needn't be as close as adjacent snap switches in a two-gang box.

- Each switch must be marked with its function. If that function is obvious, such as the overhead light, *NEC* 110.22 allows some basis for omitting this marking. However, by providing the marking you will avoid challenges.

Suppose you install a receptacle that will supply a freezer, and the owner wants to be assured that it won't be turned off inadvertently. Assume there will also be a light controlled from the house and the garage using three-way switches. Mark the three-way switch in the garage LIGHT. Run the receptacle feed through a single-pole snap switch in another box near the three-way switch, perhaps at an odd height, say 3 ft above the floor. Over the other switch place a weatherproof cover that precludes inadvertent operation, and mark it RECEP DISC. or similar.

The receptacle will have to be GFCI-protected if it is in a dwelling unit garage, and as such it will have TEST and RESET buttons. Be aware that these do not qualify the receptacle as a disconnecting means. Some GFCI-only devices without receptacle slots have OFF and ON markings in addition to TEST and RESET, which qualifies them as manual motor controllers, but not disconnects under this part of the *NEC*. In addition, even if the GFCI devices are considered disconnecting means, they have to meet the location requirements for such devices, that is, where the conductors enter the building.

Grounding requirements for outbuildings   Just as a second building must comply with all the important rules for disconnecting means as if it were a stand-alone building, so must it be in compliance regarding system grounding connections. Unless the second building contains only a single (or one multiwire) branch circuit with its own equipment grounding conductor, any grounding electrodes at the second building that would be required to be included in a grounding electrode system for a principal building must be bonded together to make a system in the second building. And similarly, unless the single-circuit provision applies, those electrodes must be connected to the wiring system in the separate building through a grounding electrode conductor, thereby establishing a local grounding reference. If no such electrodes exist, you must provide them; for the procedure refer to the page 300 discussion on providing a safe grounding system.

A key difference between the main building and a second building is that for a second building no parallel path is allowed for grounded conductor return current over a main bonding jumper to the SOURCE in an upstream building or structure. A separate equipment grounding conductor must be supplied and the grounded conductor must remain insulated from equipment grounding conductors in the second building. To understand why, suppose a main bonding jumper were installed at the second building, under the crossed-out circle in Fig. 17–22. The metal water piping system will be bonded at the first building, as covered in Chapter 16. Some of the grounded conductor current—and grounded conductors carry current routinely—will flow back to the first building over the water pipe in parallel with the grounded circuit conductor supplying the second building. Piping systems are simply not designed to be current-carrying conductors, and it can be dangerous because it maintains a voltage on uninsulated piping systems. Therefore the neutral block in the second building must remain insulated. The *NEC* objective is to keep grounded-conductor current confined to electrical conductors.

The only exception occurs for installations with no such parallel path. For example, if there were no common metallic water system, and if you went to the second building with rigid nonmetallic conduit, you could omit the equipment grounding conductor in the feeder (or branch circuit) to the second building. You would then ground the second building as if it were supplied by a service, with similar requirements for grounding electrodes and connections. The usual trade term for this practice is "regrounding" the neutral, or grounded conductor as the case may be.

Continuing with the example of one or two branch circuits arriving at a dwelling unit outbuilding, the grounding rules are essentially the following:

■ If only one branch circuit (or one multiwire circuit) is involved, ignore the establishment of a grounding electrode system if you choose; just be sure that the wiring system includes a properly wired equipment grounding conductor.

■ If two separate branch circuits (or more, or a feeder) are involved, bond all qualified electrodes into a grounding electrode system. If no qualified electrodes exist, provide one just as if you were providing a service.

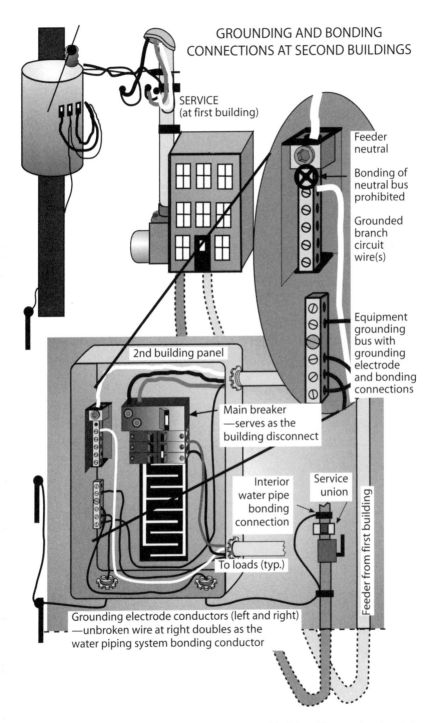

GROUNDING AND BONDING CONNECTIONS AT SECOND BUILDINGS

SERVICE
(at first building)

Feeder neutral

Bonding of neutral bus prohibited

Grounded branch circuit wire(s)

Equipment grounding bus with grounding electrode and bonding connections

2nd building panel

Main breaker —serves as the building disconnect

Interior water pipe bonding connection

Service union

Feeder from first building

To loads (typ.)

Grounding electrode conductors (left and right) —unbroken wire at right doubles as the water piping system bonding conductor

Fig. 17–22 System grounding connections at a second building. The metal water piping system common to both buildings prohibits a main bonding jumper between grounded and grounding terminals in the second building, and forces the grounded conductor to remain insulated.

■ If two separate branch circuits (or more, or a feeder) are involved, run a grounding electrode conductor between the equipment grounding conductor(s) of the supply circuit(s) and the grounding electrode system, with the size taken from *NEC* Table 250.66 based on the largest ungrounded supply conductors, just as for services. If your system qualifies and you choose to reground, make the grounding electrode connection to the grounded circuit conductor(s).

Installing a circuit for a detached garage   There are several ways to provide power to a garage that is a separate structure some distance from the house. If there is only a light and nothing else installed in the garage, and if that light is to be controlled only by a switch in the garage, or only by a switch in the house, only two wires are necessary. For simplicity, this discussion omits the additional wire that would always be present, namely the equipment grounding conductor, and focuses on the circuit conductors. In addition, this discussion omits the local disconnecting means and the system grounding connection requirements that will need to be met in any case.

Using only two wires, as in Fig. 17–23, if a receptacle is to be installed in addition to the light, and the light is to be controlled only by a switch in the garage, the receptacle can be continuously live. But if the light is controlled only by a switch at the house, the receptacle will of course be dead whenever the light is turned off.

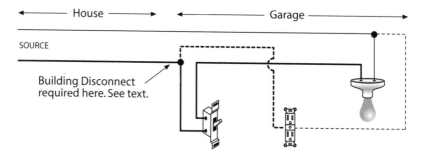

Fig. 17–23 Simple garage circuit—the receptacle is always on.

If the light and receptacle are to be controlled by three-way switches at both house and garage, three wires must be run as shown in Fig. 17–24. If the wires shown in dashed lines are disregarded, this becomes identical with the basic diagrams for three-way switches in Chapter 5. But when the *NEC*-required receptacle is installed as shown in Fig. 17–23, the receptacle will be turned off when the light is turned off by either three-way switch. This is undesirable (although minimally *NEC*-compliant) because, for example, frequently it is convenient to have a charger running all night to charge a car battery without having the light burn all night. A fourth circuit wire, as shown in dashed lines in Fig. 17–24, is necessary to make the receptacle independent of the switches. (*Caution:* A "trick" circuit that switches the grounded conductor using only three circuit wires instead of four definitely violates *NEC* requirements and is an unsafe circuit.)

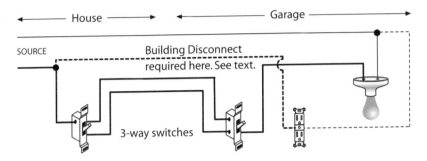

Fig. 17–24 The garage light is now controlled from either house or garage. The receptacle is always on. This requires four circuit wires between house and garage.

A pilot light at the switch in the house is also desirable. See Fig. 17–25. If the dashed lines are disregarded, the circuit becomes the same as that of Fig. 17–24. To add a pilot light, run a fifth circuit conductor as shown in dashed lines.

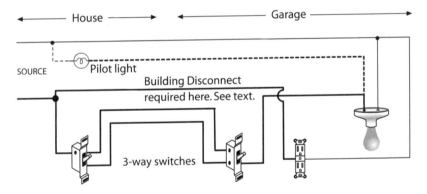

Fig. 17–25 The circuit of Fig. 17–24 with the addition of a pilot light in the house to indicate whether the light in the garage is on or off. This requires five circuit wires.

The circuit shown in Fig. 17–25 is the usual one in which three-way switches are used with a pilot light at the house end and a permanently live receptacle in the garage. Another circuit arrangement, shown in Fig. 17–26, requires only four wires instead of five. It meets *NEC* requirements and therefore may be used, though it requires more care in installation to make sure that all connections are correct. In making the circuit run to the garage, don't forget to keep all those conductors together within the same raceway, or else observe the precautions against inductive heating discussed earlier in this chapter.

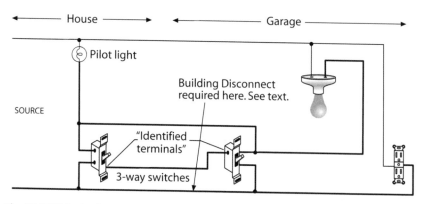

Fig. 17–26 This circuit, using only four circuit wires, serves the same purpose as the circuit of Fig. 17–25, which requires five circuit wires.

If the wires to the garage are to run overhead, they must be securely anchored at each end. Either of the insulators of Figs. 16–1 or 16–2 may be used. Where the wires enter or leave a building, either of the methods shown in Fig. 17–27 is suitable. A convenient fitting is shown in Fig. 17–28 and shown installed in *B* of Fig. 17–27. Be sure the insulators are installed at a point lower than the entrance of the wires into the building.

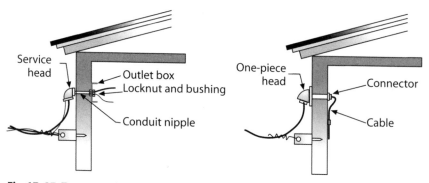

Fig. 17–27 Two ways of having wires enter the garage. The same methods can be used for other wiring, such as farm wiring.

Fig. 17–28 This wall-type entrance fitting is convenient for bringing wires into outbuildings.

CHAPTER 18

Finishing: Installation of Switches, Receptacles, and Fixtures

IN THE PRECEDING CHAPTER, the roughing-in of outlets was discussed. You are now ready to install the receptacles, switches, faceplates, and fixtures. In the trade this is referred to as "finish" wiring. This chapter begins with finish wiring procedures for conventional devices in dry, interior locations, and looks at finish wiring for grounding receptacles. The middle of the chapter covers the hanging of lighting fixtures. For this task you need to know which wire (or terminal) is which within any type of lighting fixture. Fluorescent fixtures require special consideration because of their size and the fact that their ballast channels can function in place of outlet boxes. As covered at the end of Chapter 11, flexible cord normally cannot be used for permanent wiring, but some permanently installed fixtures make legitimate use of cord, as covered in this chapter. Some fixtures sit on permanently installed poles that may, in some cases, function as raceways. Even everyday incandescent fixtures have their own conventions and rules regarding mounting hardware and positions. Paddle fans are being sold by the millions, and special mounting rules apply to these dynamic loads, a type of load not anticipated by traditional outlet box designs. It is also important to be aware of the special considerations for outdoor receptacles and switches. The chapter concludes with procedures for final testing of the entire system.

INSTALLING CONVENTIONAL DEVICES IN INTERIOR LOCATIONS

The members of a household or other end users are likely to be aware, on a daily basis, of the quality of your finishing work. While prevention of fire and shock hazards is paramount as always, durability and neatness are also important and in many ways help to support safety.

Installing switches, receptacles, and related equipment   Every device of this kind comes with a mounting strap or yoke (usually, but not always, made of metal), that has holes spaced to fit over the holes in the ears of a switch box on which it is mounted using screws that come with the device. The faceplate is then anchored to the device, not to the box; see Fig. 18–1.

For a neat installation, the yoke of the
device must be flush with the plaster or
wallboard. Because plaster or wallboard is
added after the switch boxes are installed,
the front edges of the boxes might not
always be flush with the finished surface of
the wall. Most devices have "plaster ears"
on the ends of the yokes, as shown in
Figs. 5–5 and 5–7. These ears lie on top of
the wall surface, bringing the device flush
with the surface. The metal is scored near the

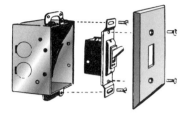

Fig. 18–1 The switch or other device is
installed on the box, and the faceplate
is installed on the device.

end of the yoke so the ears can easily be broken off if not needed. If a device does not
have plaster ears, you can insert small washers between the box and the yoke.

In the case of flush-mounted receptacles, the durability of the installation is an
even greater concern than neatness. A receptacle is subject to significant force
each time a cord is plugged into it. Many nonmetallic faceplates have enough give,
either inherently or after a crack develops in the small bridge between the two
openings of a plate for a duplex receptacle, so that the entire force of inserting a
plug bears on the yoke. The *NEC* anticipates this problem by requiring the device
yoke to be "held rigidly at the surface of the wall." If the drywall repair is at all
incomplete and the plaster ears don't seat solidly on the wall surface, use the
plaster ears or other washers under the yoke to make a solid surface to seat the
yoke against. Many devices now have extended plaster ears to help complete these
installations. This is a particular problem on nonmetallic boxes formed to
position the yoke screw to the inside of the box instead of to the outside as with
traditional metal switch boxes. Compare, for example, the lower left plastic switch
box in Fig. 10–6, on page 153, with the metal equivalents in Fig. 10–10 and 21–7,
on pages 155 and 399. The outer dimensions of the plastic box are much bigger
than the metal one in order to encompass the yoke mounting hole spacings.
Given a device yoke of constant dimension, the drywall has to be nearly perfect
for the plastic box installation in order for the yoke to have any significant bearing
surface against the wall.

If the wall consists of wood paneling or other combustible finish, you also must
comply with the *NEC* rule requiring boxes to be flush with or project from such
surfaces to protect the edge of the wall finish from any arcing resulting from a
failure within the box. Often a combustible finish gets added over an existing
wall, putting every box in the room into violation. Box inserts (Fig. 18–2)
arranged to be held in place by the yoke screws help solve this problem. Wallpaper
should not be left to cover the opening between the device and the box, although
paperhangers usually do just that, cutting through only where the device projects
from the wall. Take the time to cut the paper back to the outer perimeter of the
box—especially important if the paper has a metallic foil component.

For neatness, devices must be installed perfectly straight up and down, but often the box installation is not entirely straight. The mounting holes in the ends of the yoke are elongated rather than round to allow the device to be mounted straight even if the box tilts slightly to right or left. See Fig. 18–3.

You should have left about 10 in. of wire at each box; about 4 in. of this will be used in making the connection to the terminal of the device, as explained at the beginning of Chapter 8. Don't leave the wires too long because that would overcrowd the box. The 6 in. that is left can be neatly folded in the box leaving plenty of space for the device if the right size box was installed. The insulation of the wire must extend up to the terminal with no bare wire exposed between the terminal and the end of the insulation. Review making connections in

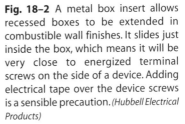

Fig. 18–2 A metal box insert allows recessed boxes to be extended in combustible wall finishes. It slides just inside the box, which means it will be very close to energized terminal screws on the side of a device. Adding electrical tape over the device screws is a sensible precaution. *(Hubbell Electrical Products)*

Chapter 8. If you're working with solid wire, it's a good idea to prefold the wires into their approximate position before making the final terminations on the device, particularly when the wires are secured to screw terminals on the side of the device. That lessens the force applied to those terminals when you push the device into the box. Another trick is to make the first fold one that runs the wire close behind the device so that as the device goes into the wall the force of folding the remaining length of wire pushes against the back of the device instead of upsetting the terminations.

Installing faceplates   After a device has been installed in a box, a faceplate must be installed using screws that come with the plate. If the device has been properly mounted at the surface of the wall, as just discussed, its yoke will be flush with the wall surface and the plate will fit snugly against both the device and the wall surface. The elongated mounting holes in the yokes of individual devices pose a problem when installing the devices in a two- or three-gang box because

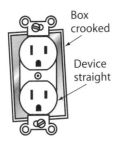

Box crooked

Device straight

Fig. 18–3 The mounting yokes of devices have elongated holes in the ends to permit vertical alignment of the device even if the box is not mounted straight.

the devices must be installed exactly parallel to each other in order for the tapped holes in their yokes to line up with the mounting holes in the multigang plate. Several adjustments of the individual yokes may have to be made.

Until you are more experienced, the approach illustrated in Fig. 18–4 can be helpful for installing ordinary switches and similar devices, though not for the interchangeable devices described on page 75. Start with a three- or four-gang metal plate, preferably with openings for switches and a receptacle, as shown at A. Saw off the edges very close to the mounting holes normally used to install the plate on the switches, as shown at B. After the switches (and receptacle if there is one) have been properly connected, install them in the box but do not tighten the mounting screws. Then temporarily install the sawed-off cover on the switches, tightening the mounting screws of the cover so that all the devices are solidly anchored to the temporary cover. All the devices will be parallel and the right distance apart. Now tighten the screws in all the individual straps, solidly mounting the devices in the box, and remove the temporary sawed-off cover. The holes in the final faceplate will fit perfectly over the various switches and receptacles.

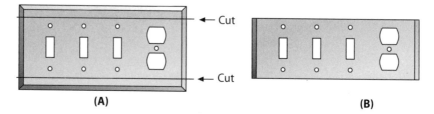

Fig. 18–4 For the beginner, this jig saves time when installing several devices in a multigang box.

Grounding receptacles   In addition to the tricks at the beginning of the chapter regarding workmanlike mounting of switches, which also apply to receptacles, finish wiring to receptacles requires careful attention to the details of proper grounding. The U-shaped opening (contact) of the receptacle is bonded to the mounting yoke of the receptacle, and to the receptacle grounding screw, which is identified by its green color and hexagonal shape. This green terminal screw of the receptacle must be effectively bonded to the box (if metal) and to the equipment grounding conductor of the supply circuit.

In the case of metal-enclosed wiring there should always be a grounding connection to any metal box. The conduit or metal armor, in turn, should already be bonded (through locknuts and bushings, or connectors) to outlet and switch boxes all the way back to the service equipment. Therefore it is necessary only to bond the U-shaped receptacle contact to the box under these circumstances. Because the yoke of the receptacle is connected to the box with steel screws, it might seem that these screws would effectively bond the grounding contact of the receptacle to the box. However, in many cases the mounting yoke is not in solid contact with the box, and the bond is dependent on 1½ or 2 threads of the small mounting screws, which is not a very effective bond.

For that reason, *NEC* 250.146 requires that an equipment bonding jumper (bare or green) be run from the green terminal of a receptacle to the box. Connect the bonding jumper (copper wire) to the box using the special metal grounding clip as shown in Fig. 18–5, or install a small screw through an unused hole in the box, as shown in the same illustration. Most outlet and switch boxes have an extra hole tapped for a 10-32 screw for the purpose. This screw hole may not be used for any other purpose.

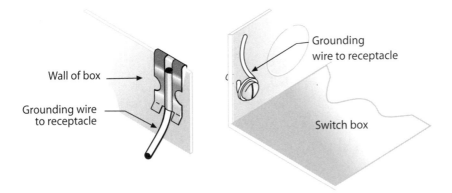

Wall of box

Grounding wire to receptacle

Grounding wire to receptacle

Switch box

Fig. 18–5 The grounding wire to the green terminal of a receptacle may be grounded to the box using either a grounding screw or a grounding clip as shown at left above.

However, many listed receptacles are now available with special screws and yoke hardware for connecting the receptacle to flush-mounted boxes, providing an effective bond between the receptacle and the box. See Fig. 18–6. The use of such receptacles does away with the need for the bonding jumper from the green terminal to the box.

When using a nonmetallic wiring method, there will be a separate equipment grounding conductor run with the circuit conductors. Connect this grounding

Fig. 18–6 Several brands of receptacles are now available with special construction that eliminates the need for running a grounding wire from the green terminal of the receptacle.

Spring-steel wire holds lobed 6-32 mounting screw in contact with the yoke.

wire as covered on page 229, depending on where the circuit originates. At each box, connect the ends of all the grounding wires together using a solderless connector such as a Wire Nut. From the junction you have made, run a short wire to the box itself as described in previous paragraphs. If the box contains a receptacle, run another short wire from the junction to the green terminal of the receptacle. All this is shown diagrammatically in Fig. 18–7. Remember that the free equipment grounding conductor length from each cable must not be less than 6 in. This allows you to remove and replace a receptacle without interrupting the grounding continuity to other outlets, as required by the *NEC*. There are special twist-on wire connectors available, green in color with a hole in the narrow end, which allow you to do this easily. Just leave one of the equipment grounding conductors about 10 in. long, and then push the extra 4 in. out the end of the connector. Insert the 6-in. end of the other equipment grounding conductor into the connector and twist it into place just as with the usual connector. You now have a 4-in. pigtail to run to the box or to a receptacle terminal.

If you are using nonmetallic boxes, the only difference is that the grounding wire does not need to be connected to the box.

Surface-mounted boxes

If you are using boxes mounted on the surface, such as "handy boxes" where the mounting yoke of the receptacle is in direct, solid contact with the surface of the box, you need not install a bonding jumper from the green terminal of the receptacle to

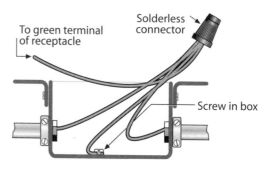

Fig. 18–7 How to connect the bare equipment grounding wire of nonmetallic-sheathed cable.

the box. In order to take advantage of this procedure, remove any nonmetallic shipping washers that keep the yoke screws from falling out of position on the yoke so nothing will prevent the yoke screws from solidly and conductively mating the yoke to the mounting ear of the box.

There is a particular type of surface receptacle application that requires a bonding jumper regardless of metal-to-metal contact, namely, a receptacle mounted in a raised cover, illustrated in Fig. 10–7. Unless the "box and cover combination" is listed as providing satisfactory grounding continuity between the receptacle and the box, and as of this writing no conventional applications are so listed, you have to install the bonding jumper. The only instances presently listed are certain explosionproof receptacle assemblies not suitable for ordinary locations. This restriction applies independently from a different but related requirement that more than a single screw must anchor the receptacle in any cover whether raised or not, unless listed otherwise. (See also Fig. 10–8, *D*, for an example of a flat cover with a receptacle.)

Tamper-resistant (safety) receptacles   Children have a natural tendency to explore, and it is not unusual for a curious child, who has seen adults push plugs into receptacles, to try to push part of a toy, or a paperclip, key, or other metal object into a receptacle slot. If the child is on a completely dry floor, probably no harm will be done. But if the child is also touching a grounded object such as a radiator, a water pipe, or the framework of a defective lamp, the result could be a violent shock. "Safety" receptacles were developed to prevent this. The correct terminology in today's *NEC* is "tamper resistant"—tacit acknowledgement that nothing is really tamperproof around children. One type has a spring-loaded cover over the slots of the receptacle; unless the receptacle is in use with a plug in it, the slots of the receptacle are concealed. Another type is designed so that inserting anything into just one of the two slots will not make electrical contact; both prongs of a plug must be inserted before contact is made. Using safety receptacles is sensible for households with small children. These receptacles are mandatory in hospital pediatric wards, as covered in *NEC* 517.18(C).

INSTALLING CONVENTIONAL LIGHTING FIXTURES

There is a very broad range of styles and sizes of conventional lighting fixtures. The wiring task is fairly consistent for all of these, but the method of hanging a fixture varies with its design. Fittings to suit a particular fixture are usually supplied by the manufacturer.

Wiring lighting fixtures   If the fixture has two terminals for connecting the wires, connect the white wire to the light-colored terminal, the black wire to the other terminal. More often the fixture has two wire leads. If these are black and white, connect them to the correspondingly colored wires. Typically you will encounter fixture wire that requires close attention, as explained in following paragraphs, so you don't reverse the connections.

Identifying grounded fixture conductors   Fixture wire, especially in the higher-temperature styles, might be available in only one color, often a relatively light shade. Look for a distinguishing tracer thread or line in the outer material of one of the wires. As explained in the page 134 white wire topic, the *NEC* generally requires a deliberately grounded conductor to be white; fixture wires, however, can rely on the tracer to be the identifying mark. For example, given two light tan fixture wires, one of which has a red tracer, the one with the red tracer is the grounded (normally white) wire. *NEC* 400.22(A) through (F) gives other identifying methods, including the use of a ridge on the exterior of a cord, as illustrated in Fig. 11–49. This method is commonly used for chandeliers. Be very careful not to reverse the intended connections, particularly if the fixture has a screw shell.

Screw-shell grounded-conductor connections in fixtures   Screw shells
used on grounded circuits must be connected to the grounded conductor. If you
reverse the wiring connections, the screw shell, instead of the center pin, will be
hot with the fixture on. Many people leave fixtures energized while relamping
them, particularly when there are multiple lampholders as on some chandeliers,
so they don't have to mark the location of a burned out bulb. In that situation a
hot screw shell is a severe shock hazard because the metal skirt of a lamp base
becomes and remains energized anytime it touches a shell. If your fingers touch
the lamp base as you are screwing it in or out, especially if your other hand is
bracing adjoining metal parts of the fixture, you will receive a severe shock from
arm to arm directly across your heart.

If a fixture is supplied by two ungrounded conductors, a hot screw shell will be
present any time the fixture is on, but the *NEC* does not permit any interior
elements including screw shells to be energized when the fixture is deliberately
off. Therefore any fixture connected line-to-line to ungrounded conductors,
whether directly to lampholders or indirectly using an electric discharge fixture
ballast, is required to use double-pole switching. For fixtures on outdoor poles
with photocell control, central control and relaying will be necessary if the
photocell is a single-pole device, which is usually the case.

Hanging traditional fixtures   There are many ways of installing lighting
fixtures, depending on the style, shape, and weight of the fixture, the particular
box involved, and the method of mounting the box in the ceiling or wall.

Simple fixtures can be mounted directly on top of outlet boxes using screws that
come with the fixture and fit into ears on the box. This method is shown in Fig. 18–8.

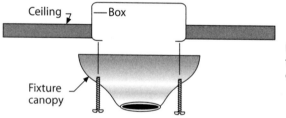

Fig. 18–8 Small, simple
fixtures are mounted directly
on the box.

For fixtures too large to permit mounting in this direct fashion, a strap is used as
shown in Fig. 18–9. The strap is first installed on the box, then the fixture is
mounted on the strap. The *NEC* requires all fixtures with exposed conductive
parts to be grounded. With the advent of nonmetallic boxes, fixture straps
commonly have grounding terminals, which makes the strap mandatory in
hanging even simple fixtures on nonmetallic boxes. Some plastic boxes, however,
do have a grounding terminal that will ground a fixture mounting screw directly.

Be aware that if you are selecting a replacement fixture for an old wiring system with
no equipment ground, you will be restricted to a very small group of fixtures that

have no exposed metal. For existing installations, however, the 2005 *NEC* now allows GFCI protection for such fixtures, in lieu of rewiring the circuit with an equipment grounding conductor.

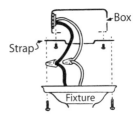

Fig. 18–9 Larger fixtures, or fixtures with nonstandard spacing between their mounting supports, are installed on a strap that has been mounted on the box.

Often a fixture stud such as shown in Fig. 18–10, also known as a "crow's foot," is used in mounting the fixture. The holes in its base line up with the mounting holes in the back center of metal outlet boxes. If one of those boxes is installed on a suitable framing member such as a short piece of framing secured to cleats between ceiling joists, you can run mounting screws directly through the crow's foot, through the box mounting holes, and then into the framing. The result is that the entire weight of the fixture will be held by the framing instead of the box ears, which allows for the installation of very heavy fixtures. Unless the box is listed for a heavier weight, the *NEC* prohibits a box from supporting any more than a 50-lb fixture.

Fig. 18–10 Fixture studs are installed in the bottoms of boxes in order to support fixtures.

In some boxes the stud is an integral part of the box. If you use a hanger of the type shown in the upper part of Fig. 10–17, the stud that is part of the hanger goes through the center knockout of the box, anchors the box to the hanger, and at the same time permits the stud to be used to support the fixture. The outside of the stud is tapped to fit ⅜-in. trade-size pipe; sometimes there is an inner female thread fitting ⅛-in. trade-size pipe.

Figure 18–11 shows the fittings generally used, called a lockup unit. It consists of an adapter fitting over the fixture stud, a short length of ⅛-in. running-thread pipe, and a nut to hold the assembly together; the nut is usually an ornamental type called a finial. Figure 18–12 shows these parts used to hold up a simple fixture. It is a simple matter to drop down the fixture while making connections, then to mount it on the ceiling.

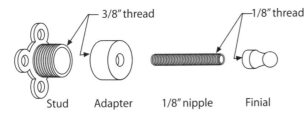

3/8" thread   1/8" thread

Fig. 18–11 A typical lockup unit for installing small fixtures.

Stud   Adapter   1/8" nipple   Finial

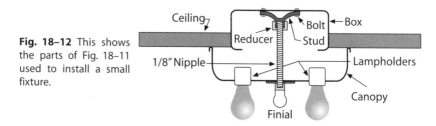

Fig. 18–12 This shows the parts of Fig. 18–11 used to install a small fixture.

If the fixture is a larger unit, the mounting is similar. The top of the fixture usually consists of a hollow stem with an opening in the side from which the wires of the fixture emerge. The top is threaded to fit on the fixture stud, and the mounting is shown in Fig. 18–13. Drop the canopy down while making the connections and then slip it back to be flush with the ceiling.

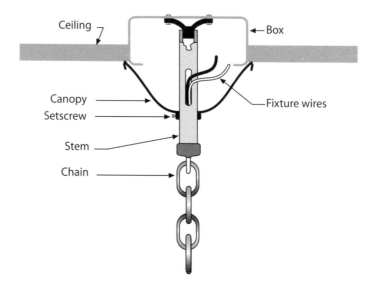

Fig. 18–13 The usual method of installing larger fixtures.

Sometimes the fixture wires emerge from the end of the stem instead of through a side opening. In that case use a hickey (Fig. 18–14) between the end of the stem and the fixture stud. Sometimes the stud is too short or the box too deep, in which case use an extension piece as shown in Fig. 18–15.

If the ceiling or wall material is combustible (such as wood or low-density wallboard) and there is an open space between the box and the edge of the fixture, *NEC* 410.13 requires the space to be covered with noncombustible material. This need not necessarily be a metallic covering, and at one time UL recognized the fiberglass insulating batting in the back of a fixture as serving this function, but no longer. There isn't a straightforward solution to this problem.

◀ **Fig. 18–14** A hickey is sometimes used between the end of the stem on the fixture and the fixture stud.

Fig. 18–15 In deep boxes it is sometimes necessary to use an extension piece between the stud and the fixture. ▶

Some manufacturers have round decorative backers (often for vinyl siding) or ceiling medallions that, depending on the canopy diameter, may fit acceptably, and ceiling boxes designed as part of surface metal raceway systems (see page 508) have round, separable bases in multiple diameters that can often be used.

Installing wall brackets   Wall brackets are often primarily decorative. The method of installing a bracket depends largely on the type of box used. Some brackets are too narrow to completely cover a 4-in. box, so a switch box is used. Sometimes a fixture stud is mounted in the bottom of the box; in that case the bracket is mounted using the lockup unit shown in Fig. 18–11, and the completed installation is as shown in Fig. 18–17. More often a fixture strap, shown in Fig. 18–16, is first mounted on the switch box, and the fixture is mounted on the strap, as in Fig. 18–18. If the wall finish is combustible, the same rules apply as for comparable ceilings.

Fig. 18–16 Fixture straps are mounted on boxes, and the fixture is then supported by the strap. These straps often have a grounding terminal for use with nonmetallic boxes.

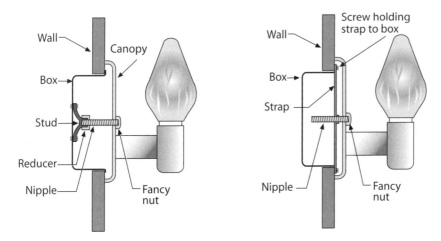

Fig. 18–17 Wall bracket installed on box by means of fixture stud in bottom of box.

Fig. 18–18 Wall bracket installed on box by means of fixture strap mounted on the box.

NEC 314.27(A) requires outlet boxes that support fixtures to be designed for this purpose. According to the product directory information supplied by the testing laboratories, boxes evaluated as acceptable for supporting fixtures have a marking on the shipping carton, which means if you get just a few boxes you won't see the marking. In general, boxes with provisions for 8-32 mounting screws are suitable, and those with 6-32 mounting provisions are not. However, many fixtures with rectangular profiles won't cover the opening made by a 4-in. octagonal box, for example, and *NEC* 370.25 requires every box to be covered as part of the completed installation. This problem was resolved in the 1999 *NEC* (amended slightly in 2002) by adding a limited exception for the use of switch boxes as fixture support. Fixtures that don't exceed 6 lb can be supported from a switch box provided the fixture is held in place by at least two 6-32 screws.

Adjusting height of fixture   A fixture with chain is adjustable to compensate for ceilings of different heights. Control the height by removing as many links of the chain as necessary to suit personal preference and location. A fixture above the dining room table should hang at least 24 in. above the table, and preferably higher. Be sure to allow for the typical 18 AWG bare equipment grounding conductor generally threaded through the chain in today's fixtures. If there is any possibility that the chain will need to be lengthened later, keep the original grounding conductor length long and tucked into the canopy.

FIXTURES THAT DON'T MOUNT DIRECTLY TO OUTLET BOXES

This group includes most fluorescent fixtures. It also includes a very limited set of circumstances in which the *NEC* allows use of flexible cord on the load side of fixture outlets. For fixtures mounted on metal poles, you must observe *NEC* requirements that apply to the poles.

Fluorescent fixtures   With the exception of the little PL folded tube and circline models designed to replace incandescent fixtures, fluorescent fixtures generally are so large that they don't "hang" from an outlet box. Instead you bring the wiring method to them after independently setting their final location. If that location is in a suspended ceiling, the ceiling members need to be securely attached to the building structure at appropriate intervals, as well as to each other. Then, after supporting the fixture within the framing, mechanically secure it to the framing with bolts, screws, rivets, or listed clips identified for use with the particular framing involved. Gravity alone is not enough. The *NEC* panel wants to be sure that in the event of an earth tremor the fixtures and the ceiling framing move as one.

For a fixture mounted to the ceiling over an outlet box, the *NEC* requires the box to be accessible without dropping the fixture. That typically means running a hole saw equal to the box opening diameter through the back of the ballast channel, and then mounting the fixture to the ceiling. Don't forget to install an equipment grounding jumper between the box and the fixture; either use a wire or drill a clearance hole in the ballast channel that lines up with a box ear and then connect the two enclosures with an 8-32 screw. If you have to mount a run of fluorescent fixtures end to end, make sure the equipment grounding continuity between the fixtures is secure. Some fixtures are designed for this and include joining hardware. In other cases you will have to supply chase nipples and locknuts to make a tight connection. After the run is complete, bring the wiring method directly to one of the fixtures. If fixtures are not listed as raceways they can't be used at will to extend circuits. However, special *NEC* provisions allow for such fixtures to carry through all the conductors of a two-wire circuit and all the conductors of the multiwire circuit to which the fixture is connected (even those legs not used for this particular fixture). In addition, one additional two-wire circuit can pass through as well if it terminates in one of the fixtures along the particular run. This is the so-called "night light" exception.

Cord-connected fixtures   The *NEC* does not generally permit flexible cord to be used as a permanent wiring method. For fixtures, however, the *NEC* allows flexible cord on the load side of a fixture outlet in five restricted instances:

- The first allowance is for a metal lampholder, such as could be attached to the end of a drop cord pendant as covered under the next heading. This permission is for a lampholder only and not a fixture.

- The second allowance is for a fixture with a part that requires aiming or adjustment after installation, such as some floodlight heads; the cord must be hard usage or extra hard usage and no longer than necessary.

The last three allowances for cord apply only to electric discharge fixtures:

- *Maintenance*—Flexible cord can be used to facilitate fixture maintenance. Fixtures equipped with flanged inlets (plugs without cords that mount directly to enclosures; see Fig. 23–7 for an example) rated at least 125% of the fixture load current can be mated to hanging cord pendants with receptacle bodies. In addition, fixtures can have flexible cord permanently mounted to them, terminating in a grounding attachment plug (or busway plug), provided they hang directly below the outlet box and the entire cord length is visible and not subject to strain or damage. Both of these allowances apply to individual fixtures, and although not clearly stated in the *NEC*, the concept is to allow fixtures to be taken down for maintenance and then rehung.

- *High intensity discharge fixtures with mogul-base lampholders*—These fixtures can be cord-connected on branch circuits up to 50 amps. The cords have to be sized according to the normal rules for such cords, namely 20-amp ratings for 40- and 50-amp circuits; 16 AWG minimum for 30-amp circuits; 18 AWG minimum for 20-amp circuits. Choose plug and receptacle ratings according to the fixture rating based on 125% of the actual current. These fixtures often hang from a busway.

- *Listed fixture assembly with strain relief and canopy*—There are specialty fixture assemblies that come with canopies and flexible cord as part of the listed fixture. These assemblies may involve substantial lighting layouts involving many fixtures placed end-to-end. Often these assemblies come with aircraft cable to carry the mechanical load or some similar arrangement, and directions to cut and terminate the aircraft cable to accommodate an uneven ceiling. In this case a length of flexible cord, listed as part of the overall assembly, may be the only practical alternative. See Fig. 18–19 for an example.

Cord-connected listed fixtures permitted to connect through a canopy furnished with the fixture, instead of a plug.

Fig. 18–19 This listed fluorescent fixture hangs on adjustable lengths of aircraft cable, and the coiled cord arrangement allows for adjustment.

In making up a drop cord, tie an Underwriters knot at the top so the weight is supported by the knot and not by the copper conductors where they are connected to terminals. This knot is illustrated in Fig. 18–20. The drop will be supported from the ceiling by a blank cover with a bushed hole in the center, mounted on an outlet box. The knot goes above the bushed hole. In general, the use of drop cords (pendant lights) can and should be avoided. They are useful, however, for repositioning of equipment in some areas, particularly in light industrial areas. The best way to wire them is using hard service or extra hard service cord supported by a suitable strain relief connector (see Fig. 18–21).

Fixtures supported on metal poles   Metal poles, such as those commonly seen in parking lots, can be used to support fixtures only if they meet *NEC* requirements. The first thing to remember about metal poles is that they are, in effect, raceways and normal raceway rules essentially apply. For example, you have to be able to get at the wiring connections inside them without taking them apart. Thus poles over 8 ft above grade must have a handhole at the base at least

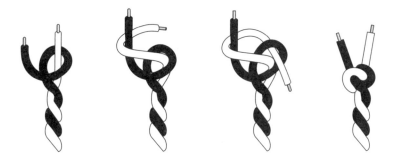

Fig. 18–20 Drop cords are not often used, but if one is installed, provide an Underwriters knot at each end.

2 in. by 4 in. with a raintight cover, or, in the case of poles not over 20 ft high, a hinged base is a suitable substitute. Poles not over 8 ft high avoid this requirement in cases where the wiring method extends all the way up the pole from grade without splices, such as when Type UF cable is pulled all the way up the pole interior. The fixture must be removable in such cases.

Poles must always be able to be grounded at the point of access to the wiring method; in the case of the short pole, that could be through the fixture connections. Vertical conductors in very high poles need support as covered in the page 503 discussion on supporting vertical runs of wire. In cases where poles support other wiring classifications, all the rules regarding wiring system separation must be observed, as covered in the page 364 topic on signaling and control wiring. For example, if you put a security camera, or a motion sensor as shown

Fig. 18–21 A strain-relief connector provides secure support for a cord drop spread over a wide area of the outer jacket. It prevents any strain from being transmitted to the terminations.

in Fig. 18–22, on a light pole, the limited energy conductors that supply that camera or sensor must run in a separate raceway within the pole in order to maintain the system separation required by the *NEC*.

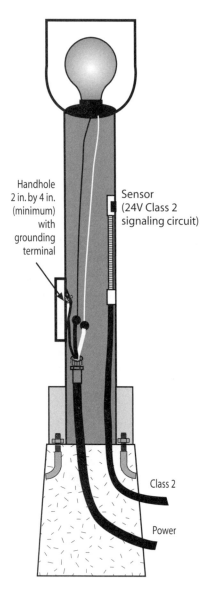

Handhole
2 in. by 4 in.
(minimum)
with
grounding
terminal

Sensor
(24V Class 2
signaling circuit)

Fig. 18–22 The inside of a fixture support pole is a raceway, and the contained wiring must be accessible and obey normal *NEC* rules for raceway installations. The motion sensor wiring, part of a Class 2 control circuit, must be completely divorced from the power wiring, as shown.

Class 2

Power

PADDLE FANS HAVE SPECIAL REQUIREMENTS

Paddle fans are popular and the ones with lighting fixtures attached below them frequently substitute for traditional overhead lights. Safely supporting paddle fans has become a major issue in the industry. Traditional boxes and their product standards have generally assumed a static load held in a particular direction. Paddle fans, especially when rotating fast and with some blade imbalance, impose a load that traditional boxes weren't designed to support. In addition to causing box failures, paddle fan loading overstresses conventional box support methods. In fact, most fan support failures probably have more to do with inadequacies in the way the box was secured to framing than with the integrity of the box itself. When you mount paddle fan boxes and supporting brackets, use very robust support hardware and follow the manufacturer's mounting instructions exactly.

The *NEC* recognizes boxes specifically listed for the support of paddle fans (see Fig. 18–23). They can generally support 35-lb fans, and heavier ones (up to 70 lb) if so listed and marked. They undergo a rigorous testing protocol involving a very heavy fan run for a long period at very high speeds with a severe blade imbalance, and with one of the screws that secure the fan to the box deliberately loosened in some cases. If your customers are even thinking of adding paddle fans, it's much easier to rough in paddle fan support boxes at the likely locations than it is to install them at existing outlets later, especially if there isn't framing in place that would allow independent fan support at the outlet location.

Fig. 18–23 These outlet boxes have been specifically listed for fan support. The one in the center is quite ingenious in terms of accounting for required wiring volume in a fan box intended for the bottom edge of a ceiling joist. *(Hubbell Electrical Products)*

With respect to framing support, the *NEC* allows paddle fans of any weight to be supported directly from structural elements of the building even at a traditional outlet box, because the building structure and not the box will be the primary support of the fan. (See Fig. 18–10 and related text describing the use of fixture studs.) This procedure has the additional virtue of allowing a fan to be mounted on any size or configuration of boxes, extension boxes, and plaster rings, some or all of which may not be available in a form listed for direct fan support. However, this procedure may also require carpentry skills as well as access to the underlying framing.

OUTDOOR AND OTHER WET LOCATIONS REQUIRE SPECIAL PROCEDURES

The required locations for outdoor receptacles at dwellings are covered in the discussion of porches, decks, and patios on page 214, and the GFCI protection requirements are explained on page 148. Following is additional information on how to finish these outlets so they are safe.

Outdoor receptacles The first question is whether the receptacle is in a damp location, as under a roofed porch, or in a wet location exposed to precipitation. For the damp location, you need to provide a cover that is weatherproof only with the cover openings closed, such as illustrated in Fig. 18–24. You may also go beyond the minimum requirements and install a cover that is weatherproof with the receptacle both in use and not in use, as shown in Fig. 18–25.

◀ **Fig. 18–24** An outdoor receptacle in a weatherproof housing. It is suitable for damp locations and wet locations only while attended. *(Leviton Manufacturing Co. Inc.)*

Fig. 18–25 This outdoor receptacle cover is weatherproof with the receptacle in use, as shown. *(TayMac Corporation)* ▶

For all wet locations (not just those outdoors), the basic rule is to use the same cover (Fig. 18–25), and this is now mandatory for all 15- and 20-amp, 125- and 250-volt receptacles in such locations. For other configurations, this type of cover is mandatory unless you can convince the inspector that the receptacle will be used only while attended. For example, a receptacle for a swimming-pool pump motor in an outdoor location would clearly be used for long periods while unattended. The covers shown can be used on a cast box (such as a Type FS shown in Fig. 27–2), and they also mate to a flush box on the side of a building. The "bubble" covers (Fig. 18–25) have largely supplanted the other covers, but be aware that while they work well outdoors when exposed to precipitation, they are not rated for a hose stream. If you have a wash-down exposure, such as with indoor dairy facilities, you will need to use your ingenuity; one solution involves a locking receptacle (see Fig. 20–6) and a cover that is designed to be watertight and receive a mating plug with a rubberized boot that seals to the flat surface of the cover while in use.

Outdoor switches. For switches exposed to the weather, use a cast Type FS box as shown in Fig. 27–2. Install an ordinary toggle switch in it, with a weatherproof cover as shown in Fig. 18–26. Outdoor wiring for farms is discussed in Chapter 22, and commercial and industrial outdoor wiring in Chapter 27.

Fig. 18–26 Install a toggle switch in a cast weatherproof box, and use this waterproof cover to control it.

TEST YOUR INSTALLATION BEFORE ENERGIZING IT

Many professional electricians do not test their installations, feeling pressured by time constraints or simply overly confident that they have done everything properly, and then regret their decision when they have to make a difficult repair that would have been easy if the walls were still open. Until you have installed a few jobs and can be sure you have made no mistakes, you should test the entire installation before the wiring is covered up by wallboard or lath-and-plaster so that any necessary changes can be made while the wiring is still accessible. You will want to know how to make the tests anyway for when a need for troubleshooting arises. For this purpose, you can use a doorbell or buzzer and a lantern battery, as shown in Fig. 18–27. Tape the bell to the dry cells or use cable ties as shown in the illustration; also secure the test leads to the cells as shown so that you can lift the entire unit by the leads without putting strain on the bell or the dry-cell terminals.

The testing must be done *before* the utility has connected its wires to the service-entrance wires. Before doing any testing, do the following:

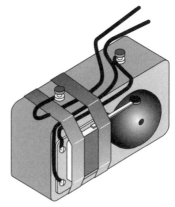

■ At each outlet containing wires that are to be permanently connected to each other, connect them now using Wire Nuts or similar connectors.

■ Where receptacles or fixtures are to be installed, let the wires project freely from the outlet.

■ At all points where a switch is to be installed, be sure that all wires that will be connected to one switch are *temporarily* connected together; the quickest way is to use a twist-on wire connector that will be removed later.

Fig. 18–27 Test outfit consisting of a 6-volt lantern battery and a bell.

■ Be sure that a transformer for doorbell or chimes has *not* been connected to a circuit.

■ Be sure that no exposed bare wire is allowed to touch any box or conduit or the armor of cable.

Testing metal wiring systems   These instructions assume that you have connected the white grounded wires of all 120-volt circuits to the grounding busbar in the service equipment. If you have used circuit breakers, turn them all to the ON position; if you have used fused equipment, make sure all fuses are in place. Then proceed as follows: At the service equipment, connect one lead of the battery-and-bell tester to the grounding busbar. Then touch the other lead of your tester in turn to the black (or red) wire of each 120-volt circuit. The bell must *not* ring. If it does, there is a short circuit, probably at a box where you have inadvertently permitted a bare wire to touch the box. Correct any such errors and retest until there is no ring.

Next, make another test for continuity. Connect a *temporary* wire jumper across the two hot service wires where they enter the service equipment. (After you have completed the tests, you must remove this jumper or you will have a total short circuit across your 240-volt service wires.) Then remove the bell from the batteries, leaving just the leads from the batteries. Connect one of them to the terminals in the service equipment across which you have placed the jumper; connect the other to the grounded neutral busbar in the service equipment. You will then have temporary 6-volt current across each 120-volt circuit that you have installed. (Remember, you will be removing that temporary jumper after you have completed the tests.)

Take the bell to each outlet where a fixture or a receptacle is to be installed. Touch it across the black and white wires; the bell should ring, just as lamps in your fixtures or floor and table lamps will later light when connected to these same wires. If the bell does not ring, check to make sure all circuit breakers are in the ON position, or that all fuses have been installed. After the check across the black and white wires of each outlet, touch the bell across the black wire and the box itself; the bell should again ring because through the conduit or cable armor all the boxes are connected together and grounded, as is the white wire. If the bell rings feebly, a poor job has been done somewhere; probably the locknut at one or more boxes has not been tightened enough. If the bell rings at each point, the wiring is all right. *Remove that temporary jumper in the service equipment.*

Note that the procedure just outlined is for 120-volt circuits only. For 240-volt circuits, touch the bell across one of the two wires at the outlet, and the box. The bell should ring. But if you touch it to the two wires of the circuit, it will not ring. To further test the 240-volt circuit, remove that jumper in your service equipment. Connect one of the battery leads to one of the two hot incoming service wires, and the other lead to the other of the two hot service wires. Then connect the bell across the two wires at the 240-volt outlet; it should ring.

Testing nonmetallic-sheathed cable installations   If you are using metal boxes and cable with the bare equipment grounding wire properly installed, proceed as in the case of metal conduit or armored cable installation, but be sure you have connected the bare grounding wires to the grounding busbar in your service equipment, and be sure they have been connected to each metal box.

If you have used nonmetallic boxes, test by touching the bell not only between the black and white wires, but also between the black wire and the bare grounding wire. The remainder of the test is as with the metal conduit system.

If all the tests check properly, *remove that jumper* and proceed to finish the installation as outlined in this chapter.

Circuit testers   Many types of commercial testers are available. The one shown in Fig. 18–28 can be used only to test the continuity of a circuit. It operates on a couple of penlite batteries and of course can be used only while the power is turned off. It can also be used, for example, to test cartridge fuses, removed from their holders, to see whether they are blown or not. Cartridge fuses look the same whether blown or not.

Fig. 18–28 A test light of this kind is convenient for tracing circuits, testing fuses, and so on. *(A. W. Sperry Instruments, Inc.)*

Another handy device is the tester shown in Fig. 18–29. It contains a tiny neon bulb that glows when energized; it can be used on either 120- or 240-volt circuits to determine whether the circuit is "hot" or not. It can also be used to test cartridge fuses while they are still in their pull-out blocks, still in the switch. On most pull-out blocks you will find tiny holes at the top and bottom of each fuse. Insert the leads of the tester into both top and bottom holes. Assuming the power is on, the neon light will glow if connected across a blown fuse, but it will not glow if connected across a fuse in good condition. Another variety of this tester has the neon glow chamber as part of one probe. That way you can see it easily as you manipulate the probe. These neon testers often glow dimly because they sense your body's capacitance when you hold one of the probe tips tightly in your hand and then touch the other probe to any source of voltage to ground. That makes them particularly handy on the fourth floor of an apartment building wired with ungrounded Type NM cable or knob-and-tube systems, and you're trying to decide which of the two conductors at the outlet, equal in overall appearance, is grounded and which is not.

Fig. 18–29 This test light is handy to determine whether a circuit is live or not. *(A. W. Sperry Instruments, Inc.)*

After the wiring is completed and the power turned on, the tester shown in Fig. 18–30 is used only to determine whether 15- and 20-amp receptacles have been properly connected. The tester has three neon lights, and when plugged into a receptacle it indicates whether the receptacle has been properly connected. If wrongly connected, the combination of lights will show the nature of the wrong connection.

Fig. 18–30 This tester tells you whether 15- and 20-amp receptacles have been properly connected.

There are GFCI testers that look similar to the tester in Fig. 18–30. They deliberately leak a certain amount of current into the equipment grounding system in order to determine if whatever GFCI device ahead of it is functioning properly. Don't use one of these testers unless it is listed, because it may give you erroneous readings in either direction. In addition, be aware that these testers cannot function on a circuit with no equipment grounding continuity. In fact, if you test a circuit with an open equipment ground, whatever is connected to that equipment ground will be energized while the tester is in place. There have been some nasty shocks as a result.

CHAPTER 19

Limited-Energy Wiring

THE CHAPTER TITLE reflects the fact that the term "low-voltage wiring," although widely used, is a misnomer because even at 6 volts enough amperes can flow through any given wire to start a fire. The *NEC* needs to assure that these circuits are not ignition capable, and therefore considers the total power available to the system. The limited-energy category does include wiring for doorbells, chimes and other signals, thermostats, and similar devices operating at low voltage, but it is the overall power limitation that technically qualifies them for consideration in this chapter. In this context, this wiring includes a Class 2 remote control and signaling circuit, which is essentially defined as a circuit originating at a listed Class 2 power supply. These supplies must be listed and cannot be constructed in the field. The power levels in such systems are held roughly to 100 VA, and the voltage is controlled to the point that the circuits do not pose either a shock or fire risk. This chapter also covers, toward the end, communications circuits that operate at somewhat higher power levels but also under strict energy limitations.

FIRE RESISTANCE OF LIMITED-ENERGY CABLING

Although limited-energy circuits don't have any inherent shock or fire hazard, the increasing volume of signaling cable in today's electronic age, even in dwellings, has become a matter of concern. After the cause of some damaging fires was traced to the contribution of these cables to smoke generation and fire transmission, the *NEC* reexamined its historical reluctance to write rules on uses of different cable constructions in limited-energy circuits.

The *NEC* established a four-level hierarchy that correlates the degree of increased fire risk with the degree of public exposure. Cables listed for any level in the hierarchy can always be used as substitutes for cables lower in the hierarchy, but never the reverse. The highest level, with the toughest listing criteria, is the level that allows exposed cabling in areas such as plenum cavity ceilings—areas through which environmental air is deliberately propelled as part of conditioning the occupied space. This level carries a "P" suffix in the cable name.

The next level down specifies cables that won't transmit fire vertically from floor to floor—these cables have an "R" suffix (for "riser"). The next level is for general commercial use and carries no suffix. The lowest level, with an "X" suffix, is for one- and two-family dwellings (or for nonconcealed spaces of multifamily dwellings). Using Class 2 cabling as an example, if you want to run these cables in a plenum cavity ceiling, use CL2P; if you want to run vertically from floor to floor, use CL2R; if you want to run from point to point in an office, use CL2; as a minimum in a house, use CL2X. In any application you can freely substitute a cable rated higher than the minimum required rating.

These categories apply to all limited-energy cabling. For example, coaxial cable connected to a community antenna television system (CATV) conforms to similar listing and labeling criteria (CATVP, CATVR, CATV, CATVX), as does fiber-optic cabling (the ultimate low-energy wiring at zero VA) and telephone (CMP, CMR, CM, CMX).

WORKMANSHIP

Although any single run of limited-energy cabling presents little fire risk, the *NEC* rules are based on the concern that a sloppy installation could be a signal of other problems. The principles of good workmanship apply to these systems as well. Exposed cabling on walls and ceilings must be supported by structural elements in order to assure that the cabling will continue to function properly under normal conditions of building use. Cable straps, staples, etc. must be the appropriate type and installed correctly so that the cable will not be damaged by them. In addition, when running the cables in new construction, be sure to route them out of the way of drywall screws, etc. The provisions of *NEC* 300.4(D) now apply to this work, as covered in the discussion on mechanical installation beginning on page 189.

In addition to these rules, the Code also addresses the effects that large numbers of these cables may have on other essential systems. The quantity and placement of cables must not present obstructions to areas intended to remain accessible, such as above suspended ceilings. To further reduce fuel loading and improve workmanship, the *NEC* now requires that accessible portions of these cables be removed if abandoned. "Abandoned" in this context is defined as "not terminated at equipment and not identified for future use with a tag."

SIGNALING AND CONTROL WIRING

For remote control and signaling, the Class 2 circuits covered here usually operate at 30 volts or less. The power for operating such low-voltage circuits is typically derived from small transformers, but there are other recognized sources in *NEC* 725.41. Under no circumstances may low-voltage wires be run in the same conduit or cable with other wires carrying full voltage. They must never enter an outlet or switch box containing full-voltage wires unless a metal barrier of the same thickness as the walls of the box separates the two types of wiring, or unless the power supply wires are introduced solely for energizing equipment to which the low-voltage wires are connected. Even in this case you must maintain a minimum of 0.25 in. of air separation between the two systems.

The lengthy list of rules in *NEC* 725.55 can be generally summarized in a single sentence: Never rely on conductor insulation alone, regardless of the voltage rating of that insulation, to establish the system separation that is required by the *NEC* between limited-energy circuits and power wiring. Although there are *NEC* provisions that allow for systems of different voltage levels to be in the same enclosure, provided the lower-voltage wiring is insulated for the maximum voltage in the enclosure, those provisions do not affect the special conditions that govern Class 2 circuits.

Transformers   If only the usual doorbell or chimes are to be operated, ordinary doorbell transformers are used. These consume a very insignificant amount of power from the 120-volt line except while they are operating the bell or chimes. One type is shown in Fig. 19–1. Mount the transformer near an outlet box. Connect the flexible (primary) leads to the 120-volt wires in the box. Similar transformers are available that are mounted on an outlet box cover that is installed directly on the box. The screw terminals on the transformer deliver the secondary or low-voltage output of the transformer. Most such transformers have a maximum capacity of about 5 watts; those for doorbells typically deliver 6 to 10 volts, and those for chimes closer to 20 volts. Some provide a choice of voltage such as 6, 12, or 18 volts; others are available in larger wattage ratings.

Fig. 19–1 A transformer for operating 24-volt signaling circuits. Many, like this one, mount conveniently to a 1/2-in. knockout. Connect the wire leads (protruding from the box in the photo) to the line voltage supply, and the transformer screw terminals to the signaling circuit. The box walls provide the required separation between systems. Note the "Class 2" rating marked on the lower end of the transformer.

Transformers for doorbells or chimes are designed so that, even if the secondary is short-circuited, the power flowing will be limited to the rating of the transformer. The transformers are rated at not over 100 VA, and the type used for residential wiring is usually from 5 to 10 VA. This limited current presents no danger of fire, and the low voltage presents no danger of shock. The usual size of the cable is 18 AWG, though sizes may range from 16 to 22 AWG. The size must be suitable for the load, the length of the run, and the voltage available. In use the cable is simply fished through walls without further protection; if run on the surface it is fastened with insulated staples.

Simple Class 2 signaling circuits   Circuits for low-voltage bells, chimes, and similar items are very simple. Consider the secondary of the transformer as the SOURCE for the circuit; the push buttons are switches. The basic circuit is shown in Fig. 19–2, which covers the installation of one bell. Push buttons can be installed in as many locations as desired, as shown by the dashed lines.

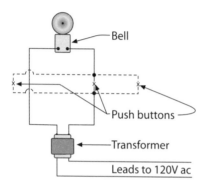

Fig. 19–2 The basic diagram for a doorbell is very simple.

It is convenient to have a different signal for the back door. Figure 19–3 shows the diagram for a combination bell and buzzer, which is a device with three terminals, one of them connected directly to the frame of the device and not insulated from it. This is the terminal from which a wire must be run to the transformer; the wires from the other terminals run to the push buttons and then to the other terminal of the transformer.

In new construction, chimes with musical notes rather than bells are almost always the choice. They come in a variety of styles. Connect them as just described for bells. But remember that most chimes require a transformer delivering from 15 to 20 volts, as compared with 6 to 8 volts for bells. If you are replacing a bell with chimes, you will have to change the transformer for satisfactory operation.

Low-voltage remote-control switching   The wiring needed to control a light from three or more points involves the expense of three-way and four-way switches, long runs of cable or conduit, and much labor. A low-voltage switching system is an economical alternative for controlling a light from any number of switch locations. Master control of all the lights in a house or for a floor of a commercial

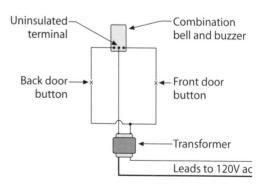

Fig. 19–3 A combination bell and buzzer has been substituted for the bell in Fig. 19–2.

building, etc., is available with systems that use a small motor rotating across contacts. Increasingly, this sort of control is available in forms that respond to commands generated through computer software.

In a low-voltage switching system the 120-volt wires end at the outlet boxes for fixtures that are to be switched; they are *not* run to switch locations. An electrically operated switch called a *relay* is installed in the outlet box for each controlled fixture. (The relay does not necessarily have to be installed in that outlet box; it could be installed near the load, or several could be installed in a centrally located cabinet. In residential work the relays are usually installed in each outlet box involved.) A relay is shown in Fig. 19–4; it will control a load up to 20 amps at any voltage up to 277 volts. From each relay three small wires are run to low-voltage switches located in as many places as you wish. Note that the relay extends through an outlet box knockout, which puts the power circuit and the Class 2 circuits on opposite sides of a grounded metal barrier, in keeping with the principle forbidding conductor insulation as the mechanism for system separation.

Fig. 19–4 The remote-control relay is an electrically operated switch. It is installed in either a master control panel or through a $1/2$-in. knockout on the side of the outlet box on which the fixture is mounted, depending on the application. (*General Electric Co.*)

The power for operating all the relays comes from a single Class 2 transformer (similar to Fig. 19–1, but with a higher output voltage) installed in the basement or some other convenient location. The transformer steps the voltage down from 120 volts to about 24 volts. As a Class 2 circuit, there is no danger of shock, and even if the wires on the secondary side of the transformer are short-circuited, the transformer delivers so little power that there is no danger of fire.

A switch typically used in remote control switching systems is shown in Fig. 19–5. It has a rocker mechanism so either the top or bottom can be pushed, and it requires a special faceplate. It is a single-pole, double-throw momentary contact switch that, when pushed, actuates one of two relay coils in the control shown in Fig. 19–4. The internal relay arrangement latches so that it remains in whatever position it was last energized, either on or off depending on the last position of one of the connected switches. The switch handle has a neutral position. Push the top end to turn a light on and the lower end to turn the light off; in either case the handle returns to the neutral position when released. Because the voltage involved is only 24 volts supplied by a special transformer with little total power, the switches need not be installed in switch boxes, but they usually are for neatness, as shown in Fig. 19–5. If you want to install them without boxes, there are frames for surface mounting. If you want to match the appearance of conventional snap switches, there are snap switches rated for line voltage with a center OFF position, and both up and down positions spring-loaded to serve as momentary contact devices.

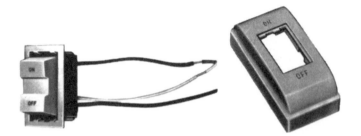

Fig. 19–5 The switch used in remote-control switching systems is equivalent to two push buttons. *(General Electric Co.)*

Because of the Class 2 circuit rating, almost any kind of wire 20 AWG or larger may be used. One type of cable used for the purpose is shown in Fig. 19–6, available in two-, three-, or four-conductor type. The four-wire type is used when a pilot light is needed. Run the cable any way you find convenient. Staple it to the surface over which it runs; in old work fish it through walls, run it behind baseboards, or run it exposed but make certain it will not be subject to physical damage. Suspended ceiling cavities must comply with *NEC* 725.5, with cabling arranged in a way that does not deny access to the cavity above. Although this is subject to interpretation, the safest course is not to run limited-energy cabling directly on the lay-in panels. Arrange a method of providing independent support for these cables through the cavity.

An installation of one light controlled by any of four switches is shown in Fig. 19–7. Install as many switches as you wish.

Connect the white wires of all the switches together, all the red wires together, and all the black wires together. The *NEC* does not require any particular color

Fig. 19–6 Because of the limited energy, wire used to connect relays to switches does not need much insulation. The special cable shown is convenient. This cable is listed as "CL2" making it suitable for general usage, but not for risers or plenum cavities.

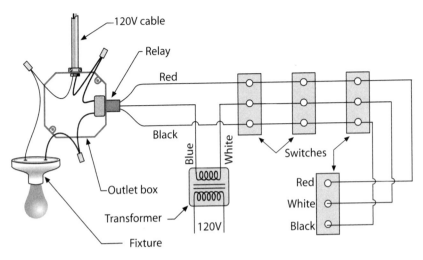

Fig. 19-7 One outlet controlled by any of four switches.

scheme for these low-voltage wires; the colors shown in Fig. 19–8 are those used by some manufacturers of this equipment. If you wish to control two or more outlets at the same time from any switch, connect the several outlets together with the usual 120-volt conductors just as if ordinary wiring were being used, and install the relay in the most convenient box.

Two installation methods   In new work, all the work can be done before the lath-and-plaster or wallboard is installed. The relays are installed in the boxes, and the low-voltage wires from the switches and transformer are connected to the relays. Leave 6 to 8 in. of slack in the low-voltage wires at the relay to allow future replacement if the relay becomes defective. An alternate method is to let 6 to 8 in. of the low-voltage wires project into the outlet box through the knockout in which the relay will later be installed. At that time, connect the low-voltage wires to the proper leads on the relay, push them out through the knockout, and push the relay into the knockout.

COMMUNICATIONS CIRCUITS

In ordinary residential work, it is too often assumed that telephone and cable television wiring can be left entirely up to the respective communications utilities (communications in this sense includes signaling). But in today's deregulated environment, it is usually the site electrician who is responsible for this work. If the communications utilities need to come in, it will be after the walls are closed and they will run their work exposed—your customer will not appreciate your avoiding this part of the job. Today for nearly all occupancies it is necessary to anticipate the effects of computers on wiring layouts. One approach is to install EMT, ENT, or some other raceway from a suitable, easily accessible location to all locations where a communications outlet may be wanted. At the very least, run appropriate cabling into each habitable room. Then install a switch box for each outlet, or at least an opening with a device plaster ring. (The box may be omitted if the particular partition does not require preservation of a fire rating.) Cover the box or opening with a special faceplate as required for the limited-energy services delivered at that point. For example, you might have a telephone jack and a cable television jack in one room, a telephone jack for a modem and a connection for a local area network in another room.

What communications wiring is the owner's responsibility?   Before deregulation of the communications utilities, their personnel typically handled both inside and outside wiring. Now there will be a network interface akin to the service point, located generally where the communications circuits enter the building. Just as a service disconnect can be extended into a building through service conductor encasement in 2 in. of concrete, the communications entrance point can also be artificially extended, in this case through using grounded rigid or intermediate metal conduit. Wherever established, the interface is where utility responsibility ends and premises wiring begins. This is the point where the grounding connection must be made (see intersystem bonding topic on page 308). If separate from the power system electrode, the communications system electrode must be bonded to the power system electrode(s) with a ground wire no smaller than 6 AWG.

Selecting wires for data transmission   Telephone circuits use pairs of wires which, if they are to be used for data transmission, should be twisted together, one pair to a twist. That keeps induced voltages on adjacent pairs in a multipair cable to a minimum. The degree of performance for data transmission is described by categories. Ordinary voice calls ("plain old telephone service" or "POTS" in the trade) do fine on Category 1. Ordinary computer modems need a higher standard, and high-speed data transmission requires the comparatively finicky Category 5, 5e (for extended), or even 6, requiring a very high standard of workmanship to maintain successfully. Even the slightest lapse can make the difference between a successful 100 Mbps ethernet connection or only a 10 Mbps connection (or today, between 1 Gbps and 100 Mbps). Alternate methods promising greater bandwidth probably represent the way of the future, such as coaxial cable service (cable modems), and optical fiber using laser-generated light transmitted over special glass or other fibers.

Color coding   Each circuit pair obeys a color coding convention, and you need to observe it or your circuits won't work properly. Outlet points use "registered jacks" with a number denoting a standard wiring configuration. For example, an ordinary two-pair outlet jack is usually an RJ-11. It has four spring contacts that mate with phone cords. The center pair of contacts connect to green and red wires in station wire with solid colored individual wires. More sophisticated station wire uses color banding with complementary colors. For example, the red wire counterpart is a blue wire with white rings, and the green counterpart is a white wire with blue rings. These wires are polarized; the red (and blue-with-white, or "blue/white") wires being the "ring" conductors and the green (and white/blue) wires being the "tip" conductors. The terms "tip" and "ring" are still used, and refer to the old-fashioned telephone operator's cords that plugged into a switchboard. Today telephone switching is all electronic, but the polarity still matters. For example, a touch-tone telephone will not operate if you reverse the connections.

An RJ-11 jack has two circuits (all pairs listed in order of tip first, then ring), green and red (or white/blue and blue/white) to the center contacts, and black and yellow (or white/orange and orange/white) to the outer contacts. That allows for two lines. More sophisticated jacks include the three-line jack (RJ-25) with the center four contacts wired like the RJ-11, and the outer two additional contacts wired to the third line using white and blue (or white/green and green/white) wires. Local area data networks (LANs) often use four-pair jacks (RJ-45), and you may be called upon to install the basic wiring for these systems even in single family homes. For example, the owner may want to install a router on a high-speed data line so that only a single internet access point is needed for all the computers in the house. The technology is constantly changing, and the most flexible basic infrastructure is always the wisest.

Listing   All equipment that will be connected electrically to a telecommunications network must be listed for this duty. This includes telephone handsets, cable jacks, recording devices, computer modems, and the wiring connecting them. Increasing power levels justify the importance of looking for listed equipment, as well as following all the directions associated with the listing.

Network-powered broadband communications systems   *NEC* Article 830 covers a new form of communications system that combines traditional cable television (CATV) with Internet access and voice/data communications over a single utility connection. The service generally involves coaxial cable, which consists of a center wire embedded in thick insulation with a braided shield around it under an outer jacket. Although this wiring is similar to the usual wiring for CATV installations, it may operate at higher power levels than ordinary CATV service. Presently there are two power levels established for these circuits, low and medium, with high power to come (perhaps) in a future Code cycle. Low-power circuits operate at up to 100 volt-amps on voltages not over 100 volts. Medium-power circuits have the same power limitation (100 VA) but can run up to 150 volts. These voltages have prompted the *NEC* to include many installation provisions that reflect requirements in power articles, such as minimum clearances to grade and above rooftops from the drop conductors, and minimum burial depths for underground broadband service laterals. In addition, medium-power circuits on buildings within 8 ft of grade must be protected by raceways or other enclosures or methods approved by the inspector.

Network-powered broadband communications circuit conductors are like service conductors in that they extend as far as a device that interfaces premises wiring and the wiring under the control of a utility. Instead of ending at service equipment, however, these circuits terminate at a Network Interface Unit (NIU). These devices split out the components of the broadband carrier, such as CATV, cable modem service for high-speed Internet access, conventional telephone, and other interactive services if available. These systems run from the NIU to communications outlets over wiring covered by other more traditional communications articles. The NIUs may (but do not necessarily) include primary and secondary protectors. Primary protectors must be placed at the point of entry, but as in other systems, the point of entry can be artificially extended using heavy-wall grounded metal conduit.

The low-power systems are intended for single occupancies, and medium-power systems are designed for apartment complexes and the like. In fact, the *NEC* grandfathers all existing CATV coaxial cable systems in place prior to the beginning of 2000 as suitable for this use, provided they are used only with low-power systems. Although the power levels may be higher for these systems, they still follow customary practice with respect to required separation from power circuits. Never rely on the cable insulation alone to provide the system separation required by the Code from power wiring. They also require their cable shields to be grounded at the point of entrance to the building (unless this point is artificially extended with conduit). Just as future construction practice will increasingly call for network and communications circuits throughout occupancies, some variety of wiring for video and other high-bandwidth applications will also be routine, and you should familiarize yourself with the customary tools and materials.

CHAPTER 20

Wiring for Multiple Circuits and Specialized Loads

UP TO THIS POINT the principal focus of the book has been applications of ordinary branch circuits for general purpose uses. This chapter focuses on more sophisticated uses of the wiring system. It covers multiwire systems and specialized equipment and applications that require a deeper understanding of electrical systems. You may want to come back and review this chapter before tackling Part 3 of this book, which goes even more in depth.

THREE-WIRE CIRCUITS

In terms of material cost, labor, and voltage drop, installing one 3-wire circuit offers many benefits over two 2-wire circuits. Any building that has a 3-wire, 120/240-volt service can have 3-wire circuits. See Fig. 20–1, which shows an ordinary 2-wire, 120-volt circuit. Assume that it is wired with 14 AWG wire and is 50 ft long, which means the current flows through 100 ft of wire. Assume the load is 15 amps. The resulting voltage drop is about 3.86 volts, or about 3.2%.

Now see Fig. 20–2, which shows two such circuits, one on each leg of the three-wire service. The voltage drop on each circuit will still be 3.86 volts. Note, however, that the two grounded wires are connected to each other at the service equipment, and they run parallel to each other. Why use two grounded wires? Why not use just one wire as in Fig. 20–3? You might think that one grounded wire serving two circuits would have to be twice as big as before in order to carry 2 × 15, or 30 amps. But that is not the case. In Fig. 20–2, each of the grounded

wires *B* and *C* does indeed carry 15 amps, but note the direction of the arrows in the illustration. The flow of current in wire *B* at any instant is in a direction opposite to the current in *C*. Therefore, in Fig. 20–3, at any given instant the single wire *BC* can be said to carry

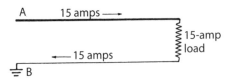

Fig. 20–1 A two-wire circuit carrying 15 amps. One wire is grounded at the service equipment.

15 amps in one direction, and 15 amps in the opposite direction, so the two cancel each other. In other words, wire *BC* has now become a neutral conductor. If the currents in the two circuits are precisely identical, the circuit would work just as well if wire *BC* were missing. But note that this is true only if the current in wire *A* is exactly the same as in wire *D*.

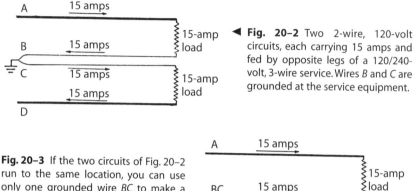

Fig. 20–2 Two 2-wire, 120-volt circuits, each carrying 15 amps and fed by opposite legs of a 120/240-volt, 3-wire service. Wires *B* and *C* are grounded at the service equipment.

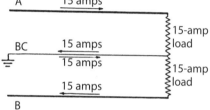

Fig. 20–3 If the two circuits of Fig. 20–2 run to the same location, you can use only one grounded wire *BC* to make a three-wire circuit. Here each half of the three-wire circuit carries 15 amps. The wire *BC* has become a neutral wire. ▶

But what about the voltage drop? In the circuits of Figs. 20–1 and 20–2, the voltage drop is 3.86 volts, based on 15 amps flowing through 50 ft of wire *A* plus 50 ft of wire *B*, a total of 100 ft. In Fig. 20–3, however, the current in each circuit flows through only 50 ft of wire, since wire *BC* carries no current. Therefore the voltage drop in the circuit involving wires *A* and *BC* is only half as great, or 1.93 volts. In the entire three-wire circuit there is only 150 ft of wire as compared with 200 ft in two separate two-wire circuits. Therefore by using a three-wire circuit we have saved 25% of the wire and halved the voltage drop.

If the two halves of a three-wire circuit are not equally loaded, there is still an advantage. An example is shown in Fig. 20–4, in which half the circuit carries 15 amps and the other half only 5 amps; the neutral carries the difference, in this case 10 amps. The voltage drop will not be reduced by 50%, as in the case of equally loaded halves, but the total losses in the three-wire circuit will still be less than in two separate two-wire circuits. If one of the halves carries no current at all, the other half will function exactly like any two-wire circuit. This is also what happens if a fuse blows in one of the two hot wires; what is left is an ordinary two-wire circuit. Therefore the neutral wire in a three-wire branch circuit must be the same size as the two hot wires.

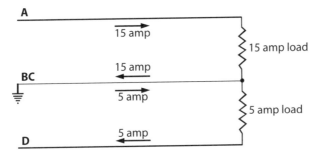

Fig. 20–4 The three-wire circuit of Fig. 20–3, one half carrying 15 amps and the other half 5 amps. The neutral *BC* carries 10 amps.

While Figs. 20–3 and 20–4 show three-wire circuits with a single load at the end of each line, three-wire circuits are not limited to such applications. Figure 20–5 shows a three-wire branch circuit with loads connected at various points. No matter how these loads are spaced, the total losses in a three-wire circuit are always lower than in two separate two-wire circuits. Note that the ungrounded wires must originate from different line buses, and therefore have full system voltage between them. If you mistakenly connect them to the same line or phase, the currents will add in the neutral, creating a severe hazard. Referring to Fig. 20–3, the current in wire *BC* would be 30 amps instead of zero.

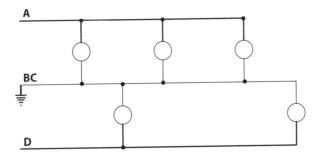

Fig. 20–5 Three-wire circuits need not carry single loads. Install loads where required.

Application of three-wire circuits   In ordinary wiring (whether house, barn, or other building) two separate two-wire, 120-volt circuits are often installed to supply two loads that are near each other. In such cases it is often more economical to install a three-wire circuit instead. Where a three-wire circuit is installed using cable, the cable will have one white wire for the grounded neutral, and one red and one black for the hot wires. When using conduit, choose the same colors. Having one hot wire black and the other red makes it easy to know which leg, or half of the circuit, you are working on. Be sure the white wire runs to each outlet, because if you were to connect a receptacle to the red and black wires, the outlet would operate at 240 instead of 120 volts. Divide the outlets on the circuit more or less evenly between the two legs. If two outlets are located near each other, making it likely that both will be in use at the same time, be sure they are on opposite legs of the circuit.

Neutrals in three-wire branch circuits   *NEC* 300.13(B) states that the continuity of the grounded neutral conductor of a multiwire circuit "shall not be dependent upon device connections, such as lampholders, receptacles, etc., where the removal of such devices would interrupt the continuity" of the conductor. This means that the neutral must be specially watched. Assume you have several duplex receptacles on a three-wire circuit. A receptacle has two brass terminal screws for the hot wires, and two light-colored terminal screws for the grounded wires. In a *two-wire* circuit it is quite all right to connect the ends of the two white wires of two lengths of cable to the two light-colored terminals of a receptacle, and then run the cable on to another receptacle where the process is repeated, then on to a third receptacle, and so on. *In a three-wire circuit this must not be done.* If using wires in conduit, you may use the loop in a continuous length of wire as shown in *B* and *C* of Fig. 17–6 on page 314, but when using cable there is no way of doing this. You must follow the procedure shown in Fig. 8–5 on page 115.

Connect the ends of the white wires of two lengths of cable to each other and to a short length of white wire, and then connect the other end of that short wire to the light-colored terminal of the receptacle. Unless this precaution is observed, removing a receptacle for replacement would (while there is no connection to the receptacle) result in a break in the neutral wire. This would place all the receptacles connected to one leg (beyond the receptacle that is temporarily removed) in series with all the receptacles on the other legs, *all at 240 volts.* If for example you have a 1000-watt appliance plugged into a receptacle on one leg, and a 100-watt appliance on the other leg, the 1000-watt appliance would have a voltage of about 20 volts across it, and the 100-watt appliance about 220 volts. Regardless of what you plugged in, everything plugged into receptacles on one leg would operate at more than 120 volts, and everything on the other leg at less than 120 volts, depending on the total wattage plugged into the receptacles of one leg as compared with the wattage plugged into receptacles on the other leg. Anything operating at more than 120 volts would be quickly damaged. Therefore, be careful to connect the white neutral so that temporary removal of a receptacle does not interrupt the neutral to the other receptacles. The same applies to the grounding wire if nonmetallic-sheathed cable is used; see Fig. 18–7.

Just as you want to avoid an open neutral with respect to energized equipment, you also want to be certain all circuit breakers protecting the legs of a three-wire circuit are open before you work on one leg of the circuit. If the other circuit breaker is closed and some line-to-neutral load is in the ON position when you break the neutral connection, full line-to-neutral voltage will pass through the connected load and, to your surprise, the white wire in your hand is carrying full voltage. The *NEC* now requires in all occupancies (no longer just residential) a common disconnecting means for both circuits when both are represented on a common yoke, such as to the two halves of a duplex receptacle. In fact, this rule now applies even to two 2-wire circuits arriving at a common device; a common disconnecting means must be arranged at the panelboard supplying the circuits.

THREE-PHASE, FOUR-WIRE CIRCUITS

Similar principles apply to a circuit originating from three transformer secondaries connected to form a neutral point, as discussed in Chapter 3 (review Fig. 3–12). In this case we have three line-to-neutral circuits using only four wires, an even better savings in the cost of wire than a three-wire circuit from a single-phase distribution. The generic term "multiwire" branch circuit is used by the *NEC*, and by this book in many cases, to refer to both three- and four-wire circuits. An important difference to keep in mind is that in a three-wire circuit from a single-phase distribution, the line-to-neutral currents from the two line conductors arithmetically cancel in the neutral, but in the three-phase, four-wire system they don't cancel unless all three legs are equal. These neutrals are subject to harmonic loading (see the page 481 discussion on counting neutrals), which may make the neutral fully loaded even when the three phase legs are balanced. In addition, many apartments in large buildings have what appears to be a conventional three-wire supply but which is actually two phase legs and a grounded conductor. You should assume those white wires are fully loaded.

NEC NOTATIONS FOR VOLTAGE SYSTEMS

Voltage systems are defined, for *NEC* purposes, in 220.5(A). On single-phase distributions, the voltage to ground appears first. On wye distributions the voltage to ground appears last, with the letter "Y" following the line-to-line voltage. Delta systems have a straight voltage designation. Here are the systems we deal with when working with the *NEC* on systems that don't run with nominal voltages above 600 volts:

120 volts:	2-wire system
120/240 volts:	3-wire system, single phase
208Y/120 volts:	3-phase, 4-wire wye-connected system
240 volts:	3-phase delta system
347 volts:	2-wire system derived from 600-volt wye connection, used in Canada
480Y/277 volts:	3-phase, 4-wire wye-connected system
480 volts:	3-phase delta system
600Y/347 volts:	3-phase, 4-wire wye-connected system, used in Canada
600 volts:	3-phase delta system

HOW THE *NEC* CLASSIFIES APPLIANCES

In addition to supplying multiple outlets, multiwire branch circuits also supply single outlets, generally located for the purpose of connecting a specific appliance. At one time the *NEC* made significant distinctions based on whether equipment was portable, stationary, or fixed, those terms increasingly involved distinctions without differences because some fixed appliances could be cord- and plug-connected, for example. Those terms still survive in the *NEC* in specialty articles such as Articles 550 and 680, but the unifying definitions that were in the 1978 version of Article 100 were dropped in the 1981 edition, never to be revived. Today there are four issues that constitute the most important considerations concerning electrical connections to specific circuits. These include whether the appliance will be *cord- and plug-connected,* or will instead be *permanently connected* ("hard wired"). The *NEC* also takes an interest in whether the appliance is *fastened in place,* or *located to be on a specific circuit,* rather than purely portable such as a hair dryer. These are the four terms introduced by a special technical committee formed to study these issues beginning in the 1978 Code cycle. In addition, installers need to track the official definition of an appliance.

■ *Appliance*—NEC Article 100 defines an appliance as utilization equipment, normally built in standardized sizes or types, that performs one or more specified functions such as clothes washing, air conditioning, etc. An appliance is installed as a unit and typically isn't used for industrial purposes.

■ *Cord- and plug-connected*—These appliances have flexible cords terminating in a plug. They are only energized when the plug is inserted into an energized receptacle at a receptacle outlet.

■ *Permanently connected*—These appliances are connected through one of the *NEC* Chapter 3 wiring methods (typically one covered in Chapter 11 of this book) that terminates directly at the unit.

■ *Fastened in place*—Appliances in this category cannot be easily moved because they have been secured at a definite location through plumbing connections or other installation conditions. Examples are water heaters and furnaces. Some equipment in this category can be connected using a cord and plug for ease of servicing, such as kitchen waste disposers.

■ *Located to be on a specific circuit*—These appliances occupy dedicated space but are not fastened. They can be moved as required after disconnecting a plug from a receptacle, but in practice are seldom moved unless the owner replaces the unit or moves and takes the appliance to the new address.

RECEPTACLES FOR APPLIANCES

Receptacles all carry voltage ratings indicating the maximum voltage at which they may be used: 125, 250, or 125/250 volts for appliances such as ranges. For nonresidential use, others are rated at 277, 480, and 600 volts. Each also carries a maximum ampere rating: 15, 20, 30, 40, or 50 amps.

The illustrations of Figs. 20–6 and 20–7 show the small and some larger types of receptacles and their most usual applications. Note that they are identified as two-pole, two-wire and two-pole, three-wire; as three-pole, three-wire and three-pole, four-wire, and so on. The number of poles indicates the number of circuit conductors carrying the current in normal use. If the number of wires is one greater than the number of poles, it means that the receptacle has an extra opening and the plug an extra prong for connection of an equipment grounding conductor, provided only for safety and never carrying current under normal conditions. Thus a two-pole, two-wire receptacle is used for an appliance with two current-carrying wires but not an equipment grounding wire; the two-pole, three-wire receptacle is used for a similar appliance having two current-carrying wires plus an equipment grounding wire.

Each receptacle's diameter in inches is indicated at the right of the receptacle drawings. On the receptacle drawings, the opening marked *G* is for the equipment grounding wire; the opening marked *W* is for the white grounded wire of the circuit; those marked *X*, *Y*, or *Z* are for other current-carrying wires.

Note that a two-pole, 125-volt receptacle may be used only for appliances operating at 125 volts or less; a similar two-pole receptacle but rated at 250 volts may be used only to supply a load operating at strictly 240 volts; a three-pole, 125/250-volt receptacle may be used only for loads operating partially or at various times at either 120 or 240 volts (an electric range is a typical example). Any of the receptacles mentioned may have an additional opening for the equipment grounding wire.

The illustrations show the more common types of available configurations. Also included, to illustrate the broad range of available configurations, are a few for higher voltages used primarily for industrial applications. The third column provides the NEMA designation for each configuration; this is a universal code that can be used to cross different manufacturers' catalogs. The charts show receptacle configurations; to convert the number into a plug designator simply change *R* to *P.* Configurations for higher voltages are not shown. Besides the receptacle types shown, there are "twist lock" styles designed so a plug cannot be inserted or removed without first twisting it to lock or unlock it.

NEC 430.109(F) requires that when a motor load uses a plug and receptacle as a disconnecting means, the receptacle must have the appropriate horsepower rating. The ratings (from the UL White Book described on page 8) that apply to the configurations on this chart are listed under the NEMA category number. Note that in most cases the straight blade slots are of unequal length. This feature polarizes the receptacle, and therefore the attachment cord; the appropriate plug can be inserted only one way even if ungrounded and therefore capable of

CONFIG-URATION	CIRCUIT WIRING	NEMA/HP	RATING/COMMENTS
		1-15 R	**15A 125V** 2-pole, 2-wire
		1/2 hp	For replacements only
		5-15 R	**15A 125V** 2-pole, 3-wire grounding
		1/2 hp	General purpose, all occupancies
		5-20 R	**20A 125V** 2-pole, 3-wire grounding
		1 hp	General purpose, 20A, all occupancies
		6-15 R	**15A 250V** 2-pole, 3-wire grounding
		1 1/2 hp	Large appliances, all occupancies
		6-20 R	**20A 250V** 2-pole, 3-wire grounding
		2 hp	Large appliances, all occupancies
		7-15 R	**15A 277V** 2-pole, 3-wire grounding
		2 hp	Commercial lighting, from 480Y/277V systems
		14-20 R	**20A 125/250V** 3-pole, 4-wire grounding
		2 hp L-L 1 hp L-N	Multiwire circuits or appliances requiring a neutral
		15-20 R	**20A 250V** 3-pole, 4-wire grounding
		3 hp	3-phase circuits on delta OR wye (208V) systems
		L5-15 R	**15A 125V** 2-pole, 3-wire grounding
		1/2 hp	Locking configuration

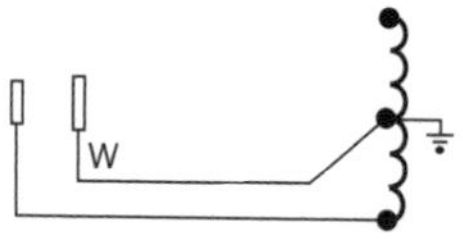

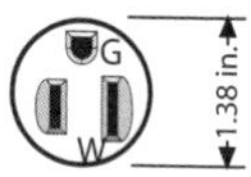

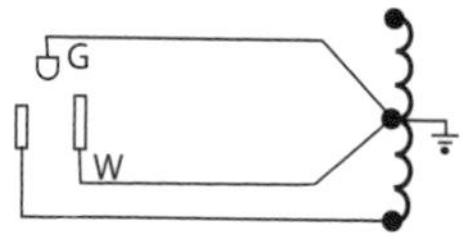

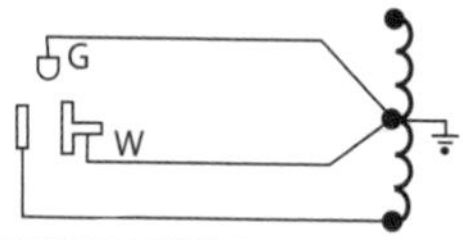

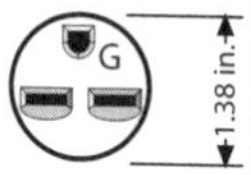

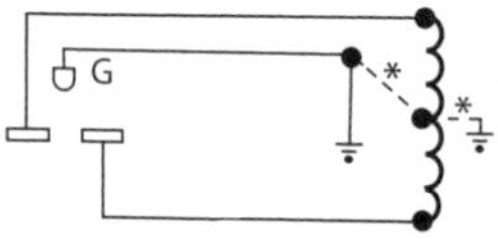

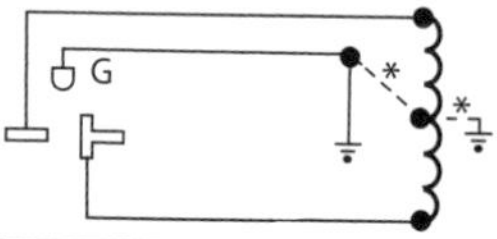

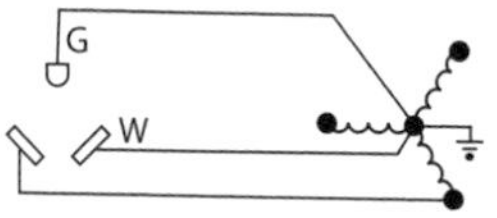

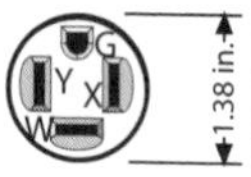

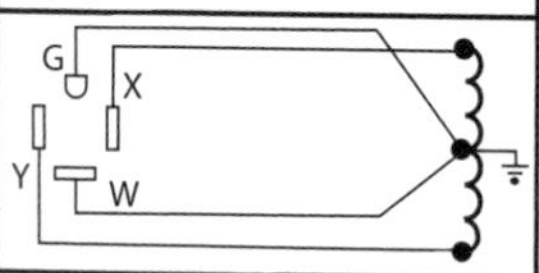

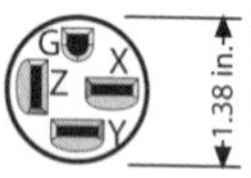

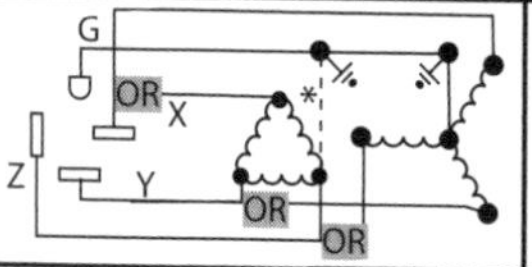

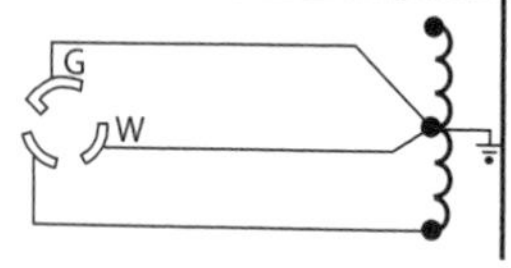

Figs. 20–6 and 20–7 A selection of the numerous receptacle configurations, designed to preserve circuit polarity and to prevent a plug from being inserted into a receptacle of a voltage class or current for which it is not rated. Note that some configurations, although providing equipment grounding, may be used on ungrounded systems. These systems omit the connections shown with an asterisk.

CONFIG-URATION	CIRCUIT WIRING	NEMA/HP	RATING/COMMENTS
		5-30 R	**30A 125V** 2-pole, 3-wire grounding
		2 hp	Large appliances and tools
		6-30 R	**30A 250V** 2-pole, 3-wire grounding
		2 hp	Large appliances and tools
		10-30 R	**30A 125/250V** 3-pole, 3-wire
		2 hp L-L 2 hp L-N	Household clothes dryers For replacements only
		14-30 R	**30A 125/250V** 3-pole, 4-wire grounding
		2 hp L-L 2 hp L-N	Household clothes dryers, other multiwire appliances
		15-50 R	**50A 250V** 3-pole, 4-wire grounding
		7 1/2 hp	3-phase circuits on delta OR wye (208V) systems
		6-50 R	**50A 250V** 2-pole, 3-wire grounding
		3 hp	Welders, heavy appliances not requiring a neutral
		10-50 R	**50A 125/250V** 3-pole, 3-wire
		3 hp L-L 2 hp L-N	Household electric ranges For replacements only
		14-50 R	**50A 125/250V** 3-pole, 4-wire grounding
		3 hp L-L 2 hp L-N	Household electric ranges, other multiwire appliances
		L21-30 R	**30A 208Y/120V, 3ϕY** 4-pole, 5-wire grounding
		3 hp	3-phase wye multiwire appliances and applications
		L16-30 R	**30A 480V** 3-pole, 4-wire grounding
		10 hp	3-phase circuits on delta OR wye (480V) systems

bypassing the position of the ground hole. Polarized receptacles are a critical safety feature because, among other benefits, they prevent Edison-base screw shells for lamps from being energized after the lamp is plugged in. More information on the importance of properly polarizing fixture screw shells is given on page 348. *NEC* 422.40 requires cord- and plug-connected appliances to have polarized attachment plugs if (1) the cord cap is not of a grounding configuration, and (2) the appliance itself contains a single-pole switch, an Edison-base lampholder, or a 15- or 20-amp receptacle.

Receptacles come in a variety of mounting methods to fit various boxes and faceplates; they also come in surface-mounting types. The 50-amp, four-wire, 125/250-volt type (NEMA 14-50R) shown third from the bottom in Fig. 20–7 is also shown pictorially in Fig. 20–8 in both flush- and surface-mounting types, along with a typical plug with a "pigtail" cord. This receptacle is used mostly in the wiring of electric ranges. An almost identical receptacle except with an L-shaped opening for the neutral (NEMA 14-30R) is rated at 30 amps, 125/250 volts and is used mostly for clothes dryers.

Receptacles and plugs are designed so a plug that will fit one particular receptacle will not fit a receptacle rated at a higher or lower ampere or voltage rating, with one exception: A plug that will fit a 15-amp, 125-volt receptacle will also fit a 20-amp, 125-volt receptacle. But a plug made specifically for the 20-amp, 125-volt receptacle will not fit the 15-amp, 125-volt receptacle.

Fig. 20–8 Typical 50-amp, 125/250-volt, three-pole, four-wire grounding receptacles, flush and surface types, and a pigtail cord connector. This is the new configuration for ranges in new installations.

Some receptacle diagrams show the configurations upside down as compared with Figs. 20–6 and 20–7, but that is of no consequence. They may be installed either way.

APPLIANCE CIRCUITS

The *NEC* in Articles 210 and 422 (and for motor-driven appliances in Articles 430 and 440) outlines conditions for branch circuits serving appliances.

If a circuit supplies one appliance and nothing else, *NEC* 422.11(E) provides that the overcurrent protection in the circuit may not exceed 150% of the appliance ampere rating, except that if the appliance is rated at less than 13.3 amps it may be connected to a 15- or 20-amp circuit. The branch-circuit wires must have an ampacity of at least 125% of the ampere rating of the appliance. To determine the ampere rating when you know the rating of the appliance only in watts, divide the watts rating by the voltage at which the appliance is to operate.

Branch circuits serving other loads   In the case of branch circuits serving other loads in addition to appliances, no portable or stationary appliance may exceed 80% of the ampere rating of the circuit. In the case of 15- or 20-amp circuits, the total rating of fixed appliances may not exceed 50% of the rating of the circuit if that circuit also serves lighting outlets or receptacles for portable appliances.

Disconnecting means   For safety when making repairs or during inspection and cleaning, every appliance must be provided with a means of disconnecting it from the circuit.

For portable appliances   A plug-and-receptacle arrangement is all that is needed. Both must have a rating in amperes and volts at least as great as the rating of the appliance. If the plug (or separable connector serving in place of a plug) is inaccessible for any reason, then the rules for permanently connected appliances apply. A connection at the rear of a range counts as accessible if you can reach it after removing a drawer.

For permanently connected appliances   If the appliance is rated at not over 300 volt-amperes or 1/8 hp, no special disconnecting means is required; the branch circuit overprotection is sufficient. Examples are range hood fans and bathroom exhaust fans.

For larger appliances, the branch circuit supplying the appliance may serve as the disconnecting means provided (1) the overcurrent protective device consists of a circuit breaker or a fuse that is part of a switch, and (2) the overcurrent device is either within sight of the appliance, or capable of being locked open. There are hasp arrangements for circuit breakers that allow them to accept a padlock in the open (off) position.

For appliances that have a built-in switch so that the entire appliance can be disconnected by the switch or switches, the disconnecting requirements are as follows:

- *If located in a single-family dwelling*—No separate disconnecting switch is required because the service-entrance disconnecting means will serve the purpose.

- *If located in a two-family dwelling where each family has access to its own service-entrance disconnecting means*—As for single-family dwellings, no separate disconnect is required.

- *If located in an apartment house with three or more separate apartments*—The disconnecting means must be within each apartment, or at least on the same floor as the apartment.

- *If located in other occupancies*—The branch circuit switch or circuit breaker can be used only if readily accessible during the process of servicing the appliance.

WIRING METHODS FOR RANGES AND OTHER SPECIAL APPLIANCES

The *NEC* does not restrict the wiring methods that may be used for connecting ranges and other special appliances. Choose conduit or cable as you wish. If the appliance is cord- and plug-connected, run your conduit or cable up to the receptacle, which may be either surface- or flush-mounted. For dwellings, the *NEC* requires receptacles for specific appliances to be located within 6 ft. In practice, considering standard cord lengths, you should always aim much closer than that.

Wiring 240-volt appliances with cable   As explained in Chapter 9, the white wire may be used only as a grounded wire. As also explained in that chapter, the white wire does not run to a 240-volt load. How can cable, which contains one black wire and one white wire, be used to connect a 240-volt load when no wire to a 240-volt load may be white? *NEC* 200.7(C)(1) allows you to reidentify the white wire by painting it black (or any other color except green) at each terminal and at all points where the wire is accessible and visible after installation. Colored tape may be used instead of paint.

Remember that an equipment grounding wire must also be installed unless you are using a suitable metal wiring system such as metal rigid conduit or armored cable.

An electric range does not operate at strictly 240 volts. On older models, when any burner is turned to "high" heat, it operates at 240 volts; when turned to "low" heat it operates at 120 volts. In other words, it is a combination 120/240-volt appliance; therefore all three wires including the neutral must run to it. Most surface range elements today are rated for full voltage with some form of stepless control, but even now the neutral is required to run the controls.

Special provisions for ranges, ovens, cooking units, and clothes dryers
NEC 338.10(B)(2) Exception and *NEC* 250.140 establish an important condition for ranges, ovens, cooking units, and clothes dryers on existing circuits only: You may use service-entrance cable with a bare neutral provided the cable runs directly from the service equipment. The bare wire serves as both the neutral and the equipment grounding conductor. (Service-entrance cable with a bare neutral may be used in any circuit provided the bare neutral is used only as the equipment grounding wire; it must not be used as a current-carrying wire in normal use except in the case of the appliances mentioned.) In ordinary residential installations it is almost always the case that these appliances are connected directly to the service equipment. But in apartment buildings there is often a feeder from the service equipment to other panelboards, from which branch-circuit wires run to various locations as required. Such panelboards are not part of the service equipment, and you may not use cable with a bare neutral beginning at such a panelboard.

In terms of new installations, the *NEC* finally decided to declare World War II over. About 60 years ago, in order to preserve copper for the war effort, the *NEC* allowed the grounded circuit conductors and the equipment grounding conductor to run as a single conductor to ranges and clothes dryers. As of the 1996 *NEC*, we are back to what always should have been the configuration: three-pole, four-wire grounding, NEMA 14-50R. Ranges with the old configuration can remain in service under a grandfathering exception.

Ranges   It is not likely that all the burners and the oven of a range will ever be turned on to their maximum capacity at the same time. *NEC* 220.55 and the associated table establish a demand factor that permits wires in branch circuits feeding ranges (also ovens and cooking units if both are served by one circuit) to be smaller than the watts rating the appliance would otherwise require. If the total rating is not over 12 000 watts, determine wire size based on 8000 watts. On that basis 6 AWG wires are usually used; for smaller ranges 8 AWG is sometimes used. If the rating is over 12 000 watts, add 400 watts for each additional kilowatt or fraction thereof. Thus for a range rated at 13 800 watts, use 8000 + 400 + 400 or 8800 watts in determining wire size. But for a circuit serving one oven, or one cooking unit, you must use the full rating of the appliance in determining wire size.

Because of the way the heating elements are connected within a range, the neutral cannot be made to carry as many amperes as the hot wires. *NEC* 220.61(B)(1) allows the neutral to be smaller, but never less than 70% of the ampacity of the hot wires, and in no case smaller than 10 AWG. The rule of thumb is to use a neutral one size smaller than the hot wires: 8 AWG neutral with 6 AWG hot wires, and so on.

Run the wires up to a range receptacle of the type shown in Fig. 20–8, flush or surface mounted. The range will be connected to the receptacle using a pigtail cord shown in the same illustration. The plug and receptacle constitute the disconnecting means. The *NEC* requires that the frames of ranges (and separate ovens and cooking units) be grounded.

Sectional ranges   In many cases, the oven is a separate unit installed in a wall or on a counter. Groups of burners in a single unit are installed in the kitchen counter. Separate components allow much flexibility in laying out a modern, custom-designed kitchen.

The *NEC* calls such separate ovens "wall-mounted ovens" and the burners "counter-mounted cooking units." Here they are referred to as ovens and cooking units. Self-contained ranges are considered cord- and plug-connected appliances located to be on a specific circuit, but ovens and cooking units are considered appliances fastened in place, and either cord- and plug-connected or permanently connected depending on the method in use.

You have a choice of two basic methods in wiring ovens and cooking units. The simplest way, preferred by many inspectors, is to provide one circuit for the oven and another for the cooking unit. An alternate way is to provide one 50-amp circuit to supply both appliances.

If installing a separate circuit for the oven, proceed as already outlined for fixed appliances in general. Use wire of the ampacity required for the load. For ovens and cooking units on separate circuits, the full nameplate rating must be used. No demand factor is permitted because the entire maximum load of either circuit is often used at the same time. The oven will probably be rated at 4600 watts, which at 240 volts is nearly 20 amps, so 12 AWG wire would appear suitable, but the minimum is 10 AWG because of the grounding requirement. In installing a cooking unit, proceed exactly as in the case of the oven, again using a minimum of 10 AWG wire, which with an ampacity of 30 amps will provide a maximum of 7200 watts, enough to take care of most cooking units—but verify the actual size needed from the watts rating of the cooking unit.

If you install a single circuit for the oven and cooking unit combined, follow the circuit of Fig. 20–9. To determine the wire size required, add together the ratings of the oven and the cooking unit, then proceed as if you had a self-contained range of the same rating. The same wire size must be used from the service equipment up to the oven and up to the cooking unit, or to the receptacles for these components. The receptacles shown are only for conven-ience and must not be used as disconnecting means for these fixed appliances. Use a circuit breaker or fuses on a pullout block in your service equipment to protect the branch circuit.

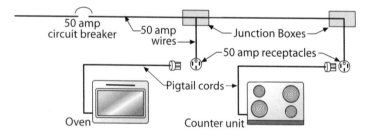

Fig. 20–9 It is best to provide one circuit for the oven and another for the cooking unit, but both may be connected to a 50-amp circuit. The receptacles shown are not required.

NEC 210.19(A)(3) Exception 1 is often misinterpreted to mean that the wire size may be reduced at the point where the circuit splits, with one set of wires to the oven and another to the cooking unit. Not so! It does permit the wires from the oven or cooking unit (as in a pigtail cord) to the receptacles, or to the splices with the circuit wires, to be smaller if large enough to carry the load, and if not longer than necessary to service the appliance, but with a minimum ampacity of 20. Install a box near the appliance fed with full-sized conductors, and then arrange a flexible connection (can include a corded pigtail) just long enough to allow for service.

As already mentioned, the receptacles in Fig. 20–9 are not required, but they may make installing and servicing more convenient. The oven and the cooking unit in the diagram may be connected directly to the circuit wires in the junction boxes. If receptacles are used, they must be rated at not less than the circuit rating. If you have used 6 AWG wire on a 50-amp circuit, the receptacles must be rated at 50 amps.

The neutral of any circuit to a range, an oven, or a cooking unit may be uninsulated only if service-entrance cable is used. Using any other method, the neutral must be insulated. Given the *NEC* change requiring neutral and equipment grounding conductor separation on these circuits, uninsulated neutrals can no longer be used for new installations.

Clothes dryers   Dryers are basically 240-volt appliances, although most have 120-volt motors in them. Their frames must be grounded. The *NEC* considers dryers to be cord- and plug-connected appliances located to be on a specific circuit.

Wire the dryer as you would a range; *NEC* 220.54 requires a circuit with a minimum 5000-watt capacity, or a capacity based on the rating of the dryer, whichever is higher. Dryers are usually installed using cord and plug, and a receptacle which must be a 30-amp, three-pole, four-wire, 125/250-volt type shown fourth from the top in Fig. 20–7 (identical with a 50-amp range receptacle except with an L-shaped opening for the neutral instead of the straight opening of the range type). Use a pigtail cord similar to the one shown in Fig. 20–8 for a range, but with smaller wires, but not smaller than 10 AWG.

Use any wiring method you choose. The rules regarding the use of the grounded conductor simultaneously as an equipment grounding conductor are the same as for ranges: never for new installations, but existing installations are grandfathered. The plug and receptacle will serve as the disconnecting means.

Automatic clothes washers   The *NEC* considers clothes washers to be cord- and plug-connected appliances located to be on a specific circuit, assuming they are connected to the plumbing with flexible hose. Install a 20-amp, 125-volt grounding-type receptacle (less satisfactory but still acceptable is a 15-amp duplex configuration, because the circuit in this case is not an individual branch circuit) on the special laundry circuit required by the *NEC* in the laundry area, as discussed in Chapter 13.

Water heaters   The power consumed by a water heater varies greatly, from 2300 to 5500 watts. A common element size is 4000 watts; electric hot water tanks have thermostatic throw-over switches that, depending on how they are wired, will either force a lower element to wait for the upper element, or allow both to cycle at the same time. Frequently the serving utility has special water heating rates in exchange for an allowance on the part of the consumer for the utility to control the availability of one or both elements. Any of these factors may shift the actual load presented by the water heater. For the remainder of this analysis, assume 4000-watt elements, top and bottom, with a throw-over thermostatic switch that requires the upper section to be satisfied before the bottom section is energized.

Number 12 AWG wire with a maximum allowable overcurrent device of 20 amps will carry 4800 watts and might appear to be suitable for most heaters, however it cannot be used because *NEC* 422.13 requires a branch-circuit rating at 125% of the nameplate rating. The nameplate rating converts to 16.7 amps; that multiplied by 1.25 is 21 amps, more than the maximum 20 amps of overcurrent protection allowed for 12 AWG wire. Meanwhile, this is a non-motor-operated appliance, and the overcurrent protection ahead of the water heater can't exceed 150% of the nameplate rating, with an allowance for the next higher standard size; 150% of 16.7 amps is 25 amps, therefore the branch circuit must be a 25-amp individual branch circuit wired with 10 AWG. A 4500-watt tank, also a common size, would be wired on a 30-amp circuit.

The branch-circuit overcurrent protective device qualifies as a disconnecting means if the water heater is within sight. Otherwise you must provide a locking arrangement unique to the water heater circuit, which allows that particular breaker to be locked open, or you can add a 30-amp, double-pole snap switch within sight of the water heater.

Room air conditioners   An air conditioner is considered a room air conditioner if it is installed in the room it cools (in a window or in an opening through a wall), if it is single-phase, and if it operates at not over 250 volts. The unit may also have provisions for heating as well as cooling. Installation requirements are outlined in *NEC* Article 440. The air conditioner may be connected by cord and plug. A unit switch and overload protection are built into the unit. The disconnecting means may be the plug on the cord, or the manual control on the unit if it is readily accessible and not more than 6 ft from the floor, or a manually controlled switch installed where readily accessible and in sight from the unit. If the unit is installed on a circuit supplying no other load, the ampere rating on the nameplate must not exceed 80% of the circuit rating. If it is installed on a circuit also supplying lighting or other loads, it may not exceed 50% of the circuit rating. Cords must not be longer than 10 ft if the unit operates at 120 volts, and not over 6 ft if it operates at 240 volts.

If the unit has a three-phase motor or operates at more than 250 volts, it is not a room air conditioner. It needs to be installed on a circuit of its own. Such a unit must meet the requirements discussed in Chapter 30 for sealed (hermetic type) motor-compressors.

Electric heat   Even though resistance electric heat is efficient in the sense that almost every watt radiated from the service location throughout the building contributes to the desired air temperature, in most areas of the country the cost of electricity makes it prohibitive as an overall heating strategy. Although it is touted as 100% efficient, that is only partially true. Looking at a metropolitan energy budget, in order to deliver the power to your door the region needs to deliver far more power than you run on your meter. This is due to transmission and distribution losses on the order of 45%, and generating capacity has to be found accordingly. Nevertheless electric heat offers some benefits: it needs no venting, it is useful for backing up other forms of heating such as wood heat, and it allows for small, highly targeted usage in some cold spots.

Multioutlet branch circuits for electric heating units must be either 15, 20, or 30 amps, and they must equal or exceed 125% of the connected heating load. Baseboard units must never be located below a perimeter receptacle outlet due to listing restrictions. If you do need the heat at that location, place a receptacle in one of the end caps as directed by the installation instructions. That way any cord plugged into the receptacle will not drape into the convection louvers over the heating element. The heating element often gets hot enough to soften conventional parallel cord from a floor lamp, resulting in an energized small conductor against the heated element. That condition combined with the quantities of dust that often accumulate in these units is a set-up for a fire.

NEC Article 424 deserves careful study if you are planning to install such specialty items as ceiling cable or large central furnaces. It isn't covered here beyond the above discussion because the large applications aren't competitive with other heating fuel sources in today's market.

SPECIAL APPLICATIONS

Chapters 6 and 7 (also 5) of the NEC cover special equipment and conditions needing provisions that modify the general rules in the first four chapters. One such article (Article 680) covers swimming pools and related environments, and another (Article 760) covers fire alarm circuits. Because these topics are not confined to large commercial occupancies covered in Part 3 of this book, they deserve some introduction here.

Swimming pools, spas, hot tubs, fountains, and hydromassage bathtubs

A complete treatment of swimming pool wiring is beyond the scope of this book, however, you should be aware of the general concepts that underlie the complicated provisions of NEC Article 680. The underpinnings of the article followed intensive research at UL regarding how electrical failures in a swimming pool environment affected persons in or adjacent to the contained body of water, whether pool or spa or fountain. Although it is generally believed that the principal hazard is the threat of electrocution given total body immersion in a conductive fluid, actually the more immediate hazard is the threat of drowning. This has to do with disorientation of the brain when the head is in the water, particularly if there is water in the ears. UL volunteers actually entered a pool under controlled conditions, and started to have problems when the voltage gradient applied across the pool exceeded about 4 volts.

That research lead directly to one of the key sections in the initial 1962 NEC version of Article 680, a set of provisions that remain in the NEC virtually unchanged to this day. A swimming pool, spa, or any artificially contained body capable of holding enough water to constitute a drowning hazard and likely to be used for bodily immersion must be wired with a bonding grid. Hydromassage tubs, which are like ordinary bathtubs and have no appreciable drowning risk, have the most relaxed requirements for just that reason. The pool bonding grid connects all metallic objects within easy reach of the pool, or part of the pool structure. In effect it creates a "Faraday cage" that assures that whatever transient voltage reaches the pool ladder, for example, also reaches the reinforcing steel on the other end of the pool at the same time, reducing any voltage gradient within the pool to near zero.

This is a bonding requirement, not a grounding requirement. Accordingly, there isn't any requirement to extend the bonding grid into a remote panelboard. Of course you are free to do so, but it will accomplish nothing other than making the pool safe for alien beings with infinitely stretchable arms who you expect, while swimming, will be reaching down through the cellar bulkhead and touching the panelboard. They would benefit, but the rest of us would not. Similarly, adding ground rods to the bonding grid at each corner of the pool or anywhere else will help assure that you don't get a shock from the azalea bush, but little else.

The other *NEC* focus still with us today is a plethora of requirements designed to assure the integrity of the grounding system, particularly for pool components that are exposed to the highly reactive chemicals in pool water. In particular, underwater lights get a great deal of attention, because you have in some cases a 120-volt source of essentially unlimited energy supplying electrical equipment surrounded by conductive pool water. Refer to *NEC* Article 680, which was completely rewritten in the 2002 *NEC* cycle.

Smoke detectors   The *NEC* does not govern the placement or selection of smoke detectors. They are covered in a different NFPA standard, the *National Fire Alarm Code*. The authority having jurisdiction over detector placement and selection is usually either a fire official or a building official, not the electrical inspector. Commonly these detectors must have battery back-up and be installed in every sleeping room, every hallway leading to a sleeping area, and at every level, preferably near stairs and definitely away from kitchens and bathrooms. Generally you use three conductors to wire these units, two for power and one that carries the alarm signal, so all will respond if one responds. Although the *NEC* doesn't control this, you will be called upon to install these life-saving devices as a part of most electrical installations.

Normally the exclusively battery-operated detectors are limited to existing occupancies; those installed as new work must be hard wired, usually with some sort of requirement that all the detectors in the building reciprocally alarm, meaning that when one goes off, they all go off. It also means that in all two-family dwellings and multifamily dwellings there will be a common alarm system, assuming the building code requires such protection, which is the usual case. This immediately raises the issue of *NEC* 210.25, which prohibits common circuits from being served by equipment that supplies an individual dwelling unit. Common alarm systems, therefore, along with common area lighting, a common well pump, and any other such load must be served from an owner's panel. This prevents the possibility that a tenant will vacate, leading to a shutoff of the power source for a common system. In the case of a fire alarm system the outcome could be catastrophic.

There are two common types of smoke detectors. In a single-family home, usually the detectors are wired to a lighting circuit that would be immediately noticed if shut off because of a tripped circuit breaker. The detectors can't be on the load side of any local switch. For quick response to flaming fires the ionization type is especially effective. It uses a small amount of radioactive material to ionize the air between two differently charged electrodes. Smoke particles, particularly charged ions from a flame, enter the area between the plates and combine with the ions produced from the radioactive source, reducing the overall amount of charged particles. This decreases the conductivity of the air, which in turn decreases the amount of current flowing between the two plates. The electronics in the detector sense the change in current and put the unit into alarm under prescribed conditions. These detectors are prone to nuisance alarms if located near ion sources such as kitchen and shower areas.

The other common type of detector is photoelectric. These have a chamber with a small, focused light source and a photocell that is completely out of the path of the light beam. When smoke particles enter the chamber, they scatter the light, inevitably deflecting some of it onto the photocell. When enough light hits the photocell the detector goes into alarm. Photoelectric detectors are better for sensing smoldering fires, and they are somewhat less likely to nuisance alarm. They also react more slowly to a flame source than an ionization type. The two types can be combined to maximize their respective strong points. There are also listed detectors that have both features.

Larger occupancies or multifamily buildings generally have actual fire alarm systems with annunciator panels, dedicated horn circuits separate from the detection circuits, heat detectors in areas of temperature extremes that make smoke detection problematic, etc. These systems are beyond the scope of this book.

CHAPTER 21
Modernizing Old Work

IN OLD WORK, which is the wiring of buildings that have been completed before the wiring is started, there are few electrical problems that have not already been covered in this book. Most difficulties can be resolved into problems of carpentry, in other words, how to get wires and cables from one point to another with the least effort and minimum tearing up of the structure of the building.

In new work it is a simple matter to run wires and cables from one point to another in the shortest way possible. In old work considerably more material is used because often it is necessary to lead the cable the long way around through available channels rather than tear up walls, ceilings, or floors to run it the shortest distance.

This chapter covers the common problems encountered in old work, but considerable ingenuity must be exercised in solving actual problems in the field. A study of buildings while they are under construction will help in understanding what is inside the wall in a finished building.

WIRING METHODS IN OLD WORK

It is impossible to use rigid or thinwall conduit in old work without practically wrecking the building. It would be used only when a major rebuilding operation is in process, and installation then would be as in new work. The usual method is to use nonmetallic-sheathed cable or armored cable. The material is easily fished through empty wall spaces. It is sufficiently flexible to go around corners without much difficulty. In some localities flexible conduit ("greenfield") is generally used. Install it as you would cable, except that the empty conduit is first installed and the wire is pulled into place later.

Connections to concealed knob-and-tube wiring   In cases where the existing wiring is concealed knob-and-tube, the *NEC* does allow it to be extended from an existing application. But that is seldom practical because the hardware is no longer available, and the existing knobs you might salvage from old jobs have internal spacings for old Type R conductor insulation that won't work on today's thinner insulation. Concealed knob-and-tube, as a wiring method, has no equipment grounding conductor carried with it. Over the generations, *NEC* provisions have changed to the point that it is almost impossible legally to wire anything without grounding it. For example, until the 1984 *NEC*, what is now 314.4 only required the grounding of metal boxes used with concealed knob-and-tube wiring if in contact with metal lath or metallic surfaces. Now all metal boxes must be grounded without exception. Meanwhile, grounding has been getting more difficult to arrange to remote extensions of concealed knob-and-tube outlets. Until the 1993 *NEC* you could go to a local bonded water pipe to pick up an equipment grounding connection, and then extend from there with modern wiring methods. Now *NEC* 250.130(C), which governs this work, requires that the equipment grounding connection be made on the equipment grounding terminal bar of the supply panelboard, or directly to the grounding electrode system or grounding electrode conductor. You are unlikely to be searching for a method of grounding concealed knob-and-tube wiring in a steel-frame building. Rather you will be attempting this in old wood-frame buildings, probably residential. In such occupancies, even if the water supply lateral is metallic, the water piping system ceases to be considered as an electrode beyond 5 ft from the point of entry. This means fishing into the basement. If you can fish a ground wire down into the basement, you can fish a modern circuit up in the reverse direction and avoid the entire problem.

It is true that some geographical areas have more extensive use of slab-on-grade construction, and here interior water piping is sometimes permitted to qualify as electrodes because the pipes extend to grade for the minimum threshold distance of 10 ft and thereby allow interior connections. But in almost every case, trying to extend knob-and-tube wiring is like trying to erect a modern structure on a rotten foundation.

Add to these problems the fact that beginning with the 1987 *NEC* this wiring method cannot be used in wall or ceiling cavities that have "loose, filled, or foamed-in-place insulating material that envelops the conductors." This effectively means that such cavities cannot be insulated, because you'd have to open all the walls to install board insulation products, and if you'd do that, you'd never consider trying to save this wiring method.

The writer considered including instructions at this point in how to perform an extension of concealed knob-and-tube wiring—this would include instructions on positioning knobs and cleats, end fittings for the new wiring method, loom, soldering, etc.—and decided against it. If you find concealed knob-and-tube wiring, try to persuade your customer to have you rewire it. Failing that, try to find some old loom from elsewhere on the job, and slip it over the individual conductors to the last knob. Loom is flexible nonmetallic tubing just big enough to slide over an individual conductor. Consider yourself lucky if the piece you find is still flexible. Cut it just long enough to enter a box at the nearest feasible point, and bring it in, one wire per ½-in. knockout (using a Type NM cable connector) or per cable knockout. Be sure the wire enters the box at least 6 in. beyond the end of the loom. If you have a steel box, snip the web between the two knockouts per *NEC* 300.20(B) so you don't create inductive heating around what is probably antique and fragile Type R insulation. Even better, use a nonmetallic box and cover for this purpose.

NEC requirements for old work   At one time the *NEC* made distinctions between old and new work regarding the minimum allowable depth of a box, but no longer. Boxes must have a minimum depth of ½ in. Figure 21–1 shows a 6-cu-in., 4-in.-diameter ceiling pan without cable clamps. This is the smallest box permitted for Type NM cable because a single 14-2 cable, with its

separate equipment grounding conductor, accounts for all of the volume (review the calculations at the end of Chapter 10). The shallowest box permitted for a flush device is ¹⁵⁄₁₆ in. For old work wiring methods, the various *NEC* cabled wiring method articles generally allow for fished cable into existing partitions. Cable is simply pulled through empty spaces in walls and ceilings; each piece must be a continuous length. Naturally it cannot be supported inside the walls or ceilings, but must be anchored to boxes with connectors or built-in clamps.

Fig. 21–1   This ceiling pan, with a volume of 6 cu in. and no required fill adjustments for an internal cable clamp, can just accommodate a single 14/2 AWG Type NM cable, or a 14, 12, or 10-2 AWG Type AC cable. *(Hubbell Electrical Products)*

CONCEALED WIRING TECHNIQUES

The walls and ceilings of the building under consideration may be of lath-and-plaster construction or may be wallboard. Illustrations for this discussion are labeled only as "wall" or "ceiling" regardless of material used.

Openings for flush-mounted devices and fixtures   To cut openings for outlet and switch boxes in walls and ceilings requires a certain amount of skill and care. Openings must not be oversized and must be neatly made. Start by marking the approximate location of the box and, if possible, allow some leeway for moving the opening a little in any direction. First make sure there is not a stud or joist in the way; usually thumping on the wall will disclose the presence of timbers. Then punch a small hole through the plaster or wallboard at the approximate location of the opening. In the case of plaster, probe until the space between two laths is found; then go through completely. Insert a stiff wire through the hole and probe to right and left to confirm that there is no stud or similar obstruction.

The sawing in a lath-and-plaster job can be done with a keyhole saw. Proceed gently in order not to loosen the bond between plaster and the lath. Many prefer to use a hacksaw blade, heavily taped at one end to serve as a handle. Have the teeth of the blade lie backward, opposite the usual position, so the sawing is done as you pull the blade out of the wall. This will help to leave a firm bond between lath and plaster, especially if you hold your hand against the plaster as you do the sawing. Unless you do this carefully, you may end up with a considerable area of plaster unsupported by laths that have become separated during the sawing operation. When the opening is in wallboard, the sawing may be done with the blade in the usual position.

The opening for a switch box must be about $3\frac{1}{4} \times 2$ in. Make a template of the opening as shown in actual size in Fig. 21–2. Trace the outline directly on thin cardboard or paste a photocopy of the outline on the cardboard, and then cut it out. This template will save you much time, especially if many openings must be cut. Lay the template against the wall where the opening is needed and trace around it. As indicated on the template, drill two $\frac{1}{2}$-in. holes at opposite corners to start your hacksaw cuts. Drill two more $\frac{1}{2}$-in. holes at top and bottom to provide clearance for mounting screws of switches or other devices installed in the box. The centers of the holes must be on the lines of the outline so that not more than half of each hole will be outside the rectangle. Do this carefully or part of the hole later may not be covered by the faceplate for the receptacle or switch. Also remember that the maximum space allowed between a flush box and the surrounding wall (or ceiling) is generally $\frac{1}{8}$ in. If your hole is sloppier than this, you will have to repair the wall opening to meet this clearance.

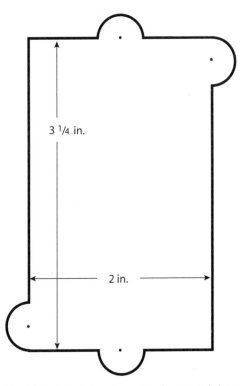

3 1/4 in.

2 in.

Fig. 21–2 Actual size of opening for a switch box. Make yourself a template of this drawing; it will save you much time.

If the wall is lath-and-plaster, remember that the ordinary switch box is 3 in. long, while two laths plus three spaces between laths measure more than 3 in. If you cut away two full laths, it will be difficult to anchor the box by its ears on the next two laths because the mounting holes in the ears will then come very close to the edges of the laths, which will split when the screws are driven in. Cut one lath completely, and remove part of each adjoining lath, as shown in Fig. 21–3. The boxes do not necessarily have to be mounted using screws; the methods outlined in the next paragraphs for drywall construction can also be used.

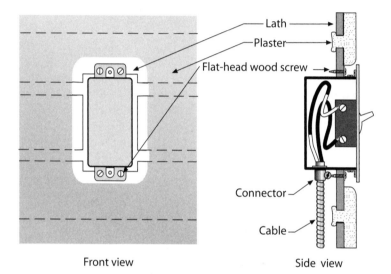

Front view   Side view

Fig. 21–3 In cutting an opening for a switch box in a lath-and-plaster wall, do not cut away two complete laths. Cut away one, and part of each of two others.

The mounting ears on the ends of boxes are adjustable to compensate for various thicknesses of walls. They are also completely reversible as shown in Fig. 21–4. In the position shown at *A* they are used for installing on lath-and-plaster walls. Reverse them as at *B* when installing the box on wallboard. Use No. 4 flathead screws for wood lath, preferably sheet metal screws that have threads the full length of the shank.

Installing switch boxes in wall   Use a deep box and internal cable clamps as shown in Fig. 21–5. Strip the jacket off the cable for about 10 in., let the ends project out of the opening in the wall, and push them into the knockout of the box, which is still outside the wall.

Then push the box into the opening; hold it in place using any of the methods shown in Figs. 21–7 through 21–9. Pull the cable into the internal clamps of the box, tighten the clamps, and you are ready to install switch or receptacle. All this is shown in Fig. 21–6. The bevel is handy, but many of these configurations

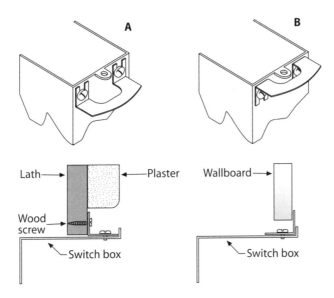

Fig. 21–4 The external mounting ears of switch boxes are reversible. In position *A* they are used when mounting boxes on lath under plaster. For wall board or plaster on gypsum lath, reverse the ears as shown at *B*.

have only about 10 cu in. of capacity; performing the calculation at the end of Chapter 10 shows that this installation on 14 AWG Type NM cable (four wires, clamps, equipment grounds, and a double allowance for the device) would require a 16-cu-in. box. If you find a beveled 3½-in.-deep box, it will probably have that much capacity and by all means use it; more likely you will be gingerly installing a conventional switch box. That process is easier if both cables enter the same side of the box. Another option is to use an angle connector to enter the rear ½-in. knockout; some of these are listed for two cables. Don't forget to identify (by tape or otherwise) the incoming wires as line and load if it will make a difference in how you wire the device.

The procedure just described assumes you are using nonmetallic-sheathed cable. If you are using one of the armored cables (Type AC or MC), your problem becomes a bit more complicated because boxes with beveled corners are not available with built-in clamps for armored cable. You have two choices. One option is to use a deep box with square corners and built-in clamps for armored cable, which will require the

Fig. 21–5 Boxes with beveled corners and internal cable clamps simplify the procedure of Fig. 21–6, but they probably won't meet conductor fill requirements. For adequate volume, use boxes with square corners.

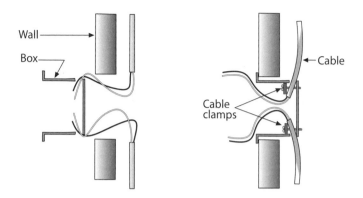

Fig. 21–6 Installing switch box in opening. The cable is loosely secured to the box by internal clamps before the box is inserted into the opening. The clamps are tightened only after box has been inserted into opening.

cable to enter at the top or bottom end of the box. After stripping off about 10 in. of the armor, let the wires project out of the opening in the wall. Back-wrap the cable bonding tape (Type AC cable only) over the redhead (anti-short bushing, see Fig. 11–40) so it won't fall out during this process. The wires (but not the armor) enter the box through the knockout, projecting into the box through the clamps. Anchor the box in the wall in the ways described. Then jiggle the wires in the box until the armor enters the clamp; tighten the clamp.

Alternately, use an ordinary box without clamps. After removing the armor, install a connector on the cable in the usual way but remove the locknut and have only the wires enter the box. After you get the box into the wall, pull on the wires to raise the male connector threads into the box. Take care not to abrade the conductors on the sharp edge of the knockout. Then install the locknut on the connector and drive the locknut home tightly. Secure the box in the wall in one of the ways described.

Three ways to install switch boxes in drywall construction   Wallboard is not sturdy enough to accept screws as in lath-and-plaster construction, so other methods must be used. In each case adjust the brackets on the ends of the box so that when finally installed it will be flush with the wall surface. Bring the cable into the box and tighten the clamps, letting about 10 in. of cable project out of the box. If using boxes without clamps, install the connector on the cable, letting it project into the box through a bottom knockout, and install the locknut only after the box is installed. (The following methods for installing switch boxes can also be used in lath-and-plaster construction.)

One way is to use the special box of Fig. 21–7. It has special clamps on the outsides of the box. After installing the cable, push the box into the opening, then tighten the screws on the external clamps. This makes the clamps collapse, anchoring the box in the wall.

Fig. 21-7 This box has exterior collapsible clamps on each side. Push box into opening, tighten screws on sides, and the box will be anchored in the wall. *(Hubbell Electrical Products)*

Another way is to use an ordinary box plus the U-shaped clamp of Fig. 21–8. Install the U-clamp with the screw holding it in place, unscrewed about as far as it will go. Slip the box into its opening; the ends of the clamp will expand outwards, and when you tighten the screw holding the clamp, it will anchor the box firmly against the inside of the wall.

A third method is to use an ordinary box, plus a pair of special straps shown in Fig. 21–9. Insert one strap on each side of the wall opening, and push the box into the opening, being careful not to lose one of the straps inside the wall. Then bend the short ends of each strap down into the inside of the box. Be sure they are bent back sharply over the edge of the box and lie tightly against the inside walls of the box so they cannot touch the terminals of a switch or receptacle installed in the box, which would lead to grounds or short circuits. To make a solid installation, if you are starting with the left clamp, let the right side of the box tip into the room slightly as you make the fold on the left clamp. Then, when securing the right clamp flush, you will add enough tension so the box won't move.

Mounting outlet boxes in ceilings   If there is open space above the ceiling on which the box is to be installed, and if there is no floor above (or there is a floor

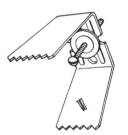

Fig. 21-8 This clamp can be used to anchor an ordinary box in the wall.

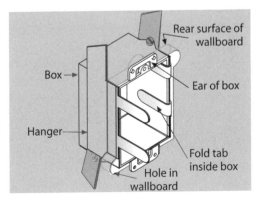

Rear surface of wallboard

Box →

Ear of box

Hanger →

Fold tab inside box

Hole in wallboard

Fig. 21-9 A pair of these hanger strips is also convenient in installing a switch box in a wallboard wall.

in which a board can easily be lifted as explained later) proceed as in new work using a hanger and the usual 1½-in.-deep box. The only difference is that you will be working from above.

If all the work must be done from below, the procedure depends on the ceiling construction, the location of the outlet, and the weight of the fixture. Let's assume you are going to install a ceiling fixture. Use a ½-in.-deep box of the general type shown in Fig. 21–1. You will probably need external cable connectors to meet box fill restrictions.

If your ceiling is lath-and-plaster, use a box that does not have a fixture stud in it. Where the fixture is to be installed, make a hole about an inch smaller in diameter than the box. Remove the center knockout from the box, and one additional knockout for each cable that is to enter the box. Run your cable(s) to this location and anchor the cable(s) to the box, letting about 8 in. project from the box. Then push a hanger (after removing the nailing ears) as shown in Fig. 10–17 through the opening in the ceiling, first removing the locknut from the fixture stud on the hanger. Let the fixture stud hang down through the opening in the ceiling into the center knockout of the box, then install the locknut on the fixture stud. Since the hanger can't be nailed to the joists, be sure to turn it to lie crosswise across several laths to distribute the weight over a wider area. Then install the fixture.

If your ceiling is made of wallboard, you can use the same procedure provided the fixture doesn't weigh more than a few pounds. For heavier fixtures, select a ½-in.-deep ceiling pan as in Fig. 21–1. Do not use a fixture stud because it must be counted in box fill. Using wood screws, install the box where it can be mounted directly on a joist. If the position on the joist is unacceptable to your customer, there are paddle fan box support systems (see Fig. 18–23 and related text for more background on why these systems have been developed) that use expanding cross-joist support bars designed for old-work applications that can be installed through a 4-in.-diameter outlet box opening. These support systems are extremely rugged. They are designed to support a rotating load and will easily support a conventional fixture.

In all cases, remember that if the ceiling is combustible (can burn), you must cover the space between the edge of the box and the edge of the fixture with noncombustible material; see page 350, last paragraph.

Installing wiring in concealed spaces   In some cases the access provided by outlet openings isn't sufficient to get the cable to where it has to run, and you have to make temporary openings in several places to make it possible to pull cable from one point to another, for example, from the ceiling into a wall. The cable does not necessarily run through the opening, and no box is installed. The opening is used only to get at the cable during the pulling process, to help it along or to get around obstructions in the wall. Such openings must be repaired when the job is finished. In other cases you may have to do the same thing in a floor.

Making temporary openings in walls (or ceilings)   If the room is papered, the paper must be carefully removed at the location of the opening and then reinstalled to look as it did before the wiring job. This is easily done. With a razor blade cut the sides and bottom of a square, but not the top. Apply moisture with a rag or sponge, soaking the cut portion until the paste has softened; lift the cut portion using the uncut top as a hinge. Fold it upward, and fasten it to the wall with thumbtacks or masking tape. These steps are shown in Fig. 21–10. Do not cut the paper at the top and fold it downward because plaster or wallboard particles will adhere to its wet surface and prevent the paper from fitting smoothly when pasted back.

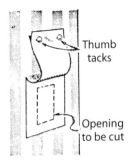

Fig. 21–10 Sections of wallpaper are easily removed temporarily. Use the top of the cut section as a hinge; this makes it easy to restore the wallpaper to its original condition after the job is finished.

If no box is to be installed (where the opening was made for access only), the opening is easily patched with a ready-mixed plaster. The same mixture may be used to fill the openings around the boxes; the *NEC* does not permit large open spaces around boxes. If openings are carefully made, no patching around the boxes need be done.

Lifting floor boards   In many cases the location of outlet and switch boxes with respect to wall or ceiling obstructions may make it necessary to lift hardwood floor boards in the floor above. This should be avoided if possible, but where necessary use extreme care so that when the boards are replaced there will be a minimum of visible damage to the floor. Ordinary attic flooring is simply lifted, but tongue-and-groove hardwood flooring presents a problem.

Before taking the next step, be absolutely sure you've got the right location. This trick works if you have any line of sight into the joist cavity, even with a mirror. Usually hardwood flooring is installed over a subfloor, which allows the ends of the boards to end randomly; the installer doesn't have to cut them to end halfway across a joist. In this case, make your best guess where you want to open the floor. At the end of one of the existing boards nearest your projected opening, drill a tiny hole (about 1/16 in.) straight down at one corner. The hole will not be noticeable because it occurs at two intersecting lines in the floor. Push a bright colored rod straight down into the joist cavity (a white coat hanger is a good choice). From below note where it came through and adjust your measurements if necessary.

It is necessary to chisel off the tongue on at least one of the boards. The thinner the chisel, the less damage that will be done to the floor. A putty knife with the blade cut off short and sharpened to a chisel edge is excellent for the purpose, very thin but sturdy because it's short. Drive it down between two boards as shown in Fig. 21–11. Do this the entire length of the opening, which should be the space across at least three joists. Having the cut section extend over this much space helps ensure the board will set solidly on at least one joist when reinstalled. In cutting the tongue, the exact location of the joists can be determined; this allows points A and B in the illustration to be identified. Bore a small hole at these two points next to the joists, and with a keyhole saw cut across the boards as shown. If you have a steady hand, you can pocket cut the board with an electric saber saw, avoiding a bored hole that will have to be plugged afterward.

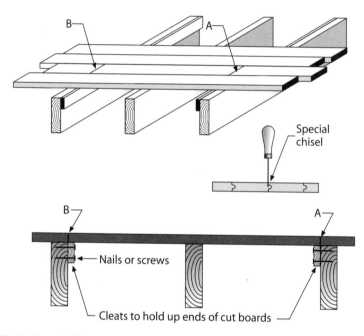

Fig. 21–11 Steps in lifting and replacing floor boards. To simplify the drawing, the subfloor is not shown.

You can now lift the board and do the wiring. To replace the board, it will be necessary to attach cleats to the joists for each end of the floor board to rest on. Anchor these cleats securely to give the cut board a solid footing. The bored holes can be filled with wooden plugs.

Another and perhaps simpler way is to use an electric circular saw set to cut a depth exactly the thickness of the flooring. Then cut the board at the exact center of the two joists. The saw will also cut part of the two adjoining boards. Save the sawdust. When the wiring has been done, replace the board and nail each end to the top of the joists with finishing nails. Then take the sawdust that you have saved, mix it with glue, and use it to patch the saw cracks.

Cable behind baseboard   Assume the bracket light on the wall at *A* in Fig. 21–12 is controlled by a switch on the fixture, but now you want to install a wall switch at point *D* on the same wall. It is a fairly easy procedure. Cut an opening for the switchbox at point *D*. Remove the baseboard at the floor and cut two holes at *B* and *C* behind the baseboard. Then cut a groove or trough (called a "chase") from *B* to *C*. If the wall is lath-and-plaster, it may be necessary to slice away parts of the laths. If wallboard is used and is quite thin, you may have to slightly notch the studs.

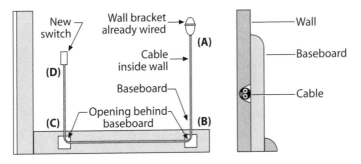

Fig. 21–12 Often cable can be concealed in a trough cut into the wall behind a baseboard.

When this has been done, prepare a length of cable long enough to reach from *A* to *B* to *C* to *D*, plus about 10 in. of wire beyond the armor or jacket at each end. Install a connector at each end. Remove the locknuts. Temporarily remove the fixture at *A*; remove a knockout in the bottom of the box on which the fixture is installed. Considering the very short lengths involved, push a short piece of stiff but flexible steel wire (or fish tape) into the knockout at *A*, down toward *B* where it can be fairly easily located. Attach the end of the cable to it, pull the cable inside the wall up to *A*, jiggle the connector into the knockout, install the locknut, and connect the wires properly to the fixture. Repeat the process at *D*, pull the cable up to *D*, jiggle the connector into the knockout in the box at *D*, install the locknut, and connect the switch. Lay the cable into the trough behind the baseboard; any slight excess length will lie inside the wall. At every point that the cable crosses a stud, protect it from future nail penetrations with $1/16$-in. steel plates before replacing the baseboard.

Cable through attic   For single-story houses, or for outlets on the second floor of a two-story house, it is generally practical to run the cable through the attic. It is a simple matter to lift a few boards of the usual rough attic flooring and lead the cable around, in that way avoiding the need for temporary openings in the living quarters. No baseboards then need to be lifted. It may require a few more feet of cable, but the saving in labor offsets the cost of a little extra cable many times over. Always explore this possibility before considering a more difficult method. For example, in Fig. 21–12 the cable would run from outlet *A,* up to the attic, under the attic floor to a point directly above outlet *D,* and then drop down to *D.*

Cable through basement   If the outlet shown in Fig. 21–12 is on the first floor, you can often run the cable through the basement, dropping straight down from *B,* into the basement, then over to the left, then upward again at point *C.* Often this will require boring holes through a sill plate, as shown in Figs. 21–13 and 21–14. Sometimes there will be obstructions in the walls, such as cross bracing, whether you go through the basement or as shown in Fig. 21–12. In such cases you will usually have to make an incision in the wall (to be patched later) in order to notch the cross braces. There are special bits available on long, flexible, spring-steel shanks many feet long that are designed to bend into a wall cavity from an opening and bore through cross bracing.

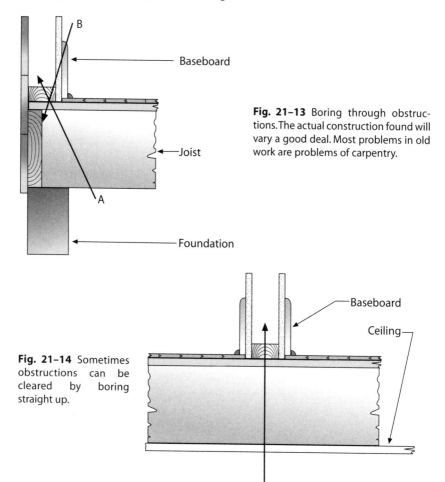

Fig. 21–13 Boring through obstructions. The actual construction found will vary a good deal. Most problems in old work are problems of carpentry.

Fig. 21–14 Sometimes obstructions can be cleared by boring straight up.

If the point where the cable is to run down into the basement is in an outer wall, the construction is likely to be as shown in Fig. 21–13. In that case bore a hole either upward as indicated by arrow *A*, or downward from a point behind the baseboard as indicated by arrow *B*. If the cable is to enter the basement from an interior wall, it is often possible to bore upward from the basement as in Fig. 21–14. The long-shank electrician's bit shown with an extension in Fig. 21–15 makes the job easier.

Fig. 21–15 Electrician's bit and extension.

Cable around corner where wall meets ceiling   Figure 21–16 shows this problem: how to lead cable from outlet *A* in the ceiling to outlet *B* in the wall, around the corner at *C*. Usually there will be an obstruction at point *C*. Any of a dozen types of construction may be used; that shown in Fig. 21–16 is typical. The usual procedure is to make a temporary opening in the wall at the ceiling at point *C*, but on the opposite side of the wall from opening *B*, as shown in the enlarged view of point *C* in the right-hand part of Fig. 21–16. Bore upward as shown by arrow *1*. Push a length of stiff wire into the hole until the end emerges at *A*. If the opening at *C* is large enough, push the opposite end of the wire downward to *B*, and pull the loop that is formed at *C* into the wall, pulling at either *A* or *B*; you will then have a continuous wire from *A* to *B* with which to pull in the cable. More likely the hole at *C* will be small; use two lengths of wire to do the fishing. Push one length from *C* to *A*, leaving a hook at *C* just outside the opening. Push another end from *C* to *B*, again leaving a small hook just outside the opening at *C*. Hook the ends together, pull at *B*, and it is a simple matter then to pull the longer wire from *A* through *C* to *A* and, with this, to pull in the cable.

If there happens to be another wall directly above point *C*, it may be better to bore downward from a point behind the baseboard as shown by arrow *2*. In that case the wire is pushed down from above through the bored hole to *B*, another length from *A* toward *C*. When the hooks on the end engage, pull down at *B* until a continuous length of wire extends from *A* through *C* to *B*. Attach cable to one end of the wire and pull into place.

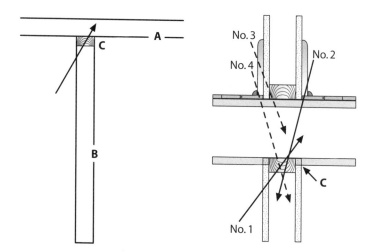

Fig. 21–16 To get cable from *A* in the ceiling to *B* in a wall, a temporary opening must often be made at *C*. Point *C* is shown enlarged at the right.

Cable from second floor to first   If the first-floor partition is directly below a second-floor partition, it is usually simple to bring the cable through by boring as indicated by arrows *2* and *3* (or *3* and *4*) in the enlarged view of Fig. 21–16. Use good judgment so that the two holes will, as far as possible, lie in the same plane, thus simplifying the fishing problem. An opening behind the baseboard is usually necessary.

If the first-floor partition is not directly below the second-floor partition, handling it as shown in Fig. 21–17 will usually solve the problem. Bore holes as indicated by the two arrows.

Fig. 21–17 Problem in bringing cable from a second-floor wall into a first-floor wall.

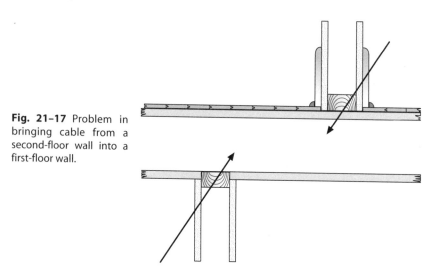

In homes where the rooms are trimmed with crown molding just below the ceiling, it is often possible to remove the molding and chisel a hole in the corner, chiseling away a part of the obstruction to provide a channel for the cable, as indicated by the arrow in Fig. 21–18. Even if the cable projects a bit, it will be concealed when the molding is put back in place.

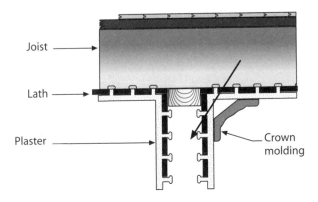

Fig. 21–18 Temporary openings can sometimes be made to advantage behind a crown molding.

SURFACE WIRING TECHNIQUES

When additional outlets are needed for fixtures or receptacles, surface wiring can be an easy and economical option if the appearance will be considered acceptable.

Extension rings   In old work it is often desirable to extend an outlet beyond an existing outlet. If the wiring must be entirely flush, it might involve a good deal of carpentry if the new outlet also must be flush. At least in certain locations (such as basements) it will be acceptable to install the new outlet on the surface. In that case use an extension ring of the type shown in Fig. 21–19. Extension rings are outlet boxes without bottoms; they are available to fit all kinds of outlet boxes. Simply mount the ring on top of an existing outlet box, and from there proceed as in any surface wiring. The extension ring must be covered with a blank cover, or with the fixture that may have been mounted on the original box.

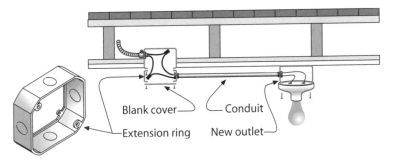

Fig. 21–19 If surface wiring is acceptable, an extension ring is handy in installing a new outlet starting from an existing outlet.

Boxes on masonry walls   In houses of masonry construction (brick, concrete, etc.), an extraordinary amount of work is involved in providing space for a box plus a channel for a raceway or cable, and serious consideration should be given to installing surface wiring. But if the decision is to install concealed wiring, space must be chiseled into the masonry to receive the box. Boxes cannot be secured directly to the masonry with screws, so it is necessary to use one of the many types of plugs or anchors available for the purpose. It will be necessary to drill holes into the masonry using one of the tools shown in Fig. 21–20.

Fig. 21–20 To drill an occasional hole in masonry, use a star drill or a masonry chisel and a hammer, as at left. This method will put a hole through a wall for a conduit, but isn't precise enough to seat most anchors. Use a carbide-tipped drill bit with an electric drill, especially in the smaller sizes (center). Use a rotary hammer for larger holes. The one shown (upper right) compresses air behind the chuck in the process of rotation, delivering explosive impacts with each revolution that allow for rapid progress.

A very common mounting method uses the well-known lead anchor expansion shields of the type shown in the top left of Fig. 21–21, which are simply inserted in a hole in the masonry and then "set" (expanded) by tapping them firmly several times with the tool shown in the upper right of the illustration. Ordinary machine screws are used to secure the box (or conduit strap, etc.) to the anchor, as shown in the bottom part of the illustration. The *NEC* prohibits using wooden plugs in holes.

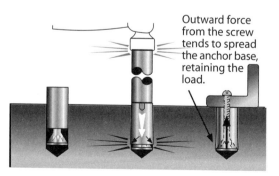

Outward force from the screw tends to spread the anchor base, retaining the load.

Fig. 21–21 One of many available varieties of screw anchors. *(Artex Rawlplug Ltd.)*

For a hollow area, such as a hollow tile wall, use toggle bolts as shown in Fig. 21–22, which also shows the installation method. Slip the collapsible wings through the opening in the wall; a spring opens the wings, which then provide anchorage when the bolt is tightened.

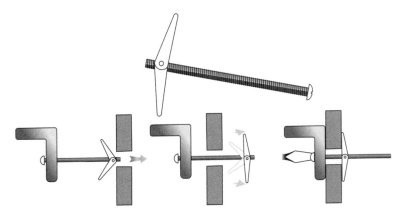

Fig. 21–22 Toggle bolts are convenient and practical in mounting equipment on hollow walls. *(Artex Rawlplug Ltd.)*

Surface wiring assemblies   An excellent method of adding additional receptacles is to use the multioutlet assemblies such as the plug-in strip shown in Fig. 27–21. Armored cable or similar wiring to supply the multioutlet assembly can usually be installed in a concealed area. This and other surface raceways are discussed in Chapter 27.

Don't overlook the combinations of nonmetallic box and wiring devices discussed in Chapter 10, and surface devices for nonmetallic cables (Fig. 5–11 shows one example). They are convenient and practical wherever surface wiring is acceptable and appearance is not a major factor. They can be used only with nonmetallic-sheathed cable.

WHEN TO REPLACE OLD WIRING

No book can outline all the problems in carpentry that will be encountered in old work. The method of construction in houses varies widely with the age of the house, the skill and integrity of the builder, the geographical area, and many other factors. Patience and common sense are great assets in doing old work.

Examine the existing wiring before you do anything. In many old houses, and some that are not so old, the wiring needs modernizing. Some, usually the oldest, need to be completely rewired. Others need only additional circuits and outlets. In most cases a larger service is needed. You must use good judgment when deciding what to do. It doesn't make sense to waste money replacing safe wiring when only additional wiring is needed. On the other hand, it is hazardous to allow unsafe wiring to remain in use.

If the insulation on the accessible ends of the old wires is brittle, if cables have been spliced to old knob-and-tube wiring, or if improper splicing (such as splicing without boxes) has been made in existing cables, the wiring should be replaced. If panelboard terminals have been overloaded for long periods and are in a deteriorated condition, the insulation on the concealed wiring might also be deteriorated and unsafe.

If the existing wiring has good insulation and has not been improperly spliced, the chances are that it can be left in service, especially if recently added appliances can be disconnected from the existing circuits and connected to new circuits installed for the purpose. Additional circuits should be added for both 240- and 120-volt appliances for which no wiring was originally provided. However, additional receptacles can safely be added to some existing circuits without overloading the circuits if the additional outlets are only for more convenient access and no additional load will be imposed. Providing an adequate number of receptacles enhances safety because it eliminates the need for extension cords, which are dangerous when householders are tempted to use them as "permanent" wiring.

For example, the living room and bedrooms may simply need new receptacles added to existing lighting circuits. An existing 20-amp kitchen small-appliance circuit might be acceptable if one or more additional circuits would be installed, with new receptacle outlets for new appliances that were acquired after the original wiring was installed, or if a new 20-amp circuit is installed for laundry equipment that was originally connected to one of the kitchen small-appliance circuits. Maybe a room air conditioner, a dishwasher, a kitchen waste disposal unit, or a freezer unit has been connected to an existing circuit that was already fully loaded instead of being provided with its own circuit. If the existing wiring is aluminum and is otherwise in good condition, you will want to replace all 15- and 20-amp receptacles and switches with devices marked CO-ALR and reconnect them properly as explained at length under the last two headings in Chapter 8. You will also want to replace any panelboard to which aluminum wiring is connected if its terminals are not marked CU-AL, unless you are going to install a larger panelboard anyway, as is often necessary where modernizing is being done.

METHODS FOR RETAINING EXISTING PANEL(S)

NEC 220.16(A) and 220.83 cover methods of computing the loads for existing dwellings, which are substantially the same as the methods explained in Chapter 13 for computing loads in new residential occupancies. Plan and lay out your circuits as you would in a new building, and then look at whether the existing panelboard can accommodate the circuits. Be sure the cooking and dining areas have at least two 20-amp circuits, and permanently installed appliances have their own circuits. The *NEC* allows smaller appliances to be connected to general-purpose branch circuits if those appliances don't exceed 50% of the branch-circuit rating, but individual circuits are a better design choice for all but the smallest permanently installed loads. The laundry area must have its own 20-amp circuit. Preferably, each freezer unit and refrigerator, especially a self-defrosting unit, should be on its own separate circuit. You can use the new allowance for a 15-amp individual branch circuit to refrigerating appliances if desired, but be sure to use a 15-amp single receptacle to finish the outlet. A homeowner can lose hundreds of dollars worth of food if a fuse blows or a breaker trips as a result of an overloaded appliance circuit that also supplies the freezer and refrigerator. An individual branch circuit is a small price to pay for insurance against that kind of loss. Don't skimp on the number of branch circuits or on the number of receptacles. Remember that most modernizing could have been avoided by more liberal planning when the house was built. The older the house, the greater the likelihood that considerable modernizing may be needed to provide for many added appliances, some of which may not have been in existence when the house was built.

Whether you must completely rewire the building or simply add new circuits and new outlets, it is likely that you will have to change the service. A 100-amp service is the *NEC* minimum for single-family dwellings, but you must make sure the service you install is at least as large as the size required by the *NEC* for the load involved. If the building must be completely rewired, a new service can be installed, and the old service and wiring can then be disconnected after the new service and wiring have been connected. But if some of the existing circuits (such as the lighting circuits) are to be retained, the new and the old circuits can be connected to the new service in one of two ways: (1) by retaining the old service equipment after altering it or (2) by replacing the old service equipment with a junction box.

Reconnecting old circuits   If the old service equipment panelboard is in good repair, if it is a 120/240-volt, three-wire panelboard, and if its grounded neutral busbar can be insulated from the panelboard enclosure by removing its main bonding jumper (by removing a screw or strap), then plan to alter and retain it. On the other hand, if the old panelboard is in poor repair, or if it is a 120-volt, two-wire panelboard, or if its grounded busbar is welded to the panelboard enclosure and it does not have a main circuit breaker or main fused switch, then plan to replace it with a junction box, or if feasible to convert it to a junction box.

Retaining the old panelboard   There are two possible ways of using the old panelboard without converting it to a junction box, depending on whether or not it has a main circuit breaker or fused switch.

Option 1: If the old panelboard has a main breaker or fused switch, whether the grounded busbar is bonded to the enclosure by a removable jumper (screw or strap) or is welded to the enclosure, this panelboard can be retained as a part of the total service equipment. The new panelboard, which will also have a main breaker or fuses, will also constitute a part of the total service equipment.

The ampere rating of the *new* panelboard will then be based on the total load *other than* that part of the load connected to the old panelboard, but the size of the service-entrance conductors must be based on the total load connected to both panelboards, because the same set of service conductors will supply both panelboards.

Where both panelboards are used as service equipment, they must be grouped closely together. Up to six main service disconnecting means are permitted, so two disconnecting means—the main breaker or fuses in each panelboard—are acceptable. Where this is done and both panelboards are supplied by the same (new) set of service-entrance conductors, install an auxiliary gutter (a short section of wireway; see pages 512 through 514) ahead of the two panelboards and connect the subset of service-entrance conductors from each panelboard to the main set of service-entrance conductors in the auxiliary gutter, as shown in Fig. 21–23.

Because the old panelboard is retained as one of the two parts of the service equipment, its grounded neutral busbar must remain bonded to the panelboard enclosure, and the grounded neutral busbar in the new panelboard must also be bonded to the panelboard enclosure, because it too is part of the service equipment. The two grounded busbars will be bonded together (as required), because both will be connected to the grounded service-entrance conductor. The ground wire connected to the old panelboard will probably not be large enough for the entire enlarged service, so install one that is large enough from the new panelboard grounded neutral busbar without using any splices to the grounding electrode system (refer to the page 303 ground wire discussion). Then, make a tap from the unbroken grounding electrode conductor and extend it into the old panelboard. Size it according to the largest ungrounded conductor feeding that panelboard, as shown in *NEC* Table 250.66. This procedure is covered in *NEC* 250.64(D).

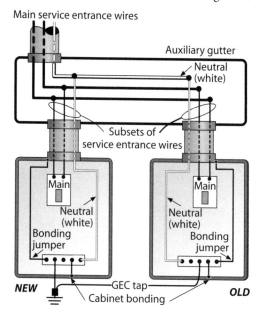

Fig. 21–23 Here the original service equipment becomes part of the total service equipment; the service equipment is now contained in two separate cabinets. Connect the service raceways and the auxiliary gutter or wireway together with bonding jumpers.

Option 2: If the old panelboard does not have a main circuit breaker or main fused switch, it can still be retained as a load-center panelboard or fuse cabinet (downstream from the service equipment) if its grounded neutral busbar can be *insulated* from the panelboard enclosure by removing its main bonding jumper (screw or strap). In this case, the old panelboard will not be a part of the service equipment, so the grounded neutral busbar *must be insulated* from the panelboard enclosure. Remove the bonding screw or strap; also remove the old ground wire after a new ground wire of the proper size has been connected to the new panelboard. Make absolutely certain the equipment grounding path from the old panelboard is secure through the new panel all the way to the main bonding jumper. Remember, the ground return path through the old panel no longer can rely on a neutral wire, because that wire is now insulated from the enclosure.

The old panelboard must be supplied by a feeder from a circuit breaker or set of fuses in the new panelboard. This feeder must have an ampacity rating at least as great as the ampere rating of the old panelboard, and the circuit breaker or fuses in the new panelboard protecting this feeder must have an ampere rating that is not greater than the ampere rating of the old panelboard. See Fig. 21–24. The two panelboards may be any distance apart, such as in different rooms or areas of a basement. So if it is desirable to install the new service in a different location, this method may be preferable to the first method described even if the old panelboard does have a main.

Removing or converting old service panelboard   This is a third option. The new service equipment must have a main breaker or main fused switch, and must have branch-circuit breakers or fuses for each old circuit and for each new circuit, and preferably a few spares for future circuits. If the old panelboard cabinet or

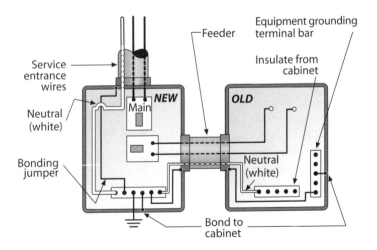

Fig. 21–24 Here the original equipment is not a part of the service equipment. It is only a "load center" for the original circuits.

enclosure will be totally enclosed (and if every knockout that has been removed but will not be used is closed with a knockout closure; see Fig. 10–1 and related text), the empty cabinet can be used as a junction box. However, if the cabinet is the very old type with the top open for connections to an old-style meter, it must be discarded. Replace it with a junction box of adequate size, at least 6 × 8 in. or preferably larger.

In either case, the old circuit wires are spliced to new circuit wires, installed from circuit breakers or fuses in the new panelboard. In other words, the old panel-board cabinet is either used as a junction box, or replaced by a suitable junction box. In either case, the new and old branch-circuit wires are spliced together, as shown in Fig. 21–25. Of course, if a new junction box is installed, the old cables must be disconnected from the old panelboard cabinet and connected to the new junction box installed in its place. A new ground wire is installed in the new panelboard in the usual manner, and the wire of the old equipment is removed.

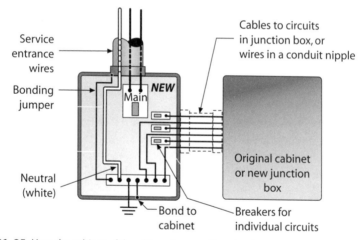

Service entrance wires

Bonding jumper

NEW
Main

Neutral (white)

Bond to cabinet

Cables to circuits in junction box, or wires in a conduit nipple

Original cabinet or new junction box

Breakers for individual circuits

Fig. 21–25 Here the cabinet of the old equipment, if retained, serves only as a junction box.

Changing over   There are several ways to make the changeover from the old to the new service. An easy way is to have the new service equipment complete, with all new circuits connected, and the old system as nearly complete as possible without working on anything while it is hot. Then have your power supplier disconnect the old service and connect the new service. After that has been done, disconnect the old service-entrance conductors from the old panelboard and, *with the circuit breakers in the new panelboard turned off,* complete the job as follows:

Disconnect the existing circuits from the old panelboard, remove the interior of the old panelboard enclosure or replace the old panelboard with a junction box, connect the old and the new cables or conduit to the old cabinet or to the new junction box, and splice the old and new circuit conductors together. Remember, the white grounded wires of the old circuits must be insulated from the cabinet.

CHAPTER 22

Farm Wiring

WHETHER IN A HOUSE in the city or one on the farm, appliances are used for the convenience they provide. But on farms, many other appliances besides those in the house are used in the business of farming, appliances bought as an investment on which the farmer expects dividends. These include water pumps, milking machines, milk coolers, hammer mills for chopping fodder, hay or crop dryers, water heaters to provide scalding water for the dairy, etc. These appliances constitute a very considerable load.

The wiring of farms involves problems discussed in previous chapters, as well as many new problems. The maximum load in use at one time is likely to be much greater for the total farm than for a city home. There is a great deal of overhead or underground outdoor wiring between the various buildings. Substantial distances are involved, which means that wires must be of sufficient size to avoid excessive voltage drop, and for mechanical strength in overhead runs. There may also be relatively poor grounding conditions. These and related factors are covered in this chapter.

The remainder of this chapter discusses electrical requirements for farms predominately in terms of the requirements of *NEC* Article 547, which applies to at least portions of most but not all agricultural buildings. Specifically, Article 547 applies to areas where excessive dust can accumulate, including all areas of livestock, fish (yes, the *NEC* also recognizes aquaculture), and poultry confinement systems where litter or feed dusts may accumulate. It also applies to similar facilities that have corrosive environments by reason of vapors from excrement, likely contact with aqueous solutions of corrosive materials as may occur in fertilizers and feedstock additives, or periodic washdown and sanitation or similar activities.

HOW TO SET UP THE ELECTRICAL DISTRIBUTION SYSTEM AT A FARM

In a typical farm installation, wires from the power supplier's transformer end at a pole in the yard (sometimes called a "maypole" to distinguish it from the power supplier's distribution pole, and because wires fan out from it in various directions).

415

From the top of the pole, wires run down to the meter, then back to the top of the pole. Sometimes there is a circuit breaker, or a switch with or without fuses, just below the meter socket. The neutral wire, the meter socket, and any breaker or switch cabinet are always grounded at the pole.

From the top of the pole, a set of wires runs to the house, and other sets to barns and other buildings. Increasingly, underground cables are being used instead of overhead wiring.

Adequacy   Adequate wiring, as discussed at the beginning of Chapter 12, applies equally to city homes and farm homes. Indeed, the farm home often requires more attention; especially on small farms, certain appliances are installed in the house instead of in other buildings as on larger farms. Adequate wiring is also required to carry on the business of farming. Special attention must be paid to the wiring in and between other buildings, and on the meter pole. Be sure to run enough circuits and large enough conductors to allow for future expansion of use. Providing more capacity than needed at the time of installation increases the labor very little and avoids later expensive alterations.

Overhead or underground?   A large number of overhead wires running all over the place gives a very untidy appearance to a farm and invites problems of various kinds. Wires to low buildings are subject to damage by vehicles and machinery passing under them. Having many overhead wires on an isolated farm increases vulnerability to lightning. In northern climates where sleet storms are common, wires can break from the weight of accumulated ice; a broken wire on the ground is dangerous. Underground wires cost little more and prevent these problems.

If you are going to use overhead wiring, consider wire size carefully. Select a wire size big enough to carry its load in amperes without excessive voltage drop (see discussion beginning on page 107). Make sure it is strong enough for the length of the span, which means that sometimes you must use wires larger than otherwise necessary for the load involved. If you are going to use underground wires, use the materials and follow the methods discussed in Chapter 16 concerning the service entrance, after reading the balance of this chapter closely.

Rod or pipe electrodes   In cities, an underground metal water piping system provides an excellent ground. On farms you must provide a substitute as explained in the grounding electrode discussion beginning on page 300.

Grounding on farms   The subject of grounding in general is covered in Chapter 9. A basic principle is that for safety the grounded circuit wire must be connected to the earth through a metal water pipe or a driven ground rod. A good ground connection offers protection not only against troubles in the wiring system but also against lightning. On a farm there is considerably more danger from lightning than in a city. Yet a good ground connection is usually much harder to obtain on a farm than in a city. A study of more than 200 farms in

Minnesota was made in a rather dry year. Only 9 of 215 (1 in 23) had a resistance to ground of 25 ohms or less—and remember that a 25-ohm ground is not a good ground. The other 22 out of every 23 had a resistance of over 25 ohms, some well over 100 ohms. In other words, perhaps one in a hundred farms had what would be considered a good ground connection in a city. The reason for the high resistance was reliance on a driven ground rod, which rarely provides as good a ground as an underground metal pipe system. These study results validate the grounding electrode system discussion in Chapter 16 and the desirability of using other grounding electrodes. The resistance to ground using a ground rod varies with changing conditions. The drier the earth, the higher the resistance, so if practicable install the rod where rain off a roof will help keep the earth wet. Usually a single rod is not sufficient; install two of them, 8 to 10 ft apart, and bond them together with 6 AWG copper wire, which may be bare, using ground clamps at each end of the wire. The minimum separation in the *NEC* is 6 ft, but the paralleling efficiency improves with greater distance. There appears to be evidence that installing more than two rods will not significantly reduce the resistance to ground unless they are widely separated.

Locating the meter pole ("distribution point" in *NEC* terminology)   There is a right and a wrong location for the meter pole. Why is there a pole in the first place? Why not run the service wires to the house and from there to the other buildings, as was sometimes done when farms were first being wired? This can be done on small farms, but isn't practical on larger farms due to the large service equipment required, the expense of running large feeders long distances, and many other complications.

Locate the pole as near as practical to the buildings that use the greatest amount of power; on modern farms, the house rarely has the greatest load. By locating the pole so the largest wires will be the shortest wires, you will find it relatively simple to solve the problem of excessive voltage drop without using wires larger than would otherwise be necessary for the current to be carried. Total cost is kept down when the large expensive wires to the buildings with the big loads are the shorter wires, and the smaller less expensive wires to buildings with the smaller loads are the longer wires.

Basic construction at pole   Of the three wires that come from the power supplier's transformer and end at the top of the pole, the neutral is usually but not always the top wire. Check with your power supplier. The neutral is always grounded at the pole.

Figure 22–1 shows a modern farm distribution point. The disconnecting means may be provided by the utility or by the owner, depending on local practice. The switch has no overload protective devices within it, thereby varying from the normal rule for services in *NEC* 230.91. The *NEC* classifies this device as a "site isolating device" to distinguish it from a service disconnect. Even if supplied and maintained by the utility, and therefore beyond the scope of the *NEC*, the *NEC* avoids needless duplication by recognizing it as a disconnecting means provided it meets the requirements in *NEC* 547.9(A). Since it is not an actual service disconnect, it follows that the wiring that leaves this device still has the status of service conductors and must meet the wiring method and clearance requirements in Article 230. Although nothing technically prevents a farm from establishing a conventional service at the distribution pole, and then routing conventional overcurrent-protected feeders to each building, the arrangement shown here is widely used for overall cost effectiveness. Note that although the switch is at the top of the pole, it can be operated from a readily accessible point through the permanently installed linkage shown in Fig. 22–1. In addition, you must install a grounding electrode conductor at this point and run it from the neutral block of the switch to a suitable electrode at the pole base.

From the pole top, go to the buildings that need to be supplied from this point. Review the discussion in Chapter 17 covering wiring to outbuildings and, in particular, the discussion about regrounded neutrals. As a general rule the farmhouse can be supplied by a three-wire service, with its neutral regrounded at the house just as if the utility had made a direct termination. The farmhouse must not, however, share a common grounding return path with the barn. If it does, as in the case of a common metallic water piping system, the house (1) has to be supplied with a four-wire service, and (2) have all instances of electrical contact between the neutral and the local equipment grounding system removed.

Although the barn, arguably, could also be wired like the house (three-wire), a three-wire hook-up would mean that the neutral and the equipment grounding system in the barn would be bonded together at the barn disconnect. That in turn would mean the neutral, in the process of carrying current across its own resistance, would constantly elevate the voltage to ground of all barn equipment by some finite amount relative to local ground, especially from the perspective of farm animals where they stand. The feet of livestock, being in close contact with moisture, urine, and other farm chemicals, are conductively rather well coupled to local earth. Most livestock are much more sensitive to voltage gradients than are people. A potential difference in the range of a fraction of a volt can take a cow out of milk production, which no farmer can afford.

The *NEC* addresses this in two ways. First, it establishes the unique rules on farm distributions being covered here. Second, it establishes an equipotential plane for these environments, discussed later in this chapter. The service to the barn is normally wired four-wire, and that is 1) customary because of the reasoning discussed in the previous paragraph, and 2) mandatory unless there are no parallel grounding return paths over water systems, etc., a necessary condition to comply with *NEC* 250.32(B)(2). There is an additional condition attached to the four-wire scheme that is unique to agricultural buildings. The separate equipment grounding conductor must be fully sized. That is, if the run to the barn is 3/0 AWG copper for a 200-amp disconnect, and the neutral is 1/0 AWG copper (both sized on the basis of load), the equipment grounding conductor is not 6 AWG as normally required by *NEC* Table 250.122; nor is it 4 AWG, the size for a grounding conductor on the supply side of a service using 3/0 AWG wires; nor is it 1/0 AWG, the size of the neutral. It must not be smaller than the largest

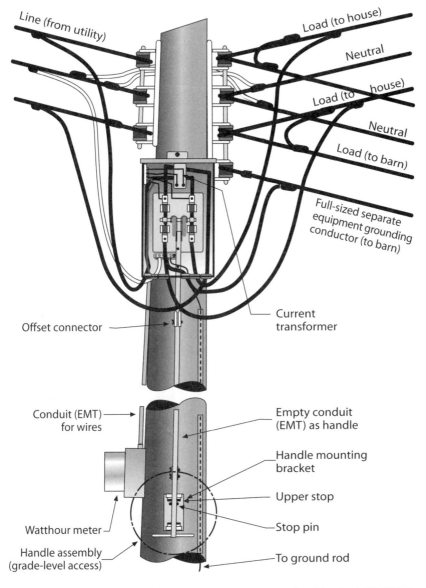

Fig. 22–1 Pole top switch and remote metering as contemplated by the 2002 *NEC* for overhead farm distribution. Leave at least half the circumference of the pole clear to allow line workers and repair workers to climb the pole without trouble.

ungrounded line conductor, or 3/0 AWG. When this wire arrives at the barn, it must arrive at a local distribution with the neutral completely divorced from any local electrodes or equipment surfaces requiring grounding.

Installing the meter socket   The meter socket is sometimes furnished by the power supplier but installed by the contractor. In other localities it is supplied by the contractor. As shown in Fig. 23–1, the meter socket is not connected across the line, but instead is connected through a current transformer (usually abbreviated CT), discussed under its own heading later in this chapter. Mount it securely at a height required by the local authorities; this is usually 4 to 6 ft above ground level.

In the case of a small service, the actual line and load conductors could come down the pole to a meter socket connected across the line. In this case the usual procedure is to wire the service disconnect immediately below the meter socket. Run service wires from the bottom jaws of the meter socket to the line terminals of a switch or circuit breaker. Then run the load wires from the load terminals of the service disconnect back up to the point of beginning at the pole top. Pass the load wires, unbroken, straight through the meter socket on their way back up. Combination meter sockets and main disconnects, usually using a circuit breaker, are commonly available and simplify this process. In sizing the service and the load wires, be aware that if you install them in a common raceway (the usual case) you must count the four hot wires as current carrying. That number of current-carrying wires run together requires their ampacities to be derated by a factor of 0.8. Today's large farm loads, however, almost always use the pole top switch and CT shown in Fig. 22–1.

Insulators on pole   Near the top of the pole, install insulator racks of the general type shown in Fig. 17–4. Provide one rack for the incoming power wires and one rack for each set of wires running from the pole to various buildings. Remember that there is great strain on the wires under heavy winds or in case of heavy icing in northern areas. Anchor the racks with heavy lag screws. Better yet, use at least one through-bolt, all the way through the pole, for each rack. Note the fourth insulator on the pole for the equipment grounding conductor run to the barn.

Ground at meter pole   For a small service using a meter connected directly across the line, the neutral wire always runs from the top of the pole through the conduit to the center terminal of the meter socket. The neutral is not necessary for proper operation of the meter, but it grounds the socket. At one time it was standard practice to run the ground wire from the meter socket (out of the bottom hub) to ground, but experience has shown that better protection against lightning is obtained if the ground wire is run outside the conduit. Run it from the neutral at the top of the pole directly to the ground rod at the bottom of the pole. It is usually run tucked in alongside the conduit as far as it goes, then to ground. In some localities it is stapled to the pole on the side opposite the conduit. For today's larger services using pole-top disconnects and CTs, just run the ground wire straight down the pole. Protect it from damage, particularly along the lower end of the pole, with U-shaped heavy plastic coverings, or run it in rigid nonmetallic conduit. If you use metallic conduit, make sure both the top and bottom ends are bonded to the enclosed ground wire or the ground wire will lose a good portion of its effectiveness (see discussion of ground-wire conduit or armor beginning on page 304).

In some areas the custom is to let the top of the ground rod project a few inches above the ground so that the ground clamp will be permanently exposed for inspection. Since the *NEC* requires such electrodes to be buried the full 8 ft, that is a violation unless you are using a 10-ft ground rod. More often the ground rod is driven about 2 ft from the pole (or building) after a trench has first been dug from rod to pole about a foot deep. The top of the rod is a few inches above the bottom of the trench. The ground wire runs down the side of the pole (or building) to the bottom of the trench, then to the ground clamp on the rod. After inspection the trench is filled in and the rod, the clamp, and the bottom end of the ground wire remain buried (see Fig. 22–2). Use the method favored in your locality. Instead of a single rod, two ground rods, as already mentioned, are recommended for a lower-resistance ground. How to determine the size of the ground wire is discussed in the ground wire topic beginning on page 303. Remember that if the ground wire runs to a rod or pipe electrode, such as a ground rod, it never needs to be larger than 6 AWG.

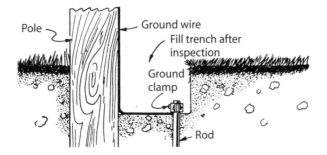

Fig. 22–2 In some localities, the top of the ground rod is below the surface of the ground. Fill the trench after inspection. If the ground rod is copper, use only a copper clamp. Per listing restrictions, insist on a DIRECT BURIAL or DB marking on the clamp.

Underground distributions   For underground distributions, extend downward from the pole-top switch shown in Fig. 22–1, entering the junction box shown in Fig. 22–3 with four wires. If no service wires will run overhead, the utility may agree to omit the pole-top switch and come straight down the pole to the metering equipment. Distribute the unprotected feeders using the same principles governing overhead distributions just described, that is, a three-wire unprotected feeder to the house, and a four-wire unprotected feeder to the barn, subject to the same constraints as three-wire feeders. However, for the underground feeder to the barn, if livestock is housed in the barn, all portions of the equipment grounding conductor run underground must be insulated or covered copper.

If several sets of underground wires are to enter the bottom of the meter socket at the pole, you will have a problem because the hub at the bottom of the socket is too small. In that case provide a weatherproof junction box just below the meter socket. Quite a number of runs can be terminated in the box with only a single set of wires from the meter socket into the junction box. A junction box for this purpose is shown in Fig. 22–3. This particular junction box has terminals for the various runs of underground wires, which eliminates the need for a lot of bulky splices in a box without terminals. A telescopic metal channel to protect a group of underground wires running up the side of a pole is shown in Fig. 22–4, installed with the junction box of Fig. 22–3.

Fig. 22–3 A rainproof junction box of this type makes it easy to install several underground runs, starting at the meter socket.

Equipotential planes—essential features of livestock confinement areas Due to the sensitivity of livestock to very small "tingle" voltages, the *NEC* now requires an equipotential plane in livestock (does not include poultry) confinement areas, both indoors and out, if they are concrete floored and contain metallic equipment accessible to animals and likely to become energized. These areas must include wire mesh or other conductive elements embedded in (or placed under) the concrete floor, and those elements must be bonded to metal structures and fixed electrical equipment that might become energized, as well as to the grounding electrode system in the building. In the case of dirt confinement areas, the equipotential plane may be omitted, but only for indoor applications. Since these systems are required for outdoor areas with metallic equipment that may become energized, and since these areas are defined in a way that mandates the use of concrete, the literal text of the 2005 *NEC* effectively disallows the use of dirt confinement areas outdoors in these

Fig. 22–4 The trough protects the aboveground portions of underground runs at the pole. The junction box is at the top.

locations. Several members of the Code-making panel, including its chair, have assured the author that this is a mistake that will be corrected in the 2008 *NEC*. Be sure to discuss this with the inspector in advance if this will affect the design of a farm system you are involved with.

Remember that the grounding system to which equipotential planes should be connected is usually (refer to the distribution point discussion of this topic) electrically separated from neutral return currents. The idea is to minimize voltage gradients, similar to the objectives of *NEC* Article 680 in swimming pools (review that topic in Chapter 20). Due to the well-grounded environment, the *NEC* also requires all general purpose 15- and 20-amp, 125-volt receptacles in the area of an equipotential plane to have GFCI protection. This GFCI protection requirement also applies to similar receptacles in all damp or wet locations, including outdoors, and for dirt confinement areas whether indoors or out. Receptacles for specified (not general purpose) loads are not covered by this requirement.

Calculating wire sizes—pole to buildings   Each building must have a service entrance quite similar to those discussed in Chapter 16, except without a meter. Some buildings, however, particularly those housing livestock, will be using the four-wire service arrangements described previously, and you must carefully isolate the neutral wire from any equipment grounding wires. Never make any equipment grounding connection to an isolated neutral, or you will destroy the effectiveness of the special grounding arrangements the *NEC* requires to prevent stray voltages from harming farm animals. Wires from the pole to any building must be of the proper size. Part V of *NEC* Article 220 shows how to compute the load for service conductors and service equipment for each farm building (except the house, which is computed as any other house) and the total farm load.

Assume that the building will have a three-wire, 120/240-volt service. The total load in amperes at 240 volts must be determined. For motors, see *NEC* Table 430.148. For all other loads, start with the load in volt-amperes.

For incandescent lights you can determine the total load beginning with the wattage of the lamps you intend to use. (For fluorescent lights, add about 15% to the watts to get volt-amperes, because the watts rating of a fluorescent lamp defines only the power consumed by the lamp itself; the ballast adds from 10 to 20% depending on the ballast power factor.)

For receptacles, if you use the value of 180 VA for each, based on *NEC* 220.14(I), you will probably be on the safe side; they will not all be used at the same time.

This principle of applying demand factors to theoretical connected loads is well established in the *NEC* as well as in routine engineering practice. Don't make the mistake of assuming that your wiring system must accommodate every last ampere of branch circuit capacity. A feeder to a farm building with ten 20-amp, 120-volt branch circuits probably won't ever carry more than half that load at any given time.

Divide the total watts by 240 (volts), and you will have the amperes at 240 volts.

For each building to which wires run from the pole (except the house, which is figured as explained in Chapter 13), first determine the amperes at 240 volts of all the loads that have any likelihood of operating at the same time. Enter the amperes in *a* of the tabulation below. Then proceed with steps *b* through *f:*

a. Amperes at 240 volts of all connected loads that in all likelihood will operate at the same time, including motors if any: ____amps

b. If *a* includes the largest motor in the building, add here 25% of the ampere rating of that motor (if two motors are the same size, consider one of them the largest): ____amps

c. If *a* does not include the largest motor, show here 125% of the ampere rating of that motor: ____amps

d. Add the amps in *a* to the amps in *b* or *c,* whichever is applicable, and write the total here: ____amps

e. Amperes at 240 volts of all other connected loads in the building: ____amps

f. Total of *d* + *e:* ____amps

Now determine the minimum service for each building by one of the following steps:

▪ If *f* is 30 or less and if there are *not more than two* circuits, use a 30-amp switch and 10 AWG wire.[1]

▪ If *f* is over 30 but under 60, use a 60-amp switch and 6 AWG wire.

If *d* is less than 60 and if *f* is over 60, start with *f.* Add 100% of the first 60 amps plus 50% of the next 60 amps plus 25% of the remainder. For example, if *f* is 140, add 60, plus 30 (50% of the next 60 amps) plus 5 (25% of the remaining 20 amps), for a total of 60 + 30 + 5 or 95 amps. Use 100-amp switch and wire with ampacity of 95 amps or more.

▪ If *d* is over 60 amps, start with 100% of *d.* Then add 50% of the first 60 amps of *e,* plus 25% of the remainder of *e.* For example if *d* is 75 amps and *e* is 100 amps, start with the 75 of *d,* add 30 (50% of the first 60 of *e*), and add 10 (25% of the remaining 40 of *e*), for a total of 75 + 30 + 10 or 115 amps. Use a switch or breaker of not less than 115-amp rating and wire with corresponding ampacity.

The wire sizes determined above will be the minimum permissible by the *NEC* for each building. You would be wise to install larger sizes to allow for future expansion and to reduce voltage drop.

---

1. The *NEC* in 230.42(B) and 230.79(B) requires a minimum ampacity of 30 if the building has not more than two 2-wire branch circuits, and that will be a 10 AWG based on termination restrictions (review Chapter 7); if run overhead farther than 50 ft, *NEC* 225.6(A)(1) will require 8 AWG for mechanical reasons.

Overhead spans   *NEC* 225.6(A)(1) requires a minimum of 10 AWG wire for overhead spans up to 50 ft, and 8 AWG for longer spans. While the *NEC* is silent on the subject, it is recommended that if the distance is more than 100 ft, use 6 AWG; if it is more than 150 ft, it is best to install an extra pole. If the wires are installed in northern areas on a hot summer day, remember that a copper wire 100 ft long will be a couple of inches shorter the following winter when the temperature is below zero. Leave a little slack so the insulators will not be pulled off the buildings during winter.

Triplex cable   This material is shown in Fig. 16–3; it consists of two insulated wires wrapped spirally around a strong, bare neutral wire. It requires only one insulator for support. One triplex cable is usually considered preferable to three separate wires. The *NEC* covers it as Article 321, "Messenger Supported Wiring," and the utilities have been using it for service drops for years.

Calculating total load   The calculations just described determine the size of the wires from the pole to the separate buildings. To determine the size of the wires on the pole (from top, to meter socket, back to top of pole), proceed as follows using the load in amperes at 240 volts, as determined above, for each building (excluding the house until step 6):

1. Highest of all loads in amperes for an individual building: ____amps at 100% ....____amps

2. Second highest ampere load: ____amps at 75% ....____amps

3. Third highest ampere load: ____amps at 65% ....____amps

4. Total of all other buildings (except house): ____amps at 50% ....____amps

5. Total of above ....____amps

6. House computed as discussed in other chapters, at 100% ....____amps

7. Grand total including house ....____amps

Important: If two or more buildings have the same function, consider them as one building for the purpose above. For example, if there are two brooder houses requiring 45 and 60 amps respectively, consider them as one building requiring 105 amps and enter 105 in line 1 above.

The total of line 7 above is the minimum rating of the breaker or switch (if used) at the pole, and the minimum ampacity of the wires that you must use on the pole. Note that in listing the amperes for any building, the load in amperes used is the *calculated* load, not the rating of the switch or breaker used. For example, if for a building you calculated a minimum of 35 amps but you use a 60-amp switch (because there is no size between 30 and 60 amp), use 35 as the load, not 60 amps.

To connect the various wires at the top of the pole to each other, use solderless connectors of the general type shown in Figs. 8–19 and 8–20. Unless these connectors have snap-on insulating covers, the splices must be taped.

Use current transformers for modern high-capacity installations   If the service at the pole is rated at 200 amps or more (line 7 of the preceding calculation), very large wires would run from the top of the pole down to the meter and back again to the top. That is both expensive and clumsy. It is not necessary to run such large wires down to the meter—instead use a current transformer (CT).

An ordinary transformer changes the voltage in its primary to a different voltage in its secondary. In a current transformer, the current flowing in its primary is reduced to a much lower current in its secondary. Most current transformers are designed so that, when properly installed, the current in the secondary will never be more than 5 amps. A typical current transformer, shown in Fig. 22–5, has the shape of a doughnut 4 to 6 in. in diameter. It has only a single winding: the secondary. The load wire (in which the current is to be measured) is run through the hole of the doughnut and becomes the primary. Assuming that the transformer has a 200:5 ratio, for use with a 200-amp service, the current in the secondary will be 5/200 of the current in the primary. If the current in the primary is 200 amps, 5 amps will flow in the secondary; if it is 100 amps in the primary, 2½ amps will flow in the secondary. For a 400-amp service, the transformer would have a 400:5 ratio.

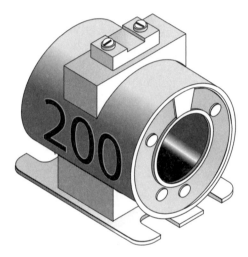

Fig. 22–5 A current transformer.

One current transformer is installed at the top of the pole, with the two hot wires running through the hole. Four small 14 AWG wires run from the top of the pole to the meter: two from the secondary of the current transformer, and two from the hot wires for the voltage. The meter must be the type suitable for use with a current transformer. The meter operates on a total of not more than 5 amps but the dials of the meter will show the actual kilowatthours consumed. A wiring diagram is shown in Fig. 22–6.

Caution: The transformer's secondary terminals must always be short-circuited while any current is flowing in its primary. If connected to a kilowatthour meter, the meter constitutes a short circuit. As purchased, the transformer will probably have a short-circuiting bar across its secondary terminals; this must not be removed until the transformer is connected to the meter. If you were to touch the terminals of such a meter that is not short-circuited, you would find a dangerous voltage of many thousands of volts.

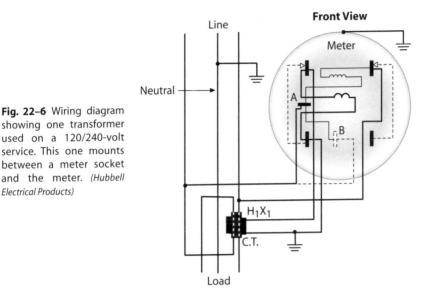

Line

Front View

Meter

Neutral

Fig. 22–6 Wiring diagram showing one transformer used on a 120/240-volt service. This one mounts between a meter socket and the meter. *(Hubbell Electrical Products)*

H_1X_1

C.T.

Load

In an installation using a current transformer at the top of the pole, if you also want a switch at the pole, install it at the top. Such pole-top switches are operated by a handle near the ground level, connected through a rod to the switch at the top as shown in Fig. 22–1. Note that the current transformer is installed in the same cabinet with the switch. Such switches are also available in the double-throw type, as required if a standby generating plant is installed for use during periods of power failure.

Surge (lightning) arresters   While lightning-caused damage to electrical installations is quite rare in large cities, it is rather frequent in rural areas. The more isolated the location, the greater the likelihood of damage. It occurs frequently on farms and to a lesser extent in suburban areas and smaller towns. It is more frequent in southern areas, especially in Florida and other Gulf states.

Lightning does not have to strike the wires directly; a stroke *near* the wires can induce very high voltages in the wires, damaging appliances and other equipment as well as the wiring. Sometimes the damage is not apparent immediately, but shows up later as mysterious breakdowns.

While proper grounding greatly reduces the likelihood of damage, a surge arrester correctly

Fig. 22–7 A surge arrester should be installed at the pole. Often they are installed in the service equipment of individual buildings. This one mounts conveniently between a meter socket and the meter. This is appropriate if the meter is placed directly across the line. *(Hubbell Electrical Products)*

installed reduces the probability of damage to a very low level. Three leads come out of it; connect the white wire to the grounded neutral busbar, the other two to the hot wires.

One arrester should be installed at the meter pole. If the feeders from pole to building are quite long, install another at the building. Install it on or in the service equipment, either letting the neck of the arrester project into the enclosure through a knockout, or by using one designed to mount within the enclosure as shown in Fig. 22–8.

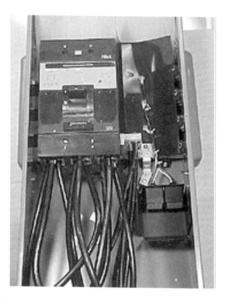

Fig. 22–8 A surge arrester installed on a service switch cabinet. Be sure its white wire is connected to the grounded neutral busbar in the cabinet. Take care to terminate its wiring at appropriate terminals; most main circuit breaker lugs are not listed for the combination of the small wires from the arrester and the much larger feeder or service conductors. Consult with the panelboard manufacturer as to the availability of appropriate terminations for the surge arrester. *(Square D)*

HOW TO WIRE BRANCH CIRCUITS AND OUTLETS IN FARM BUILDINGS

The basic wiring principles covered in previous chapters apply to farm installations as well. In addition, your wiring design must take into account the special conditions that exist in farm buildings, particularly in buildings that house animals.

Cable for barn wiring   A record of failures using Type NM cable in the 1930s led to the development of what at first was called "barn cable," and which has now been standardized as Type NMC nonmetallic-sheathed cable. Later, underground feeder cable Type UF was developed (see page 192). This too is suitable for use wherever Type NMC (or Type NM) is otherwise used. Type UF may even be buried directly in the ground. (For further information, review the discussion on selecting underground service wires beginning on page 284.)

This led to the development of what at first was called "barn cable," and which has now been standardized as Type NMC nonmetallic-sheathed cable. Later, underground feeder cable Type UF was developed (see page 192). This too is suitable for use wherever Type NMC (or Type NM) is otherwise used. Type UF may even be buried directly in the ground. (For further information, review the discussion on selecting underground service wires beginning on page 284.)

In the wiring of barns and other farm buildings housing livestock, use only Type NMC or Type UF. There is no reason why the same types should not be used in all the buildings on a farm.

If you were to wire a barn using metal conduit or armored cable, the same corrosive conditions that rot away ordinary Type NM cable would also attack the metal of the raceway or cable. Remember that the raceway or armor serves as the equipment grounding conductor; thus as the metal rusts away, the equipment grounding conductor is destroyed. Suppose that a metal system had been used, and that the metal had been destroyed at some point. If an accidental ground fault occurs in the raceway, armor, or a metal box at a point beyond the break, and then a person or an animal touches the metal, the circuit is completed through the body to the ground. The result is a shock, unpleasant or dangerous depending on many factors. Figure 22–9 illustrates the point; the ground through the body is equivalent to touching both wires and is equally dangerous. Many farm animals have been killed through just such occurrences. Cattle and other farm animals cannot withstand as severe a shock as a human being, and are killed by a shock that would be only unpleasant for a person.

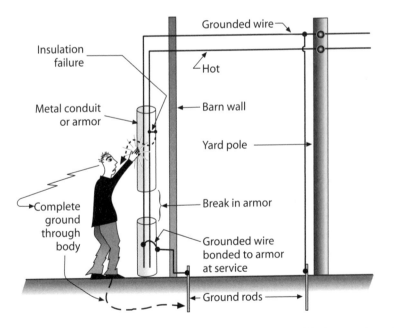

Fig. 22–9 If there were no break in the conduit or armor, ground-fault current would normally blow a fuse. Because the conduit or armor does have a break, a dangerous condition exists.

For these reasons never wire a barn using metal conduit, armored cable, or metal boxes, although the *NEC* now recognizes Type MC cable if it has an impervious nonmetallic jacket. Use only nonmetallic-sheathed Type NMC or Type UF cable, or rigid nonmetallic conduit with nonmetallic boxes (refer to the nonmetallic box discussion on page 153). You can also use Type SE cable, provided it is made up with copper conductors. The more usual aluminum type cannot be used. If you need a flexible wiring method, liquidtight flexible nonmetallic conduit can be used. Regardless of the wiring method selected, a separate copper equipment grounding conductor must be installed or be a constituent of the wiring method, and, if run underground, the wire must be insulated or covered. For cabled wiring methods, always secure them within 8 in. of the boxes or cabinets in which they terminate.

In the farmhouse, Type NM cable or any other wiring system may be used. If Type NMC cable is hard to find, as is the case in many localities, Type UF cable (which costs only slightly more than Type NMC) may always be used wherever Type NM or Type NMC may be used.

Poultry, livestock, and fish confinement systems   These areas, a subset of agricultural buildings generally, have been plagued by accumulations of litter and feed dusts, which are often corrosive and likely to infiltrate electrical components. These areas require dustproof and weatherproof enclosures in addition to the restricted list of wiring methods. Also, motors must be totally enclosed, or designed to prevent dust, moisture, and corrosive particles from getting inside.

In the installation of nonmetallic boxes there is a precaution to observe. Wood swells and shrinks in locations where dampness and humidity levels vary. Steel boxes, if solidly mounted on supporting timbers, do not present a mechanical problem as the timbers swell with moisture, because the steel can give a little if required. But nonmetallic boxes, if screwed down tightly on dry timbers, have been known to break out their bottoms as timbers swelled with increasing moisture. So if you mount nonmetallic boxes on the surface of dry timbers, leave just a little slack; don't drive the mounting screws down completely tight.

Wiring of barns   The physical makeup of the circuits, the combination of cable and boxes and wiring devices, is not different from that discussed in other chapters. The chief points to observe are practicability and common sense. Locate switches and receptacles where they cannot be bumped by animals in passing. But locate switches at convenient heights and locations that allow the farmer to operate them by the elbow when hands are full. A great convenience is to have three-way switches to control at least one light from two different entrances to the barn. Never use metal sockets; always use plastic or porcelain.

Barns come in all sizes and descriptions. Provide lighting outlets and receptacles in proportion to the need. Figure 22–10 shows some suggested wiring diagrams. It is wise to locate outlets for lamps between joists so the lamps do not project too far into the aisle between stalls where they might easily be damaged. Most barn ceilings are dark and dusty and reflect practically no light. About half the light from an exposed lamp is directed toward the ceiling and is lost. Provide a reflector for every lamp; the area underneath will be lighted almost twice as well as without a reflector. One type of reflector is shown in Fig. 22–11. They are reasonable in cost, and using them is a good investment. Keep reflectors clean.

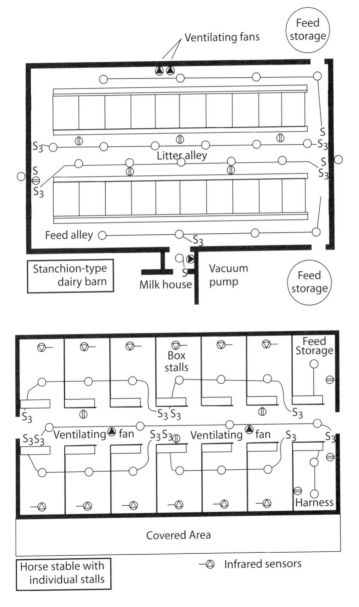

Fig. 22–10 Suggested wiring diagrams for farm buildings.

Fig. 22–11 A 60-watt lamp with a reflector is often as effective as a 100-watt lamp without a reflector. Reflectors must be kept clean.

Provide a light to illuminate the steps to the haymow. In the haymow itself, inspectors often require so-called vaporproof fixtures of the type shown in Fig. 22–12. A vaporproof fixture is simply a socket for a lamp with a tight-fitting gasketed glass globe that covers the lamp, and frequently a metal guard that protects the glass. Hay and the dust that arises in haymows are easily ignitable, even explosive. If an unprotected lamp is accidentally broken, the lamp burns out instantly, but during that instant the filament melts at a temperature above 4000°F. This flash can set off an explosion, hence the requirement for fixtures that "minimize the entrance of dust, foreign matter, moisture, and corrosive material."

Cable in haymows should always be installed where it cannot possibly be accidentally damaged. In haymows, cover the cable with strips of board at points where it might be punctured by hayforks. Where the cable passes through a floor, *NEC* 334.15(B) requires that it be protected by conduit or other metal pipe (or Schedule 80 PVC) to a point at least 6 in. above the floor. Many inspectors sensibly require this protection for about 6 ft, especially where there is danger of forks damaging the cable. Some inspectors require conduit or EMT or rigid nonmetallic conduit to extend from the haymow fixture to a point outside the haymow area. In such cases simply pull the cable through the raceway. This length of pipe or conduit does not make the system a conduit system; the pipe is used only for protection against mechanical damage to the cable. Be sure to terminate the pipe at an outlet where it can be grounded by bonding it to the equipment grounding conductor in the cable. The *NEC* does, however, allow "short" sections of metal pipe used for physical protection of an enclosed cable to omit a grounding connection. The inspector is the judge of the meaning of the word "short" in any particular application.

Cable in barns and other farm buildings should not be run along or across the bottoms of joists or similar timbers because this exposes the cable to mechanical injury. The cable will receive good protection if it is run along the side of a joist or beam. You have to purchase more cable in order to run it from the side of an aisle out to the middle for an outlet, then back to the side of the aisle, but consider the extra cost as insurance against damage. Figure 22–13 shows the details of recommended practice.

Fig. 22–12 A typical vaporproof fixture often used in haymows.

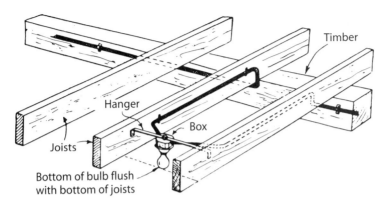

Fig. 22–13 Keep exposed cables away from centers of aisles. To avoid damage, it is best to keep lamps from projecting below the bottoms of joists.

Poultry houses—good lighting design   One point that should be particularly noted is that special wiring is frequently required for lighting designed to promote egg production. It is well known that hens produce more eggs during winter months if light is provided during part of the time that would otherwise be dark. It is best to provide light at both ends of the day, morning and evening. Opinions vary as to the ideal length of the "day" but a 14- or 15-hour period seems reasonably acceptable. Of course that means the length of time the lights must be on varies from season to season; adjusting the period every two weeks seems a reasonable procedure.

If all the lights are turned off suddenly in the evening, the hens cannot or will not go to roost, but will stay where they are. Therefore it is necessary to change from bright lights to dim lights to dark. This can be done by manually operated switches, but an automatic time switch of the type shown in Fig. 22–14 costs so little that manually operated switches should not be considered. In early morning, the time switch turns the bright lights on at a predetermined hour, then turns them off at daylight. In the evening, again at a predetermined hour, the switch turns the bright lights on. Later the bright lights go out and the dim lights come on; during this interval of dim lights the hens go to roost. Shortly afterward all lights are automatically turned off. The wiring for such switches is simple, and wiring diagrams are furnished with the switches.

Be sure that both the bright and the dim light fall on the roosts; if the roosts remain in darkness when the lights come on, the hens may not leave their roosts. Neither will they be able to find the roosts in the evening if the roosts are in darkness while the dim lights illuminate the rest of the pens. One 40- or 60-watt lamp for every 150 to 200 sq ft of floor area is usually considered sufficient for the "bright" period. Normal spacing is about 10 ft apart. For the "dim" period 15-watt lamps are suitable, but only about half as many as the number of "bright" lights

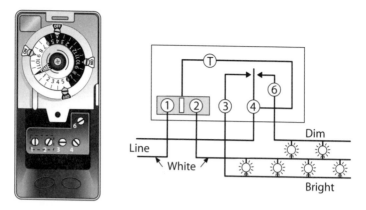

Fig. 22–14 A time switch automatically controls poultry-house lights, dim and bright, for forcing egg production.

are needed. It is a good idea to provide each light with a shallow reflector; otherwise a good share of the light falls on the ceiling and is lost.

Water pump—fire safety considerations   Every farm will have a water pump. It not only serves to provide water for all usual purposes but, in addition, is a tremendous help in case of fire. But in case of fire, quite often power lines between buildings fail and fuses blow so that the pump cannot run just when it is needed most. That failure can be avoided by adding an independent feeder to the pump, as if it were a fire pump. Simply arrange another feeder from the distribution point directly to the pump, preferably using an underground wiring method for reliability. Fire pumps are exhaustively covered in *NEC* Article 695, and given the fact that this pump would be routinely used for other purposes, it could not for many reasons technically qualify as such. However, the second feeder to the pump location would be permissible under *NEC* 225.30(A)(1), on condition of the provision of reciprocal labeling in the building served (see end of page 59).

Motors   Stationary motors must be wired as discussed in Chapters 15 and 30. For a portable motor of significant size, provide a receptacle with a rating in amperes and volts at least equal to the rating of the motor, and with a suitable horsepower rating, as covered in the page 378 appliance receptacle topic. One of the types shown in Figs. 20–6 and 20–7 will be suitable. If the receptacle must be installed outdoors, use a cast Type FS box and one of the covers for it, of the proper size for the receptacle involved; refer to the page 358 discussion of outdoor receptacles.

Yard lights   Every farm will have at least one yard light. It could be installed on the meter pole or on an outside wall of one or more buildings. The light should be controlled by three-way switches at house and barn. Note that the wires to the light must not be tapped off the service wires on the pole; the light may be mounted on the pole, but must be fed by separate wires from any branch circuit at house or barn. The yard light itself might be the old style of Fig. 22–15, or the modern type shown in Fig. 22–16. When using the type with an exposed lamp, the lamp itself will be one of the Type R or PAR discussed in Chapter 29. Such lamps have internal reflectors and are available in either floodlight or spotlight type, depending on whether a large or a small area is to be lighted. Be sure the lamp you use has a "hard glass" bulb which will not be damaged when cold rain hits the hot bulb. Lamps with bulbs made of ordinary glass shatter when hit by cold rain or snow and therefore should be used only indoors or where protected by an effective reflector.

Fig. 22–15 Old-style yard light.

A different and very modern yard light uses one of the types of high-intensity discharge (HID) lamps described in Chapter 29. Such lamps produce many times as much light per watt as ordinary lamps. They cost about five times as much as ordinary reflector lamps of the hard-glass variety, but last at least twelve times longer. They require special fixtures. When used for yard lighting, they are often equipped with photoelectric controls that auto-matically turn the light on in the

Fig. 22–16 Photo-cell controlled, high pressure sodium yard lights contribute to convenience and safety on the farm.

evening and off at daylight. In arranging photocell control, be sure the switching system complies with *NEC* 410.54(B) and opens all ungrounded conductors to the lamp ballast. If the photocell operates only as a single-pole device, it can operate a relay but it cannot be used to control one of these fixtures connected line-to-line. In addition, note that there isn't any height limitation on the grounding provisions in *NEC* 410.18(A). Unless the inspector judges the instal-lation inaccessible to unqualified personnel, you will need to provide an equipment grounding conductor to this light.

Outdoor switches and receptacles   On farms, switches and receptacles often must be installed outdoors exposed to the weather, and special means must be used to protect them. The correct methods are explained in the page 358 discussion of outdoor and other wet locations.

CHAPTER 23

On-Site Engine Power Generation and Supply of Premises Wiring

YOU WILL FREQUENTLY ENCOUNTER engine generators in many guises. When a utility failure occurs, some loads have to stay on, and an on-site generator is what usually serves the purpose. This chapter looks first at how an engine generator fits into a safe electrical system designed for the convenience of its owner. The chapter then looks at how the electrical system differs when used for emergency or legally required purposes, and concludes with a brief look at an engine generator used in parallel with the utility. On-site power sources that don't involve engine generators, such as solar photovoltaic and wind power, are beyond the scope of this book.

ENGINE GENERATORS AND OPTIONAL STANDBY SYSTEMS

Isolated power plants are used in locations where electric power is not available from commercial power lines. Examples are remote production or camp areas, repair or construction crews, boats, and mobile equipment such as used on fire-fighting vehicles. Consider the terminals of the generator as the SOURCE, and wire the premises substantially as if served by a commercial power line; observe the safety requirements practiced in ordinary wiring. Some but not all engine-driven plants could be used as standby plants or emergency plants as well as isolated plants.

NEC Article 702 covers *optional standby plants,* which generally means plants installed to serve economic purposes rather than primarily for the safety of people. An interruption of power for a few hours in a home can be a considerable nuisance, but is not necessarily an emergency. In a chicken hatchery, a few hours of interruption might ruin a large load of eggs. In a greenhouse with electrically controlled heating equipment, loss of power could drop the temperature very low with consequent great loss. Broadcasting stations need standby protection. On farms, power outages are often measured in days rather than hours; without power, the milking machines, pumps, water heaters, and building heating equipment are usually out of operation. Continuous power is important in many hundreds of circumstances, and the only guarantee against an interruption is a standby generating plant (or an emergency generating plant, as defined later in this chapter). A typical installation of an engine-driven generating plant is shown in Fig. 23–1.

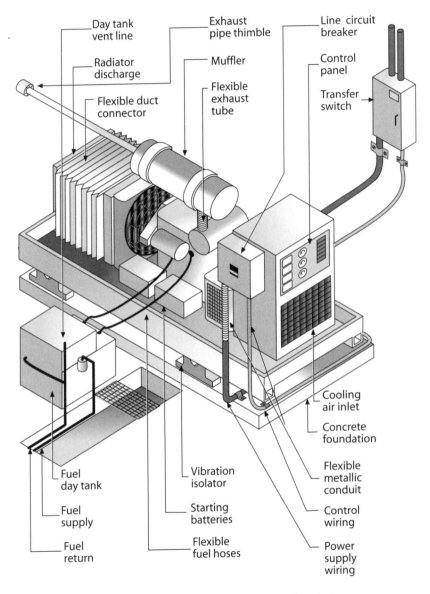

Fig. 23–1 Typical installation of a large engine-driven generating plant.

Power source and location of installation   The most common generating plants are powered by diesel or gasoline engines, the latter also being available for operation on natural or other gas. The plants typically produce 60-Hz power (or 50-Hz if required for foreign countries), in capacities from 500 watts to 500 kW or more. They are available in any required voltage, single- or three-phase. Figure 23–2 shows some of the enormous variation in plant sizes available.

Fig. 23–2 Standby generating plants developing 12 kW and 750 kW of power, single-phase or three-phase, of any standard ac voltage. *(Onan Corporation)*

The plant must be installed where the engine will cool properly. A small room is not considered a good location unless means are provided for adequate circulation of air to carry away the heat created by the engine. The fuel supply system must be considered. If gasoline is used, an underground tank outside the building might provide the most convenient storage, but environmental regulations could require a double-walled tank or make it otherwise problematic. If you have natural or other gas in your building, use it for fuel in order to avoid storing gasoline. Depending on the richness of the gas in your particular area, the output of the plant might be reduced somewhat. In the case of an emergency system the fuel supply must be on site unless the inspector judges that the simultaneous failure of both the electric utility and the public utility gas system is unlikely. This approval may not be granted in, for example, earthquake-prone areas.

Starting methods   Engine-driven plants must be provided with means for starting the engine. The hand-starting method using a rope or crank is practical only with the smallest plants. Usually a 12-volt, heavy duty automotive-type battery is used to crank the engine. Smaller plants have a special starting winding in their ac generators, which also contain a dc winding for keeping the starting battery charged. On larger plants the starting method is substantially the same as in an automobile.

Start-stop push buttons on the plant allow control. Usually additional control buttons may be installed at a distance from the plant, up to several hundred feet away, connected using inexpensive control wire similar to that used in wiring door chimes, thermostats, etc. The voltage in the wire is never more than the voltage of the cranking battery. There is also a "full automatic" plant (only in the smaller sizes up to 10 kW) that starts when a light or an appliance is turned on and stops when everything is turned off. This type of control is especially popular on boats.

Load transfer equipment—essential to safety   A generating plant must never under any circumstances be installed in a way that allows its power to feed back into the utility company's line at any time. That can be extremely dangerous to line crews who are trying to repair the utility lines. Several deaths of line workers have been traced to their working on such supposedly dead lines that were energized by haphazardly installed standby plants. Remember that the same transformer that reduces 2400 volts to 120 or 240 volts also steps 120- or 240-volt power from a generating plant up to 2400 volts if the line is not disconnected while the plant is in operation.

In every case install a double-throw switch. In one position the switch connects the load to the power line; in the other position it disconnects the load from the power line and transfers it to the generating plant. If you have 120/240-volt service, you may usually use a three-pole, two-wire, solid-neutral switch, as shown in Fig. 23–3. In some localities this is prohibited and you must use a three-pole, three-wire switch that does not have a solid neutral; the neutral conductor is then switched along with the hot wires as shown in Fig. 23–4. The type with all poles switched is frequently preferable (see discussion in later paragraphs). Such switches can be either hand-operated, as shown in the illustrations, or electrically operated in the case of automatic equipment. Trace the wiring in Figs. 23–3 or 23–4 and you will see there is no possible way for the generating plant to feed power back through the transformer into an otherwise dead line.

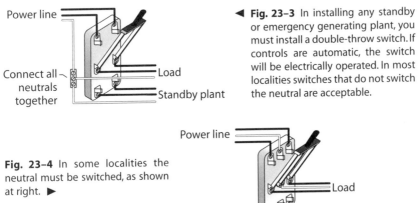

Power line

Connect all
neutrals
together

Load

Standby plant

◀ **Fig. 23–3** In installing any standby or emergency generating plant, you must install a double-throw switch. If controls are automatic, the switch will be electrically operated. In most localities switches that do not switch the neutral are acceptable.

Power line

Fig. 23–4 In some localities the neutral must be switched, as shown at right. ▶

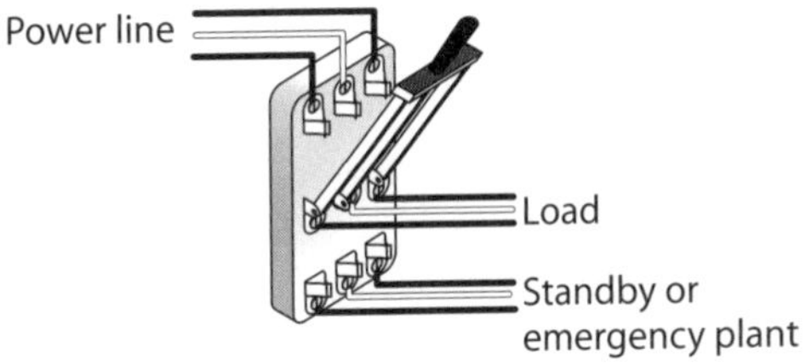

Load

Standby or
emergency plant

Match capacity to load   Where an optional standby system is to be wired, the *NEC* requires only that it have enough capacity to handle all the load the owner chooses to have remain operational at one time. The owner decides what those loads are to be, but the decision should be reflected in the permanent wiring on site. This is done by creating a standby panelboard that includes the loads the owner has determined are essential to remain energized. Then, knowing what loads will be represented and using load calculation provisions generally as presented in *NEC* Article 220, determine the required electrical demand for the standby panelboard. Compare this result with the proposed rating of the generator and the transfer equipment and make sure they both are up to the job. If not, either decrease the anticipated load or increase the intended plant and equipment accordingly.

Following are the steps in planning the installation, beginning with deciding where to locate the transfer switch and whether it will be manual as shown diagramatically in Figs. 23–2 and 23–3, or whether it will be automatic. The owner needs to make a cost versus convenience versus reliability decision at this point.

Advantages of automatic control   In many applications, including all those that the *NEC* classifies as "emergency" installations (covered later in this chapter), there is no choice—the transfer switch must be automatic. If the lights in a hospital operating room were to be out for even a few minutes during a surgical operation, the delay could be serious. If installed to provide power for heating systems, the automatic control provides power even at night or over the weekend when no one is on the premises. Automatic installations are often made in isolated locations such as in microwave transmitting stations where there may be no caretaker present for weeks at a time.

A standby plant is rarely large enough to deliver all the power needed in the place where it is installed. In a chicken hatchery, it might be big enough to take care of specific loads (as with emergency plants). In other cases, automatic electric switches can be installed to disconnect selected loads from the line when the power fails, but not reconnect them automatically when the voltage is restored by the generating plant. The standby plant then supplies only the remaining loads.

In addition to the basic advantage of being completely automatic, such transfer equipment also provides other conveniences. Among them are automatic charging of the cranking battery; test switches to test the equipment and take over the load or not, as desired; "exercisers" to run the plant for a short time daily or on selected days during the week without taking over the load in order to keep the engine always ready to start instantly.

Should you transfer the neutral?   The next step in planning the installation is to decide if the neutral will be transferred or not. Transferring all conductors is widely done on very large systems with permanently installed prime movers because it prevents any fault currents that develop from entering stray pathways in the parallel system. To use a small-scale example, suppose the basement freezer is selected to be on the standby system, and its cord gets pinched and shorts to ground on normal power. That fault current returns to the standby panel and then to the service panel over the following path (see Fig. 23–5): equipment grounding conductor in the branch circuit, to transfer switch, to normal feeder to the transfer switch, to normal source through the main bonding jumper in the service panel.

If the transfer switch does not transfer the neutral and if the generator is permanently connected, there are two possible outcomes, both of which adversely affect grounding performance under fault conditions, especially on large systems. Referring to Fig. 23–5, if the generator bonding connection is removed, any fault current flowing *during generator operation* must leave the transfer switch over normal system connections and pass through the main bonding jumper, returning on the normal feeder neutral to the transfer switch, and then to the generator source, all of which adds impedance to the fault return path. If the generator bonding connection is not removed, the result is even worse *during normal operation.* Some of that fault current can take another route through a parallel path, diverging from the previously described path in the transfer switch enclosure, as follows (see Fig. 23–5): equipment grounding conductor of the standby system feeder, to bonding jumper in the generator, to standby feeder neutral back to the transfer switch, and then over the normal system neutral conductor to the utility source.

The effect is to return some of the fault current over a downstream feeder neutral conductor instead of over the normal equipment grounding path. This is a significant problem on large systems that utilize special sensors at the service equipment to determine if a severe ground fault is in progress, because any fault return current that arrives over the normal system neutral tends to fool the sensors. Fault current returning on a neutral looks like ordinary load current to these sensors. If they don't respond appropriately under fault conditions, the result is likely to be a fire. For more information, review the discussion of ground-fault protection beginning on page 544.

For a typical simple farm or one-family dwelling, particularly when the generator isn't permanently connected, there might be little risk in not transferring the neutral. However, local or utility policies might preclude the solid neutral connection. Consult your local authorities early in the planning process. Always remember that if the neutral is not transferred, the generator connection utilizes the existing grounding system in the building. Referring again to Fig. 23–5, that grounding connection will be at a remote location, in this case at the service equipment. If someone unfamiliar with the system attempts to work on the service equipment with the generator running and opens the feeder neutral connection, the consequence is that the generator system now runs without any

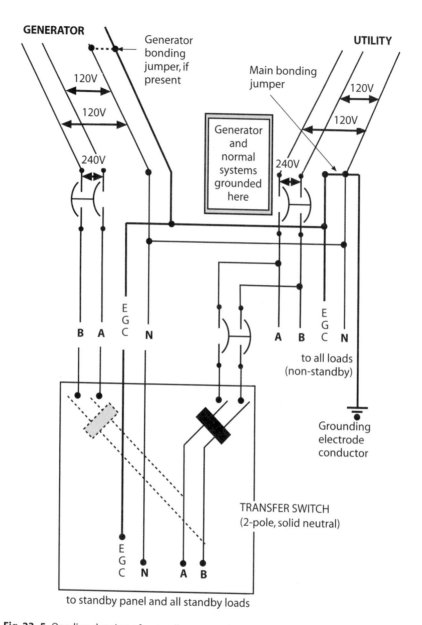

Fig. 23–5 One-line drawing of a standby system that is not a separately derived system.

ground reference, an extreme hazard. For this reason the *NEC* requires a sign that identifies all standby and normal sources making use of the grounding electrode conductor connection at a given point. For the installation depicted in Fig. 23–5, that sign would be placed at the service equipment.

If the neutral is transferred, there are no conductors in common between the generator source and the normal utility source. This is a separately derived system, and the *NEC* requires separately derived systems to have all the essential components of a normal system. Ahead of the transfer switch you would need to install a main disconnect complete with a main bonding jumper (see page 132 discussion of main bonding jumper; this jumper serves a similar function) and an independent connection to the building grounding electrode system. Many generators include some or all of these functions, but the grounding electrode system connection obviously needs to be done on site. See Fig. 23–6.

Decide between permanently installed or portable   The final planning step is to determine how both the normal and the standby feeder conductors are going to arrive at the transfer switch. For example, is the standby plant permanently installed and hard wired? Or is it on wheels and connects through a cord? Very common users of optional standby systems are municipal public works departments that have to keep sewage lift stations operational during utility outages. If those stations have an inlet for a cord connection from a generator, the crews can bring a generator from station to station and empty the ejector pits.

Optional standby systems with either permanently installed prime movers or portable alternate power supplies are covered in *NEC* Article 702. A failure to install transfer switches for all optional standby systems, including those not permanently installed, will endanger the lives of utility line crews as just described. The owner cannot be relied on in all cases to remember to shut off the normal main breaker before operating the standby plant. For these reasons, a transfer switch is now required whether or not the standby source is portable.

Wiring for standby generator installation   The connection from source to building disconnect involves different considerations for permanently installed plants and those that are cord- and plug-connected. Careful labeling of the disconnecting means is important. Finally, the load transfer switch is wired. The critical role of this transfer switch is discussed earlier in the chapter.

Wiring from source to disconnect   If the standby plant is to be hard wired, select any appropriate wiring method and install a feeder from the generator output terminals to the building disconnect (discussed under the next heading). If it will be cord- and plug-connected, special rules apply to that connection beginning with exactly what is plugged into what. For providing wiring access on the outside of a building we normally think in terms of a receptacle. But this would be a receptacle energized only when the plug fed from the generator was plugged in—the reverse of the usual situation. And the *NEC* absolutely forbids using the energized blades of a plug as the source of supply to a receptacle; the potentially exposed, energized blades of an attachment plug present a serious shock hazard. The solution is to use a "flanged inlet" as in Fig. 23–7. This is a recessed male plug that accepts a receptacle cord body. In effect, it allows you to

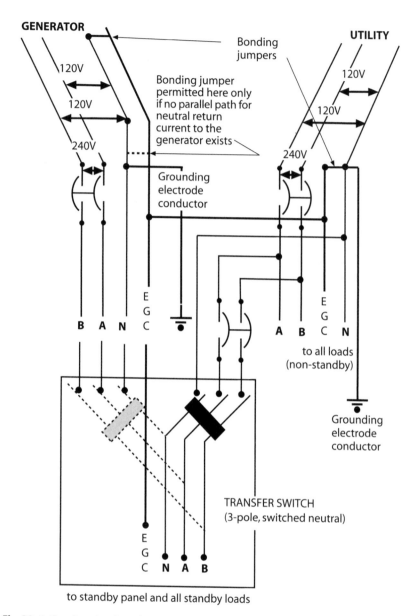

Fig. 23–6 One-line drawing of a separately derived standby system.

plug in the house. Size the cord to 115% of the current the generator is capable of supplying (100% if it meets the exception in *NEC* 445.13) based on its nameplate rating, and do the same for the flanged inlet and receptacle cord body.

Fig. 23–7 Flanged inlet and the mating receptacle cord body that could energize it. *(Pass & Seymour/LeGrand)*

Disconnecting means   The generator supply constitutes a source of power to the premises, and as such, all the rules that apply to multiple supplies to a building also apply here. (See the paragraph on principles applying to building disconnects on page 60 and the topic on garages and outbuildings on page 334.) Install a main switch or circuit breaker sized per system requirements and located at a readily accessible point nearest the point of entry of the conductors or the flanged inlet. Provide a label on this device explaining its function. If it is separated from the service, add to the label the fact that it is one of two power source disconnects and specify the location of the service source; at the service disconnect provide a reciprocal label identifying the existence and location of the standby source disconnect. If the generator will be functioning as a separately derived system, bring a grounding electrode conductor to this point from the grounding electrode system in the building and connect the main bonding jumper. If the generator has a disconnecting means on it that is listed as suitable for use as service equipment, and if it is located close to the building, the inspector may be willing to qualify it as a building disconnect. If it is thus qualified, and the system will be separately derived, extend a grounding electrode conductor to the generator disconnect and make the connection there. Don't forget the reciprocal labeling.

Wiring the transfer switch   Select any appropriate wiring method and install the feeder from the generator supply to the transfer switch. Add a similar feeder from a multipole breaker in the normal system panelboard to the transfer switch. Then add the feeder from the transfer switch to the standby panel. The standby panel will normally require individual protection in accordance with its rating. The feeder protective devices in both sources should provide that protection, but if for any reason either source feeder won't protect the panel while that source is functioning, use a panel that includes a main breaker.

Tractor-driven generators   For standby use on farms, tractor-driven generators are often used. They are available in sizes up to 25 000 watts. Naturally they cost less than generating plants complete with their own engines. They do have the disadvantage of requiring time to line up the tractor with the generator after spotting the generator in the right location; they also preclude the use of the tractor for other purposes during the power failure, which on farms sometimes lasts for extended periods.

This point is so important it bears repeating: Never install a generator, whether with its own engine or tractor-driven, without installing a double-throw transfer switch. That is an absolute necessity for safety purposes.

In matching the generator with your tractor, be sure your tractor has sufficient horsepower at the power takeoff shaft *at the speed required by the generator*. It should have at least 2 hp for each kilowatt of generator capacity; 2½ hp for each kilowatt will enable you to take better advantage of the generator's overload capability. If you don't have enough horsepower in the engine, it will slow down and so will the generator. The voltage, the frequency, and the capacity of the generator will drop off rapidly.

Battery-powered systems   Many years ago when few farms had commercial power available, many of them had their own engine-driven dc generator that charged a battery, usually 32 volts. Such systems necessarily had a very limited capacity to operate only a small number of lights and very few, if any, appliances. Today the use of such plants is rare. But in nonfarm applications there are some uses where direct current is the normal power, such as for telephone and other communications systems and for signaling purposes. Storage batteries of proper voltage are used, and they are kept charged from the usual ac supply. To protect against failure of the ac power, an ac standby generating plant is used to power the battery-charging apparatus.

EMERGENCY AND LEGALLY REQUIRED STANDBY SYSTEMS

The term "emergency" is one of the most misunderstood electrical terms applied to circuitry. Although all emergency systems are standby in character, there are also many standby systems that are not emergency systems within the meaning of the *NEC* even though their owners might consider them to be emergency systems. These systems, whether legally required or optional, are subject to far different requirements and typically include many features that, while perfectly acceptable for their intended use, would never qualify for inclusion in a true emergency system.

A standby power source is either required or not required by a statute or legally enforceable administrative regulation. Emergency systems and legally required standby systems are the same to the degree that both are legally required. If a standby system is not legally required, then it is not an emergency or legally required standby system, no matter how potentially devastating the failure of the system might be. The owner's engineers must design adequate performance into the system; the *NEC* will not do it for them.

If you know that the standby system is required by law, classify it as either an emergency system or simply a required standby system—though determining which is sometimes problematic. In some cases, the applicable regulation stipulates the specific *NEC* article, making the job easy. In some other cases, it is difficult to decide which article to follow because there is some overlap in the *NEC*. The most important criterion in distinguishing between the two types is the relative length of time that power can be interrupted without undue hazard. Here are some admittedly overlapping considerations. First, what sort of occupancy is involved? Does it involve large congregations of people in a setting conducive to panic? High-rise buildings and large auditoriums typically require emergency systems. Second, how critical is a particular load to safety? Exit signs and egress lighting are usually objects of emergency-system mandates. Third, how much danger to personnel would be involved if a load were interrupted? For example, if cooling were interrupted to a chemical reaction that would quickly turn violent in the absence of coolant, the coolant pumps might be classified as emergency loads, assuming a regulatory authority had issued the appropriate orders.

Emergency plants   Per *NEC* Article 700, an emergency power plant (as distinguished from a legally required standby plant) is installed primarily for protection of *people,* although it may serve other purposes as well. Emergency plants are installed in locations such as hospitals, where a loss of power in a critical location such as an operating room could be a catastrophe. In a windowless workplace, in a theater, in a store, or similar location, complete darkness due to power failure might easily lead to panic with consequent injuries or even death. In *NEC* language, emergency systems are "generally installed in places of assembly where artificial illumination is required for safe exiting and for panic control in buildings subject to occupancy by large numbers of persons, such as hotels, theaters, sports arenas, health care facilities, and similar institutions."

Where an emergency generating plant is required, the *NEC* specifies that the plant must provide full power within 10 seconds of the power failure; and the plant once started by a loss of power must continue to run for 15 minutes after power is reestablished (after power has failed it is often restored for only a few seconds or minutes before it fails again—this provision prevents the plant from starting and stopping repeatedly within those 15 minutes). Naturally every installation must include automatic means for starting the plant and transferring the load to it upon failure of commercial power.

The output of an emergency generating plant must be used only to supply special emergency circuits to selected loads to be operated during the emergency, entirely independent of the usual wiring. Wires of the two systems must never enter the same raceway, cable, box, or cabinet of the usual wiring, except in exit or emergency lights supplied by both sources. Naturally wires from both systems must enter the automatic transfer equipment, however, the transfer switch may supply only emergency loads. The owner who wants to oversize the generator to also supply nonemergency loads may do so, but those loads must be supplied through a different transfer switch, and automatic load shedding must be established so those loads drop off line in the event the generator approaches overload from the emergency system demand. The generator can, on the same conditions, also be used to reduce the load taken from the utility.

Legally required standby systems   The key to separating legally required standby systems from emergency systems is the length of time an outage can be tolerated. The *NEC* includes a very useful note that gives a good sense of the typical loads it anticipates will be supplied by such systems. "Legally required standby systems are typically installed to serve loads, such as heating and refrigeration systems, communication systems, ventilation and smoke removal systems, sewerage disposal, lighting systems, and industrial processes, that, when stopped during any interruption of the normal electrical supply, could create hazards or hamper rescue or fire fighting operations." In contrast to emergency systems, legally required standby systems may take a full 60 seconds to restore power. In addition, legally required standby system circuits need not be isolated from raceways containing normal circuits.

Although less critical in terms of restoration time, these systems may be very critical in terms of long term environmental consequences. They are also directed at the performance of selected electrical loads instead of the safe egress of personnel. For example, numerous regulations governing large sewage treatment facilities require standby power. The facility must remain in operation in order to prevent environmental problems, and incidental to that requirement, some lighting may need to be assured. However critical, this is far different than lighting for emergency egress, and for that reason a longer time delay is allowed between loss and restoration than is allowed for emergency systems.

INTERCONNECTED POWER PRODUCTION

Engine generators are also used to operate in parallel with the utility, that is, to do deliberately what must never be done inadvertently. This process is usually known as "cogeneration." Cogeneration facilities are the subject of *NEC* Article 705. These facilities produce both electric power and some other useful form of energy, such as heat. For example, why run a generator as simply a standby generator when you could run it all the time, or on the call of the local utility, and capture the waste heat as domestic hot water or some other useful energy storage? Cogeneration facilities run in parallel with the utility and reduce the total plant power costs while reducing the need to expend energy from outside sources for the other uses. The design of these systems is beyond the scope of this book, but to get an overall picture you should be aware of some of the more important requirements. First, these systems must be arranged so that if the primary source fails, they must be immediately disconnected from the primary source until that source is restored. This need not involve transfer equipment if the cogenerator will simply stay down until the utility returns. Some cogeneration equipment involves ac induction generators that must not be operated as an island, and the control equipment for these simply takes them off line completely until utility power returns. However, if you chose to install equipment capable of independent generation, you would need transfer equipment to prevent parallel operation when power fails.

Cogeneration output must normally come into the premises system "at" the service—the rule purposely avoids saying whether on the line or load side. Utilities differ as to their preference on this point. For large systems the point of entry can be at downstream locations. Large cogeneration systems in this case consist of those operating at over 1000 volts, or those with a capacity above 100 kW. There must be qualified personnel available for service and operation, and suitable protective equipment and safeguards must be in place. The generator output must be controlled so that it is "compatible" (not necessarily identical) with the utility system voltage, wave shape, and frequency.

CHAPTER 24
Manufactured Homes, Recreational Vehicles, and Parks

UNTIL IT ARRIVES ON SITE, the interior wiring of the type of housing discussed in this chapter is entirely under the control of its manufacturer, who has to meet national standards before shipping the products in interstate commerce. In order to benefit this industry, which would be adversely affected by having to meet different standards in every state and major municipality, the U.S. Congress invoked the supremacy clause in the U.S. Constitution and declared that the regulations made by the U.S. Department of Housing and Urban Development would supersede those of any locality. Those national regulations are based on the *NEC* in large degree, but not entirely, and certainly not contemporaneously with local adoption of a specific edition. Since the interior wiring is not done on site, it is not covered here. This chapter explains what this housing is and is not, so you can properly classify what you're working with. Nonresidential uses of these items are discussed at the end of the chapter. The main portion of the chapter focuses on wiring that is done by field electricians: service wiring where allowed, feeders and power outlets at the various specific sites, and overall park wiring.

DEFINING THE TERMS USED IN THIS CHAPTER

It is important to distinguish carefully among the *NEC* definitions for the types of housing covered in this chapter, because different requirements apply to each. Although manufactured buildings are not the focus of the chapter, their description is included here because they must never be confused with manufactured homes.

Manufactured building—Any building that is of closed construction and is made or assembled in manufacturing facilities on or off the building site for installation, or assembly and installation, on the building site, other than manufactured homes, mobile homes, park trailers, or recreational vehicles. (*NEC* 545.3)

Manufactured home—A structure, transportable in one or more sections, that is 2.5 m (8 body ft) or more in width or 12 m (40 body ft) or more in length in the traveling mode or, when erected on site, is 30 m² (320 ft²) or more; which is built on a chassis and designed to be used as a dwelling, with or without a permanent foundation when connected to the required utilities, including the plumbing, heating, air-conditioning, and electrical systems contained therein. (This is only the first sentence; the definition in *NEC* 550.2 now includes additional details regarding how these dimensions are to be measured.)

Mobile home—A factory-assembled structure or structures transportable in one or more sections that is built on a permanent chassis and designed to be used as a dwelling without a permanent foundation where connected to the required utilities, and includes the plumbing, heating, air-conditioning, and electric systems contained therein. (*NEC* 550.2) For the purpose of the latest *NEC* and unless otherwise indicated, the term "mobile home" includes manufactured homes.

Park trailer—A unit that meets the following criteria: (1) Built on a single chassis mounted on wheels, and (2) having a gross trailer area not exceeding 400 ft^2 (37.2 m^2) in the set-up mode. (*NEC* 552.2)

Recreational vehicle—A vehicular-type unit designed primarily as temporary living quarters for recreational, camping, or travel use, which either has its own motive power or is mounted on or drawn by another vehicle. The basic entities are travel trailer, camping trailer, truck camper, and motor home. (*NEC* 551.2)

Recreational vehicle park—A plot of land upon which two or more recreational vehicle sites are located, established, or maintained for occupancy by recreational vehicles of the general public as temporary living quarters for recreation or vacation purposes. (*NEC* 551.2)

"Manufactured homes" were called "mobile homes" when the concept was first addressed in the 1965 *NEC*, and the term still remains in Article 550. When this type of housing originated, the buildings were somewhat smaller and more mobile than they are today, and for a long time the Code-making panel didn't want a service brought on to something it considered to be capable of comparatively easy movement. The federal government, seeing these units as a way to generate low-cost home ownership, set about removing certain popular stigmas; it changed the terminology by regulation and declared that services didn't have to be off the buildings, as had always been the case with mobile homes. In fact today these housing units are rarely moved. Manufactured homes still, however, include the term "chassis" in their definition, and they significantly differ from "manufactured buildings," even manufactured buildings designed as dwelling units, popularly referred to as "prefabricated."

A "manufactured building" has no chassis and cannot be transported on running gear. It is delivered to the site in sections on flatbed trucks and the sections are put into place with a crane. In terms of regulations, it must meet the building and electrical codes of its destination municipality. *NEC* Article 545 includes no special rules to govern receptacle placements or calculate loads. Probably the most significant concession from normal wiring protocol is the *NEC* 545.13 allowance for field-installed splicing devices that do not have to be in a box and accessible after installation. This allowance allows circuiting to proceed in a logical way between sections that must be set in place separately.

"Recreational vehicles" (called RVs in the rest of this chapter) are, on the other hand, truly transient. In addition to the popular term "motor home" this category includes camping trailers and travel trailers, which are towed to the campsite, and truck campers, which travel in the bed of a pick-up truck. The 1996 *NEC* added a new article on "park trailers" to cover a larger version of travel trailers. Travel trailers don't require special highway movement permits and can't exceed 320 sq ft. Park trailers can be as large as 400 sq ft; nevertheless, they are designed for seasonal use only, and they can't be used for permanent habitation or for commercial purposes. *NEC* Article 552 has no rules governing sites for park trailers, some 10 000 of which are coming off assembly lines each year and are presumably being set up in RV parks. No demand provisions in *NEC* 551.73 address unit feeders of indefinite capacity above 50 amps 120/240 volts, however, and park trailers, like mobile homes, can be supplied with a straight feeder instead of a cord-and-plug connection.

MANUFACTURED (MOBILE) HOMES AND PARKS

Just as for conventional residential installations, the minimum required service for manufactured/mobile homes is 100 amps. A mobile home park supplies 120/240-volt, single-phase power to the homes located within it. Explanation is provided for determining minimum feeder size, taking into account the demand factor based on the number of homes served by each feeder.

Three ways of supplying service to manufactured homes   Manufactured homes are permitted to have and now usually do have service equipment right in or on the home, provided the home is on a permanent foundation. The equipment has to comply with *NEC* Article 230 as interpreted by the local inspector, and the home manufacturer must provide means, such as an empty conduit, to route a grounding electrode conductor out of the service equipment and to the exterior of the structure. Together with the service equipment, or in lieu of it if fed from a cord or a feeder, the manufactured home includes a distribution panel with a single disconnecting means and main overcurrent protection, plus additional overcurrent protection for all the branch circuits.

In cases where the service is not mounted in or on the home, normally the home would be supplied by a feeder connected to a similar distribution panel and rated in accordance with the load. The neutrals of these panels must never under any conditions be grounded to the cabinets. They must be fully insulated just as the hot wires. The service equipment itself (consisting of main overcurrent equipment and disconnecting means, such as a circuit breaker or fuses on a pull-out block) would then be installed away from the home in a raintight housing. It may be installed on a pole and fed by overhead wires, or installed on a metal pedestal and fed by underground wires. The service equipment must be 120/240 volts and rated at not less than 100 amps. It must contain provision for connecting the feeder to the home by a permanent wiring method, and may and usually does contain receptacles into which cords from the home are plugged.

Smaller homes that do not require more than 50 amps at 120/240 volts may be equipped with a permanently attached cord with a plug to fit the 50-amp receptacle in the service equipment. The cord must be from 21 to 36½ ft long (no extension cords allowed). Whichever method is used for service not mounted in or on the home, the feeder from remote supply equipment to the home must consist of four wires: two hot wires, an insulated white neutral wire, and an insulated green equipment grounding wire.

Minimum service equipment is 100 amps   A typical service equipment is shown in Fig. 24–1. It is rated at 100 amps and contains one 50-amp receptacle and one 50-amp breaker as already described, also a 20-amp, two-pole, three-wire, 125-volt receptacle of the grounding type (either 15- or 20-amp plugs will fit) protected by its own circuit breaker and GFCI. Note there is space to install a 100-amp, two-pole, 125/250-volt breaker at a future date if the home's requirements increase. In that case the 50-amp receptacle and 50-amp breaker would be disconnected, and the home would be served by a direct connection using no cords.

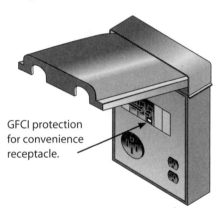

GFCI protection for convenience receptacle.

Fig. 24–1 Basic service-entrance equipment for a mobile home. Such equipment must have minimum rating of 100 amps (the terminals must accept wires with an ampacity of at least 100). The receptacle must have GFCI protection.

Note that while the service equipment must be rated at a minimum of 100 amps (having terminals suitable for wire with an ampacity of at least 100 amps), any receptacle and the breaker protecting it may not exceed a 50-amp rating. (Larger breakers may be installed if the home is not served by cords but by direct permanent connection to terminals in the service equipment.) Each service equipment must have a grounding busbar for the neutral of the cord, the grounded wire of the smaller receptacle(s), and the green grounding wire of the cord. It must be bonded to its cabinet, and a ground wire run from it to an underground piping system if available, or to a ground rod.

Service equipment with optional meter socket   Figure 24–2 shows this type of equipment; note that it contains a meter socket (which is optional on most service equipment for individual mobile homes). It is installed on top of a metal pedestal to be supplied by an underground feeder; the pedestal is optional on all similar units. The particular equipment shown is rated at 200 amps. It contains the usual 50-amp receptacle for one cord and a 50-amp breaker to protect it, and a 20-amp, two-pole, three-wire, 125-volt grounding-type receptacle protected by a separate breaker and a GFCI. It contains space for a 200-amp, two-pole circuit breaker that can be installed at a future date for a home with a permanent connection and no cords. Though not visible, the unit contains terminals for "loop feed." Wires from the underground feeder to the service equipment end at a set of terminals in the equipment, and wires from another set of terminals are then run to the next service equipment, eliminating splices within the equipment.

Parks for mobile homes   A mobile home park is a parcel of land used for the accommodation of occupied mobile homes. The park contains a park office, and often facilities such as laundry equipment provided for common use. A mobile home lot is that space within a park for one mobile home and accessory buildings or structures for the exclusive use of the occupant.

Service equipment for entire mobile home park The service equipment for the entire mobile home park must be single-phase, 120/240-volt. The details vary with the size of the park. In a very small park it might be located in the park office, with feeders to groups of individual lots. In most parks the supply to the park may be at a voltage higher than 240 volts, but with this higher voltage running only to a number of separate transformers each delivering 120/240 volts. Each transformer serves a feeder for a number of homes, just as in city wiring where one transformer serves a number of houses. The number of feeders required is determined by the number of homes in the park. To minimize voltage drop, it is best to install separate feeders for relatively small groups of homes. The feeders might be overhead, in which case service drops to individual lots would be much as in wiring a group

Fig. 24–2 A larger service entrance rated at 200 amps. It is equipped with a meter socket and supported by a metal pedestal for use with underground wiring.

of farm buildings (refer to the discussion on locating the meter pole beginning on page 417), but underground feeders are more desirable.

Calculating size of feeders for mobile home parks   How big must each feeder be? Start with an assumed load of 16 000 watts for each home, at 120/240volts. As an alternative, if you know that park policies preclude homes larger than a certain size, you can do a calculation in accordance with *NEC* 550.18 and use those results on a per-home basis instead. Then apply the demand factors of *NEC* 550.31.

The demand factor for a single home is 100%, but it drops off rapidly as the number of homes increases. For example, the demand factor for 5 homes on 1 feeder is 33%; the minimum ampacity of the wires in the feeder for 5 homes would be

$$\frac{(16\,000)(5)(0.33)}{240} = \frac{26\,400}{240} = 110 \text{ amps}$$

Similar calculations show that the minimum ampacity for a feeder for 10 homes would be 180 amps.

Now suppose you have a park with 20 homes and 4 feeders, one for each 5 homes. Will the service conductors for the entire park have to be 4 times the 110 amps of each feeder or 440 amps? Not at all. Look at *NEC* 550.31 again. The demand factor for 20 homes is 25%. So the minimum ampacity of the main service conductors must be

$$\frac{(16\,000)(20)(0.25)}{240} = \frac{80\,000}{240} = 333 \text{ amps}$$

Actually they should be somewhat larger than that to take care of the park office and other requirements in the park for equipment that perhaps is not installed in the individual homes.

RECREATIONAL VEHICLES AND PARKS

RVs may have ac or dc systems, or they may have both for maximum flexibility. RV parks may now have both three-phase and single-phase power, although individual sites will always be served with a single-phase feeder. The individual RV accesses the park power supply through a power outlet that must meet special wiring and siting requirements. Explanation is provided for determining minimum required feeder size—with a handy trick for doing the necessary three-phase calculations.

Three ways of supplying service to recreational vehicles   RVs equipped with wiring for use at 120 volts have electric power only while in an RV park. These RVs are usually equipped with a permanently attached cord that must be a minimum length of 23 ft for a side entry or 28 ft for a rear entry. The cord must be sized according to the electrical load in the RV, and be equipped with a plug of the type and rating based on the load.

Some RVs are equipped only with a battery system, usually 12 volts and never over 24 volts, entirely apart from the battery of the vehicle on which it is mounted or towed. The amount of power from such a system is very limited. The battery is kept charged from the usual 120-volt power. RVs equipped with both a 120-volt system and a battery system can have power available whether in or away from the RV park. The wiring of the two systems must be kept entirely separate; wires from the battery circuits must under no circumstance ever enter the same outlet boxes or similar fittings containing 120-volt wires, unless entering a fixture or appliance listed for dual voltage operation.

RVs equipped with independent engine-driven generating plants delivering 120- or 120/240-volt, 60-Hz power are completely independent of parks if the owners wish. If an owner wants to take advantage of the power available at a park instead of running the RV plant for long hours, equipment must be installed in the vehicle so that commercial power from the park and power from the generating plant cannot possibly be connected to each other or to any piece of equipment at the same time.

Parks for recreational vehicles   A *site* is the space in a park occupied by one RV, and the *park* is an area intended to take care of a number of RVs.

The requirements for RV sites and parks are similar to those for mobile homes, with some important differences. Each RV is equipped with a distribution panelboard with an insulated neutral busbar, and provided with a cord that plugs into separate equipment installed near the RV. The cord is sized according to the electrical load in the RV, and may be either three-wire, 120-volt (one hot wire, the white grounded wire, and the green equipment grounding wire) if the load does not exceed 30 amps at 120 volts, or for larger or luxury RVs it may be four-wire, 120/240-volt (two hot wires, the white neutral wire, and the green equipment grounding wire) and rated at 50 amps. RV sites are supplied by feeders ending in a "power outlet" that meets special requirements and from which the RV is supplied. Groups of power outlets are served by feeders from the service equipment that supplies the park as a whole.

Figure 24–3 shows a typical power-outlet assembly for one RV. It contains one 20-amp, two-pole, three-wire, 120-volt grounding receptacle, and one 30-amp, two-pole, three-wire, 120-volt grounding receptacle, each protected by a circuit breaker. The type of construction shown is by far the most popular type of equipment for RV parks using underground wiring.

Unlike service equipment in mobile home lots, the special power outlets for RVs must *not* be grounded at each site. The grounded wire (neutral in the case of 120/240-volt installations) must be insulated from the cabinet of the power outlet at each site, while the green grounding wire must be bonded to each cabinet. The white grounded wire must be continued from site to site until it reaches the service equipment for the entire park, where it must be grounded, as must the green equipment grounding wire. But if the feeder to the individual power outlets at the various sites is in underground metal conduit, that conduit serves as the equipment grounding conductor.

Determining where to locate power outlets

In order to minimize the use of extension cords, the *NEC* in cooperation with the RV industry has standardized the locations of power outlets at RV sites. As part of this process, the *NEC* also distinguishes between a conventional ("back in") site and a "pull through" site that accommodates an RV with another vehicle or boat in tow. For a conventional site, draw a line 6 ft from the anticipated driver's (road) side of a parked vehicle and parallel to the longitudinal centerline of the stand (desig-

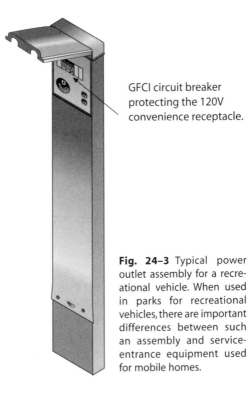

GFCI circuit breaker protecting the 120V convenience receptacle.

Fig. 24–3 Typical power outlet assembly for a recreational vehicle. When used in parks for recreational vehicles, there are important differences between such an assembly and service-entrance equipment used for mobile homes.

nated parking location). On this line, or within 1 ft on either side of it, install the power outlet at any point from the rear of the stand to a point 15 ft forward of the rear line. If the outlet is for a pull-through site, it goes on the same line but in a zone beginning 16 ft from the rear of the stand and extending forward, ending at the center point between the two access roads serving the stand.

Receptacles in RV power outlets

Grounding receptacles of the two-pole, three-wire, 20-amp, 125-volt type (either 15- or 20-amp plugs will fit) and those of the 50-amp, three-pole, four-wire, 125/250-volt type are the general-purpose type shown in Figs. 20–6 and 20–7. But note that the 30-amp, two-pole, three-wire, 125-volt type, as shown in Fig. 24–4, is not a general-purpose type. It is used only in RV power outlets and must not be used for other purposes. The general-purpose type of the same rating may not be used in RV power outlets.

Three-phase and single-phase distribution in RV parks

In contrast to a park filled with manufactured homes, most of the individual sites in an RV park involve 120-volt outlets only (discussed under the next heading). This allows for a 208Y/120-volt distribution, with equal load allocations across all three phases. Because electric power on the primary side of utility distribution transformers is always generated as three-phase, utilities often prefer to serve large loads in this

Fig. 24–4 The special-purpose 30-amp, 125-volt, two-pole, three-wire receptacle that may be used only in recreational vehicle power outlets.

Receptacle   Plug

way, and the *NEC* permits it in this case. This also allows the park administration ready access to three-phase power for large power applications at a central facility if they choose to provide it. However, any sites that have 120/240-volt, 50-amp outlets must be fed from a single-phase distribution only.

Calculating size of feeders for RV parks   In the following calculations, note that the *NEC* recognizes that not every RV site needs to have electric power, and that the park owners often designate some sites as dedicated tent sites with 120-volt, 15- or 20-amp receptacles. None of those sites count in the percentages that follow. The *NEC* specifies that every power-outlet assembly for an individual site designated for an RV must contain one 20-amp, 125-volt receptacle, and that 70% of all the assemblies must contain both the 20- and the 30-amp, 125-volt receptacles. Additional configurations, if appropriately supplied, are allowed but not required. The *NEC* also requires that the RV park include the 50-amp, 125/250-volt receptacle (in addition to the 20- and 30-amp types) in at least 20% (just increased from 5%) of the power outlets to accommodate luxury motor homes that require a large amount of power.

How big must each feeder be? Allow 3600 VA per lot served by both 20- and 30-amp receptacles, 2400 VA for those supplied only by a 20-amp receptacle. Add 9600 VA for each site with a 120/240-volt, 50-amp receptacle, and add 600 VA per dedicated tent site with electric power (20-amp supply maximum). Then apply the demand factor of *NEC* Table 551.73, reproduced here as Table 24–1, based on the highest rated receptacle at each site. Let's assume one section of a park has 150 sites, of which 25 are without power and 25 are dedicated tent sites. How large must the 208Y/120-volt feeder be that supplies this section of the park, assuming a Code-minimum outlet distribution? First, exclude from percentage calculations the unpowered sites and the tent sites, leaving 100 sites of which 70 end up with 30-amp supplies, 20 with 50-amp supplies, and the remaining 10 sites with 20-amp supplies. The 50-amp sites cannot be supplied by the three-phase feeder. That leaves the following in the load calculation, based on a 0.41 demand factor from the table:

$$\{(70 \text{ sites})(3600 \text{ VA/site}) + (10 \text{ sites})(2400 \text{ VA/site})$$
$$+ \ 25 \text{ sites } (600 \text{ VA/site})\}(0.41) = 119 \text{ kVA}$$

This has to be converted to amperes. The procedure is explained under the next two headings.

Three-phase calculations and square root of three   There is a trick that will allow you to understand and do these three-phase calculations easily every time. But first, here is some background information.

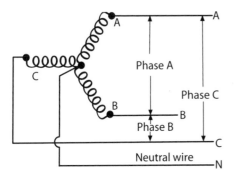

Fig. 24–5 The current flowing in the transformer bank between A and B must be added vectorially, not arithmetically, to the current flowing between C and A in order to determine the actual current flowing in the wire connected at A.

Referring to Fig. 24–5, the load is presumed to distribute equally across each phase winding. That means coils *A* and *B* see one-third of the load, coils *B* and *C* see one-third, and coils *C* and *A* see one-third. However, wire *A* does not see "one-third plus one-third equals two-thirds" of the load, nor do wire *B* or wire *C*, because this is a three-phase distribution, and the voltages and currents in each phase winding occur at different instants in time. The load in wire *A* (or *B* or *C*) is the vectorial summation of the load on the two simultaneously connected windings.

A vectorial quantity is a scalar quantity (a quantity with size but no direction) with direction added. For example, 25 mph is a scalar quantity called speed; 25 mph due north is a vectorial quantity called velocity. If you can swim 2 mph and you set out across an eastbound river that is moving 1 mph, your velocity will not be 1 nor 2 nor 3 mph. Your velocity will be the vectorial summation of 2 mph north plus 1 mph east. The resulting velocity is the hypotenuse of a right triangle with a base of 1 mph east and an altitude of 2 mph north, or $\sqrt{5}$ mph northeast (actually, bearing about 26½ deg east of due north).

In three-phase work one plus one doesn't always equal two for the same reason.[1] Looking at the drawing, if you do some simple geometry on coil *C* plus coil *A* (angle = 120 deg), the resultant quantity from adding the two presumed equal quantities together is the basic quantity times the square root of three. This is why the square root of three shows up repeatedly in three-phase calculations. (For an example, see the voltage drop calculation for three-phase circuits on page 110.) In every instance, when one wire is effectively connected to two different transformer phase windings at the same time, the electrical result contains a square-root-of-three multiplier. Review Figs. 3–5 and 3–6 for how these windings combine.

---

1. Technically, ac polyphase systems have "phasors" because the displacements are in terms of time. But directional vectors are easier to visualize and the numerical results come out the same.

Three-phase calculations—here's the trick   We know that the total volt-amperes on any winding is one-third of the total kVA from the load calculation. Applying what we just learned about three-phase systems means that a wire connected to adjacent phase windings carries one-third of the kVA load on one winding times the square root of three. On our 208Y/120-volt system, coils *A-B* carry 119/3 kVA, as do coils *C-A*. Using the three-phase "algebra" we just learned, wire *A* must carry $(119/3)(\sqrt{3})$ kVA. Now rearrange the terms: $(119)(\sqrt{3}/3)$ kVA. We want to calculate how many amperes we need to size wire *A* for—which means dividing by the voltage, 208 volts. As explained in Chapter 3, 208 volts is simply the line-to-neutral voltage—120 volts—times the square root of three. Therefore instead of dividing by 208 volts, divide by $120(\sqrt{3})(\sqrt{3})$ volts, or 360 volts.

Here is the ampere calculation:

$$\frac{\{119(\sqrt{3}/3)\ \text{kVA}\}}{120(\sqrt{3})\ \text{V}} = \frac{119\ \text{kVA}}{120(3)\ \text{V}} = \frac{119\ \text{kVA}}{360\ \text{V}} = 331\ \text{amps}$$

When doing three-phase load calculations on wye systems, divide the volt-ampere result by three times the line-to-neutral voltage. Thus, on 208Y/120-volt systems, divide kVA results by 360, and on 480Y/277-volt systems, divide kVA results by 831.

A 400-amp feeder would serve the 120-volt loads in this part of the park.

The twenty 50-amp sites figure at 9600 VA each, with an allowable demand factor of 0.45.

   (20)(9600 VA)(0.45) = 86.4 kVA

   convert to amperes at 240 V: (86 400 VA)/240 V = 360 amps

The 120/240-volt, single-phase, 50-amp sites would require a 400-amp feeder.

Table 24-1 [*NEC* Table 551.73]* DEMAND FACTORS FOR FEEDERS AND SERVICE-ENTRANCE CONDUCTORS FOR PARK SITES

Number of recreational vehicle sites	Demand factor, %	Number of recreational vehicle sites	Demand factor, %
1	100	10–12	50
2	90	13–15	48
3	80	16–18	47
4	75	19–21	45
5	65	22–24	43
6	60	25–35	42
7–9	55	36 plus	41

*Reprinted with permission from NFPA 70-2005 *The National Electrical Code,* © NFPA 2004.

REQUIREMENTS FOR NONRESIDENTIAL USES OF MOBILE HOMES AND RVs

If a mobile home or an RV is used for other than the normal residential purposes covered in the definitions at the beginning of the chapter, but is instead used for nonresidential purposes such as portable office, merchandise display, or dormitory on construction site, *and* if it is intended to be supplied from outside 120- or 120/240-volt power, it must meet all the requirements discussed, except the number and size of branch circuits need not be as normally required by the *NEC*.

CHAPTER 25

Wiring Apartment Buildings

ALTHOUGH A SINGLE APARTMENT within an apartment building presents no new wiring problems of any consequence, the building as a whole does present such problems. Detailed discussions of some of the problems in the wiring of larger apartments are beyond the scope of this book. The information in this chapter should be considered only as a general guide and foundation. You must also study *NEC* Article 230.

FROM SIMPLE TO COMPLEX—START WITH A SINGLE APARTMENT

For an individual apartment, the minimum number of circuits required is determined in essentially the same way as for a single-family house, as outlined in Chapter 13. There are some allowances if the building management provides centralized laundry services, for example, but in principle the *NEC* treats dwelling units alike, regardless of the mechanics of ownership. For similar reasons, a condominium and an apartment are the same in the eyes of the *NEC*. Remember that the *NEC*-required minimum is often inadequate for convenience and practicality, so additional circuits may be needed. Load calculations for determining wire size are also done in a similar way as for a house, although the 100-amp minimum size rule does not apply to individual apartments.

Determining minimum number of branch circuits To determine the minimum number of branch circuits for any single apartment, proceed as outlined in the Chapter 13 discussion on installing branch circuits for a single-family dwelling. For lighting, allow 3 VA per sq ft. For example, a small apartment of 800 sq ft will require (800)(3) or 2400 VA for lighting and general purpose receptacle outlets, which means two circuits. To this must be added the two 20-amp small-appliance circuits for the dining and cooking areas, just as for a single-family house. (Laundry equipment is usually not installed in individual apartments, but if it is, a separate 20-amp circuit must be provided for it.) The *NEC* now requires a separate circuit for bathroom receptacles, but no specific load allowance needs to be included for this circuit. A separate circuit must be provided for each special appliance such as range or dishwasher. The number of receptacles is determined as in separate houses. Don't overlook the GFCI-protection requirements for receptacle outlets on kitchen counters, within 6 ft of sinks, and in bathrooms.

Determining wire size   The wires to the overcurrent devices in individual apartments are feeders. But if actual service equipment is installed in each unit (see Fig. 25–3) instead of a common service at a common location (see Figs. 25–1 and 25–2), the supply wires are subsets of service-entrance wires. In either case, the ampere load (and therefore the wire size) is determined as described for service-entrance wires in the page 230 topic on selecting service-wire size. Allow 3 VA per sq ft of area for lighting including the usual receptacles; add 3000 VA for the two required small-appliance circuits; if laundry equipment is installed in the individual apartment, add 1500 VA. Total all the above, count the first 3000 VA at 100% demand factor, the remainder at 35% demand factor. Then add the volt-amperes rating of individual circuits to appliances such as range, air conditioner, dishwasher, and you will have the total computed volt-amperes. For an 800-sq-ft apartment, your calculation would look like this:

	Gross computed volt-amperes	Demand factor, %	Net computed volt-amperes
Lighting, 800 sq ft at 3 VA	2400		
Small appliances, (minimum)	3000		
Total	5400		
First 3000 VA		100	3000
Remaining 2400 VA		35	840
Total			3840

At 120 volts, this is equivalent to 3840/120 or 32 amps. This will require two 15-amp, general-use lighting (and receptacle) circuits and two 20-amp, small-appliance circuits.

According to *NEC* Table 310.16, 8 AWG wire is the smallest that may be used for 32 amps, but only if the wiring is as shown in Figs. 25–1 or 25–2, where the wires to the apartment are feeders. If installed as shown in Fig. 25–3, they are subsets of service-entrance wires, and *NEC* 230.42(B) and 230.79(D) require that they be a minimum of 6 AWG (technically, having an ampacity of 60; however, since no termination on this equipment would likely be listed for use above 75°C, this becomes the basis for the wire size; also see the page 106 discussion on termination temperatures).

All the above applies to the two-wire, 120-volt supply to the apartment. If the main fuse protecting that apartment blows, the entire apartment will be without power and in darkness. It would be better to install a three-wire, 120/240-volt supply. The 3840 volt-amperes at 240 volts would equal 16 amps, so it might appear that 12 AWG wire would be large enough, but *NEC* 230.42(B) and 230.79(D) would still require a minimum of 6 AWG for a subset of service-entrance wires per Fig. 25–3; if the installation is as shown in Figs. 25–1 or 25–2, then any conceivable *NEC* load calculation would require a minimum of 10 AWG for the feeder.

In actuality, there will be few apartments having only the circuits discussed. Usually there are additional circuits to supply appliances such as range, air conditioner, dishwasher, and kitchen waste disposer. To add a range, follow *NEC* 220.55 and add 8000 VA (if the range is rated at more than 12 000 watts; review the discussion of special appliance loads beginning on page 231) to the 3840 VA already calculated, making a total of 11 840 VA. Since a range is a combination 120/240-volt load, a three-wire supply must be installed. The 11 840 VA at 240 volts is equal to 11 840VA/240V, or 49.3 amps, for which 6 AWG wire is suitable.

However, as stated in the discussion on determining size of the grounded neutral in the service beginning on page 235, the neutral serving a range cannot be made to carry as many amperes as the hot wires carry. Consider only 70% of the allowance for the range. In the example above, the neutral would be figured this way: The range load is 8000VA/240V, or 33.3 amps, and 70% of 33.3 is 23.3 amps. Add this 23.3-amp load to the 16-amp lighting load originally figured, for a total of 39.3 amps, for which an 8 AWG wire is suitable for the neutral.

REQUIREMENTS FOR SERVICE ENTRANCE FOR THE BUILDING

In practically all cases there is but a single service for the entire building. In many cases each tenant pays for the power consumed within the tenancy, so there is a separate meter for each apartment, plus another meter for "house loads" such as hall lights, water heaters, office loads, and heating-system motors. The service equipment must then supply a number of separate meters and disconnecting means. *NEC* 230.72(C) requires that each occupant have access to his or her disconnecting means. Although *NEC* 230.40 Exception No. 1 allows a set of service conductors to extend to a remote location to serve either one unit or a group of units, this chapter assumes a single service disconnect. If the *NEC* exception does apply to your situation, complete the calculations in this chapter based on the load presented by the group of occupancies served by the service conductors in question. Don't confuse this with the more common case of a protected feeder extending to a remote group of apartments, for which the load calculation is identical, but for which the wiring method need not meet service requirements for being outside the building (or utilizing equivalent protection—review the page 57 topic on the unique overcurrent protection problems of service conductors).

Common access to service equipment   In general, service equipment must be grouped in an area to which all occupants have access, as shown in Figs. 25–1 and 25–2, unless special permission is given otherwise (Fig. 25–3). Branch-circuit protective devices are located within each occupancy for convenience and to reduce the wiring costs of providing a home run all the way to the basement for every circuit. Keeping the branch circuit protection in local panelboards also reduces voltage drop.

Determining service-entrance wire size   The minimum size of the service-entrance wires for the entire building is determined by the probable total load at any one time. The method is similar to that used for a single-family dwelling—the lighting (including receptacles), the small-appliance circuits, and the special loads such as ranges are simply computed for the entire building.

The greater the number of individual apartments in a building, the less likelihood there is that all occupants will be using maximum power at the same time. Therefore the *NEC* has established demand factors for lighting and small appliances in 220.42, for clothes dryers in 220.54, and for ranges in 220.55. The greater the number of apartments, the lower the demand factor.

Looking first at the lighting and small-appliance loads for the building as a whole, allow 3 volt-amperes per sq ft of area in all the apartments; then add 3000 volt-amperes for the small-appliance circuits in each apartment. Apply the demand factors of *NEC* 220.42 as follows:

First 3000 VA at   100%

Next 117 000 VA at   35%

All above 120 000 VA at   25%

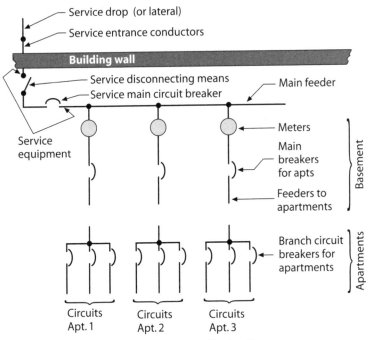

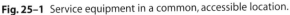

Fig. 25–1 Service equipment in a common, accessible location.

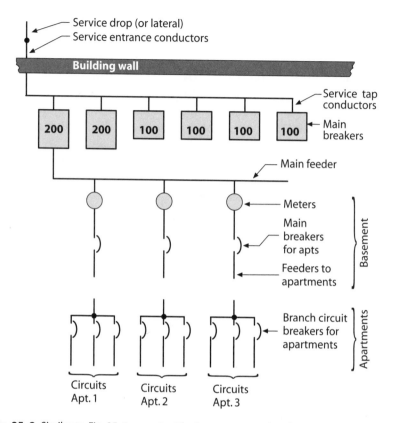

Fig. 25–2 Similar to Fig. 25–1, except with six separate service disconnecting means, the maximum number permitted in one location.

Now looking at ranges, it is not likely that all occupants will be using their ranges at maximum capacity at the same time. Decreasing demand factors are shown in *NEC* Table 220.55. Note that the table specifies that its Column C must be used at all times with the exception of those cases mentioned in the third Note following the table. Assuming that each range is rated at not over 12 000 watts, allow 8000 watts for one range, 11 000 watts for two ranges, 14 000 watts for three ranges. For four or more ranges, see Table 220.55.

Thus to determine the ampacity of the service-entrance wires, add together the volt-amperes required for the lighting and small-appliance circuits (after application of the demand factor) and the watts required for the ranges. To this total must be added the "house load," which can be very considerable, to include hall and basement lights, laundry circuits at 1500 volt-amperes each, clothes dryers (the first four dryers must be included at full nameplate rating; if there are more than four, see *NEC* Table 220.54), motors on heating or other equipment, water heaters if electric, and so on. *NEC* 220.53 allows a demand factor of 75% to be applied to

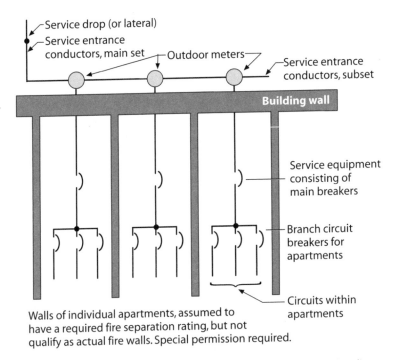

Walls of individual apartments, assumed to have a required fire separation rating, but not qualify as actual fire walls. Special permission required.

Fig. 25–3 If there is no common location accessible to all occupants, service disconnects may, by special permission only, be located in each apartment. Usually the utility will insist on grouped metering.

the nameplate ratings of appliances (other than ranges, dryers, air-conditioning or space-heating equipment).

Service entrance for three-apartment building   The maximum probable load can be computed in accordance with the following calculation sequence:

	Gross computed volt-amperes	Demand factor, %	Net computed volt-amperes
Lighting, 2400 sq ft at 3 VA	7200		
Small appliances, 3 apartments at 3000 VA	9000		
Total gross computed VA	16 200		
First 3000 VA		100	3000
Remaining 13 200 VA		35	4620
Total net computed VA			7620

The 7620 VA covers only the apartments. The "house load" will probably come to 6000 VA, considering the need for the 20-amp laundry circuits. Adding the

6000 VA to the 7620 VA determined for the apartments makes a total of 13 620 VA, which at 240 volts is 56.7 amps. Accordingly a 60-amp service would be acceptable, and per *NEC* Table 310.16, 6 AWG Type THW or RHW wire or 4 AWG Type TW would be acceptable. But remember that *NEC* 230.42(B) and 230.79(C) require a 100-amp service for a single-family house. It is likely that most people would consider a 60-amp service quite skimpy for such an apartment building; prudence suggests that a 100-amp or larger service be installed.

If an electric range consuming not over 12 000 watts is added in each apartment, column C of *NEC* Table 220.19 shows that 14 000 watts must be added for the three ranges in calculating the service. Adding 14 000 watts to the 13 620 watts already determined produces a total of 27 620 watts, which at 240 volts is equivalent to about 115 amps. The minimum would be a 125-amp service, using a circuit breaker as the disconnecting means; if fused equipment is installed, the minimum switch size would be 200 amps, since there are no switch sizes between 100 and 200 amps. The minimum ampacity of the service wires would in any case be 125 amps. A 150- or 200-amp service would be better, because it would allow for some expansion for future use.

Optional computing methods   If the apartment building contains three or more individual apartments, *NEC* 220.82 permits an optional method of computing loads in individual apartments, provided the apartment feeder is at least 100 amps; follow the procedure for a single-family dwelling using the optional calculation in Chapter 13. In *NEC* 220.84 there is an optional method for the total building, but only if all the following conditions are met: (1) All apartment loads considered by this calculation are served by the same feeder or service conductors, and (2) each apartment's cooking equipment is electric, and (3) each apartment has electric space heating, or air conditioning, or both. If all these conditions are met, proceed as follows, with all calculations based on the entire load across all apartments connected to the feeder or service conductors being calculated:

1. For each square foot of area, allow 3 VA for lighting and general-purpose receptacles .... _____VA

2. Add for each special appliance circuit (two small-appliance circuits minimum plus laundry circuit if present) 1500 VA per circuit: .... _____VA

3. Total nameplate ratings of all appliances fastened in place or located to be on a specific circuit (review Chapter 20 for discussion of these terms); includes ranges, clothes dryers, space heaters, water heaters, etc....... _____VA

4. Air-conditioning or electric heating equipment (larger of the two loads) at full nameplate rating .... _____VA

5. All motor and "low power factor" loads, at nameplate rating .... _____VA

6. Total of steps 1, 2, 3, 4, and 5 .... _____VA

7. Multiply total in step 6 by demand factor permitted in *NEC* Table 220.84

8. Divide result in step 7 by 240, giving you the ampacity required for the hot wires of the feeder. The neutral might be smaller per *NEC* 220.61; see discussion in Chapter 13.

9. Add the house loads as computed in the traditional fashion .... ____VA

WIRING THE SERVICE TO A LARGER APARTMENT BUILDING

Based on the information already discussed, you should be able to determine the details of wiring larger buildings with more apartments. Study *NEC* Appendix D, Examples D4(a) and D4(b). The final installation of the service equipment in a larger building might well be as shown in Fig. 25–4. This shows a four-meter module containing four meters plus four circuit breakers, one for each of four apartments. The modules are gangable, so for example you could put three together to handle 12 apartments. Since this is more than the maximum of six permitted disconnecting means, a main circuit breaker serving as the disconnecting means is provided ahead of all in a separate section.

Computing electric range and clothes dryer loads   For such a large apartment building, the utility will probably require a 208Y/120-volt service in order not to unbalance their three-phase distribution. Study *NEC* Appendix D, Example D5(a) in conjunction with this process. The key difference involves how electric ranges (and clothes dryers if present) are calculated under the normal procedure. *NEC* 220.55 (and 220.54) requires these loads to be computed based on twice the maximum number of such appliances connected between any two phase legs.

If there is an electric range in each apartment, with four apartments on each pair of phase legs, this means applying the demand table based on $4 \times 2 = 8$ ranges, or 23 kW. This is a per phase demand of 11.5 kW; there being a total of three phases, the total kW to put into the load calculation is 34.5 kW. This is significantly higher than the Table 220.55 result of 27 kW for 12 ranges.

Review the technical discussion on three-phase calculations at the end of Chapter 24. Remember that at the very end we are selecting a wire, and everything we do must ultimately relate to the load that the electrical system will impose on the wire we select. The reason the *NEC* rules are correct has to do with how single-phase loads add together on three-phase distributions. The building will have 4 ranges across the Phase A-B connection, 4 across B-C, and

Fig. 25–4 A typical group metering panel for use in larger apartment buildings. *(Square D)*

4 across C-A. Each wire sees a total of 8 actual and $4\sqrt{3}$ effective range connections, because the loads are out of phase. Using the demand table for 8 ranges, that should be 23 kW, or 2875 W per range. Each range is connected at 208 volts, and dividing watts by volts gives 13.8 amps per range. The effective current on each wire then is the number of effective ranges $(4\sqrt{3})$ times the current: $(4\sqrt{3})(13.8 \text{ amps}) = 96$ amps.

The *NEC* is correct if it generates the same result. Take double the highest number of ranges across any phase pair, or 8 ranges. Again we are selecting wires, and each wire sees 8 range connections. That is the relevant parameter with which to enter the demand table. The demand for 8 is 23 kW, or 11.5 kW for each winding, and that would be 34.5 kW in the total load calculation. At the very end of the calculation, to figure wire size convert kVA to amps. As in the RV park at the end of Chapter 24, divide by three times the line-to-neutral voltage, or 360 volts. The result checks: 34,500 kW/360 volts = 96 amps.

Limited-energy signaling circuits   Doorbell and door opener wiring will usually be Class 2 control circuits. In a multifamily dwelling, where this wiring is run concealed (usual condition), it must be listed as CL2 or better (such as CL2R or CL2P) for this use.

Doorbell and buzzer system   The usual door-bell and buzzer system (or chimes) involves limited-energy circuits and will be installed in accordance with the principles outlined in Chapter 19. However, in buildings of considerable size you must take into consideration the resistance of the wires in the longer lengths involved.

Sometimes the installation is made as shown in Fig. 25–5. Theoretically this diagram is correct, but the more distant bells will ring faintly because of the substantial voltage drop in the long run of small wire. A drop of 2 volts on a 120-volt circuit is not serious, but when the starting voltage is likely to be under 20 volts, a 2-volt drop is at least 10%. If the transformer voltage is stepped up high enough for the more distant bells to ring properly, the nearby bells will ring too loudly and the distant bells will still not be loud enough.

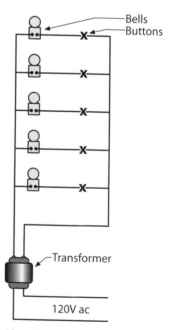

The solution is to use the circuit shown in Fig. 25–6. A little more wire is needed, but the number of feet of wire involved for any one bell

Fig. 25–5 Using this wiring diagram, the distant bells ring too faintly and the nearby ones ring too loudly.

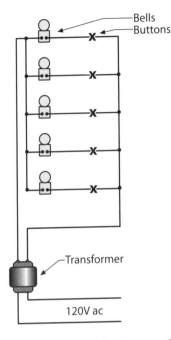

Bells
Buttons

Transformer

120V ac

Fig. 25-6 Using this diagram, all bells ring with equal volume. It requires only a little more wire than the previous diagram.

is exactly the same as for any other bell. Accordingly, all will ring with equal volume; a transformer voltage that is correct for one is correct for all.

Door openers   In many apartment buildings the door leading into the inside hall is equipped with a door opener. A key is needed to enter the building. To enter without a key it is necessary to buzz or call an individual apartment from the entryway; the occupant pushes a button in the apartment to release a latch in the opener allowing the door to open. A typical door opener is shown in Fig. 25–7. It consists of an electromagnet energized by the same transformer serving the bells, and that releases the latch when the button in an apartment is pushed. The opener is mortised into the door frame opposite the lock in the door, as shown in Fig. 25–8, which also shows the wiring diagram.

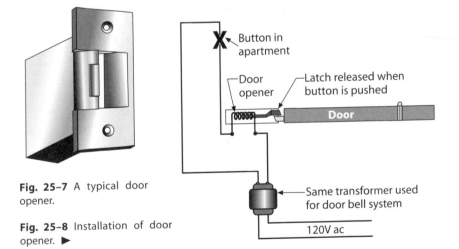

Button in apartment

Door opener

Latch released when button is pushed

Door

Same transformer used for door bell system

120V ac

Fig. 25-7 A typical door opener.

Fig. 25-8 Installation of door opener. ▶

CHAPTER 26

Sizing Conductors for All Load Conditions

ONE OF OUR PRIMARY RESPONSIBILITIES in the electrical trade is to select electrical conductors. Now that you understand from previous chapters how wire sizes are standardized and how they relate to each other, and you are familiar with some of the common types of insulation available for wires, it is time to learn how to select the minimum allowable size of a wire, assuming you know the load to be carried. Determining the load comes up constantly in many different forms, and it is discussed repeatedly throughout the remainder of this book. After you finish this chapter you will be prepared to pick the right size wire at any point that, as you consider various rules for calculating loads, you are able to draw a conclusion about the actual load.

Each time you select a wire size, you will be considering several principles and how they relate to each other. Great deliberation and mental discipline are required to apply these principles correctly and consistently.

MIDDLES AND ENDS OF WIRES REQUIRE SEPARATE CALCULATIONS

The key to applying the rules that follow is to remember that the end of a wire is different from its middle. Special rules apply to calculating wire sizes based on how the terminations are expected to function. Entirely different rules aim at assuring that wires, over their length, don't overheat under prevailing loading and conditions of use. These two sets of rules have nothing to do with each other—they are based on entirely different thermodynamic considerations. Some of the calculations use, purely by coincidence, identical multiplying factors. Sometimes it is the termination requirements that produce the largest wire, and sometimes it is the requirements to prevent conductor overheating. You can't tell until you complete all the calculations and then make a comparison. Until you are accustomed to doing these calculations, do them on separate pieces of paper.

Current is always related to heat   Every conductor has some resistance (see page 31), and as you increase the current, you increase the amount of heat, all other things being equal. In fact, as we've seen, you increase the heat by the square of the current. The ampacity tables in the *NEC* reflect heating in another way. As the excerpt from *NEC* Table 310.16 (see Table 26–1) shows, the tables tell you how much current you can safely (meaning without overheating the insulation) *and continuously* draw through a conductor under the prevailing conditions—which is essentially the definition of ampacity in *NEC* Article 100: The current in amperes that a conductor can carry continuously under the conditions of use without exceeding its temperature rating.

Table 26–1 ALLOWABLE AMPACITIES OF SELECTED INSULATED CONDUCTORS, BASED ON *NEC* Table 310.16[†] [See Appendix Table A1 for full reproduction]

Size	Temperature Rating of Conductor (See Table 310.13)		
	60C (140F)	75C (167F)	90C (194F)
AWG or kcmil	Types TW, UF	Types RHW, THHW, THW, THWN, XHHW, USE, ZW	Types TBS, SA, SIS, FEP, FEPB, MI, RHH, RHW-2,THHN, THHW, THW-2, THWN-2, USE-2, XHH, XHHW, XHHW-2, ZW-2
COPPER			
10*	30	35	40
8	40	50	55
6	55	65	75
4	70	85	95
3	85	100	110
2	95	115	130
1	110	130	150
1/0	125	150	170
2/0	145	175	195
3/0	165	200	225
4/0	195	230	260
CORRECTION FACTORS			
Ambient Temp. (C)	For ambient temperatures other than 30C (86F), multiply the allowable ampacities shown above by the appropriate factor shown below.		Ambient Temp. (F)
36–40	0.82	0.88	0.91	96–104
41–45	0.71	0.82	0.87	105–113
46–50	0.58	0.75	0.82	114–122

*See 240.4(D).
[†]Reprinted with permission from NFPA 70-2005 *The National Electrical Code,* © NFPA 2004.

Ampacity tables show how conductors respond to heat The ampacity tables (see Table 26–1) do much more than what is described in the previous paragraph. *They show, by implication, a current value below which a wire will run at or below a certain temperature limit.* Remember, conductor heating comes from current flowing through metal arranged in a specified geometry (generally, a long flexible cylinder of specified diameter and metallic content). In other words, for the purposes of thinking about how hot a wire is going to be running, you can ignore the different insulation styles. As a learning tool, let's make this into a "rule" and then see how the *NEC* makes use of it:

A conductor, regardless of its insulation type, runs at or below the temperature limit indicated in an ampacity column when, after adjustment for the conditions of use, it is carrying equal or less current than the ampacity limit in that column.

For example, a 90°C THHN 10 AWG conductor has an ampacity of 40 amps. Our "rule" tells us that when 10 AWG copper conductors carry 40 amps under normal-use conditions, they will reach a worst-case, steady-state temperature of 90°C just below the insulation. Meanwhile, the ampacity definition tells us that no matter how long this temperature continues, it won't damage the wire. That's not true of the device, however. If a wire on a wiring device gets too hot for too long, it could lead to loss of temper of the metal parts inside, cause instability of nonmetallic parts, and result in unreliable performance of overcurrent devices due to calibration shift.

TERMINATION RESTRICTIONS PROTECT DEVICES

Because of the risk to devices from overheating, manufacturers set temperature limits for the conductors you put on their terminals. Consider that a metal-to-metal connection that is sound in the electrical sense probably conducts heat as efficiently as it conducts current. If you terminate a 90°C conductor on a circuit breaker, and the conductor reaches 90°C (almost the boiling point of water), the inside of the breaker won't be much below that temperature. Expecting that breaker to perform reliably with even a 75°C heat source bolted to it is expecting a lot.

Testing laboratories take into account the vulnerability of devices to overheating, and there have been listing restrictions for many, many years to prevent use of wires that would cause device overheating. These restrictions now appear in the *NEC*. Smaller devices (generally 100 amp and lower, or with termination provisions for 1 AWG or smaller wire) historically weren't assumed to operate with wires rated over 60°C such as TW. Higher-rated equipment assumed 75°C conductors but generally no higher for 600-volt equipment and below. This is still true today for the larger equipment. (Medium-voltage equipment, over 600 volts, has larger internal spacings and the usual allowance is for 90°C, but that equipment is beyond the scope of this book.) Today, smaller equipment increasingly has a "60/75°C" rating, which means it will function properly even where the conductors are sized based on the 75°C column (Table 26–1).

Figure 26–1 shows a "60/75°C" marking on a 20-amp circuit breaker, which means it can be used with 75°C conductors, or with 90°C conductors used under the 75°C ampacity column. Both the panelboard and the device at the other end of the wire must make the same allowance for a 75°C temperature assumption to be acceptable. Otherwise the 60°C column applies.

Fig. 26–1 A 20-amp circuit breaker marked as acceptable for 75°C terminations.

In the case of a circuit breaker, using the allowance for 75°C terminations requires that both the CB as well as its enclosing panelboard carry this designation. Always remember, though, that wires have two ends. Successfully using the smaller (higher ampacity) wires requires equivalent markings on both the circuit breaker and panelboard (or fused switch) and the device at the other end. Refer to Fig. 26–2 for an example of this principle at work.

Every wire has two ends. This wire must be sized as if it had 60°C insulation because ...

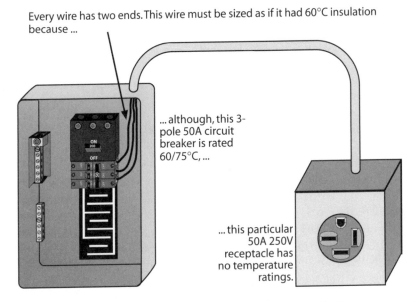

... although, this 3-pole 50A circuit breaker is rated 60/75°C, ...

... this particular 50A 250V receptacle has no temperature ratings.

Fig. 26–2 Always consider both ends of a wire when making termination temperature evaluations.

Splices are terminations   Not all terminations occur on electrical devices or utilization equipment. Some terminations occur in the middle of a run where one conductor is joined to another. The same issue arises when we make a field connection to a busbar that runs between equipment. Busbars, usually rectangular in cross section, are often used to substitute for conventional wire in applications involving very heavy current demands. When you make a connection to one of these busbars (as distinct from a busbar within a panel), or from one wire to another, you only have to be concerned about the temperature rating of the compression connectors or other splicing means involved. Watch for a mark such as "AL9CU" on the lug. This one means you can use it on either aluminum or copper conductors, at up to 90°C, but only where the lug is "separately installed" (*NEC* text).

Lug temperature markings usually mean less than they appear to mean. Many contactors, panelboards, etc., have termination lugs marked to indicate a 90°C acceptance. Ignore those markings because the lugs aren't "separately installed." Apply the normal termination rules for this kind of equipment. What's happening here is that the equipment manufacturer is buying lugs from another manufacturer who doesn't want to run two production lines for the same product. The lug you field install on a busbar, and use safely at 90°C, also works when furnished by your contactor's OEM. But on a contactor you don't want the lug running that hot. The lug won't be damaged at 90°C, but the equipment it is bolted to won't work properly.

Protecting devices under continuous loads   The NEC defines a continuous load as one that continues for three hours or longer. Most residential loads aren't continuous, but many commercial and industrial loads are. Consider, for example, the banks of fluorescent lighting in a store. Not many stores always stay open less than three hours at a time. Although continuous loading doesn't affect the ampacity of a wire (defined, as we've seen, as a continuous current-carrying capacity), it has a major impact on electrical devices. Just as a device will be affected mechanically by a heat source bolted to it, it also is affected mechanically when current near its load rating passes through it continuously. To prevent unremitting thermal stress on a device from affecting its operating character-istics, the NEC restricts the connected load to not more than 80% of the circuit rating. The reciprocal of 80% is 125%, and you'll see the restriction stated both ways. Restricting the continuous portion of a load to 80% of the device rating means the same thing as saying the device has to be rated 125% of the continuous portion of the load. If you have both continuous and noncontinuous load on the same circuit, take the continuous portion at 125%, and then add the noncon-tinuous portion. The result must not exceed the circuit rating.

Suppose, for example, a load consists of 51 amps of noncontinuous load and 68 amps of continuous load (119 amps total) as shown in Fig. 26–3[1]. Calculate the minimum capacity we need to allow for our connected equipment as follows:

$$
\begin{aligned}
&\text{Step 1: } 51\,A \times 1.00 = \quad 51\,A \\
&\text{Step 2: } 68\,A \times 1.25 = \underline{\quad 85\,A} \\
&\text{Step 3: Minimum} = 136\,A
\end{aligned}
$$

A device such as a circuit breaker that will carry this load profile must be rated no less than 136 amps, even though only 119 amps actually passes through the device. In the case of overcurrent protective devices, the next higher standard size would be 150 amps. In general, for overcurrent protective devices not over 800 amps, the NEC allows you to round upward to the next higher standard overcurrent device size. In this case, suppose a standard wire size had an ampacity of 140 amps (there doesn't happen to be one, but this is only for discussion). The

---

1. This example, and the further examples in this chapter derived from it, formed the basis of the new Example D3(a) in Annex D of the 2005 NEC. This material appeared in the 18th edition of *Practical Electrical Wiring*, and has been slightly modified in this edition to correlate with the actual text of the NEC.

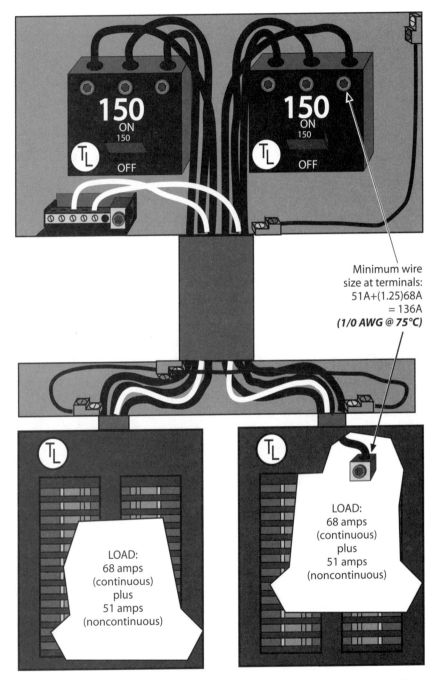

Fig. 26–3 Continuous loads require increased wire sizes so terminating devices will not overheat. This is true even if the size based on the ampacity required over the run seemingly allows a smaller wire.

NEC normally allows a 150-amp fuse or circuit breaker to protect such a conductor.

Having gone this far, it's easy to make two mistakes at this point. First, although you can round up in terms of the overcurrent device rating, you can't round up in terms of conductor loading, not even one ampere. Number 1 AWG conductors in the 75°C column can carry 130 amps. If your actual load runs at 131 amps, you have to use a larger wire. Second, when continuous loads are a factor, you have to build in additional headroom on the conductor sizes to assure that the connected devices perform properly. This last point continually results in confusion because it may seem to contradict what was said about conductor ampacity tending to be the factor that determines minimum wire size.

In our work we handle conductors, and we worry about conductors getting overheated. Device manufacturers don't worry about conductors in this sense; they worry about their devices getting overheated and not performing properly. Continuous loads pose real challenges in terms of heat dissipation from the inside of mechanical equipment. Remember that when you bolt a conductor to a device, the two become one in the mechanical as well as the electrical sense. Device manufacturers rely on those conductors as a heat sink, particularly under continuous loading. The *NEC* allows for this by requiring conductors carrying continuous loads to be oversized according to the same formula that applies to the device, namely an additional 25% of the continuous portion of the load.

Our 10 AWG THHN conductor, for example, will carry 40 amps for a month at a time without damage to itself. But under those conditions the conductor would represent a continuous 90°C heat source. Now watch what happens when we (1) size the conductor *for termination purposes* at 125% of the continuous portion of the load, and (2) use the 75°C column for the analysis. This calculation assumes the termination is rated for 75°C instead of the default value of 60°C:

> Step 1:   1.25 x 40 A = 50 A
>
> Step 2:   Table 310.16 at 75°C = 8 AWG

We go from a 10 AWG conductor to an 8 AWG conductor (6 AWG if the equipment doesn't have the allowance for 75°C terminations). Now that is just one customary wire size, but look at it from the device manufacturer's perspective. Number 10 AWG carrying 40 amps continuously is a continuous 90°C heating load. What about the 8 AWG? Use the ampacity table in reverse, according to our "rule." Forty amps happens to be the ampacity of an 8 AWG, 60°C conductor. Therefore, any 8 AWG wire (THHN or otherwise) won't exceed 60°C when its load doesn't exceed 40 amps. By going up just one wire size, the termination temperature dropped from 90°C to 60°C. The *NEC* allows manufacturers to count on this headroom.

To recap, if you have a 40-amp continuous load, the circuit breaker must be sized at least at 125% of this value, or 50 amps. In addition, the conductor must be sized to carry this same value of current *based on the 75°C ampacity column (or 60°C if not evaluated for 75°C)*. The manufacturer and testing laboratory count on a relatively cool conductor to function as a heat sink for heat generated within the device under these continuous operating conditions.

In the feeder example, including the 125% on the continuous portion of the load brings us to a 136-amp conductor, and the next larger one in the 75°C column is a 1/0. Remember to use the 75°C column here because the 150-amp device exceeds the 100-amp threshold (below which the rating is assumed to be 60°C). Remember, only 119 amps (68 amps + 51 amps) of current actually flows through these devices. The extra 17 amps (the difference between 119 amps and 136 amps) is phantom load. You include it only so your final conductor selection is certain to run cool enough to allow it to operate in accordance with the assumptions made in the various device product standards.

There are devices manufactured and listed to carry 100% of their rating continuously, and the *NEC* recognizes their use in exceptions. Typically these applications involve very large circuit breaker frame sizes in the 600-amp range (although the trip units can be smaller). Additional restrictions accompany these products, such as on the number that can be used in a single enclosure and on the minimum temperature rating requirements for conductors connected to them. Learn how to install conventional devices first, and then apply these 100%-rated devices if you run across them, making sure to apply all installation restrictions covered in the directions that come with this equipment. The warning about wires having two ends applies here with special urgency; be aware that one of these devices at one end of a circuit doesn't imply anything about the suitability of equipment at the other end.

THE MIDDLE OF THE WIRE—PREVENTING CONDUCTORS FROM OVERHEATING

None of the preceding discussion has anything to do with preventing a conductor from overheating. That's right. All we've done is to be sure the device works as the manufacturer and the test lab anticipate in terms of the terminations. Now we have to be sure the conductor doesn't overheat. Again, ampacity is by definition a continuous capability. *The heating characteristics of a device at the end of the run don't have any bearing on what happens in the bowels of a raceway or cable assembly.*

To reiterate, you have to compartmentalize your thinking at this point. We just covered the end of the wire; now we'll get to the middle of the wire. Remember being asked to do these on separate pieces of paper? Lock the first one up, and forget everything you just calculated. It has absolutely no bearing on what comes next. Only after you've made the next series of calculations should you retrieve the first sheet of paper. And only then should you go back and see which result represents the worst case and therefore governs your conductor choice. If you have trouble making this distinction, and many do, apply an imaginary pull box at each end of the run (Fig. 26–4).

Review the ampacity definition. Conductor ampacity is its current-carrying capacity *under the conditions of use.* For *NEC* purposes, two field conditions affect ampacity, and they are mutual heating and ambient temperature. Either or both may apply to any electrical installation. Both of these factors reduce ampacities from the table numbers.

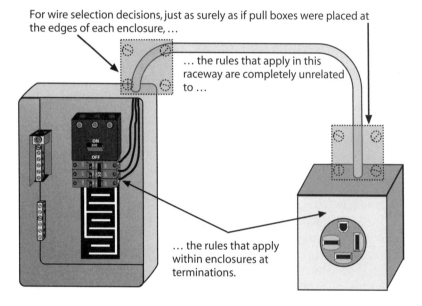

For wire selection decisions, just as surely as if pull boxes were placed at the edges of each enclosure, ...

... the rules that apply in this raceway are completely unrelated to ...

... the rules that apply within enclosures at terminations.

Fig. 26–4 These imaginary pull boxes at each end of the run illustrate how to separate raceway/cable heating calculations from termination calculations.

Mutual heating   A conductor under load dissipates its heat through its surface into the surrounding air; if something slows or prevents the rate of heat dissi-pation, the temperature of the wire increases, possibly to the point of damage. The more current-carrying conductors there are in the same raceway or cable assembly, the lower the efficiency with which they can dissipate their heat. To cover this mutual heating effect, the *NEC* imposes derating penalties on table ampacity values. Penalties increase with the number of current-carrying conductors in a raceway or cable assembly. *NEC* Table 310.15(B)(2)(a) limits the permissible load by giving derating factors that apply to table ampacities. For example, if the number of wires exceeds three but is fewer than seven, the ampacity is only 80% of the table value; if the number exceeds six but is fewer than 11, 70%; more than ten but fewer than 21, 50%, and so on. However, if the raceway is not over 24 in. long (classified as a nipple), the *NEC* assumes heat will escape from the ends of the raceway and the enclosed conductors need not have their ampacity derated.

Count only current-carrying conductors for derating calculations. Grounding conductors (covered in Chapter 9) are never counted. A neutral wire that carries only the unbalanced current of a circuit (such as the neutral wire of a three-wire, single-phase circuit, or of a four-wire, three-phase circuit) is not counted for derating purposes in some cases. But remember: Grounded wires are not always neutrals. (Grounded wires are covered in Chapter 9; suffice it to say for now that in most modern systems one conductor is deliberately connected to ground, and neutrals are almost always grounded.) The grounded ("white") wire of a two-wire circuit carries the same current as the hot wire and therefore is *not* a neutral. If you install two such two-wire circuits in a conduit, they must be counted as four wires.

How (and when) to count neutrals   Although neutral conductors are counted for derating purposes only if they are actually current carrying, it is increasingly common in commercial distribution systems derived from three-phase, four-wire, wye-connected transformers to find very heavily loaded neutrals. If the circuit supplies mostly electric-discharge lighting (fluorescent, mercury, and similar types discussed in Chapter 29), you must always count the neutral. In such circuits, the neutral carries a "third harmonic" or 180-Hz current produced by such fixtures. The explanation of why such fixtures produce such a 180-Hz current is far beyond the scope of this book. But be aware that the neutrals of such circuits or feeders must be counted because those harmonic currents add together in the neutral instead of canceling out, and therefore require derating. The remainder of the illustrations in this chapter covering feeder ampacity (Figs. 26–5, 26–6, and 26–8) assume (due to typical industrial load profiles) that the neutrals must be counted as current-carrying conductors.

In addition, harmonic currents from nonlinear loads (see page 47) often add instead of canceling in three-phase, four-wire, wye-connected neutrals as well. In fact, there are conditions where such loads cause the neutral to carry *more load* than the ungrounded conductors. There are now Type MC cable assemblies manufactured with oversized neutral conductors to address this problem. Finally, any time you run just two of the three phase conductors of a three-phase, four-wire system together with the system neutral, that neutral always carries approximately the same load as the ungrounded conductors and must always be counted. This arrangement is very common on large apartment buildings where the feeder to each apartment consists of two phase wires along with a neutral, but the overall service is three-phase, four-wire.

However, the neutral of a true single-phase, three-wire system (review last page of Chapter 2) need not be counted, because harmonic currents fully cancel in these systems. The overwhelming majority of single-family and small multifamily dwellings and most farms have these distributions, which greatly simplifies your wire selection calculations.

How (and when) to count control wires   Many types of control wires, such as motor control wires and similar wires that carry only intermittent or insignificant amounts of current need not be counted, provided the continuous control current does not exceed 10% of the conductor ampacity. For example, a motor starter with a 120-volt coil might require 100 VA to operate, or less than 1 amp. Since that is an intermittent load, and in any event it is less than 10% of the ampacity of 14 AWG, you could use 14 AWG wires for this control circuit and run them in the same raceway as the power wires without counting them for derating purposes. Since derating applies to every current-carrying wire in the conduit, for a typical three-wire control circuit this special allowance has a major impact on the ampacity of those power wires. For a normal three-phase motor circuit running in the same conduit, it allows the power wires to be counted at 100% of their table value instead of 80%.

Conductor ampacity derating   Now that you know how to count the number of current-carrying wires in a conduit, it's time to learn how to apply *NEC* rules to the result. Using the *NEC* directly means going from the ampacity table to the derating factor (by which you multiply), and comparing the result to the load. That's great for the inspector who checks your work, but it doesn't help you pick the right conductor in the first place. You want to go the other way: Knowing the load, you want to select the proper conductor. Figure 26–5 shows an example, using the feeder with 51 amps of noncontinuous load and 68 amps of continuous load.

Refer to Fig. 26–5. Suppose you have two of those feeders supplying identical load profiles and run in the same conduit. That would be (as previously noted we are counting the neutrals) eight current-carrying conductors in the raceway. *For this part of the analysis, ignore continuous loading and termination problems.* Remember, you should be using a fresh sheet of paper for this calculation.

Start with 119 amps of actual load (51 amps + 68 amps), and *divide* (you're going the other direction, so you use the opposite of multiplication) by 0.7 [the factor in Table 310.15(B)(2)(a)], to get 170 amps in this case. In other words, any conductor with a table ampacity that equals or exceeds 170 amps will mathematically be guaranteed to carry the 119-amp load safely. A 1/0 AWG THHN conductor, with an ampacity of 170 amps, will carry this load safely under the conditions of use, and it might appear to work. Whether it represents your final choice depends on the outcome of the analysis explained in the later topic on choosing a conductor.

Ambient temperature problems   High ambient temperatures hinder the dissipation of conductor heat just as in the case of mutual heating. To prevent overheating, the *NEC* provides ambient temperature derating factors at the bottom of the ampacity tables. Suppose, for example, circuit conductors go

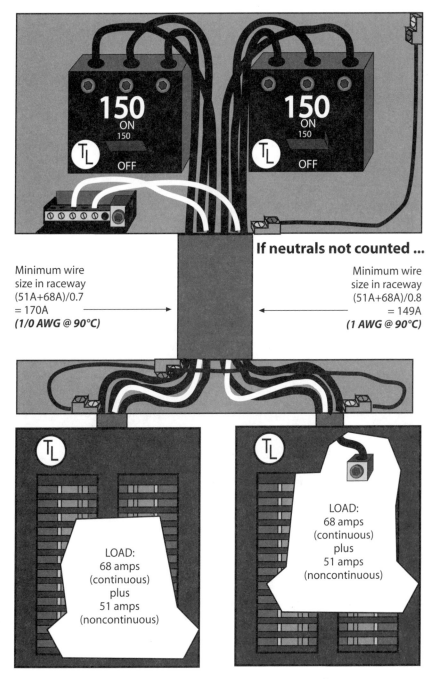

Fig. 26–5 An example of more than three current-carrying conductors in a common raceway. Assume the raceway is longer than 24 in. and that the neutrals must be counted due to harmonic load content.

through a 35°C ambient. Their ampacity goes down (for 90°C conductors) to 96% of the base number in the ampacity table. Here again, start with 119 amps, and *divide* by the 0.96 to get 124 amps. Any 90°C conductor with an ampacity equal to or higher than 124 amps would carry this load safely. What happens if, as shown on the right side of Fig. 26–6, you have *both* a high ambient temperature and mutual heating? Divide twice, once by each factor. In this case:

$$119 \div 0.7 \div 0.96 = 177 \text{ A}$$

A 2/0 AWG THHN conductor (ampacity = 195 amps) would carry this load without damaging itself. Again, this would be true whether or not the load was continuous, and whether or not the devices were allowed for use with 90°C terminations. Don't cheat; the termination calculations are still supposed to be locked up in another drawer.

When diminished ampacity applies only to a small part of the run   You will run into installations where most of a circuit conforms to *NEC* Table 310.16 but a small portion requires very significant derating. For example as illustrated in Fig. 26–7, your circuit may be 207 ft long, with 200 ft in normal environments and 7 ft of it passing through a corner of a boiler room with a very high ambient temperature. The *NEC* generally observes the weak-link-in-the-chain principle and requires the lowest ampacity anywhere over a run to be the allowed ampacity. However, for very short runs where the remainder of the circuit can function as a heat sink, the *NEC* allows the higher ampacity to be used.

Specifically, any time ampacity changes over a run, determine all points of transition. On one side of each point the ampacity will be higher than on the other side. Now measure the length of the wire having the higher ampacity (in this example, the portions *not in* the boiler room) and the length having the lower ampacity (in this example, *in* the boiler room). Compare the two lengths. *NEC* 310.15(A)(2) Exception allows you to use the higher ampacity value beyond the transition point for a length equal to 10 ft or 10% of the circuit length having the higher ampacity, whichever is less.

In this case of the 200-ft run beyond the 7 ft in the boiler room, 10% of the circuit length having the higher ampacity would be 20 ft, but you can't apply the rule to anything over 10 ft. Since 7 ft is less than or equal to 10 ft (and less than the 10% limit of 20 ft), the exception applies and you can ignore the ambient temperature in the boiler room in determining the allowable ampacity of the wires passing through it. In the words of the exception, the "higher ampacity" (which applies to the run outside the boiler room) can be used beyond the transition point (the boiler room wall) for "a distance equal to 10 ft or 10% of the length figured at the higher ampacity, whichever is less."

CHOOSING A CONDUCTOR

Now you can unlock the drawer and pull out the termination calculations. Put both sheets of paper in front of you, and design for the worst case by installing

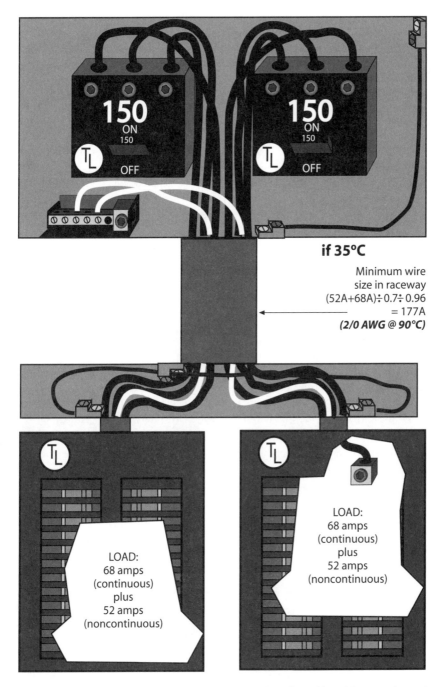

Fig. 26–6 The two feeders in Fig 26–5 as they would be affected by adding an elevated ambient temperature.

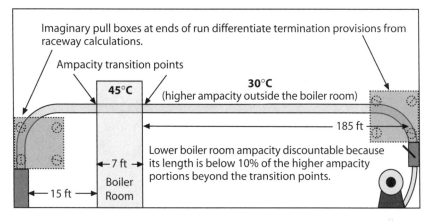

Fig. 26–7 There is a limited exception to the weak-link-in-the-chain principle.

the largest conductor that results from those two independent calculations. The termination calculation (Fig. 26–3) came out needing conductors sized, under the 75°C column, no less than 136 amps even though the actual load was only 119 amps. You could use 1/0, either THHN or THW.

Suppose you put two feeders (eight conductors) in a conduit, as in Fig 26–5. The termination calculation comes out 1/0, and as we've seen, the raceway derating calculation also comes out 1/0 AWG THHN. Here, the termination rules happen to agree with the raceway analysis. If the same raceway also goes through the area with high ambient temperature, however, you will need 2/0 THHN. Now the raceway conditions are limiting and you size accordingly. Figure 26–8 shows the calculations under both conditions. Finally, suppose for a moment that the neutrals do not carry heavy harmonic loading and therefore need not be counted, and suppose the ambient temperature does not exceed 30°C, as shown on the right side if Fig. 26–5. With only six current-carrying conductors, the derating factor drops to 0.8, so the required ampacity becomes 119A/0.8 = 149A, seemingly allowing 1 AWG THHN (90°C) conductors. Now the termination rules (shown on the right side of fig. 26–3) would be the worst case and would govern the conductor selection process.

Remember, the terminations and the conductor ampacity are two entirely separate issues. Just because you need to use the 75°C column for the terminations doesn't mean you start in that column to determine the overall conductor ampacity. Go ahead and make full use of the 90°C temperature limits on THHN, and its resulting ampacity, to solve derating problems.

Don't get confused by the fact that the derating factor for continuous loads (0.8) is the same as the one for four to six conductors in a raceway (0.8). This is only a coincidence. One applies to devices and terminal heating at the end of a conductor, and the other applies over the interior of the raceway. They never apply at the same point in a circuit because the technical basis for each is

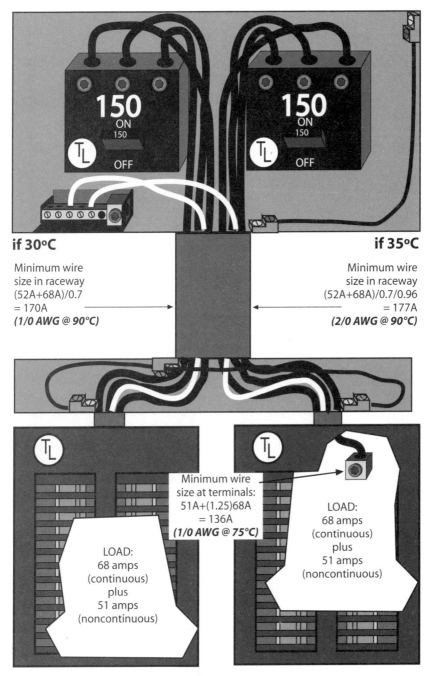

if 30°C

Minimum wire
size in raceway
(52A+68A)/0.7
= 170A
(1/0 AWG @ 90°C)

if 35°C

Minimum wire
size in raceway
(52A+68A)/0.7/0.96
= 177A
(2/0 AWG @ 90°C)

Minimum wire
size at terminals:
51A+(1.25)68A
= 136A
(1/0 AWG @ 75°C)

LOAD:
68 amps
(continuous)
plus
51 amps
(noncontinuous)

LOAD:
68 amps
(continuous)
plus
51 amps
(noncontinuous)

Fig. 26–8 Calculations for termination limitations must be divorced from run calculations. In some cases the termination restrictions produce the worst case, and in others the run calculation will be the limiting factor. As previously, these calculations assume a raceway length greater than 24 in.

entirely different. The middle is never the end, so never apply the rules for one to the other.

If you still aren't sure, review again Figs. 26–3, 26–5, and 26–8, which illustrate different instances of similar connected loading. In one case the termination requirements force an increase in conductor size. In another case the termination requirements agree with the results of the ampacity calculations; and in yet another case the ampacity calculation forces you to use a larger conductor than what the termination rules would predict.

The conductor must always be protected   Don't lose sight of the fact that the overcurrent device must always protect the conductor. For 800-amp and smaller circuits, *NEC* 240.4(B) allows the next higher standard size overcurrent device to protect conductors. Above that point, *NEC* 240.4(C) requires the conductor ampacity to be no less than the rating of the overcurrent device. As a final check, be sure the size of the overcurrent device selected to accommodate continuous loads protects the conductors in accordance with these rules; if it doesn't you will need to increase the conductor size accordingly. Refer to the discussion of noncontinuous loads, in the paragraph after the next, for an example of where, even after doing both the termination and the ampacity calculations, this consideration forces you to change the result.

Small conductors   Small conductors (14, 12, and 10 AWG) present an additional wrinkle. The *NEC* imposes special limitations on overcurrent protection beyond the values in the ampacity tables for these small wires. Generally, the overcurrent protective device for 14 AWG wire can't exceed 15 amps; for 12 AWG wire, 20 amps; for 10 AWG wire, 30 amps. The higher ampacities of these conductors remain what the table says they are, however, and there are cases, notably including motor circuits, where this restriction doesn't apply. In general, perform all ampacity calculations as described previously, based on the ampacity table limits. But at the very end, make sure your overcurrent device doesn't exceed these specific ampere limits unless you fall into one of the exceptions specifically tabulated in *NEC* 240.4(G).

Noncontinuous loads   Suppose none of the load is continuous on our 150-amp feeder in Fig. 26–3, and suppose the character of the load on the neutrals does not require them to be classified as current-carrying conductors. Looking at the two feeders in Fig. 26–5, suppose the ambient temperature doesn't exceed 30°C. The termination need not include any phantom load allowance, but it still needs to assume 75°C termination restrictions. A 1 AWG conductor will carry the actual 120-amp load without exceeding 75°C, and therefore would seem to be usable until you consider the mutual effects of multiple conductors in the common raceway. Suppose you went to a 1 AWG THHN conductor, ampacity 150. Will it carry the 120-amp load safely? Yes, because 150 A × 0.8 (the derating factor for six conductors) = 120 A. Will it overheat the breaker terminations? No, because 1 AWG copper is 1 AWG copper no matter what insulation style it has around it, and it won't rise to 75°C until it carries 130 amps. But its final derated ampacity within the conduit is 120 amps. The next higher standard-sized overcurrent device is 125 amps. The 150-amp breaker does not protect this wire under these conditions of use, and has to be reduced to 125 amps, or else you need to increase the wire size to 1/0 AWG.

WIRES IN PARALLEL

Instead of using a very large wire, *NEC* 310.4 permits two or more smaller wires to be connected in parallel for use as a single wire. There are seven conditions, all of which must be met. All the wires that are to be paralleled to form the equivalent of one larger wire must (1) be 1/0 AWG or larger, (2) be of the same material (all copper, all aluminum, all copper-clad aluminum, etc.), (3) have precisely the same length, (4) have the same cross-sectional area in circular mils, (5) have the same type of insulation, (6) be terminated in the same manner (same type of lug or terminal on all), and (7) be arranged so the same number of conductors run in each paralleled raceway or cable.

Keep the wires to be paralleled as equal as possible   The seven conditions noted in the previous paragraph must be met as exactly as possible. A very slight difference in length, for example, produces a slight difference in resistance and therefore impedance. Even slight differences in impedance can cause very significant inequalities in the way current divides between the conductors in the parallel circuit. Any inequality in current division is a potential hazard because paralleled runs don't, by definition, have individual overcurrent protection on each conductor. Consider a 500-amp parallel make-up with two legs sized for 250 amps each. If one of the legs carries 300 amps and the other 200 amps, the overloaded leg will burn up without any immediate response from the overcurrent device.

It is a wise practice to cut all paralleled conductors of any given phase at the same time, prior to installation, when they can be laid out and cut to exactly the same length. (Note that Phase A wires can be longer than Phase B, etc., as long as all wires comprising any given phase have the same characteristics.) Consider leaving them long enough to loop at terminations if the enclosure is large enough. Then, after pulling them into the conduits, each loop can be adjusted in size easily to take up any additional length resulting from the geometry of the paralleled run and the position of the terminals relative to the conduit entry point. If you must cut a conductor, measure the amount that you cut and then trim the others to the same length.

As an example of a practical application of paralleled circuiting, consider a task of providing a three-wire circuit carrying 490 amps. According to *NEC* Table 310.16 (75°C column), you will need 800-kcmil wire. In place of each 800-kcmil wire, you can use two smaller wires each having half the required ampacity, or 245 amps. According to *NEC* Table 310.16, two 250-kcmil wires each with an ampacity of 255 would appear to be suitable.

But if you plan to run all six wires through one conduit, 250-kcmil wire is not suitable because you will have six wires in one conduit, and therefore must derate the wire's normal ampacity of 255 to 80%. The ampacity of the 250-kcmil wire then becomes 80% of 255, or 204. Use 350-kcmil wire with an ampacity of 80% of 310, or 248.

Use parallel raceways as well as wires   Instead of running all wires through one conduit as described in the previous paragraph, use two separate conduits with three wires in each (or instead of two conduits you may use three or even more conduits provided all the requirements outlined are met); then derating is not required and 250-kcmil wire is suitable. But note that the conduits or other raceways through which the wires run must have the same physical character-istics; all might be steel rigid conduit, or all might be EMT. It would not be permissible to use one steel and one aluminum conduit.

When running paralleled wires through a raceway, special care must be used to meet the requirements of *NEC* 300.20. One wire of each phase must be run through each conduit. See Fig. 26–9 which shows six wires in two conduits. Wires *A* and *a* are paralleled with each other, but they must not run together in the same conduit. Likewise wires *B* and *b* are paralleled, and again must not run in the same conduit. The same applies to *C* and *c*. If you ran, for example, *a, b,* and *B* in one conduit, and *c, A,* and *C* in the other, each conduit would contain wires of only two of the three phases. In that case considerable induced and eddy currents would be set up in the conduits themselves, resulting in excessive heating and power losses.

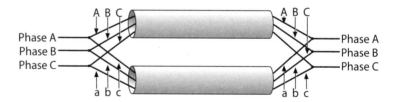

Fig. 26–9 Where paralleled wires are run through conduit, each conduit must contain one wire from each phase. The installation shown does not include a neutral.

There is one common exception to the rule about running one of each of the phase conductors in each raceway. These are referred to as "isolated phase installations" in *NEC* 300.5(I) Exception 2. If you're using nonmetallic raceways in an underground application, you can run all the *A* phase conductors in one conduit, all the *B* phase conductors in another, and so on. You have to be sure that when you enter the switchboard or other enclosure that you meet *NEC* 300.20, such as by providing a nonmetallic window through which all the conduits enter, preventing inductive reactance. Often these large installations enter a pit in a concrete floor below the switchboard, for example, and having all of one phase together simplifies landing the conductors on their terminations. It also facilitates making all the conductors the same length.

The neutral wire of a feeder or circuit might often be smaller than the hot wires. If that is the case in your installation, the neutral also must be paralleled even if its ampacity is less than that of the hot wires, just so that their combined ampacity is equal to that which would be required if the neutral were not paralleled. But the 1/0 threshold still applies, and the paralleled neutrals cannot be reduced below that size. This is shown in Fig. 26–10, which is the same as Fig. 26–9 except that the neutral has been added, labeled *N* and *n*. If the wires are installed in a nonmetallic raceway, the raceway cannot serve as the equipment grounding conductor, and in that case an equipment grounding conductor must usually be installed in each raceway.

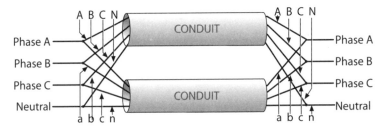

Fig. 26–10 Same as Fig. 26–9 except a neutral has been added. Each conduit must contain one wire of the neutral. In each illustration, while three ungrounded wires are shown, there might be only two (or even one if a grounded wire is also installed).

Conductors in parallel are quite common where using one wire would require a very large wire. The smaller wires have much higher ampacity per thousand circular mils than the larger one. Thus much less copper by weight is used when two smaller wires are paralleled to replace one larger one. The total installation cost is reduced. However, when you use two smaller wires in place of one larger wire, the combined circular-mil area of the two smaller wires will be less than that of one larger wire. That will result in a somewhat greater voltage drop than you would have using one wire of the larger size. If the circuit is long, you might have to increase the size of the paralleled wires to offset the difference in voltage drop.

Special rules for sizing grounding conductors in parallel make-ups   If you use separate grounding conductors for any reason, as in the case of rigid nonmetallic conduit, refer to the size of the overcurrent protective device and choose an equipment grounding conductor no smaller than required by *NEC* Table 250.122. Run this full-size grounding wire in each conduit. The *NEC* imposes this requirement because there are cases when a single such equipment grounding conductor will be called upon to return fault current contributed by all of the paralleled conductors of the faulted phase due to backfeed into a fault through the downstream connections. Note, however, that the 1/0 AWG threshold that normally applies to paralleled conductors does not apply to equipment grounding conductors. For example, if you run two nonmetallic conduits with 3/0 AWG THWN (200 amp) conductors in each for a 400-amp feeder, the equipment grounding conductor in each run must be at least 3 AWG (*NEC* Table 250.122). This is larger than the 6 AWG required for 200-amp circuits, but smaller than the 1/0 AWG minimum that applies to other paralleled wiring.

This rule also applies to cabled wiring methods run in parallel. Type MC cable is very commonly used in large feeders. Suppose you need to run a 600-amp feeder, and you decide to run two parallel 350-kcmil cables (75°C ampacity = 310 A each). Be sure you inspect the grounding conductor in the cable. The manufacturer probably assumed that his 350-kcmil cable would be used on a 300-amp circuit, and *NEC* Table 250.122 lists a 4 AWG equipment ground as adequate in such cases. Number 4 AWG is far below the 1 AWG required for 600-amp circuits. Each 350-kcmil cable run must include the 1 AWG grounding conductor. The proper cable is available, but usually only on special order. There is an elaborate exception in the *NEC* on this point, but it is entirely academic because the equipment required to comply with it is not now in production, and it requires a degree of expert supervision generally applicable to installations beyond the scope of this book.

CHAPTER 27

Nonresidential Wiring Methods and Materials

THIS CHAPTER OPENS with the uses of enclosures such as conduit bodies, pull and junction boxes, and outlet boxes in configurations seldom used in residential applications. It also covers the support of large conductors in long vertical runs, and special grounding rules for terminating raceways on systems over 250 volts to ground. The wiring methods discussed in the rest of the chapter could be (and occasionally are) applied in residential occupancies, but it would rarely be cost effective to do so. While studying this chapter, consider that the wiring methods covered in Chapter 11 are not exclusively residential applications. Conduit, EMT, Type MC cable, and the various flexible raceways are commonly used for industrial work. Think of this chapter as working with Chapter 10, "Outlet and Switch Boxes," and Chapter 11, "Wiring Methods," to round out the coverage of nonresidential electrical enclosures and wiring methods.

HOW TO CONNECT WIRING TO ENCLOSURES

For exposed runs of conduit, ordinary outlet boxes may be used, but cast conduit fittings of the type shown in Fig. 27–1 are sturdier, neater, and more frequently

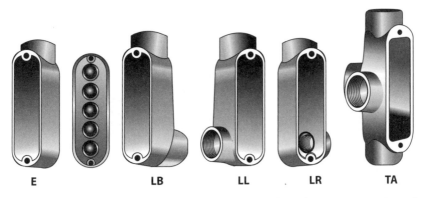

| E | LB | LL | LR | TA |

Fig. 27–1 For exposed runs of conduit, fittings of the type shown here are commonly used. There are dozens of different shapes or types.

used. They are known by many trade names such as Electrolets, Condulets, Unilets, etc. These devices are actually specialized forms of outlet boxes, but instead of having removable knockouts, they have one or more threaded (or setscrew for EMT, or solvent-welded for rigid nonmetallic conduit) openings or hubs. With a very few basic body shapes, dozens of different combinations of openings are available. Each opening is sized for the trade size of conduit for which the fitting is designed. Note that the regular threaded fittings can be used with EMT because the thread of an EMT connector will fit the threaded opening of the fitting.

A few of the more common types are shown in Fig. 27–1. While the illustration shows only one cover, naturally you must install a cover on each fitting after the installation is completed. Type E with the cover at its right is used at the end of a run to some types of motors, transformers, or similar equipment not having provisions for a conduit connection. The other fittings shown avoid awkward bends in conduit. Type LB is commonly used where conduit must run through a wall or ceiling; it is equally useful in going around a beam or similar obstruction. Types LL and LR are handy for 90-degree turns where the mounting structure is not suitable for an LB. However, the wires are more difficult to feed into an LL or LR than into an LB. There are very many kinds of combinations, some as complicated as the Type TA with four openings, which obviously is not used very often; Type T is similar but has only three openings, without the opening in the back or bottom side (not bottom end) of it.

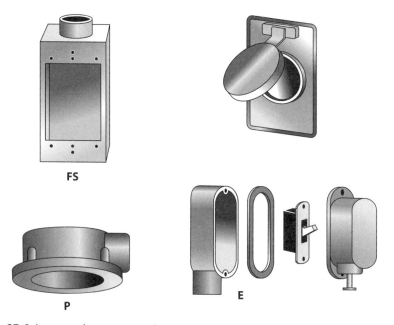

Fig. 27–2 Larger enclosures are used to support switches, receptacles, fixtures, etc.

Don't confuse conduit bodies with boxes that have threaded hubs, such as the Type FS and Type P boxes in Fig. 27–2. These are boxes and are subject to the same volume calculations for boxes as covered at the end of Chapter 10. Also shown in Fig. 27–2 is a Type E with a special gasketed cover that permits installation of a toggle switch outdoors. On exposed runs of conduit, lighting fixtures can be mounted on round Type P fittings.

Most conduit bodies can be used for two functions: (1) to simply redirect and provide access to a raceway interior and (2) to function as a box that contains splices or supports devices or fixtures. One group, the "short radius" conduit bodies (Fig. 27–3), may be used only for the first function and must never be used as boxes. These include such items as handy ells, capped elbows, and service-entrance elbows. They have no fill restrictions beyond those of the entering raceways (review Chapter 11). They cannot be used in any other way, and since they don't meet minimum requirements for wire bending space, they cannot be used for conductors larger than 6 AWG.

Conventional conduit bodies must have a cross-sectional area at least double the area of the largest conduit or tubing to which they are connected; they must be marked with their volume, and they can be used as boxes if the normal volume calculations for boxes (see Chapter 10) allow it. Relatively few conduit bodies have enough volume to allow use as boxes. For example, a typical ¾-in. tee conduit body has a volume of about 10 cu in., and splicing a run of just two 14 AWG wires would require 12 cu in. (six wires—two in, two off to the side, and two continuing on, at 2 cu in. each, for a total of 12 cu in.).

Fig. 27–3 "Short radius" conduit bodies only enable raceway installation and do not function as boxes. Due to restricted size, they cannot be used for 4 AWG or larger conductors.

Raceway support of boxes—general requirements   Although boxes are normally supported by structural elements of the building, under some circumstances they can be supported from the wiring method itself without relying on any other support. You can use only rigid or intermediate metal conduit for this purpose, no other wiring method, and the *NEC* includes a number of restrictions on this procedure. In general, conduit bodies come under the same rules, although exceptions assure that as long as the required distances to conduit supports are met, conduit bodies can be supported by their entering raceways regardless of size and wiring method.

However, to qualify for these exceptions, the conduit bodies must be no larger than the largest trade size of the entering raceways, and the conduit bodies must not support devices or fixtures. In other words, as long as the conduit body is the same trade size as the supporting raceway, and not called on to support anything besides itself, the *NEC* doesn't ask for any additional support. If otherwise, the *NEC* treats it as a box, and to avoid independent support it must meet all of the requirements for boxes supported on a raceway, including the restriction to rigid or intermediate metal conduit as the wiring method. Conduit bodies are very difficult to support directly; they have no mounting feet, etc., so these considerations must be kept in mind. The following general rules apply to raceway-supported boxes, and to raceway-supported conduit bodies if they have to be installed as boxes due to being larger than the raceway trade size or due to device or fixture support:

- The box size can't exceed 100 cu in.
- The box must have threaded entries. These are usually cast into the box (as in the FS box, Fig. 27–2), but hubs of the sort pictured in Fig. 27–4 can also be added to a sheet metal box, and once added they qualify the box for this procedure. This rule prohibits under any circumstances the use of locknuts on the end of conduit runs as the means of supporting boxes.

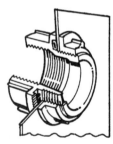

Fig. 27–4 This fitting adds a threaded hub in the field. The two pieces have serrated edges that bite into the enclosure wall when tightened, making a very secure entry. It is available with an O-ring seal, allowing it to be used in wet locations.

- The conduits must be threaded wrenchtight into the box. There are connectors, similar to those shown in Fig. 11–18 for EMT, that allow rigid (and intermediate) metal conduits to be terminated at boxes without having to thread the conduit. These connectors do not qualify under this rule. If the problem, because of job-site geometry, is an inability to twist the conduits into the box, use a short nipple connected to a union. Unions allow threaded conduits to be joined without rotation, and they don't reduce the strength of the overall run.

- There must always be two or more conduits used for support. A box on the end of a single conduit can work loose and twist off. Two or more conduits make that impossible. The only exception is the Type E conduit body shown in Fig. 27–1. Aside from the obvious impossibility of arranging a double entry to a fitting with only one entry, Type E conduit bodies aren't much wider than their supporting conduits. If they are installed wrenchtight as required, they are unlikely to loosen because any impact on their edges provides very little torque.

■ The conduits must be supported within 36 in. of the box, and they must enter the box on two or more sides. Alternatively, the two or more required conduits can enter the same side of the box if the support distance is reduced to 18 in. If the box supports a fixture or a wiring device, the support distance must be reduced to 18 in. regardless of the direction of conduit entry.

Cantilevered boxes   Figure 27–5 shows a common use of boxes supported by a raceway, and it obviously doesn't conform to the foregoing list of conditions. This discussion also applies to fixtures with wiring compartments that serve in

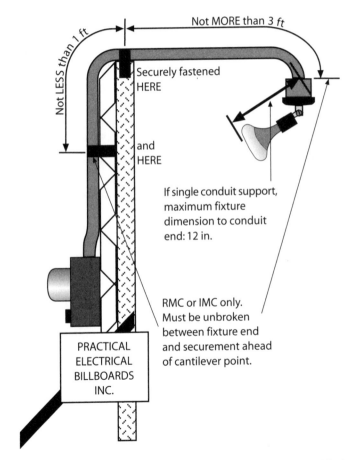

Fig. 27–5 The last point of support is, effectively, a fulcrum for a lever. The *NEC* rules assure that the worst-case mechanical advantage will be 3:1, and the worst-case loading will be 60 lb as imposed on the point of support next back from the last point of support. This is the most vulnerable point in the support system.

lieu of boxes; many outdoor fixtures are of this type. The *NEC* now includes comprehensive rules for these applications:

- The conduit must be securely fastened with the maximum extension not exceeding 3 ft.

- The conduit arm (rigid or intermediate metal conduit only) must be unbroken.

- The length from the end support (the effective fulcrum of this system) to the next point back where the conduit is securely anchored must not be less than 1 ft.

- The conduit arm and fixture must not be where passersby might easily twist, hit, or hang on them. Unless inaccessible to unqualified persons, the minimum height is 8 ft and the minimum horizontal reach is 3 ft from balconies, etc. Some billboards qualify for bottom-mounted fixtures by having fences that restrict unqualified access.

- No fixture supported by a single conduit can exceed 12 in. in any direction from the point of conduit entry. This reduces wind loading and minimizes possible torque on the conduit threads at the fixture end.

- No single conduit can carry more than 20 lb; with the 3:1 worst-case leverage, the highest static load on the supply end is 60 lb.

- The conduit must be threaded wrenchtight into the box (or conduit hub like the one in Fig. 27–3); a connector, such as a setscrew connector, cannot be used.

Pendant boxes   Figure 27–6 shows another application, also common, that doesn't comply with the usual rules for raceway support of boxes. As in the case of the cantilevered box, this discussion also applies to fixture wiring compartments that serve in lieu of boxes. Here are the rules:

- The pendant must be rigid or intermediate steel conduit, threaded wrenchtight into the box or fixture wiring compartment. Field-installed hubs (Fig. 27–4) are permitted, and the pendant may have more than one conduit length coupled together.

- If the stem is over 18 in. long, the connection at the supply end of the conduit must be flexible. The threaded area on a piece of conduit is its weakest part by far, and this rule minimizes the danger of broken conduit threads resulting from mishaps such as someone running into the fixture with a ladder.

- If only a single conduit is involved, the conduit connection at the box must be protected from being easily or inadvertently loosened. Protection can be accomplished by adding setscrews to the conduit joint, or by installing the fixture so its lowest point is at least 8 ft above the floor or standing platform and it is at least 3 ft horizontally from a reaching exposure.

- The fixture must not exceed 12 in. in any horizontal dimension if there is only a single support. This controls the mechanical advantage imposed by the edge of a fixture farthest from the conduit entry, lessening the danger to the conduit threads at the point of conduit entry.

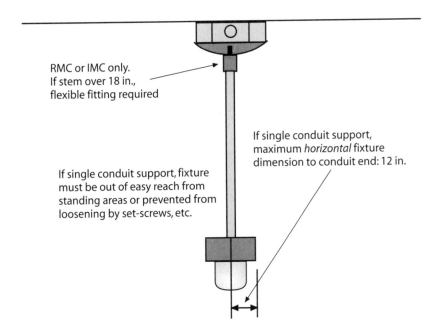

RMC or IMC only.
If stem over 18 in.,
flexible fitting required

If single conduit support,
maximum *horizontal* fixture
dimension to conduit end: 12 in.

If single conduit support, fixture
must be out of easy reach from
standing areas or prevented from
loosening by set-screws, etc.

Fig. 27–6 Pendant boxes on raceway stems don't need any special weight limitations beyond the normal rule that fixtures over 50 lb must be supported directly to structural elements of the building unless the box (at the upper end in this case) is listed for a greater capacity.

Flexible cords also support pendant boxes, such as the ones often used to enclose the controls to an electric hoist. These boxes must be supported by the cord in a way that protects the circuit conductors from strain. There are strain-relief connectors designed for this purpose, as shown in Fig. 18–21, and they should be threaded into a box equipped with a hub.

Junction and pull boxes   Wires of the smaller sizes are sufficiently flexible to be readily pulled into conduit. If the runs are long, boxes of the type described in Chapter 10 as well as the conduit fittings shown in Figs. 27–1 and 27–2 may (within their capacity limits) be installed as pull boxes. All boxes must be installed where they will be permanently accessible without damaging the structure of the building.

The larger the wire, the more difficult the pulling process becomes. For wires 4 AWG and larger, and particularly in the large kilocircular-mil (kcmil) sizes, it becomes increasingly difficult to pull them into conduit, especially if there are long runs or several bends. Then it becomes necessary to install pull boxes in strategic locations; many times they are used in place of conduit elbows. A pull box, as the name implies, is a box located so that the wires can be pulled more easily into the conduit from one outlet to the next. A single box is often used for a number of runs of conduit, as shown in Fig. 27–7. Boxes are also used where wires have to be spliced. In many cases, a box is used as both a splice box and a pull box. All such boxes are called junction boxes.

For wires 6 AWG and smaller, boxes discussed in Chapter 10 may be used. If the conduit contains 4 AWG or larger wires, or if cables are used containing 4 AWG or larger wires, *NEC* 314.28 specifies the minimum dimensions of junction boxes. (If cables are used, determine the size of the conduit that would be required to contain the wires in the cable as if they were separate wires; to determine the required size of

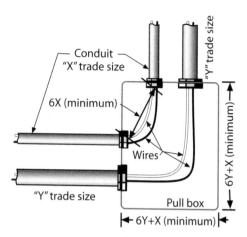

Fig. 27-7 Junction boxes are used with conduit to make it easier to pull wires through long lengths of conduit, or for splicing wires, or for both purposes in the same box.

the junction box, pretend that that size of conduit is being used instead of the cable.) For straight pulls (conduit entering one wall of box and leaving by opposite wall) the length of the box must be at least eight times the trade size (not the outside diameter) of the largest conduit entering the box; several lengths of conduit may run into and out of the box. If the largest size of conduit is the 3-in. size, the box would have to have a minimum length of 24 in.

For angle or U pulls, if only one conduit enters any one wall of the box, the distance to the opposite wall must be at least six times the conduit size. But if more than one conduit enters the box, the minimum distance to the opposite wall must be six times the size of the largest conduit plus the sizes of each of the other conduits entering the box. If there are multiple rows of conduit in that wall, choose the row that calculates as requiring the largest dimension. If, for example, only one 2-in. conduit enters the box, its minimum length must be 12 in. But suppose two 2-in. conduits plus two 1½-in. conduits enter the box. Then the minimum length of the box is 12 + 2 + 1½ + 1½, or 17 in. On the other hand, if the two 1½-in. conduits entered on a second row, the dimension would drop to just 14 in.

What about a conduit entering the back of a pull box, with its wires exiting one of the sides—does the depth of the box have to be six times the raceway diameter, since those conductors are part of an angle pull? The answer is no. An exception waives the six times rule when wires enter opposite a removable cover. In this case the box depth is the same as that for a conduit body, that is, the minimum bending radius per *NEC* Table 312.6(A). For example, if a 3-in. conduit enters the back of a pull box with 500 kcmil wires, the box has to be 6 in. deep, or 8 in. deep for 600 kcmil.

There is one other requirement. The distance from the center of the conduit by which a set of wires enters a junction box to the center of the conduit by which it leaves the box must be at least six times the trade size of the conduit. This rule applies to entries for the same conductor in the physical sense, not the electrical sense. That is, if you splice the conductors entering the box from two conduits, although the spliced conductors become one in the electrical sense, the conduits can still enter with no minimum spacing between them. Nevertheless, a pull box used for splices must still comply with the overall dimensional requirement for enclosures used for angle pulls.

All dimensions discussed are *NEC* minimums. It is often good practice to use boxes larger than the minimum size (especially if a large number of conduits enter the box) to avoid crowding, which increases labor and leads to possible damage to insulation.

Handhole enclosures

The 2005 *NEC* officially recognizes for the first time a special pull and junction box designed for use at grade level. Although these boxes have been used for utility applications for many years, they also apply to normal electrical distribution. Figure 27–8 shows one in use. The majority of these enclosures have no bottom, but their covers are often designed for substantial

Fig. 27–8 A handhole enclosure, in this case without a bottom, in use. This one is unusual because it encloses another box instead of housing the open wires directly, but under the 2005 *NEC* it could do so provided both the wires and the splices were listed as suitable for wet locations. In this case the appropriately rated inner box allows for conventional splicing methods.

load. The one in the photo is a good example; its cover is designed for vehicular loading. These enclosures are sized the same way as regular pull boxes if they contain large conductors, except a conduit entry through the bottom would be measured from the end of the conduit instead of using a wall-to-wall dimension. The cover must prominently identify the electrical function of the enclosure, and require the use of tools to open (unless it weighs at least 100 lb). If the cover is conductive (the one in the photo is not), an equipment bonding conductor must be run to the cover and connect to the equipment grounding system in the enclosure. This is crucial. There have been a number of fatalities and injuries resulting from covers energized by hot conductors with worn insulation.

Concrete boxes   Walls and ceilings are often of reinforced-concrete construction. The conduit and the boxes must be embedded in the concrete if the devices in the boxes are to be flush with the surface of the walls. Ordinary outlet boxes may be used, but special concrete boxes are preferred and are available in depths up to 6 in. One of these is shown in Fig. 27–9. The conduit and the boxes must be in position before the concrete is poured. These boxes have special ears by which they are nailed to the wooden forms for the concrete. When installing them, stuff the boxes tightly with paper to prevent concrete from seeping in. When the forms are removed, the conduit and the boxes are solidly embedded; the interiors of the boxes are clean and ready to use. Figure 27–10 shows an installed box.

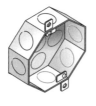

▲
Fig. 27–9 A concrete box designed to be embedded in concrete as it is poured.

Fig. 27–10 Concrete boxes are nailed to wooden forms before the concrete is poured. **▶**

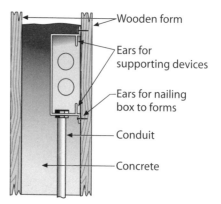

Wooden form

Ears for supporting devices

Ears for nailing box to forms

Conduit

Concrete

Deflection of wires   If wires are bent sharply and press against a protruding surface (as that of a conduit bushing) where they emerge from a conduit, the insulation might be damaged and ground faults might develop. This is especially true of larger wires because of their weight. *NEC* 300.4(F) requires an insulating bushing (or the equivalent) where 4 AWG or larger wires enter or leave the conduit, such as at a box, cabinet, or similar enclosure. (The design of a conduit fitting such as those shown in Figs. 27–1 or 27–2, or a threaded hub in a box, eliminates the need for this extra protection.) A metal conduit bushing with an insulated throat (molded insulating material on the inside) is shown in Fig. 27–11. Such bushings are also available with a terminal for a bonding jumper, as shown in the same illustration. A bonding jumper is a commonly used means of assuring the required grounding continuity between a service conduit and the service equipment cabinet, or in cases of higher voltages to ground (discussed later in this chapter).

A bushing made entirely of insulating material is shown in Fig. 27–12. When using such bushings, before the bushing is installed you must install two locknuts, one on the outside of the box or cabinet and another on the inside. This ensures grounding continuity between the conduit and the box or cabinet.

An acceptable substitute for a bushing with an insulated throat is an insulating liner, which is also called a bushing; this is similar to the fiber bushings ("red

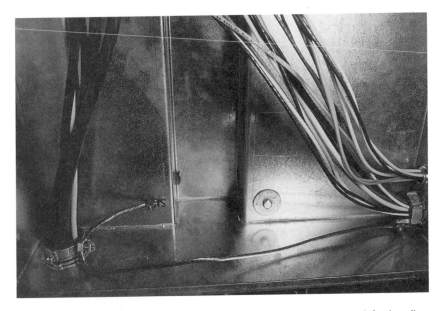

Fig. 27-11 Metal grounding bushing with insulated throat and terminal for bonding jumper.

heads") used with armored cable but of much larger size. They must be of a construction that enables them to be snapped into place and not easily displaced. Usually this type is used only on existing installations where an insulating bushing as described in preceding paragraphs was overlooked in the initial installation.

Fig. 27-12 Conduit bushing made entirely of insulating material.

Locate cabinets so the incoming runs of conduit will be placed in a way that requires minimum deflection of wires where they emerge from the conduit, and so that in vertical runs the weight of the wire will not be supported by a bend in the wire at the end of the conduit run. Figure 27–13 shows the wrong method, and Fig. 27–14 the right method. Let the bends in the wire be sweeping and gentle rather than abrupt.

Supporting vertical runs of wire   Terminals on panelboards and similar equipment are not designed to support any substantial weight. Where there is a vertical run of wire, the weight of the wire is quite considerable, especially in the larger sizes. If such vertical runs are connected directly to terminals, damage may result. *NEC* 300.19(A) requires that wires be supported at the top of a vertical raceway, or as close as practical to the top, and at additional intervals as shown in Table 27–1. If the vertical portion of a raceway is less than 25% of the

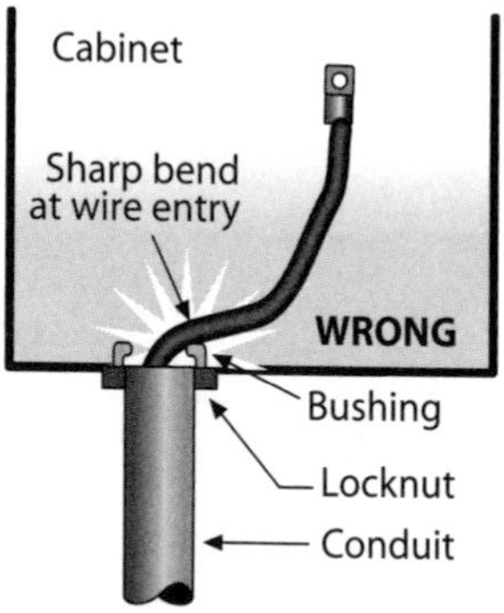

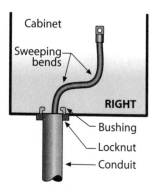

Fig. 27–13 The bend where the wire emerges from the conduit is too abrupt. The wire rests on the conduit bushing, which could lead to damage to the insulation.

Fig. 27–14 The bend in the wire should be gentle and sweeping. This helps to prevent ground faults where wires emerge from conduit.

spacing shown in Table 27–1, no supports are required.

A common way of accomplishing the required supports is to use the fitting shown in Fig. 27–15. Boxes or cabinets are installed at the required intervals, and the fitting is installed in the conduit as it enters the bottom of the cabinet. Be sure one is installed at the topmost cabinet.

Fig. 27–15 This fitting is very effective in supporting vertical runs of wire. ▶

Table 27–1 FROM *NEC* Table 300.19(A)* SPACINGS FOR CONDUCTOR SUPPORTS			
	Conductors		
AWG or circular-mil size of wire	**Support of conductors in vertical raceways**	**Aluminum or copper-clad aluminum**	**Copper**
18 AWG through 8 AWG	Not greater than	100 feet	100 feet
6 AWG through 1/0 AWG	Not greater than	200 feet	100 feet
2/0 AWG through 4/0 AWG	Not greater than	180 feet	80 feet
Over 4/0 AWG through 350 kcmil	Not greater than	135 feet	60 feet
Over 350 kcmil through 500 kcmil	Not greater than	120 feet	50 feet
Over 500 kcmil through 750 kcmil	Not greater than	95 feet	40 feet
Over 750 kcmil	Not greater than	85 feet	35 feet

*Reprinted with permission from NFPA 70-2005 *The National Electrical Code* © NFPA 2004. (Metric measurements not shown.)

NONRESIDENTIAL APPLICATIONS MAY INVOLVE DIFFERENT GROUNDING RULES

Nonresidential occupancies frequently involve higher voltages to ground that require increased precaution regarding equipment grounding in order to avoid shock hazards. In addition, these occupancies often use electronic equipment that is very sensitive to even small voltages on its equipment grounding connections. The *NEC* has special rules to address these concerns.

Continuity of ground   Preceding chapters show how the various runs of conduit or armored cable tie outlet boxes or other equipment into one continuously grounded system. However, if the voltage of one or more of the wires in a feeder or branch circuit is above 250 volts to ground, *NEC* 250.97 requires special bonding means. One approach is to use any of the methods for services, such as grounding bushings (see Fig. 27–11) or threaded hubs (see also the top and bottom of Fig. 16–13 which shows both of these common approaches used in order to comply with this rule.)

Another approach, which can be used only if the knockout is not oversized and is not the inner portion of a concentric or eccentric knockout, is to use the double-locknut method shown in Fig. 27–16. Install one locknut on the outside, another on the inside of the cabinet, plus the usual bushing on the inside. If EMT or armored cable is used, the shoulder on the connector serves in place of a locknut on the outside of the cabinet. With respect to entering the

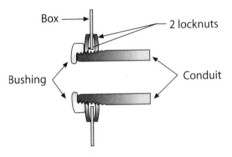

Fig. 27–16 Where the voltage to ground is over 250, use the double-locknut construction shown here instead of the ordinary construction shown in Fig. 10–13.

inner portion of concentric and eccentric knockouts, the *NEC* now allows for test labs to list such knockouts as providing acceptable grounding continuity. Boxes that are so listed will have their smallest unit shipping cartons marked. Most of the common outlet boxes such as 4-in.-square boxes and others are now being listed accordingly, but larger pull boxes and cabinets usually have not been evaluated, and in the case of an existing box there is no way of knowing.

Isolated ground receptacles and equipment   Equipment grounding conductors serve as inadvertent antennas for spurious injections of electronic noise from other parts of the wiring system. Occasionally those random currents will be enough to confuse the ground reference for sensitive electronic equipment, causing data errors. To address this problem, the *NEC* allows an additional insulated equipment grounding conductor to be run to sensitive equipment. These grounding conductors remain insulated from the normal equipment grounding return paths in the building as far back as desired. However, these conductors must ultimately combine with the normal equipment grounding return path at the service, or at the main disconnect for a second building supplied from another building, or at the source of a separately derived system as covered in Chapter 28 under that heading.

In the case of cord- and plug-connected equipment, use an isolated ground receptacle, which is identified by an orange equilateral triangle on its face. Frequently these receptacles are entirely orange in color, but the only Code rule is for the orange triangle, which could appear on receptacles of different colors and still mean the same thing. These receptacles have their yokes insulated from the U-ground holes in the receptacle face, and the U-ground holes alone connect to the green grounding screw on the receptacle body. That means that a green-insulated equipment grounding conductor connected to one of these receptacles will return fault current from whatever is plugged into the receptacle, but will not accumulate other electronic noise on its way back to the system's point of origin. Figure 27–17 shows how to circuit one of these receptacles in a second building fed from a parent building.

The NEC does not and never has allowed these conductors to return to a separate grounding electrode. To have them do so is extremely hazardous, though some wiring specifications still ask for it. Fault current in such cases will need to pass through the earth between the separate electrode and the normal system electrode in order to complete the circuit at the normal power source. The inherent impedance in earth connections makes it probable that under such conditions a fault will leave a dangerous voltage on the equipment. For example, suppose the main electrode system has an impedance of 10 ohms to ground, and the new electrode has an impedance of 25 ohms, which is being very optimistic. The net impedance would be something over 35 ohms, which at 120 volts would increase the total current by just 3.4 amps, an amount not likely to cause the overcurrent device to respond.

If you are using metal conduit or tubing as the wiring method, the normal grounding connections to the device box are the same as always (refer to Chapter 18), however an additional green-insulated conductor must be brought into the box to connect to the receptacle grounding terminal. If the wiring method is entirely nonmetallic, you need either one or two equipment grounding conductors depending on what type of faceplate will be used. If the faceplate will be nonmetallic as well, the only equipment grounding conductor required is the insulated one. If the faceplate will be metal, you need two grounding conductors and a way to attach an equipment ground to the faceplate. One way that makes sense, although isn't specifically in the NEC, is to rely on a grounding terminal of an adjacent conventional device, if this is a multi-gang application. The other way is to use a special nonmetallic box with a grounding terminal designed to ground any metal device yoke screwed into it.

Cabled wiring is more problematic. Type MC cable is available with two equipment grounding conductors for just this application, and Type AC cable is available with an additional insulated equipment grounding conductor in

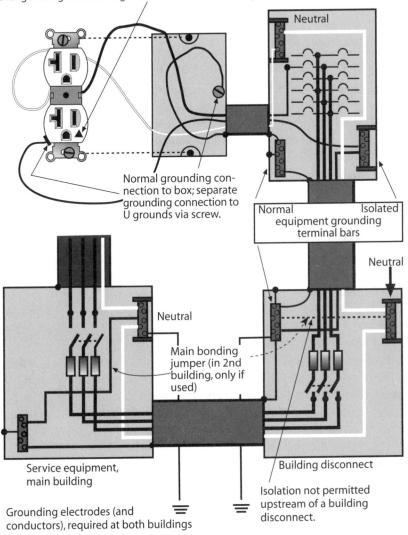

Orange triangle means U-grounds insulated from the yoke.

Neutral

Normal grounding connection to box; separate grounding connection to U grounds via screw.

Normal   Isolated
equipment grounding
terminal bars

Neutral

Neutral

Main bonding jumper (in 2nd building, only if used)

Service equipment, main building

Building disconnect

Isolation not permitted upstream of a building disconnect.

Grounding electrodes (and conductors), required at both buildings

Fig. 27–17 Although insulated from other equipment grounding conductors, the equipment grounding return path from the insulated grounding terminal on the receptacle follows all the normal rules for equipment grounding conductors. It runs with its associated power conductors at all points, and it returns fault current to the power source over a solid connection that doesn't rely on the earth. The device box is independently grounded to the normal equipment grounding system.

addition to the cable armor, but both of these cables need a metal box. Type NM cable is not really suitable because its grounding conductor is uninsulated; however, if the inspector agrees that only qualified individuals will service the system, a three-conductor cable could be used with one of the insulated conductors reidentified as green at all points of access. Remember that the duplicate set of equipment grounding conductors involved in many of these applications requires a duplicate equipment grounding deduction in calculating box fill, as covered at the end of Chapter 10.

WHAT OTHER WIRING METHODS ARE YOU LIKELY TO USE?

Chapter 3 of the *NEC* covers all the recognized wiring methods or systems, including many not covered in this book. In fact, a verbal shorthand to describe recognized wiring methods covered in the *NEC* is simply to refer to a "Chapter 3 wiring method." This book covers the most important methods, either in Chapter 11 or in this chapter.

In nonresidential occupancies (and in high-rise apartment buildings), the wiring is most often installed in raceways such as rigid metal conduit, intermediate metal conduit, EMT, or rigid nonmetallic conduit, and special cables such as Type MC cable. Raceway methods, when properly installed, provide considerable flexibility in that circuits may be changed, wires added, and breakdowns repaired by pulling new wires into existing raceways. Type MC or TC cables in a cable tray offer similar benefits, but might not be practical in many locations. Where exposed, raceways are neat and afford ample protection for wires. Cabled methods typically involve lower initial installed cost and are increasingly used, but raceway methods will continue to be extremely important, particularly in nonresidential applications due to the reusability factor.

Surface metal raceways   In nonresidential buildings, the wiring is usually in conduit embedded in the concrete. To add new outlets or move existing outlets is almost impossible if the conduit is to be concealed, and conduit that is installed on the surface is unsightly and rarely acceptable. Yet changes are often necessary, especially in offices, stores, and other buildings where layout changes are frequent. Instead of conduit, use metal surface raceways; they are attractive and blend well with the background surfaces. (Nonmetallic surface raceways are also available.) Such raceways can also be used for telephone, data transmission, and similar circuits, but they must never be installed in the same channel with power wires.

Two styles are available: the one-piece type (which actually consists of two pieces reassembled at the factory) and the two-piece type. The one-piece type is installed empty; the wires are then pulled into it just as in conduit. With the two-piece type, the base member is installed first, the wires are then laid in place, and finally the cover is installed. The *NEC* does not list the number of wires that may be installed in any given size and type of surface raceway, but *NEC* 386.21 (and 388.21 for the nonmetallic counterpart) states that the size of any wire may not be larger than that for which the raceway was designed, and 386.22 (and 388.22 for the nonmetallic counterpart) states that the number of wires installed may not exceed the number for which the raceway was designed. Test labs evaluate the raceway for the "designed" size and number of wires and list the product only if suitable. You only have to consult the tables furnished by manufacturers for such information.

Splices and taps normally are made only in junction boxes, but *NEC* 386.56 (and 388.56 for the nonmetallic counterpart) permits them at any point of a metal surface raceway of the removable-cover type. The wires including the splices and taps must not occupy more than 75% of the cross-sectional area of the raceway at the point where the splice is made.

NEC 386.12(5) (and the same location in Article 388 for the nonmetallic variety) limits both one- and two-piece types to *exposed* runs in dry locations, although the material may be run through (but not inside of) dry walls, partitions, and floors if the raceway is in a continuous length where it passes through the wall. The voltage between any two wires must not exceed 300 volts in the metallic product [386.12(2)] unless the material in the raceway is at least nominally 0.040 in. thick, in which case the voltage between wires may be up to 600 volts. The nonmetallic product [388.12(3)] has no absolute voltage limit in the *NEC*, however if the raceway will be used for wires running over 300 volts it must be listed for that duty.

The one-piece type, in which the wires are pulled in after the raceway is installed, is shown in Fig. 27–18. This figure also shows the dimensions of the smallest size; larger sizes are available. It is held in place by clips, couplings, or straps shown in the same illustration. Many appropriate kinds of elbows, adapters, switches, and receptacles are available to fit each size of raceway. Some of these are shown in Fig. 27–19.

The two-piece type is shown in Fig. 27–20, which also shows the dimensions of the smallest size; larger sizes up to 4¾ in. wide by 3⁹⁄₁₆ in. deep are available.

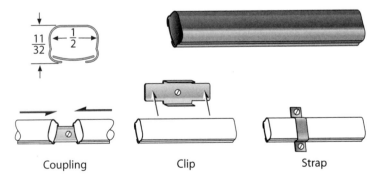

| Coupling | Clip | Strap |

Fig. 27–18 Example of one-piece metal surface raceway. Wires are pulled into place after the raceway is installed.

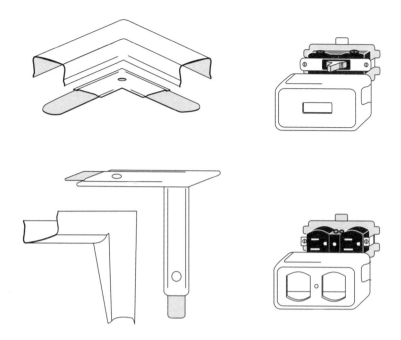

Fig. 27–19 Fittings used with the raceway shown in Fig. 27–17.

Many fittings are available similar to those shown in Fig. 27–19 but designed for this type of raceway. One of the larger sizes of raceway is available with a metal divider in the base, providing two separate channels. Use one channel for power wires, the other for telephone or other nonpower purposes. This makes a neat installation and meets the *NEC* requirement that power wires and other wires must occupy separate channels.

The *NEC* states that metal surface raceway must not be installed where "subject to severe physical damage unless otherwise approved." There are many cases where the material must be run on the floor, certainly a location where physical damage might be expected. For this purpose there is available a "pancake" raceway shown in Fig. 27–21, which also shows the dimensions of the smallest size available. As with the other two-piece type, first install the base, then lay the wires into place, and install the cover. As with other surface raceways, channels

Fig. 27–20 Example of two-piece metal surface raceway. Install the channel, lay the wires in place, then snap the cover into place.

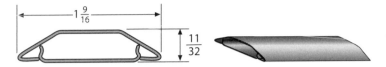

Fig. 27–21 "Pancake" type of metal surface raceway for installation on floors.

may contain power wires, or telephone and similar wires, but never in the same channel. As with other types, fittings such as elbows and so on are available.

Multioutlet assemblies   Especially in stores, schools, laboratories, and similar locations, there is often need for many closely spaced receptacles. Such receptacles are not necessarily all in use at the same time, but having them available at reasonably close intervals is very convenient, avoids the need for extension cords, and permits, for example, many floor lamps in a store to be plugged in at the same time so that any one of them can be turned on instantly to show to a customer. Much installation labor is involved if many such receptacles are installed using ordinary wiring methods.

To reduce the cost considerably, use the special raceway shown in Fig. 27–22. The covers are prepunched for receptacles so that special fittings for the receptacles are not needed. This type of raceway, called a multioutlet assembly in the *NEC*, is available with prewired receptacles in place, or may be obtained empty, with the receptacles prewired on separate long lengths of wire. Several spacings are available, with receptacles from 6 to 60 in. apart. A typical installation is shown in Fig. 27–23.

Fig. 27–22 Raceway with multiple receptacle outlets at regular intervals. If many receptacles are needed, this saves much time over other installation methods.

NEC Article 380 covers multioutlet assemblies. The material may be installed like other types of metal surface raceways, but if it runs through a wall or partition, no receptacle may be within the wall or partition. Moreover, it must be installed so the cover can be removed from any portion outside the wall or partition without disturbing the sections inside the wall or partition.

This kind of material is used not only in commercial establishments, but is equally useful in homes, especially in the small-appliance circuits discussed at the beginning of Chapter 13. It is equally suitable for home workshops or other locations where numerous receptacles are desirable.

Except for homes or guest rooms of hotels and motels, each 5 ft of a multioutlet assembly must be considered as one outlet, taken for load calculation purposes as not less than 180 VA. In locations where it is likely that a number of appliances may be used at the same time, however, each foot must be taken as a 180 VA load. See *NEC* 220.14(H).

Power poles A special type of multioutlet assembly, called a "Tele-Power Pole" by the manufacturer, is essentially a rectangular raceway in cross sections from 3/4 × 11/2-in. to 23/4 ×

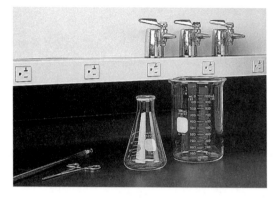

Fig. 27–23 A typical installation of the multioutlet assembly shown in Fig. 27–21. *(Wiremold)*

27/8-in. in 10- to 15-ft lengths, steel or aluminum, with a variety of finishes to go with most office decors. The pole is installed vertically between floor and ceiling wherever desired, and supported at floor and ceiling with a variety of appropriate fittings available. The pole has an internal divider providing two separate channels, one for power wires, the other for telephone or similar low-voltage wires. All wires are fed into the channels from the top. Usually the power wires are factory-installed in the power channel, with two or more duplex receptacles near the bottom end. Figures 27–24 and 27–25 show the material as purchased and in an installation. The definition of "multioutlet assembly" in Article 100 was recently amended to allow for freestanding uses of these assemblies, thereby clarifying that this is the appropriate classification.

Wireways Often it is necessary to run large wires, or a large number of wires, for a considerable distance to a central location or several locations. Instead of using large conduit, it is quite common practice (and more economical) to use a wireway. *NEC* Article 376 defines metal wireways as "sheet metal troughs with hinged or removable covers for housing and protecting wires and cables in which the conductors are laid in place after the wireway has been installed as a complete system." Article 378 covers the nonmetallic equivalents in much the same way as the traditional metal counterparts. Figure 27–26 shows a short section of wireway and some of the fittings used with it: elbows, crossovers, end sections, and so on. Wireways are available in lengths of 1, 2, 3, 4, 5, and 10 ft, and in cross sections from 21/2 × 21/2 in. up to 12 × 12 in. Raintight types are available for outdoor installations. The various sections and fittings are assembled in place as a complete raceway system, then the wires are laid in place.

If the splice is made with a power distribution block (a fixed assembly of multiple terminal lugs), the block must be listed and the wireway size must, in addition to the 75% rule, equal or exceed the minimum size specified in the installation instructions, and the wiring space must meet Table 312.6(B) sizing rules for the

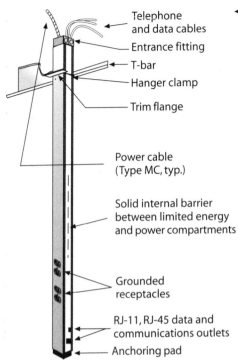

Telephone
and data cables

Entrance fitting

T-bar

Hanger clamp

Trim flange

Power cable
(Type MC, typ.)

Solid internal barrier
between limited energy
and power compartments

Grounded
receptacles

RJ-11, RJ-45 data and
communications outlets

Anchoring pad

◀ **Fig. 27–24** "Tele-Power Pole" contains, in two separate channels, both power wires and telephone or similar wires. It can be moved easily as office layouts change.

Fig. 27–25 Typical installation of Tele-Power Pole. *(Wiremold)*
▼

Fig. 27–26 Wireways of this kind are very useful, especially when installing large sizes of wire.

conductors landing on the block. These blocks are a very convenient way to splice equal-sized conductors in a circuit set, or to make smaller taps from a larger feeder. The power distribution block must be arranged so it can be completely insulated before the wireway cover is replaced.

A wireway may be run any distance, and must be supported every 5 ft unless listed for support at 10-ft intervals. An exception allows vertical runs to be supported at up to 15-ft intervals where there is only one joint between supports. Of course all joints must be made up tight. Nonmetallic wireways require much closer supports (3 ft horizontally and 4 ft for vertical runs). A wireway may be run through a wall if in an unbroken section. Splices and taps must be made where permanently accessible and must not exceed 75% of the wireway cross section where made.

Wireways must not contain so many wires that more than 20% of their internal cross-sectional area is occupied by conductors, except as allowed for splices where made. Within that 20% limitation, however, metal wireways have a very useful permission: if no more than 30 conductors at any particular point are current carrying, the derating penalties that normally apply to more than three current-carrying conductors in raceways (see page 105 for adjusting *NEC* Table 310.16 values) need not be applied. This allowance does not apply to nonmetallic wireways, which don't function as a heat sink as effectively.

Wireways are sometimes used in place of pull boxes, and the *NEC* imposes similar rules. When conductors enter a wireway and change direction at that point, or if the wireway itself changes direction by more than 30 deg, then the wireway cross-section dimension at that point must at least equal the minimum size as covered in *NEC* 312.6 for the conductors involved. In addition, the distance between raceway entries enclosing the same conductor must have the same minimum spacing of six times the raceway trade diameter as applies to a pull box. See Fig. 27–27.

Auxiliary gutters   Sometimes it is necessary to run large wires for a relatively short distance but with many taps in the wires. For an example, see Fig. 25–2 where the service-entrance wires must run to six switches. Unless very special switches are used (which is rarely done), it is not permissible to run wires to the first switch, make splices within the switch and run them on to the second, make more splices there and run on to the next, and so on. Instead, install an auxiliary gutter, covered in *NEC* Article 366. An auxiliary gutter is similar in appearance to a wireway, and a short section of a wireway is often used as an auxiliary gutter. But the function of an auxiliary gutter is more like that of a long junction box.

Auxiliary gutters may contain only wires or busbars; they must not contain switches, overcurrent devices, or similar equipment. They must not run more than 30 ft beyond the equipment they supplement. If necessary to go beyond 30 ft, use a wireway. Gutters must be supported every 5 ft (3 ft in the case of nonmetallic equipment). The gutter is first installed; then the wires are laid into place. Gutters that have hinged covers simplify the installation. The restrictions

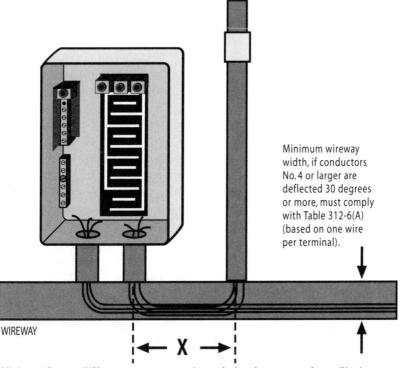

Minimum wireway
width, if conductors
No. 4 or larger are
deflected 30 degrees
or more, must comply
with Table 312-6(A)
(based on one wire
per terminal).

WIREWAY

X

Minimum distance "X" betweem raceway entries enclosing the same conductor (No. 4 or larger) not less than six times the trade diameter of the larger raceway.

Fig. 27–27 Wireways used as pull boxes need to meet comparable requirements.

regarding the number of wires and the space they may occupy are the same as for wireways.

Because of the endless range in size and shape required, gutters are often made to fit a particular space. But a section of wireway, if available in a size to suit the purpose, is entirely suitable for use as an auxiliary gutter. In fact, most wireways carry a dual listing and are simultaneously evaluated for both functions.

Strut-type channel raceway   For exposed use, and only in dry locations, a subspecies of U-shaped framing channel akin to that used for constructing heavy shelving, trapeze hangers for electrical raceways, supports for panelboards that have to be racked off a wall, etc. is now a Chapter 3 wiring method, covered in *NEC* Article 384. Support rules follow those for circular steel raceways, that is, every 10 ft and within 3 ft of terminations. The basic wire fill provisions and derating thresholds are the same as for metal wireways, but there is an additional provision that allows you to use a more dense fill than the 20% for wireways, all the way to a 40% fill for unobstructed strut and 25% fill for strut with internal joiners. However, if you pass the 20% mark, the usual raceway derating factors apply, even if you stay under the 30 current-carrying conductor threshold.

Flat cable assemblies   This is another spin-off from the strut industry, used primarily for factory and warehouse lighting. It uses $1\frac{5}{8}$-in. × $1\frac{5}{8}$-in. U-shaped solid strut, open side down. After installing the strut, the cable assembly consisting of a flat layout of four (two- or three-conductor versions are also available) 10 AWG conductors is drawn into the strut. The cable wires have a special stranding and the insulation consists of a special self-healing polymer that, taken together, can be safely tapped at any and various points through the use of special fittings. These fittings use special pins to puncture the insulation and supply the fixtures that mount to the strut, as well as making any taps that might be required for other purposes. They can be repositioned as often as

required. The cables originate in special terminal blocks with a mandatory color code of first white, then black, then red, and then blue. If the strut is at least 8 ft above the floor, the channel opening need not be covered. See Fig. 27–28.

One of these assemblies can electrically support an entire 480Y/277-volt multi-wire branch circuit. For fluorescent lighting, which doesn't have heavy duty lampholders and therefore qualifies for not more than a 20-amp branch circuit under *NEC* 210.21(A), this circuit would support over 13 kVA of lighting load (0.8 × 20 amps × 831 volts). To review where these numbers come from, refer to the discussion on protecting devices under continuous loads beginning on page 476, and the calculation trick on page 460. In the case of HID fixtures with heavy duty lampholders on a 30-amp circuit, the comparable calculation would be 24 amps × 831 volts, for just barely under 20 kVA on one multiwire branch circuit.

Fig. 27–28 Cutaway view of Type FC cable installed in metal strut designed to be compatible with the wiring method. The photo also shows a tap device designed to penetrate a specific phase conductor in the multiwire branch circuit shown, and then both supply and support a lighting fixture. *(Sentinel Lighting)*

Busways   A busway consists of a grounded metal enclosure of any length depending on manufacture, which rigidly encloses factory-mounted solid conductors configured as bare or insulated copper or aluminum bars, rods, or tubes. The sections are bolted together in the field using components provided by the manufacturer to make a run of any length. At one time busway was found only in heavy industrial applications involving thousands of amperes, but now it shows up in high-rise buildings of all sorts, and in commercial applications as well. Instead of involving only very high current densities, busways now range anywhere from 50 to 6000 amps. They can also form, as branch circuits, the basis for movable equipment connected to rolling power take-off arrangements (the "trolley bus") and for support of fixtures.

Busways also come in plug-in configurations that accept branch circuit extensions. Each plug-in module includes provisions for branch-circuit overcurrent protective devices, and the modules can be positioned at regular intervals, allowing a very flexible layout. Although this busway and the switches on it generally run overhead well above the normal upper reach limit for switches (6 ft 7 in. to the center of the operating handle), *NEC* 404.8(A) includes an exception for busway switches, provided there are chains, hooksticks, or other suitable means provided to operate the switches from the floor.

Busway extensions can be in raceway, or in a special type of listed cord called "bus drop cable." Although it appears similar to flexible cord, in the UL directories it is listed in the miscellaneous category under the general heading of "Wires"— completely separated from the heading "Flexible Cord." It will be marked as bus drop cable and not as one of the types of flexible cord. It is covered in *NEC* Article 368 and not Article 400 for that reason.

Busways must be installed out in the open as a general policy. However, they can run above a suspended ceiling (or other accessible areas) with access panels, provided the busway uses a totally enclosed, nonventilating construction, with all joints accessible for maintenance. If the suspended ceiling is an air-handling ceiling, further restrictions apply. The individual busbars must be insulated, and the busway must not have any plug-in provisions.

If busways run vertically and penetrate two or more dry floors in other than industrial occupancies, *NEC* 368.10(C)(2)(b) requires you to provide curbing around the penetrations. The curbing must be at least 4 in. high, and not more than 12 in. from the floor opening. Other electrical equipment must be located so it won't be damaged by liquids retained by the curb. This rule reflects the fact that busway risers often pass through janitors' closets with floor-level slop sinks and hoses. Normally the floor penetration would be firestopped in accordance with *NEC* 300.21. Depending on the configuration of the firestopping, however, floor-level liquids might easily reach and flood the vertical busway. In addition, the other electrical equipment that is in the room and suitable for ordinary locations shouldn't be on or near the floor where it might get flooded by whatever fluids get held back by the curbing.

The rule on curbing vertical penetrations needs to be read in its larger context. *NEC* 368.10(C)(2)(a) permits only floor penetrations through dry floors, and both paragraphs apply to this work. Don't draw the inference that because this paragraph refers to dry floors, you could do something different or less restrictive on a wet floor. The first paragraph also continues the rule requiring unventilated, totally enclosed busways up to the 6-ft level to prevent physical damage.

Cablebus   This wiring method is your opportunity to make your own busway out of insulated conductors. It comes in the form of a completely enclosed, ventilated, protective metal housing. In accordance with the manufacturer's specifications, you add individual conductors to the enclosure, which has supports arranged to maintain good air separation around the conductors on all sides. In this form, the conductors qualify for consideration in *NEC* Table 310.17 as being in free air. This allows some impressive ampacities, such as 620 amps for 500 kcmil at 75°C, instead of the usual 380 amps.

Underfloor raceways   When laying out large open areas prior to construction, areas that will be used for office activities, or some research or very light industrial production, underfloor raceways provide an excellent way to route power circuits to individual desk or work station locations. These wiring methods allow for access to their interior to relocate outlets at almost any floor location. There are three types. Cellular metal floor raceways (*NEC* Article 374) use a series of parallel cells, made of sheet metal, that serve as partial concrete forms for a poured-concrete floor, especially in steel-framed buildings. Cellular concrete floor raceways (*NEC* Article 372) consist of parallel voids that are precast in structural concrete floor slabs. Underfloor raceways as shown in Fig. 27–29 (*NEC* Article 390) use actual metal raceway sections in a rectangular grid pattern, frequently with header ducts crossing at right angles to allow ease of cross-circuiting as required.

In today's offices signaling and communications wiring are as important as power wiring, and frequently these systems will be duplicated or partitioned to allow universal access to limited-energy wiring as well as power. All the underfloor methods use panelboards and limited-energy cabinets on perimeter walls arranged to connect to the underfloor wiring through either a special raceway attachment or a conventional raceway segment. To facilitate finding inserts, the pre-set elements must line up perfectly, and the *NEC* requires the floor system to be anchored accordingly prior to pouring the concrete. The final insert on each run must be marked as well, typically using a brass flat-head screw.

Just as underfloor systems allow for outlets to be added along the length of the run at specified intervals due to floor plan rearrangements, so also must these systems allow for outlets to be abandoned and the flooring restored. This raises practical concerns about splices in an underfloor system, since splices should usually be accessible. The *NEC* only allows splices in the header, which has a removable cover; however, it does recognize "loop wiring" for receptacles in tombstone floor outlets. This involves skinning the insulation off the middle of a length of wire and looping it over the terminal screw, as in *C* of Fig. 17–6. If you subsequently abandon this outlet, however, you can't just reinsulate the wire and

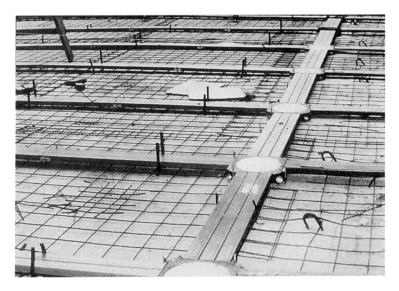

Fig. 27–29 Underfloor raceway prior to the pouring of a concrete floor, showing junction boxes and preset inserts for power and limited-energy circuits. *(Wiremold)*

push it back into the underfloor raceway. The next person who comes along with a metal fish tape from a downstream outlet point could inadvertently catch the repair, and the resulting arc would do a great deal of damage.

Therefore the *NEC* expressly forbids reinsulating abandoned loop wiring; the entire run from the header or panelboard out must be replaced. One way to avoid this is to put each tombstone on its own pair of wires, and make the circuiting connections in the header. Then, if you abandon an outlet, simply attach a pull string to the end of the wires before pulling them out. That way you didn't reinsulate any wires and, if you subsequently reactivate the outlet, a pull wire is in place from the header (or other legal splice point) and the abandoned outlet location. Be aware, however, that this arrangement adds quite a few conductors to the run, which can lead to problems due to mandatory derating for mutual conductor heating. To create an incentive for doing this work correctly, however, some jurisdictions don't impose derating penalties provided there aren't more than 30 current-carrying conductors at a given underfloor raceway cross-section, similar to the *NEC* allowances for wireways. Check with your inspector.

Mineral-insulated (Type MI) cable   The letters "MI" stand for mineral insulation. The cable consists basically of solid bare wire, two or more conductors properly spaced, with a seamless copper sheath around the outside. The wires are separated from each other and from the copper shield by a tightly packed mineral insulation (magnesium oxide), which is a good insulator and not affected by heat. See Fig. 27–30, which shows both the cable and the special connectors used with it. If the cable is supported and run in accordance with the strict requirements of

UL (requirements which, for this purpose, are considerably more severe than the *NEC*'s), Type MI cable can qualify as an "electrical circuit protective assembly" in the *UL Fire Resistance* directory, with a two-hour rating. This means it will withstand a two-hour fire exposure without compromising the contained electrical circuiting. It will also withstand a hose-stream test afterward.

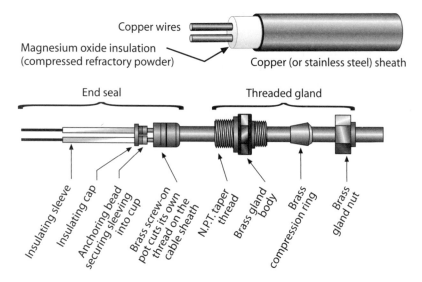

Fig. 27–30 Type MI cable and the special fittings used with it.

It is installed like other cable, and being of relatively small diameter it can be readily installed in crowded quarters. The outer copper sheath is waterproof so the material may be installed in wet locations, and even underground if protected against physical damage and corrosion. However, immediately after cutting the copper sheath, a seal must be installed to prevent the entrance of moisture into the highly absorbent insulation. Since it is completely filled, it cannot transmit vapor along its core, making it the only cable accepted without special additional conditions for hazardous (classified) applications.

It must be supported every 6 ft unless fished through existing wall spaces. In bending, the radius of any bend must be at least five times the diameter of the cable for cable up to ¾-in. in diameter; ten times the cable diameter for ¾-in. to 1-in. diameters. If run as single-conductor cable, the circuit conductors must be grouped, and the termination of the group at a ferrous enclosure must be such that inductive heating is avoided. Refer to *NEC* 300.20.

The grouping of single conductors (required to prevent induced voltages from building up on the sheath) raises an interesting question: Is the cable group ampacity based on single conductors (Table 310.17) or a cable assembly (Table

310.16)? In fact, such a grouped configuration does not allow for the same degree of heat radiation as a separated arrangement. On the other hand, even if the ampacities in Table 310.17 are used (and the cable temperatures within the group exceed 90°C as a result), the inherent heat resistance of this wiring method is so great that it can run well above 90°C without damage. The 2002 *NEC* recognizes these realities by allowing Table 310.17 to be used, providing the arrangement is such that the temperature limitations applicable to the end seals are not exceeded. Since the maximum temperature ratings for end seals typically don't exceed 90°C, this means that as the cable group approaches a termination, the conductors must be spread out to allow the additional heat to dissipate. Since the *NEC* does not as yet provide a distance, discuss this with the engineering department of the cable manufacturer. The distance over which the cables must be spread out generally runs in the 2-ft range. Finally, the terminations within the enclosure must use conventional Table 310.16 limits. The MI cable manufacturers are aware of this, and have developed step-up butt splice connectors that provide an increased pigtail size that translates the likely smaller sizes of their cables into the larger sizes required to provide the heat sink functionality required by 110.14(C). See the termination restriction topic on page 474.

Cable trays   Cable trays blur many distinctions. Fundamentally they are cable support systems and not wiring methods, but under some conditions they become quasi-raceways. Some tray configurations are acceptable equipment grounding conductors, at least through a specified range of overcurrent protective device sizes. Figure 27–31 shows one of the traditional uses of this method, with a very heavy-duty tray open at the top and supported on trapeze hangers. Since the hanger supports run to the ceiling on both sides, the cable has to be pulled into the tray; the electrician is guiding cable into position as it passes through a 90-degree turn at a tee. Modern tray systems also involve center-support trays, not as heavy duty, but which allow for the cables to be slipped over the side, which is much faster.

Fig. 27–31 Metal-clad cable, Type MC, being installed in a ventilated trough cable tray. *(MP Husky Corp.)*

In the case of ventilated tray, as shown in Fig. 27–31, the free circulation of air allows for increased ampacity allowances, provided spacing between cables is maintained. This type of tray (or ladder or ventilated channel), where used in industrial occupancies under qualified supervision, can support individual conductors marked for cable tray use and not smaller than 1/0 AWG. This is where cable tray becomes a quasi-raceway. These restrictions are often ignored, creating hazardous conditions. This allowance does not make cable tray into a wireway without a cover, even in an industrial occupancy. This allowance does not allow cable tray to support open conductors in any non-industrial occupancy, not a hospital nor an exhibition hall nor an airport, etc., no matter how large the individual conductors.

In general, cable trays are supposed to be installed as a complete system, but the *NEC* provides some leeway that acknowledges long industrial experience. For example, cable trays need not be completed between two discontinuous tray layouts, provided that when cables go from one tray to the other, they are supported in accordance with their applicable *NEC* article. For example, if the cable in Fig. 27–30 changed trays, the other tray would have to be not over 6 ft above or below, to allow for support of the wiring method. In the event you have a legitimate application of single conductor cables, for example, and you want to drop from the tray to a cabinet, the cabinet could not be over 6 ft distant.

Type TC cable   Although cable trays can support most any cabled wiring method, there is one Chapter 3 method developed specifically for cable trays— Type TC cable (*NEC* Article 336). This is a nonmetallic wiring method that either runs in cable tray or exits the tray to specific loads within a raceway wiring method. In addition, for this method to be used in the open in lengths of up to 50 ft, there is a limited allowance in industrial occupancies with qualified mainte- nance and supervision. It must be listed and found to meet the crush and impact requirements for Type MC cable to be used in this way, and it must be supported every 6 ft.

CHAPTER 28
Planning Nonresidential Installations

ALTHOUGH THE SAME BASIC PRINCIPLES covered in Parts 1 and 2 of this book also apply to nonresidential projects, there are some additional problems. Larger currents, higher voltages, three-phase power in addition to single-phase power, and different kinds of materials and equipment may be involved.

This book cannot possibly cover all the details of larger projects. Some typical moderate-size projects are discussed as a guide to help you in your career. The information in this chapter through the remainder of the book should be considered a preview of the kinds of problems you may encounter. This chapter discusses the distribution systems often used to supply nonresidential occupancies, either directly from a utility, or indirectly through the use of separately derived systems. Since separately derived systems usually originate at transformers, the chapter covers how transformers must be installed and protected. Included are the special grounding arrangements that must be made for these systems. Any of these systems, whether originating at a transformer or a service, ultimately supplies branch circuits. A description of branch circuits permitted to be installed is provided, along with general restrictions on the use of lighting fixtures on these circuits. The chapter concludes with a look at special *NEC* rules for temporary wiring.

WHAT ARE COMMON NONRESIDENTIAL DISTRIBUTION SYSTEMS?

Service-entrance wires where they enter a building may end at one or more switchboards or at one or more panelboards. All necessary disconnecting equipment and overcurrent protection, covering the building as a whole, are located in this equipment. Compare such a system to a big tree. The trunk is the service-entrance wire, the point where the trunk breaks up into half a dozen or more branches is the service equipment protecting the building as a whole, and the points where the larger branches in turn split into smaller branches are feeder or branch-circuit distribution points.

Equipment used for distribution—*NEC* definitions What is the difference between a switchboard and a panelboard? And between a service-distribution panelboard and a branch-circuit panelboard? The *NEC* in Article 100 defines panelboards and switchboards as follows:

- *Panelboard*—A single panel or group of panel units designed for assembly in the form of a single panel; including buses, automatic overcurrent devices, and with or without switches for the control of light, heat, or power circuits; designed to be placed in a cabinet or cutout box placed in or against a wall or partition and accessible only from the front. (See Fig. 4–4.)

- *Switchboard*—A large single panel, frame, or assembly of panels on which are mounted, on the face or back or both, switches, overcurrent and other protective devices, buses, and usually instruments. Switchboards are generally accessible from the rear as well as from the front, and are not intended to be installed in cabinets. (See Fig. 4–5.)

Generally speaking, if the equipment is installed in or on a wall, it is a panelboard, which can range in size from the small fuse cabinet of Fig. 28–1, to the larger cabinet with many circuits shown in Fig. 28–2, to still larger ones. If the equipment is installed away from a wall and is accessible from either the front or back, it is a switchboard. From this statement and the *NEC* definition, you could easily infer that switchboards are never enclosed, but most switchboards now being installed are of the metal-enclosed type. Switchboards are usually installed only in very large buildings.

▲
Fig. 28–1 This simple fuse cabinet is a small panelboard.

Fig. 28–2 A much larger panelboard containing many circuit breakers. *(Square D)*
▶

The distribution riser diagrams of Fig. 28–3 show a one-story building at *A* and three-story buildings at *B* and *C*. These might be factories, office buildings, stores, or any other large building. These diagrams are typical of installations where a separate switchboard is installed.

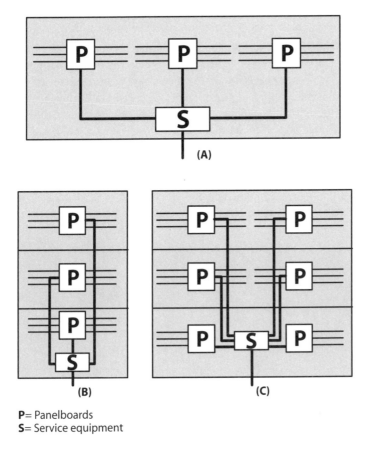

P= Panelboards
S= Service equipment

Fig. 28–3 Typical distribution riser diagrams for large buildings.

Interior distribution   In most houses and many other buildings of moderate size, the branch-circuit wires run directly from the service equipment. In such cases there are no feeders between the service equipment and the branch circuits, although the short copper (or aluminium) busbars within the service equipment technically are feeders. In larger installations, it impractical to start all branch circuits from the service equipment due to the distances involved, the size of the total load, and the number of branch circuits. Instead, in large buildings install various panelboards, which might be (1) service equipment panelboards, or (2) distribution panelboards, or (3) branch-circuit panelboards. A single panelboard could combine two of these functions or all three in a single cabinet. Branch circuits may start from any kind of panelboard, and if all three kinds are installed in a building, the branch circuits could start from one, or two, or all three of them.

If the panelboard you install is a service equipment panelboard (whether or not its cabinet also contains feeder or branch-circuit overcurrent protection), it must be of a type suitable for service equipment, and be so marked.

A building may have up to six separate disconnecting means and main overcurrent protection equipments; each of these can be a service equipment panelboard which may also contain overcurrent protection for feeders or branch circuits. From this equipment, run feeders to branch-circuit panelboards located in various parts of the building; the branch circuits start from there. If the building is quite large, one or more distribution panelboards can be installed between the service equipment and the branch-circuit panelboards. In other words, from the service equipment run feeders to the distribution panelboards located where you wish. From each distribution panelboard run additional feeders to the branch-circuit panelboards. See Fig. 28–4. Each single line in that diagram represents whatever number of wires might be involved in the run.

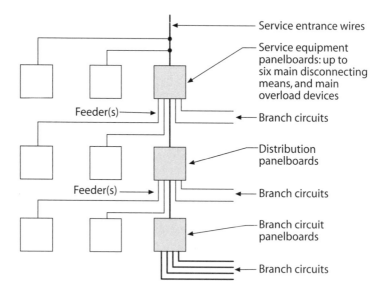

Fig. 28–4 The relative positions of three different kinds of panelboards.

Lighting and appliance branch-circuit panelboards This is the kind of panel-board that you will encounter most often, and it has special restrictions. It is defined in *NEC* 408.14 as "having more than 10 percent of its overcurrent devices protecting lighting and appliance branch circuits." A lighting and appliance branch circuit is "a branch circuit that has a connection to the neutral of the panelboard and that has overcurrent protection of 30 amps or less." No more than 42 overcurrent devices (in addition to the main overcurrent devices) may be installed in any lighting and appliance branch-circuit panelboard. Each fuse and each pole of a circuit breaker is counted as a separate overcurrent device. In other words, each two-pole breaker is counted as two overcurrent devices, and each three-pole breaker as three overcurrent devices.

A physical means must be provided to prevent the installation of more overcurrent devices than the panelboard is designed, rated, and approved for. If "standard" width breakers can be replaced by the half-size "thin" breakers, this is taken into account in the total number of spaces allowed in the panelboard and in its ampere rating.

Each lighting and appliance branch-circuit panelboard must be protected on its supply side by not more than two sets of main overcurrent devices (two breakers or two sets of main fuses). There are two exceptions to this rule, only one of which is relevant to nonresidential applications. This exception (which applies to all occupancies) provides that if the overcurrent device (breaker or fuses) protecting the feeder to the panelboard has an ampere rating not greater than the ampere rating of the panelboard, no further main overcurrent device is required for the panelboard. In some cases, a feeder supplies two or more panelboards, in which case it is necessary to have main overcurrent protection at each panelboard. The overcurrent device may be inside as an integral part of the panelboard, or a separate circuit breaker or fused switch may be installed ahead of the panelboard. If a feeder supplies only one panelboard, the usual practice is to that the overcurrent device that protects the feeder will have an ampere rating not greater than that of the panelboard; in that case the overcurrent device also provides the required overcurrent protection for the panelboard.

Power panelboards   If the panelboard is not a "lighting and appliance branch-circuit panelboard," it is a "power panelboard," defined in an exactly complementary fashion. There is no restriction on the number of overcurrent devices it may contain, and they may be of any size. No specific overcurrent protection is required. In effect, power panelboards are protected by their load calculation.

Beginning with the 1999 *NEC*, however, the Code-making panel decided to rein in at least the smaller power panels that used circuit sizes (and likelihood of changes being made by untrained persons) similar to their lighting and appliance branch-circuit panelboard cousins. A power panelboard with more than 10% of its overcurrent device poles protecting at 30 amps or less requires individual line-side protection by a single protective device that is either part of the panelboard or is remote.

Such panelboards may also be used as service equipment, provided there are not more than six disconnecting means and the panelboard is suitable for service equipment. The small power panelboard provisions include an exception for just this application, since service equipment panelboards are more likely to have qualified inspection after modifications are complete.

Types of distribution systems   Previous chapters cover the manner of computing the size of services, feeders, etc., including the size of the neutral conductor, and how to install them. As the response beginning on page 131 to the question "Is it a neutral wire?" clarifies, a grounded wire is not always a neutral wire. Where it is not a neutral wire, it carries the same current and must be the same size as the ungrounded wires. However, you need to know more about arranging feeders and branch circuits of various systems.

Lighting circuits illustrate connection of single-phase loads on three-phase systems   In larger buildings the service is usually three-phase, but the two wires that run to any one lighting fixture are necessarily single-phase. How then is the lighting to be handled? There are four ordinary ways, as follows:

1. A separate 120/240-volt, single-phase service may be installed in addition to the three-phase service, to supply all the usual 120- or 240-volt loads such as lighting receptacles and small single-phase motors.

2. If the service is wye-connected, three-phase, four-wire, 208Y/120-volt, then 120-volt, single-phase, two-wire circuits are directly available by connecting to the neutral of the service and one of the phase wires, as shown in Fig. 28–5. In Fig. 28–6, the diagram at left is the sort usually supplied with such a three-phase panelboard; at right is a diagram that may be easier for beginners to understand. Figures 28–5 and 28–6 illustrate a three-phase service used only for 120-volt lighting, which is rarely the case. More often the three-phase service supplies both single- and three-phase circuits; then you would install one or more three-phase feeders from the service equipment to one or more three-phase panelboards used only for lighting. This is explained in more detail later in this chapter.

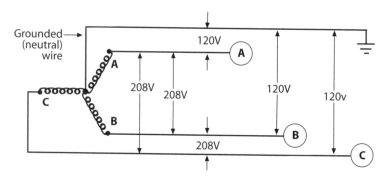

Fig. 28–5 A 208Y/120-volt system can supply both 120-volt, single-phase circuits and 208-volt, three-phase circuits at the same time.

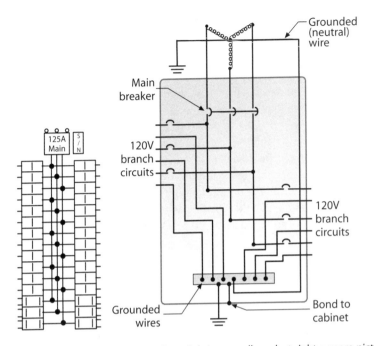

Fig. 28–6 At left, diagram for a three-phase lighting panelboard; at right, a more pictorial diagram.

In Fig. 28–5, one of the wires is marked "grounded wire." That wire while grounded remains just a grounded wire; it becomes a neutral only if it becomes part of a four-wire circuit.

3. When 120-volt current isn't directly available from the service equipment, transformers must be installed to step the higher voltage down to 120 or 120/240 volts. All the wiring beginning with the secondary of such a transformer constitutes a "derived system," the proper installation of which is discussed later in this chapter. The primaries of such transformers are connected to two of the phase wires, whether their voltage is 277, 480, or 600 volts. Of course the three-phase transformer, or transformer bank, could be used, each of the secondaries then supplying 120/240-volt current for lighting, receptacles, or other single-phase loads such as small single-phase motors.

4. If the service is wye-connected, three-phase, four-wire, 480Y/277-volt, fluorescent lighting operating at 277 volts is usually installed. The wiring would be as shown in Figs. 28–5 and 28–6, except that each single-phase circuit would operate at 277 volts instead of 120 volts. Of course step-down transformers would still have to be installed to provide 120/240-volt power for incidental 120-volt lighting (incandescent lighting is usually prohibited by NEC 210.6 on 277-volt circuits), for receptacles for office machines and similar 120-volt loads, and to operate small single-phase motors.

Economic benefits of 277-volt lighting   Increased lighting levels over the years have meant consumption of more watts per square foot of floor area. That in turn requires more circuits or larger wires. Instead of using 120-volt circuits, why not use a higher voltage, thus reducing the amperage of any given load? Then any size of wire will carry more watts than at a lower voltage. For example, 14 AWG wire, allowed on no more than a 15-amp circuit and operating at 120 volts, can carry 15×120 or 1800 volt-amperes; if operating at 277 volts the load rises to 15×277 or 4155 volt-amperes. (If the loads are continuous, requiring derating as already discussed, the figures would be 1440 and 3324 volt-amperes.) Fewer circuits would be required for any given load, with consequent great savings of cost in the number of conduits needed, the amount of wire needed, and especially in installation labor. Install the 277-volt lighting system as you would a 120-volt system, but be sure switches and branch-circuit breakers are listed for the higher voltage.

In all 277-volt wiring, a word of caution is in order regarding switches to control lighting. Each two-wire circuit for 277-volt fixtures of course consists of a grounded wire connected to the neutral of the 480Y/277-volt system, and one of the three hot or phase wires of the system. If the circuit is three-wire, it will include two hot wires. In an installation of any size, some of the circuits will contain the hot wires from phase A, others the hot wire from phase B, and still others the hot wires from phase C. Now the voltage between the grounded wire and any hot wire is 277 volts, but between any two hot wires it is 480 volts. If switches to control the lighting are grouped in a multigang box, *NEC* 404.8(B) requires that the voltage between any two adjacent switches must not exceed 300 volts. If the switches cannot be arranged to accomplish this, you must provide permanent barriers in the box between any two switches where the voltage between them would be not 277 but 480 volts. The same rule also applies, as of the 2002 *NEC*, to other devices. For example, a 125-volt receptacle placed next to a 277-volt switch would usually need a barrier. The voltages add vectorially, and given the phase shifting that takes place in the separately derived system transformation, the actual voltage will not be an arithmetic sum of the two voltages. However, it would usually exceed 300 volts in these cases.

Wye distributions are increasingly popular   These systems provide a solidly grounded central ground reference that is of equal potential to each of the phase conductors. In cases where line-to-neutral loads are a significant percentage of the total load, that makes load balancing among phases comparatively simple. In addition to the 208Y/120- and the 480Y/277-volt systems, some industrial users have moved to 600Y/347-volt systems. Although uncommon in the U.S., Canadian industry uses this system routinely. In the interest of promoting universal applicability of the *NEC* and reducing trade barriers, the *NEC* now covers this system, although it is not addressed further in this book.

Circuit arrangements—208Y/120-volt systems   Such systems can provide 120-volt, single-phase and 208-volt, three-phase circuits; either voltage alone, or both at the same time, as shown in Fig. 28–5. Figure 28–6 shows the system providing only 120-volt circuits for lighting. Now see Fig. 28–7 which shows how the system can supply five different kinds of circuits at the same time, as follows:

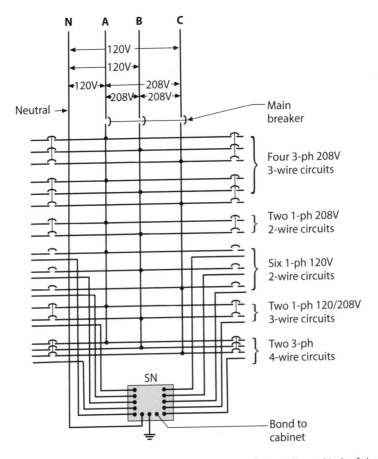

Fig. 28–7 This shows how a 208Y/120-volt system can supply five different kinds of circuits.

1. Two-wire, 120-volt, single-phase circuits can be supplied by simply connecting the grounded wire of each circuit to the neutral N, and the ungrounded or hot wire of each circuit to either A, B, or C. As far as possible, the number of ungrounded wires should be divided equally between A, B, and C. Note that the grounded wire of any such 120-volt circuit is not a neutral even if it is connected to a neutral wire. (If all four wires were connected in a circuit, the grounded wire would then be a neutral.) Each such two-wire circuit is of course a single-phase circuit.

2. Two-wire, 208-volt, single-phase circuits can be supplied by simply connecting the two ungrounded wires of the circuit (whether feeder or branch circuit) to any two of wires A, B, or C. As far as possible, connect the same number of circuits to wires A and B as you do to B and C, or to A and C. Such 208-volt circuits are used to supply, for example, water heaters rated at 208 volts.

3. Three-wire, 120/208-volt, single-phase circuits (feeders or branch circuits) can be supplied by connecting the grounded wire of each such circuit to the neutral N, and two ungrounded wires to either A and B, or B and C, or A and C. If there are to be a number of such circuits, they should as far as practical be connected so that wires A, B, and C carry equal loads. Such three-wire, 120/208-volt circuits are quite common because they make available the usual 120-volt current for lighting, receptacles, and so on, as well as 120/208-volt current for ranges designed for that voltage. The grounded wires of such 120-volt circuits are not neutrals, because the grounded wire in this case carries the same current as the ungrounded wires.

4. Three-wire, 208-volt, three-phase circuits (feeders or branch circuits) can be supplied by simply connecting the three ungrounded wires to the three wires A, B, and C. Such circuits are for operating three-phase motors, three-phase heating loads, and so on.

5. Four-wire, 208Y/120-volt, three-phase circuits (feeders or branch circuits) can be supplied by connecting the grounded wire (which is now a neutral) to the neutral wire N, and three ungrounded wires to wires A, B, and C. Such feeder circuits are often used to supply one or more three-phase panelboards in commercial and industrial buildings (as well as in larger apartment houses); each such panelboard can supply branch circuits or other feeders operating at any of the voltages discussed in circuits 1 through 4 above.

Often such four-wire, three-phase feeders are installed as "risers," each extending from the service equipment (which is usually in the basement) to several upper floors. At each floor, a three-wire, single-phase subfeeder, as described in circuit 3 above, is installed to supply 120- or 208-volt loads. In installing such subfeeders, be sure to observe the 10- and 25-ft tap wire rules already discussed. Remember too that the grounded wire in such a circuit derived from a four-wire source is not a neutral, and the grounding busbar in the cabinet must be insulated from its cabinet.

Circuit arrangements—480Y/277-volt systems Everything that has been said about 208Y/120-volt systems is also correct for 480Y/277-volt systems, except that wherever "120 volts" appears you must change it to "277 volts," and where "208 volts" appears you must change it to "480 volts," and where "208Y/120 volts" appears, change it to "480Y/277 volts." Note, however, that water heaters and ranges are not available for voltages higher than 240 volts. As discussed elsewhere, 277-volt circuits are used primarily for fluorescent and other forms of electric discharge lighting.

Today 277-volt fluorescent lighting is very common. If 277 volts seems a peculiar voltage, remember that if a building is served by a three-phase, wye-connected, 480Y/277-volt service, the voltage between the neutral and any phase wire is 277 volts. Just as 208 is to 120 volts as $\sqrt{3}$ is to 1, so also is 480 to 277 volts as $\sqrt{3}$ is to 1. The three-phase, 480-volt power is still available for three-phase loads.

Delta distributions are still important for industrial applications   Unlike wye distributions, which must have their star points grounded, delta systems may or may not have a grounded circuit conductor. In their ungrounded form, they offer the advantage of power continuity, particularly in cases where a sudden, random stop could cause significant harm to a continuous process. This means they can be arranged so a ground-fault doesn't necessarily trip an overcurrent device. A fault merely shifts the system ground reference to the faulted phase, and the maintenance staff can schedule an orderly shutdown and then fix the problem later. This makes the systems attractive in industries where a random, non-orderly shutdown could cost millions of dollars in lost production.

Circuit arrangements—delta-connected, three-wire, three-phase systems   If the service is at 240 volts, you will have 240-volt, three-phase power available by connecting to all three of the wires, or 240-volt, single-phase by connecting to any two of the three wires. If the service is at 480 volts, you will have 480-volt, single-phase power by connecting to any two of the three wires. Single-phase, 480-volt power is not often used, but could be used to serve the primary of a transformer of a derived system, as discussed later in this chapter.

Such ungrounded distributions must now have ground detectors that monitor whether or not a phase leg is grounded somewhere. The simplest detectors involve lights connected from each phase to ground; if a light goes out, the phase must be grounded and maintenance personnel should start arranging for an orderly shutdown. Far more sophisticated arrangements are more common today. The first ground on an ungrounded system does no harm; it merely corner-grounds the system at that point. The danger comes from the next fault. When the second fault occurs, the odds are two to one that the faulted conductors are supplied by different phase legs. If that happens, the result is a phase-to-phase short circuit across the equipment grounding system in the building. Given the high voltage and extensive components in the path (for example, think of all the locknuts, any one of which might be a little loose), quite a bit of damage can be done until an overcurrent protective device opens. In addition, the first fault, if undetected, makes the system as vulnerable to random outages due to faults as the grounded alternative.

Delta systems often have one of the wires grounded, and it is then known as a "corner grounded" system because the grounding point is at one "corner" of the delta. The grounded wire is not a neutral; it is just a grounded circuit wire. Such a grounded wire must never have a fuse or breaker installed in it. Like other grounded circuit wires it must be white, or for wires larger than 6 AWG (which are not available in white) it must be painted white or marked with white tape at all connection points. This does not change the voltage relationship between wires. It simply means that the grounded wire will be at zero voltage to ground, and the other two will have a voltage of 240 (or 480 or 600) volts to ground.

The circuit breakers on a corner-grounded system will be two-pole, however, those two-pole breakers must potentially handle a full-voltage fault that will involve only a single pole of the breaker. For example, if you apply a two-pole breaker on a 480Y/277-volt system and the circuit faults to ground, the breaker pole initiating the clearing sequence will open a 277-volt short; if the short is between the phases, both poles of the breaker split up the fault current, in effect making it easier to clear. On a corner-grounded system, however, there is no such thing as a 277-volt fault, and the 480-volt fault will often have to be cleared with a single pole of the breaker absorbing the clearing energy. Breakers that are capable of this duty are specially marked "1ϕ–3ϕ" and must be used for this application.

Watch split-voltage ratings on circuit breakers   Most commonly available circuit breakers have voltage ratings that reflect grounded single-phase or wye-connected systems. For example, a 480/277-volt circuit breaker may be used on either 277-volt line-to-grounded conductor loads such as fluorescent lighting, or on 480-volt loads, provided the 480-volt system is wye-connected and grounded at some point. If, however, the system is one that actually (or potentially in the case of an ungrounded circuit) operates at 480 volts to ground, as in the case of all delta-connected, thee-wire, three-phase systems, you must only use circuit breakers with single voltage ratings not less than the line-to-line voltage (for example, 480 instead of 480/277 and 240 instead of 240/120). Be aware that these breakers have a more robust internal construction and are made in much smaller volume. This means they are usually special order items at two or three times the cost of conventional breakers.

Circuit arrangements—impedance grounding   There is a way to combine the best features of the ungrounded systems in terms of reliability, and the best features of the grounded systems in terms of their ability to dissipate energy surges due to their grounding connection. That procedure is the high-impedance grounded neutral approach, covered in *NEC* 250.36. See Fig. 28–8.

These systems behave like ungrounded systems in that the first ground fault will not cause an overcurrent device to operate. Instead, alarms required by *NEC* 250.36(3) will alert qualified supervisory personnel. Remember, a capacitor is two conductive plates separated by a dielectric. A plant wiring system consists of miles and miles of wires, all of which are separated by their insulation. This means that a plant wiring system is a giant though very inefficient capacitor, and it will charge and discharge 120 times each second. The resistance is set such that the current under fault conditions is only slightly higher than the capacitive charging current of the system. Since a fault will often continue until an orderly shutdown can be arranged, the resistor must be continuously rated to handle this duty safely.

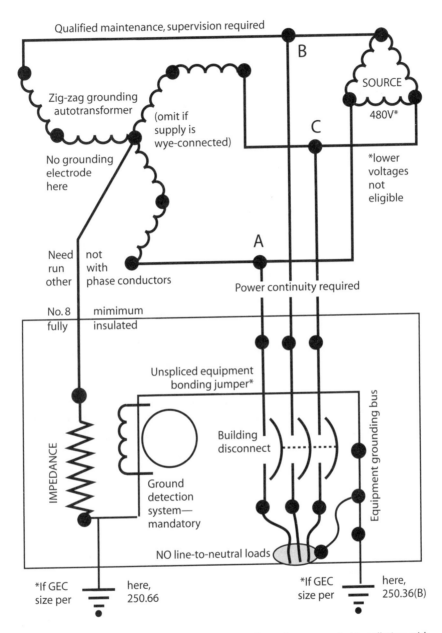

Fig. 28–8   This drawing shows a likely retrofit on a formerly ungrounded installation, with its neutral conductor created through a zig-zag grounding autotransformer. Note that the size of the equipment bonding jumper now (2002 *NEC*) depends on the end to which the grounding electrode conductor is connected. A connection at the line (left) end makes the bonding jumper a functional extension of the grounding electrode conductor, and it must be sized accordingly. A connection at the load (right) end makes the bonding jumper a functional extension of the neutral, normally sized at 8 AWG.

As shown in Fig. 28–8, the grounding impedance must be installed between the system neutral and the grounding electrode conductor. Where a system neutral is not available, the grounding impedance must be installed between the neutral derived from a grounding transformer and the grounding electrode conductor. The neutral conductor between the neutral point and the grounding impedance must be fully insulated. Size it at 8 AWG minimum. This size is for mechanical concerns; the actual current is on the order of 10 amps or less.

Contrary to the normal procedure of terminating a neutral at a service disconnecting means enclosure, when the system is high-impedance grounded, the grounded conductor is prohibited from being connected to ground except through the grounding impedance. In addition, the neutral conductor connecting the transformer neutral point to the grounding impedance is not required to be installed with the phase conductors. It can be installed in a separate raceway to the grounding impedance. The normal procedure (usually performed by the utility) of adding a grounding electrode outside the building at the source of a grounded system (should one be used as the energy source for an impedance-grounded system) must *not* be observed, because any grounding currents returning through the earth to the outdoor electrode will bypass and therefore desensitize the monitor.

An equipment bonding jumper must be installed unspliced from the first system disconnecting means or overcurrent device to the grounded side of the grounding impedance. The grounding electrode conductor can be attached at any point from the grounded side of the grounding impedance to the equipment grounding connection at the service equipment of the first system disconnecting means.

Circuit arrangements—delta-connected, four-wire, three-phase, 240-volt systems
Such a system as previously discussed is basically a three-wire, 240-volt, three-phase system with one of its transformer secondaries tapped at its midpoint. That makes 120/240-volt, single-phase power available as shown in Fig. 3–12. The wire from the midpoint of that secondary must, under *NEC* 250.20(B)(3), be grounded. It is a neutral as long as it continues throughout the system in three-wire circuits only. At any point where a 120-volt circuit is led off from the three wires, it ceases to be a neutral and becomes merely a grounded wire.

Power at 240 volts, single-phase is obtained by running wires from any two of the three wires, not including the grounded wire. Power at 240 volts, three-phase is available by connecting to all three of the wires, again not including the grounded wire.

In such installations, the transformer with the tap at its midpoint is usually larger (of a higher kilovolt-ampere rating) than the other two, because it must provide all the extra power from the single-phase circuits. The wiring diagram is shown in Fig. 28–9, in which the grounded wire is marked *N* and the three phase wires *A*, *B*, and *C*.

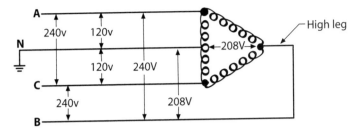

Fig. 28–9 In a 240-volt, delta-connected, three-phase system, the voltage from the neutral to the "high leg" is 208 volts. Use extreme care never to connect between the neutral and the high leg what you want to be a 120-volt circuit.

High leg—orange identification   In Fig. 28–9, the voltage between N and either A or C is 120 volts. But the voltage between N and B is 208 volts ($120\sqrt{3} = 208$). The wire B is then called the "high leg" (high-voltage leg, or the wire with the higher voltage to ground; it is also called by other names such as "wild leg" or "stinger"). The high leg does not necessarily have to be wire B; it is whichever wire is not connected to the transformer secondary that has a tap at its midpoint. However, *NEC* 408.3(E) requires that it be the middle, "B" phase in new switchboards and panelboards.

In wiring 120-volt circuits on such a four-wire delta system, it has often happened that instead of connecting the two wires of the circuit to wires N and A or C, they are accidentally connected between N and wire B; then what was intended to be a 120-volt circuit becomes a 208-volt circuit, which would of course ruin 120-volt equipment connected to it. Therefore *NEC* 215.8 requires that the wire in the high leg must be orange-colored. Instead of an orange color throughout you may instead identify the wire at any accessible point if the neutral is also available at that point, which you can do by tagging it or using other effective means such painting it orange at the accessible points or wrapping orange-colored tape around it at those points.

One great disadvantage of the four-wire, delta-connected, three-phase system (besides the danger of accidentally connecting 208 volts to what is intended as a 120-volt circuit) is that one of the three transformers must be larger than the other two, and the wires from the transformer secondary with the midpoint tapped must be larger than the third or high-leg wire. That makes it difficult to properly protect all three wires. The simplest solution is to use the same size of wire throughout, thus making wire B larger than needed. The other solution when using a smaller wire from the high-leg B than from A and C requires a very special circuit breaker with a smaller trip coil in the one wire than in the other two, or if fused equipment is used, with a smaller fuse (sometimes requiring an adapter that permits only a fuse of a smaller rating than the other two for the wire from B, in the switch). In some localities this type of system is no longer installed, and many utilities now actively discourage or prohibit it for new installations.

Open-delta, three-wire, three-phase system   Suppose one of three transformers in a three-wire, three-phase system fails and must be replaced. While one transformer is missing, the two remaining constitute what is called an open-delta system (sometimes called a "V" system), as shown in Fig. 28–10. The open-delta system with two transformers will still deliver three-phase current, but only 57% ($\frac{1}{2}\sqrt{3}$) of the normal power available from all three transformers. That reduced power may be an inconvenience to the power user, but it usually does not totally disable that user's business.

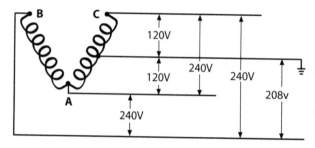

Fig. 28–10 A 120/240-volt, "open delta," three-phase system. The transformer supplying single-phase power should be larger than the other transformer.

But this characteristic can be advantageous in some installations. If the power need is likely to be on the small side for some time, the utility may install only two transformers, connected open-delta. If each of the two transformers is rated at 50 kVA, they can deliver 85.5 kVA of power (85.5 kVA is 57% of the power that three 50 kVA transformers could deliver). When the power needs of the installation increase, the power supplier boosts the total available power from 85.5 to 150 kVA by adding one 50 kVA transformer.

One of the two transformers of an open-delta system may have its secondary midpoint tapped, as shown in Fig. 28–10, to supply 120/240-volt, single-phase power. That transformer should of course have a higher capacity than the other one in the system.

Neutrals of feeders—single-phase   In a two-wire feeder, both wires must be the same size. If a three-wire feeder serves only 120-volt loads, all wires must be the same size. But if a three-wire feeder serves both 120-and 240-volt loads, *NEC* 220.61(B) permits the neutral in some circumstances to be smaller than the hot wires. The "feeder" under discussion might be the service-entrance conductors serving the entire installation, or a feeder serving part of the load.

See Fig. 28–11 which shows a feeder serving both 120- and 240-volt loads. Assume now that the 120-volt loads are disconnected; in your mind erase them from the circuit. The neutral does not run to the 240-volt load; it is not necessary to the operation of the 240-volt load. Therefore the 240-volt load does not need to be considered as far as the size of the neutral is concerned. The neutral needs to be large enough only to carry the "maximum unbalance," which in *NEC* 220.61(A) is defined thus: "The maximum unbalance shall be the maximum net computed load between the neutral and any one ungrounded conductor."

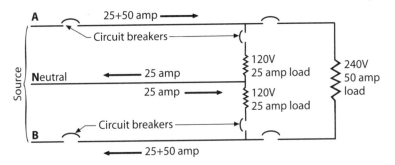

Fig. 28–11 A feeder serving both 120-volt loads and 240-volt loads.

Reverting now to Fig. 28–11, if the 120-volt loads are 25 amps each, and the 240-volt load is 50 amps, each hot wire must carry 25 + 50 or 75 amps. If the two 120-volt loads are identical, for example 25 amps as in the illustration, the neutral carries no current at all, but if one of the 120-volt loads is turned off, the neutral must carry the current for the other 120-volt load, in this case 25 amps, and must be sized accordingly.

In ignoring the 240-volt load, an electric range will give you a problem because it is not a 240-volt appliance, but a 120/240-volt appliance. Ranges with fixed intermediate burner positions operate line-to-neutral in the lower heat positions. Newer ranges with infinite heating controls tend to operate line-to-line, but still have 120-volt controls, oven lights, etc. For the purpose of determining feeder size, the *NEC* essentially classifies ranges as 30% 240-volt, and 70% 120-volt. To calculate the feeder, proceed as follows:

A. Total load in watts (including the range and/or dryer at the value normally required by the *NEC*)......_____watts

B. Deduct all loads operating at 240 volts, but not including the range and/or dryer....._____watts

C. Deduct 30% of the watts that you have included in *A* for the range and/or dryer....._____watts

D. Remainder (*A* minus *B* minus *C*)....._____watts

Divide the total wattage of *A* by 240 to get the ampacity of the minimum size of wire acceptable for the hot wires. Divide the remaining watts of *D* by 240 to get the required ampacity for the neutral, assuming the 120-volt loads are evenly balanced between the two hot wires. If for some reason it is impossible to divide the 120-volt loads into two substantially equal portions, determine the wattage of the larger of the two groups and divide by 120, which gives you the maximum unbalance and determines the minimum ampacity of the neutral.

If the maximum unbalance is over 200 amps, under *NEC* 220.61(B)(2) count only 70% of the amperage above 200. For example, if the total unbalance is 300 amps, the total to use is 200 amps, plus 70% of the remaining 100 amps or 70 amps, for a total of 270 amps. If, however, part of the load consists of fluorescent lighting, or other nonlinear loads (review nonlinear loads on page 47), the 70% factor may not be applied to that portion of the load. Follow the example of later paragraphs concerning three-phase feeders.

It must be noted that all the above is correct only if the service wires come (a) from a single-phase transformer as in B of Fig. 3–10 or (b) from a three-phase delta-connected transformer as in Fig. 3–13, and then only if the wires come from that leg of the transformer which has the center tap, so that the voltage between either of the hot wires and the neutral is 120 volts, or (c) from a three-phase, four-wire, wye-connected transformer and all four wires including the neutral are used.

There are circumstances, as in the case of individual apartments in large multi-family dwellings, where the premises are served by the neutral and only two of the hot wires of a three-phase, four-wire, wye-connected system. In that case, as explained in the page 481 discussion on how to count neutrals, the neutral must always be the same size as the hot wires because it carries the same amount of current.

Neutrals of feeders—three-phase   Now assume a three-phase, four-wire feeder of 120/208 or 277/480 volts, serving both three-phase loads connected to the three hot wires, and one or more single-phase lighting loads. Any one single-phase load will of course be connected to the neutral and one of the hot or phase wires. For the purpose of determining the neutral, ignore the three-phase loads such as motors.

Now consider each of the three single-phase loads (lighting, appliances, single-phase motors, etc.) separately. Determine the amperage of each, but exclude the amperage for any nonlinear load, including electric-discharge lighting. Select the amperage of the largest of these three loads; assume it is 350 amps, excluding the electric-discharge lighting.

If the unbalance is over 200 amps, *NEC* 220.61(B)(2) permits a demand factor of 70% for the unbalance over 200 amps whether the feeder is single-phase or three-phase, but in the case of three-phase not for any portion of the load that consists of nonlinear loads including electric-discharge lighting. Assume that the unbalance for the electric-discharge lighting load determined separately is 90 amps. To determine minimum neutral size, proceed as follows:

First 200 amps of load other than
electric-discharge lighting, demand factor 100% ....200 amps

Remaining 150 amps of load other than
electric-discharge lighting, demand factor 70% ....105 amps

Maximum unbalance, excluding electric-discharge
lighting ....305 amps

Unbalance for nonlinear loads,
demand factor 100% in every case ....90 amps

Final maximum unbalance ....395 amps

The neutral must have an ampacity of at least 395 amps.

Calculating different occupancies With the above general discussions it should be possible to calculate almost any type of building, taking into consideration the lighting demand requirements in *NEC* Table 220.11. However, some types of buildings are covered separately in some detail in later chapters.

GROUNDING NONRESIDENTIAL SYSTEMS

The theory and importance of grounding are discussed in Chapter 9. The actual method of grounding several small projects such as outlets and fixtures is covered in Chapter 17. Review the subject of grounding in those two chapters before proceeding with this chapter. You must consider separately (a) system grounding and (b) equipment grounding. System grounding refers to grounding one of the current-carrying wires of an installation. Equipment grounding refers to grounding exposed, non-current-carrying components of the system.

System grounding This subject is principally covered by *NEC* 250.20. Alternating-current systems must be grounded if this can be done so that the voltage to ground[1] is not over 150 volts. In practice this includes single-phase installations at 120 volts, 120/240 volts (also 240-volt installations if the service is derived from a transformer that is grounded, as is usually the case), also three-phase installations at 208Y/120 and 480Y/277 volts.

Grounding is required on 277/480-volt systems, but only if one or more circuits operate at 277 volts; it is not required if only the 480-volt power is used. It is also required on all 240-volt, three-phase, delta-connected installations if one of the phases is center-tapped as in Fig. 3–12.

Neutral busbar: grounded or insulated? Every piece of service equipment must have what is usually called a grounding busbar. This must be bonded to the cabinet as already explained in other chapters. This busbar contains connectors for the neutral of the service wires, and the ground wire, and usually smaller connectors for the grounded wires of all the feeders or branch circuits. All the equipment grounding wires must also be connected to a grounded busbar which must be bonded to the grounding busbar; often the connectors for these equipment grounding wires are installed along with the other connectors on one grounding busbar.

But a panelboard that is not used as service equipment but is connected on the load side of ("downstream from") the service equipment cabinet, whether an

---

1. "Voltage to ground" in the case of grounded systems means the maximum voltage between the grounded neutral wire and any other wire in the system; in the case of ungrounded systems, the maximum voltage between any two wires.

inch or a hundred feet away, must have the busbar for all the grounded wires insulated from the cabinet. But if the wiring is by means of nonmetallic-sheathed cable containing a bare equipment grounding wire, the cabinet must contain a separate busbar for all the grounding wires, and that must not be insulated from the cabinet, but rather must be bonded to the cabinet. Such a separate busbar is shown in Fig. 28–12.

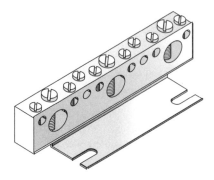

Fig. 28–12 Sometimes grounded circuit wires must be insulated from the cabinet, but if equipment grounding wires are required, they must be bonded to the cabinet. In such cases use a grounding bar, of the general type shown, for all the equipment grounding wires, and bond it to the cabinet.

Occasionally a building has only 240-volt, single-phase loads, or only three-phase loads. Such loads do not require a neutral for proper operation. Nevertheless *NEC* 250.24(B) requires that if there is a ground at the transformer or transformers serving the building (as there usually is), the grounded wire must be brought into the building. It ends at the service switch, where it is grounded in the usual way. This grounded wire, ending at the service switch, is sized in most cases equal to the size of ground wire required for the system. However, since its function is to return fault currents to the supply transformer, on very large systems it will have to be larger than the grounding electrode conductor. The rule is that when the service phase conductors comprising the largest phase connection (most services size all three phases the same, but technically that doesn't always happen) exceed 1100 kcmil copper or 1750 kcmil aluminum, then the minimum size grounded circuit conductor brought to the service disconnect must be at least one-eighth the cross-sectional area of the largest phase connection. In a parallel hook-up, the usual rules apply (review the page 489 topic on wires in parallel), which means that these conductors must be run in parallel also, subdivided in size according to the number of parallel runs, but never reduced below 1/0.

Referring to Fig. 28–13, suppose a power load supplied from Service Enclosure No. 2 develops a ground fault. Whether or not a grounded conductor enters the enclosure, the system is still grounded, and it will attempt to behave accordingly. The fault return current will still seek out the transformer neutral point to complete the circuit, probably through Service Enclosure No. 1. Current would flow over the two grounding electrode conductors and through part of the grounding electrode, although with uncertain results given that the path was never designed for this duty. The fault wouldn't clear quickly as a result, as well as because of the increased reactance that comes from the return path being at a significant distance from the supply conductors. The result is the worst of both worlds—an "ungrounded" system with no ability to ride peacefully through an initial fault without an outage, and a "grounded" system without the ability to properly clear the same initial fault.

Grounded Systems Require Grounded Circuit Conductor(s) Run to Every Service Disconnect

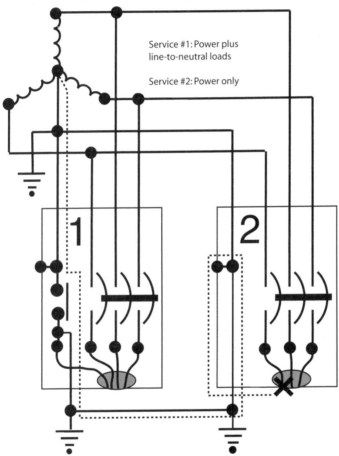

Assume an arcing fault at "X" from Service #2; dashed line shows likely fault current path in the event the grounded circuit conductor were not run to the second service disconnect. This path does not run with the supply conductors and the resulting higher impedance will retard the operation of overcurrent devices.

Fig. 28–13 A grounded circuit conductor, large enough for any load and large enough to return fault currents, must enter every service disconnect enclosure when the supply system is grounded at any point.

The 2005 *NEC* further develops this concept by requiring the grounded conductor in any feeder (if present) to be sized no less than the minimum size equipment grounding conductor for the feeder in question (see next topic). The point is to be sure that a short circuit from a phase conductor to the grounded conductor doesn't end up destroying the grounded conductor, which otherwise could be sized in terms of the connected (and very possibly minimal) load.

Sizing equipment grounding conductors   In the smaller branch circuit sizes typically addressed in Chapter 9, when an equipment grounding conductor consists of a separate wire as opposed to the enclosing metal raceway or cable armor, the equipment grounding conductor is sized the same as the branch-circuit conductors. This is true for 15-, 20-, and 25- or 30-amp circuits. When the overcurrent protection exceeds 30 amps, this is no longer the case. *NEC* Table 250.122 bases equipment grounding conductor sizes on the rating (or setting) of the overcurrent device, and above 30 amps the wire size reflects the fact that equipment grounding conductors carry current only during brief periods of circuit failure. For example, the equipment grounding conductor for a 100-amp feeder need not be larger than 8 AWG. If multiple circuits run in a common raceway, only one equipment grounding conductor needs to be run, with its size based on the largest overcurrent device protecting any one of the circuits within the common raceway (or cable assembly).

This rule is subject to modification in some cases. For example, the equipment grounding conductor need not be larger than the ungrounded conductors. This addresses the fact that short-circuit and ground-fault protection for motor circuits routinely exceeds the ampacity of the circuit conductors, often by several times, because overload protection is provided separately. (For more information you may want to look ahead to the page 589 discussion of motor branch-circuit and ground-fault protection.) For the same reason the *NEC* also allows the rating of the overload protective device to be used for the sizing decision. In addition, if the circuit conductors are increased in size for any reason, such as to avoid excessive voltage drop, a proportionate increase must be made in the size of the equipment grounding conductor.

Ground-fault protection of equipment (GFPE) on 480-volt services (and feeders)   As discussed in Chapter 3, when we refer to an ac voltage, we are referring to its RMS or effective voltage, and not its peak voltage. Peak voltage on an ac sine wave is the effective voltage times the square root of 2. Thus the peak voltage on a 120-volt system is $(120 \text{ volts})\sqrt{2} = 170$ volts, and the peak voltage of a 277-volt circuit is $(277 \text{ volts})\sqrt{2} = 392$ volts. This is significant because the peak voltage determines the likelihood of an arc restriking, once initiated. The difference between 170 volts and 392 volts is enough to explain why arcing failures on 480Y/277-volt systems have been so destructive. The arc, which naturally tends to go out 120 times a second at each zero crossing point, keeps restriking at each voltage maximum. The result is what is euphemistically called the "arcing burn-down." Actually, don't call it anything, just call the fire department instead. If such an event is in progress, and unprotected service equipment is anywhere but in a vault, an enormous amount of damage will unavoidably result.

There is another problem. Electrical arcs, once in progress, have substantial impedance, enough to cut the current flowing in a 277-volt arc to a few thousand amperes. To a 1000-amp overcurrent device in a large switchboard, a few thousand amperes looks like a heavy load trying to start and not necessarily a reason to trip immediately. And a few thousand amperes is probably optimistic. These problems combine to almost nullify the effectiveness of overcurrent protective devices in controlling a ground-fault in a 480Y/277-volt system. Nor is this engineering conjecture. Significant numbers of very serious fires resulted from this type of failure.

GFPE was the answer to the problem. It divorces short circuits from ground faults in the protective system. Review Fig. 28–14. If everything is quiet, the vectorial summation of all current leaving on the phase conductors and returning on the neutral is zero. No current should be flowing in an extraneous water pipe or over a run of conduit. On the other hand, if a ground-fault initiates, that current will not be flowing on the neutral until it gets to the main bonding jumper. Just as in the case of a GFCI for personnel protection (review Fig. 9–14), this type of current imbalance can be measured using a magnetic window set around all the circuit conductors because it creates a net magnetic flux.

The *NEC* requires all 480Y/277-volt and 600Y/347-volt services, and all feeders at those voltages unprotected by service-level GFPE for whatever reason, to have GFPE protecting every disconnect rated 1000 amps or higher. In the case of a fused switch the rating goes to the largest fuse that can be installed; in the case of a breaker with an adjustable setting, it means the highest possible setting on the breaker. The maximum setting permitted to be the triggering point for a main breaker trip is 1200 amps, and the time delay for that trip to occur is no more than one second with a 3000-amp ground fault in progress. Actual settings are typically more sensitive than the *NEC* maximum limits as users come to understand the awesome destructive power involved in these systems. Ultimately there is an engineering judgment to be made, taking into account requirements for continuity of power and the desired level of protection. The *NEC* requires the systems to be acceptance tested when first installed, and they should be retested periodically thereafter.

Figure 28–14 demonstrates how current returning from a fault in progress passes through the service portion of a switchboard, and how critical the relative position of components really is. The current comes in from the utility transformer, through the metering current transformer without break, and through the service disconnect. From there it passes through the magnetic window and continues to the load and the fault location. Returning, it passes over the load-side equipment grounding conductors and arrives at the equipment grounding

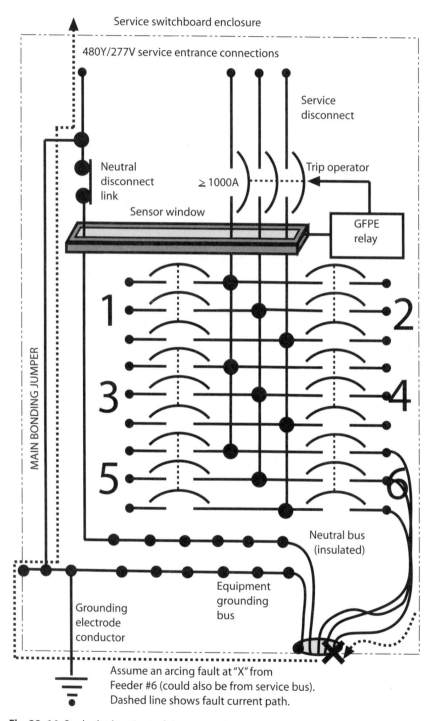

Fig. 28–14 Study the locations of the essential elements of this switchboard. The relative locations of the sensor and the main bonding jumper are critical.

bus. From there it bypasses the sensor and the neutral disconnect linkage, and then rejoins the neutral for the trip back to the system source.

The drawing shows the most popular style of GFPE, a "zero sequence" sensor that senses net current from including a sensor around all wires. Had the connection been made on the load side instead of the supply side of the service disconnect, the fault current would have passed directly through the sensor, fooling it completely since there wouldn't be any unbalanced current.

WHAT IS A SEPARATELY DERIVED SYSTEM AND HOW IS IT GROUNDED?

If there is no way to provide 120/240-volt, single-phase circuits for lighting, receptacles, and other small 120- or 240-volt loads from the service installed in a building, as is the case with 480- or 480Y/277-volt services, some other way must be found to provide those circuits. The usual way is to install what is called a "separately derived system." This consists of a transformer (or several as required) whose primary is connected to two 277- or 480-volt wires of the installation. The secondary delivers the 120/240-volt power that is wanted. See Fig. 28–15.

The secondary only of the transformer, and all wiring connected to it, constitutes the separately derived system. The terminals of the secondary of the transformer become the SOURCE for all wiring of the separately derived system. All the wiring in the separately derived system must be installed under the same conditions, as if the power for the system came from a separate 120/240-volt service entrance.

Note in Fig. 28–15 that the primary of the transformer is supplied by two 480-volt wires. If those wires are in metal conduit, or in armored cable, the separate

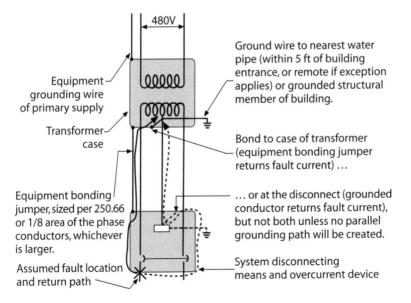

Fig. 28–15 Proper method of installing and grounding a "derived system."

equipment grounding wire shown in the diagram would not be installed, because the conduit or the armor serves as the grounding wire. The same would be true if the primary were supplied by the grounded neutral and one wire of a 277-volt circuit. But the grounded neutral in the circuit serving the transformer would of course have to be fully insulated all the way, and insulated from the non-current-carrying parts such as the transformer case. The equipment grounding wire shown would be installed only if the primary were supplied by nonmetallic-sheathed cable or wires in a nonmetallic raceway. In any of these cases, if a ground fault developed between the wires serving the transformer (or the primary itself) and the metal enclosure of the transformer, the fault current would travel all the way back to the service equipment and disconnect the circuit.

Note, as shown in Fig. 28–15, that any grounded wire in the basic higher-voltage system does not extend to the derived system; the grounding connection is lost because there is no electrical connection between the primary and the secondary of the transformer. Therefore the separately derived system must be grounded as provided in NEC 250.20(D) and 250.30. The size of the ground wire can be determined by NEC Table 250.66, and the installation rules are the same as for services.

Grounding separately derived system transformer secondaries   Since the secondary of the transformer delivers 120/240-volt power, its midpoint is a neutral. The neutral of the secondary of the transformer must be bonded to the transformer case, or the case of the main disconnecting means for the separately derived system, but never both locations. A bond in both locations would create a parallel current path for neutral return current over nonelectrical building components, which is as objectionable in these systems, in the viewpoint of the NEC, as when there is a second building (refer to garages and outbuildings topic beginning on page 334).

With bonds in both locations now comparatively unusual, the 2002 NEC set about clarifying how to handle fault return currents on the supply side of a separately derived system disconnect. First, identify how and where fault current returning on an equipment grounding conductor would reach the grounded conductor connection at the transformer secondary, based on the location of the bonding connections. Then make sure that conductor is sized for its responsibility as a potential fault return path, as shown in Fig. 28–15.

You also bond the neutral (or other grounded system conductor, as in the case of a corner-grounded delta system) to the nearest water pipe (as a local bonding connection to a water piping system, not necessarily as an electrode) and the nearest metal structural member of the building (if present, as an electrode). In addition, if the building has an interconnected steel frame and that frame is exposed, it must be bonded just as in the case of water piping systems. If the only electrode is the water piping system, then you must make an additional connection to it as an electrode within 5 ft of where it enters the building, in addition to the local bonding connection. There have been many examples of other trades working on the water system and disrupting the continuity of the path to ground. An exception may apply, however, in cases where there is qualified maintenance and supervision in an industrial or commercial occupancy if the interior water pipe is exposed over its entire length. Note that the *NEC* definition of "exposed" includes a run above a suspended ceiling. In such cases, the remote bonding connection also works as the grounding electrode connection. Whether that local connection qualifies as the connection to a grounding electrode becomes a major concern, for example, if your separately derived system originates on the tenth floor of the building.

The *NEC* has a practical alternative in these cases. Now you can run a common grounding electrode conductor for an entire group of separately derived systems, sized at not less than 3/0 AWG. Run taps (sized for the individual system requirements) from each such system to the common grounding conductor using hydraulic compression connectors or exothermic welding (see Figs. 8–23 and 8–24). In addition, for this purpose you can land the common grounding electrode conductor on a grounding busbar at least ¼ in. by 2 in. (but still leaving that conductor uncut) and terminate separately derived system grounding tap conductors on the busbar. The grounding electrode tap runs must meet the usual rules for grounding electrode conductors at services, as covered in the grounding system topic beginning on page 300.

The secondary of the transformer must be treated, as has already been mentioned, just as if it were a separate service entrance. In other words, it must have a disconnecting means and overcurrent protection; the neutral from the secondary must be grounded to the grounding bar of the cabinet containing the disconnecting means and the overcurrent protection, and that in turn must be bonded to that cabinet. But remember, if panelboards are installed downstream from that first cabinet (or transformer enclosure, if that is where the bonding jumper is connected), the grounded busbar must be insulated from those cabinets, just as in other wiring.

Instead of one or more transformers providing 120/240-volt, single-phase power, sometimes a three-phase transformer bank is installed to supply a separate 208Y/120-volt supply, as for example when the service to the building is supplied by a 480- or 480Y/277-volt service. Proceed basically as outlined, and install any kind of circuit that you could install from a 208Y/120-volt service entrance.

Not all separately derived systems use transformers   The separately derived system is not necessarily fed from one or more transformers. It could be supplied by a motor-generator: a motor operated from the available voltage, driving a generator producing the desired voltage. In that case the generator and all wires leading from it constitute the derived system. It could be fed from a converter: a device that looks like a motor, but has in a single case the motor itself fed from the available voltage, and other windings producing the desired voltage. In that case the windings producing the desired voltage, and all wiring from its terminals, constitute the derived system.

An emergency or standby generator, if feeding entirely separate circuits in no way connected to the ordinary wiring of the building, constitutes a separately derived system (review Chapter 23). Solar photovoltaic systems and fuel cell generating equipment also serve as sources for separately derived systems.

HOW TO INSTALL AND PROTECT TRANSFORMERS AND THEIR CONDUCTORS

Install transformers where they won't be subject to damage. Transformers are extremely efficient, converting more than 99% of the incoming (primary) power into useful power at the secondary voltage. However, given the large amounts of power passing through them, even a 1% loss within the transformer results in substantial heating. Make sure their ventilating openings aren't obstructed, and that the required clearances marked on the transformer are respected.

All liquid-filled transformers (high power transformers immerse their coils in special dielectric fluids that more efficiently remove heat from the core) must be readily accessible to maintenance personnel. The *NEC* includes many other special requirements for liquid-filled transformers, including requirements for location in fireproof transformer vaults in the case of transformers using combustible cooling fluids or operating at very high voltages. Beyond this mention, such installations are not within the scope of this book.

Dry transformers, which use convection currents of air for cooling, have requirements that are more lenient. They don't have to be readily accessible as long as they are "in the open." This allows for common industrial locations on columns and walls. In addition, if not over 50 kVA, they can go above suspended ceilings provided they have suitable ventilation (see previous paragraph) and provided they are completely enclosed. Although this is the usual case today, unenclosed transformers can also be used in these locations provided they meet the separation rules that apply to dry transformers generally. Dry transformers not completely enclosed and rated over 112½ kVA must be separated by at least 12 in. from combustible material, or separated by a fire-resistant and heat-insulated barrier. Larger dry transformers must be in a transformer room with a minimum construction rating of 1 hour, unless their insulation system is classified at 155 or higher (generally meaning suitable for operation in a 40°C ambient with a 115°C temperature rise). Dry transformers with higher class insulation systems don't need a dedicated room provided there is a thermal barrier similar to that for smaller transformers, or provided there is a separation of 6 ft horizontally and 12 ft vertically from combustible material.

Providing overcurrent protection for transformers   Just as wires and motors and appliances need overcurrent protection, so do transformers. The *NEC* offers two general approaches. You can closely protect just the primary side, or you can more loosely protect the primary if the secondary side provides equivalently close overload protection. Specifically, if you protect the primary side within 125%

of its rated current (next higher standard overcurrent protective device allowed), you need no other protection. However, often a transformer will be one of several supplied by one feeder, and designers would like to avoid individual protection on any one transformer primary. You can raise the primary protection to 250% of its rated current (next higher standard overcurrent device *not* allowed) if the secondary overload protection doesn't exceed 125% of its rated secondary current, rounded up if necessary to the next higher standard overcurrent device size. The secondary protection may comprise up to six grouped devices, with a combined rating no higher than that allowed for one device. Refer to *NEC* Table 450.3(B) for additional allowances for small transformers.

Transformers supplying nonlinear loads require special design

Nonlinear loads impose heating, in transformers as well as ordinary wires, beyond that which would normally be predicted by just looking at the numbers of amperes flowing at any given time. Nonlinear loads, as described on page 47, involve some combination of higher frequency harmonic currents with the 60-Hz fundamental. These high frequency currents tend to flow disproportionately over the outer margin of a conductor (this "skin effect," discussed on page 305, is much more pronounced at elevated frequencies). Since the current doesn't use the entire conductor cross section, the part it does use gets hotter, just as surely as if you forced ordinary 60-Hz current through a comparatively small wire. In addition, these currents tend to add in the neutral, sometimes to the point where they exceed the current flowing in any of the phase wires.

A transformer coil is particularly vulnerable to this effect because the same current passes through many wires all bunched together, and if there is any problem with skin effect heating, the problem multiplies by each turn on the core. For this reason, the testing laboratories declare that ordinary transformers are not suitable for installations "where a significant nonsinusoidal current is present." Possible examples cited in the product directories include electronic ballasts, uninterruptible power supplies, data processing equipment, and solid state motor speed controllers. The industry has developed K-rated transformers with special core conductor geometry and shielding to deal with this problem.

"K" factors[2] reflect a transformer's ability to cope with greater harmonic loads. Standard K-rated transformers include factors of 4, 9, 13, 20, 30, 40, or 50. By way of

2. The "K" factor required is based on an analysis of how much current is flowing at each harmonic frequency. Look at the per-unit rms current (effective current—review in Chapter 3) at each given harmonic h and square it, and then multiply each result by the square of the harmonic order involved. The K factor is the summation of those results over the range of applicable harmonic orders. For example, suppose for a unit 1 amp rms load, a harmonic analysis shows that it comprises 0.91 amps at 60 Hz (first order), 0.34 amps at 180 Hz (third order), 0.22 amps at 300 Hz (fifth order), 0.11 amps at 420 Hz (seventh order), and 0.05 amps at 540 Hz (ninth order). The even order harmonics generally cancel. Remember that these currents are not in phase, and do not add arithmetically (hence seeming to add to more than 1). Square each of these numbers, multiply each one by the square of the harmonic order (1, 9, 25, 49, 81 respectively), and add the results together. The result (3.97) points to a K-4 rating for this application.

comparison, linear loads (incandescent lighting, resistance heating, and motors connected without variable-frequency speed controls) are K-1 and most electric discharge lighting (covered in Chapter 29) and solid-state lighting control comes out as a relatively modest K-4. On the other hand, telecommunications equipment powered from building power circuits often requires K-13 transformers, as do branch circuits supplying academic classrooms and modern health care facilities. Exclusive data processing loads and variable speed drives require K-20 ratings or upwards.

Protecting wires supplied from transformers   The existence of transformer protection has no bearing on whether the wires connected to the transformer are safely protected. You have to make an independent evaluation of the status of conductor protection, which is covered in an entirely different *NEC* article. Review the end of Chapter 6 for some context for what follows; that chapter covers taps from feeders, leaving six connections from transformer secondaries for this chapter. The six *NEC* provisions taken together amount to five specific allowances for protecting conductors connected to transformers, as listed below; note that many of them have strict length limitations, which are intentionally absent from the transformer provisions. This is another example of when you have to make two tentative design analyses based on two entirely different *NEC* articles, and then correlate the results at the end.

Transformer with potentially unprotected primary and secondary conductors
This is a special case of the 25-ft tap rule, where a feeder is tapped and run to the primary, and other wires extend from the secondary to a remote overcurrent protective device. This arrangement is permitted, as long as the following conditions are true:

1. The total wire length from the feeder tap to the primary plus the secondary length doesn't exceed 25 ft. The *NEC* doesn't count, however, any portion of the primary side protected at its ampacity. This means that if you make the run to the primary with a fully sized and protected feeder, you have the entire 25 ft left over to use in positioning the secondary-side protection.

2. The primary-side overcurrent protection, as translated into protection on the secondary side by the transformer winding ratio, doesn't exceed three times the secondary-side ampacity. For example, if you have a 480- to 240-volt transformer with a 100-amp fuse on the primary size, that would translate into effectively 200-amp secondary protection. As long as the secondary conductor ampacity is at least one-third of that (67 amps) it is considered to have adequate protection.

3. The primary-side wires must have an ampacity at least one-third that of the protective device rating ahead of the feeder from which they are tapped.

4. The conductors connected to the transformer must be suitably protected from damage and the secondary conductors must arrive at a single overcurrent device rated not less than their ampacity (no next higher standard size allowance permitted).

Twenty-five foot secondary rule   The second allowance allows you to omit overcurrent protection, in industrial occupancies only, at the point of supply if all four of the following conditions are met:

1. The secondary wire is not over 25 ft long.

2. The secondary wire has an ampacity at least equal to the rating of the transformer secondary.

3. The secondary wire ends in one or a group of overcurrent devices (if more than one device, they must be grouped) having a combined ampere rating not greater than the ampacity of the secondary wire. Since you cannot tap a tap, the path to the overcurrent devices must be at least equal in ampacity to that of the wires connected to the secondary. This would be practical in the case of a large transformer feeding an industrial switchboard.

4. The secondary wire is protected against physical damage.

Ten-foot secondary rule   The third allowance is the transformer equivalent of the "10 ft rule" permitted by *NEC* 240.21(B)(1). A secondary wire may be unprotected if it meets all the following conditions:

1. The secondary wires must not be not over 10 ft long.

2. The secondary wires must have an ampacity (a) not less than the combined computed loads on the circuits supplied by the tap conductors, (b) not less than the ampere rating of the panelboard or other device supplied by the tap wires, and (c) not less than one-tenth the rating of the primary-side protective device, with that value multiplied by the transformer primary/secondary winding ratio. This allowance, however, does not supersede the individual protection requirement for lighting and appliance branch-circuit panelboards covered earlier in this chapter.

3. The secondary wires may not extend beyond the switchboard, panelboard, or control device they supply.

4. The secondary wires must be installed in a raceway (such as conduit) from the tap to the panelboard or other overcurrent-device enclosure.

Outdoor feeder secondary rule   The fourth allowance allows outdoor field wiring to be run under similar rules as for comparable utility practice. As long as you stay completely outside the building (except at the termination), the tap can be any length, and any relative size. At the load end, however, the secondary wire must arrive at a single overcurrent device having an ampere rating not greater than the ampacity of the secondary wire. Such taps supply electrical equipment in buildings or at structures. Remember, the rule would lose its technical validity if it were to result in conductors of unlimited length within a building or structure. For that reason, the tap must end within the building or structure disconnecting means or immediately adjacent thereto, and the disconnect must be outdoors, or inside nearest the point of conductor entry. However, that point of entry can be artificially extended, just as in the case of taps from feeders (review the discussion of outdoor feeder taps beginning on page 97), through 2-in. concrete encasement or other methods covered in *NEC* 230.6.

Rule for secondaries protected by the winding ratio   Secondary conductors that are not part of a multiwire secondary can rely on the primary protection as translated to the secondary side through the winding ratio of the transformer. For example, a 100-amp primary fuse on a 480-volt primary would protect a 200-amp secondary conductor connected at 240 volts. This is permitted only for two-wire/two-wire single-phase or three-wire/three-wire delta-delta transformations. If the 480/240-volt transformer actually had a 120/240-volt multiwire secondary, an ungrounded secondary conductor, presumably sized for 200 amps, could end up carrying 400 amps before the primary protection opened. This would happen if all the load were connected line-to-neutral on one line. The transformer would see that as a simple 480/120-volt (4:1) transformation resulting in not 200-amp but effectively 400-amp protection instead.

Industrial occupancies have additional tap allowances   Very large industrial installations benefit from additional allowances in Part VIII of *NEC* Article 240, requiring sophisticated engineering supervision. These installations are beyond the scope of this book.

WHAT BRANCH CIRCUITS CAN BE USED?

Branch circuits are covered by *NEC* Article 210. The ampere rating of a branch circuit is determined by the rating of the overcurrent protection (breaker or fuses) protecting the circuit. The ampacity of the wires in the circuit must be such that the wires are protected by the overcurrent protective device, with exceptions notably including taps and motor circuits. Circuits supplying two or more outlets are rated as 15, 20, 25, 30, 40, and 50 amps. If a single load consumes more than 50 amps, it must be supplied by a separate circuit of any required ampere rating; the circuit must usually supply only one outlet. However, for industrial occupancies with qualified maintenance and supervision, multioutlet branch circuits are allowed for nonlighting loads. An individual branch circuit may always supply any load for which it is rated.

Fifteen-ampere branch circuits   A 15-amp circuit may serve any kind of load in any kind of occupancy: incandescent or fluorescent lighting, receptacles, and so on. Either ordinary or heavy-duty lampholders may be installed on the circuit. Only 15-amp receptacles may be installed. The ampere rating of any one cord-and plug-connected appliance may not exceed 80% of the rating of the circuit; for appliances fastened in place the limit is 50%. See *NEC* 210.23(A).

Compute all known loads, such as lighting, special appliances such as office copying machines, air conditioners, etc., and add a liberal allowance for unknown loads, such as maintenance equipment (floor polishers, vacuum cleaners, etc.). Receptacle outlets for which the load is unknown must be calculated at 180 VA.

Each heavy-duty lampholder, if its exact load is unknown, must be figured at 600 VA, about 5 amps at 120 volts. This is covered by *NEC* 220.14(E).

Twenty-ampere branch circuits   Everything said about the 15-amp circuit is also correct for a 20-amp circuit, except that either 15- or 20-amp receptacles may be installed on it.

Thirty-ampere branch circuits   These circuits may be used for appliances in any occupancy; the rating of any portable or stationary appliance may not exceed 80% of the rating of the circuit. In dwellings, they may not be used for lighting. In other occupancies they may be used for lighting provided they serve only fixtures with heavy-duty lampholders (defined in the lighting fixture topic that follows). Fluorescent lampholders are not of the heavy-duty type, therefore they may not be used on 30-amp circuits, but only on 15- or 20-amp circuits. On the other hand, HID lighting commonly uses heavy-duty lampholders and qualifies for these branch circuits. Only 30-amp receptacles may be installed.

Forty- and fifty-ampere branch circuits   In dwelling occupancies, these circuits may be used only to supply cooking appliances. If such an appliance is otherwise permitted to be connected by cord and plug, such as a range, that does not change its classification as "fixed." In other occupancies, these circuits may be used to supply cooking appliances as in dwelling occupancies, and may be used for infrared heating appliances. They may be used to supply fixtures with heavy-duty lampholders.

Receptacles on 40-amp circuits may be either 40- or 50-amp type; on 50-amp circuits they may be only the 50-amp type. At present there is no configuration for a 40-amp receptacle.

WHAT LIGHTING FIXTURES CAN BE USED ON VARIOUS CIRCUITS?

Lighting fixtures inevitably have a lampholder as one of their components. A lampholder is any device by which current is carried to a lamp. As discussed near the beginning of Chapter 5, most people simply call them sockets. Some lampholders with no other components, such as the pull-chain lampholder pictured in Fig. 5–3, can be installed independently on outlet boxes. There are other types of lampholders besides those designed for use with screw-shell lamp bases; some lampholders for use with other kinds of lamp bases are described and illustrated in Chapters 14 and 29. Note that ordinary medium screw-shell bases are not permitted on lamps larger than 300 watts; they must have mogul screw-shell (or other) bases. Heavy-duty lampholders are defined in *NEC* 210.21(A) as those rated at 750 watts or more. (If the socket is the "admedium" size, it is a heavy-duty type if rated at 660 watts or more. But admedium bases and sockets are now practically museum pieces.)

In nondwelling occupancies, per *NEC* 210.6(B)(1), the voltage between conductors in a circuit may not exceed 150 volts if the circuit supplies one or more medium-base screw-shell lampholders, the kind used with ordinary incandescent lamps. The restriction therefore does not apply to fluorescent lighting.

NEC 210.6 requires that branch circuits supplying lampholders, fixtures, or receptacles supplying loads not over 1440 volt-amperes (or less the ¼ hp) must not exceed 120 volts between conductors in residential applications. In other occupancies including industrial establishments the 120-volt limitation applies to lampholders within their voltage ratings, ballasts, and receptacles, but that is not the upper limit in all cases. These other occupancies can also use line-to-neutral branch circuits on 480Y/277-volt systems to supply HID and fluorescent fixtures and their ballasts, mogul-base screw-shell lampholders or other lampholders within their voltage ratings, and utilization equipment. These rules make possible the routine use of 277-volt branch circuits for fluorescent lighting, and they make it impossible to run 277 volts to a medium-base lampholder. Rounding out the picture, 480- and 600-volt circuits may supply ballasts for electric discharge lamps mounted at least 18 ft above grade in permanently installed fixtures in tunnels, and at least 22 ft up on poles or similar structures.

Electric-discharge lighting loads   This category includes fluorescent lighting of all kinds, as well as other types defined in the next chapter. When calculating the load on a circuit, remember you must include not only the rated watts of the lamps but also the reactive power produced by the ballasts. For example, a fluorescent fixture with four 40-watt lamps consumes 160 watts in the lamps. For the power consumed by the ballasts, which varies with the type and size of the lamps, adding 15% would be a fair average. The 160 watts for the lamps, plus 15% of that or 40 volt-amperes in the ballast, comes to 184 volt-amperes.

Electric discharge lighting, which includes ballasts with reactive components, inherently has significant inrush currents which decline rapidly after the circuits are turned on. In commercial and industrial applications where the lighting is turned on and left on for entire shifts, or even weeks at a time, many designers omit conventional snap switch control and simply leave the circuit breaker as the means of control. Be sure the circuit breakers you use have been listed for this application. In the case of fluorescent lighting, the breaker must be marked "SWD" if used on 120-volt or 277-volt circuits. In the case of metal-halide and other HID applications (see Chapter 29), the inrush currents are even more severe and the breaker must be marked "HID" for this service. Any breaker marked "HID" will also be suitable for fluorescent lighting control.

Taps on branch circuits   Many times it is not practical or necessary to run wires of the same size as the circuit wires to individual loads on the circuit. This subject is covered in *NEC* 210.19(A)(4). A common example is wires to individual lighting fixtures supplied by the same branch circuit. Such taps must not be more than 18 in. long (in the case of lighting fixtures, measured from the point where the wires emerge from the fixture), and must always have an ampacity sufficient for the load. In any event, taps must have an ampacity of not less than 15 if on 15-, 20-, 25-, or 30-amp circuits, and not less than 30 if on 40- or 50-amp circuits.

In wiring recessed lighting fixtures, as already discussed, ordinary wire may be used in wiring the circuit up to a junction box, from which a tap not over 6 ft long, ending at the fixture, is permitted and often required, as covered in *NEC* 410.67(C). As noted in the page 332 discussion of high temperatures relating to recessed fixtures, this allowance may no longer have many practical applications.

WHAT ALLOWANCES DOES THE *NEC* PROVIDE FOR TEMPORARY WIRING?

Large construction projects often involve the arrangement of temporary power. As covered in *NEC* Article 590, most of the normal rules apply to these installations, but the circuitry can use multiconductor cable or flexible cords. This allowance is contingent on adequate support to protect them from damage. Lamps require lamp guards for protection from accidental contact or damage. Receptacles must be on separate circuits from lighting circuits.

Most receptacles now require GFCI protection   The requirements of *NEC* 590.6 are not limited to receptacles on constructions sites, but now cover all receptacles used for building "construction, remodeling, maintenance, repair, or demolition." This section now reaches receptacles in industrial plants, for example, that are permanent in character but used for temporary maintenance activities. First, any 125-volt, 15- 20-, or 30-amp receptacles must have GFCI protection. The only exception is for industrial establishments with qualified maintenance and supervision. Beginning with the 2002 *NEC*, even this exception is now further conditioned on instances where a nuisance trip would cause a greater hazard such as with a vacuum-assisted core-boring machine, or the use of equipment incompatible with GFCI protection. Examples of this include some small MIG welders and some equipment with heating elements that have high ground leakage until they come up to temperature.

Perhaps not just any GFCI, either   The GFCI protection provided as part of the temporary wiring, as opposed to being permanently installed and just supplying the temporary wiring, must be "identified for portable use." There are extension cords with GFCI protection built into the plug or located in-line and very close to the plug. In addition, there are portable distribution boxes intended for construction site use that include GFCI-protected receptacles for cord- and plug-connected equipment. These GFCI devices (see Fig. 28–16) have "open neutral" protection (meaning the white wire even though at this point in the circuit it is not actually a neutral), which means that if the grounded supply conductor opens for any reason, a normally open relay (held closed whenever the cord set is energized) opens all circuit conductors. This is in contrast to conventional GFCI devices, which fail in the closed position if the grounded conductor opens.

If the grounded conductor opens with a conventional GFCI receptacle in the untripped mode, anything plugged into the receptacle stops because the circuit is

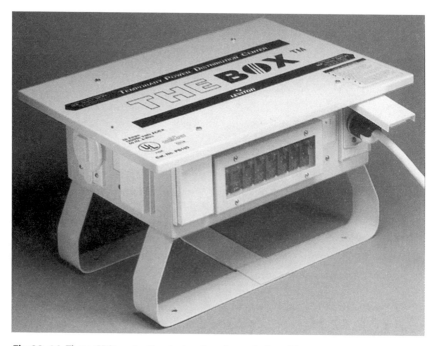

Fig. 28–16 These GFCI protective devices have been designed for construction site environments, and include means to deenergize the ungrounded conductors in the event the grounded conductor opens. *(Leviton Manufacturing Co. Inc.)*

no longer complete. This may create an illusion that the appliance has been disconnected. However, the inside of the appliance (saw, drill, etc.) is still energized at 120 volts to ground. If the hot wire insulation fails and energizes the case, the result is an extremely hazardous voltage (construction sites are notoriously wet and full of grounded surfaces) on an item that appears to be on a dead circuit, and the GFCI will not offer any protection. To see why, look at the drawing in Fig. 9–14 of the inside of a GFCI. Without 120 volts available, the electronics have no operating voltage and the device stops responding to current imbalances. However, the normal circuit conductors pass straight through the device unbroken, allowing the ungrounded conductor to remain energized.

Don't try to make up a GFCI extension cord in the field Although *NEC* 240.5(B)(4) normally allows you to make up an extension cord in the field using "separately listed and installed components," you cannot field-construct a GFCI extension cord to comply with this requirement, because GFCI receptacles with "open neutral" protection are not available as separate items. Put another way, you can legally make extension cords using 4-in.-square boxes and receptacle raised covers (refer to Figs. 10–4 and 10–7) using strain relief connectors, and you could even install a GFCI receptacle in the box. However, you would still need to install a GFCI receptacle at a permanent location (or install a GFCI circuit breaker in the panel) in order to comply. Conventional GFCI devices lack "open neutral" protection because the *NEC* and UL both take the position that a grounded circuit connection failure in the permanent wiring of the branch circuit is so statistically remote that the open circuit protection isn't needed in such cases, and the experience of the last 25 years supports that conclusion. Temporary wiring applications are another matter entirely.

AEGCP—a program that will require closer attention   Industrial occupancies that don't qualify for the GFCI exception for certain applications as described previously, and all occupancies using circuits of higher amperages or voltages, must use the highly complex Assured Equipment Grounding Conductor Program (AEGCP). This program requires you to periodically recheck the equipment grounding continuity (and proper connection) of all temporary cord sets, tools, and receptacles at first use, when there is evidence of damage, after repairs, and at intervals not over three months. This testing and recordkeeping doesn't involve just the electrician; it involves every tradesperson/independent contractor on the job, and the records are to be made available to the inspector for review. Nor is it an option in all cases simply to use GFCI, since it isn't available over 60 amps, and won't work on a 277-volt circuit. There are residual current detectors available for 480Y/277-volt applications, but they operate at far too high an amperage (typically 30 mA) to provide personnel protection. Therefore, if the inspector enforces these rules on applications that effectively mandate AEGCP, the paperwork will be required even on a 200-amp, 480-volt, pin-and-sleeve receptacle.

CHAPTER 29

Nonresidential Lighting

THE FUNDAMENTALS OF LIGHTING and the kinds of incandescent and fluorescent lamps used mostly for residential and some nonresidential lighting are discussed in Chapter 14. The same types are also used in large nonresidential occupancies, but in a greater variety and often in higher wattage ratings. Other types of lighting often used in nonresidential occupancies are covered in this chapter.

GROUP RELAMPING

A good example of the scale involved in nonresidential occupancies involves the simple task of relamping fixtures. It often becomes an expensive procedure to replace lamps one at a time as they burn out due to the labor costs involved in setting up, including assembling ladders and tools and material at one fixture. If a hundred were replaced at a time, the cost per lamp would be much lower. Many large establishments make a practice of replacing all lamps in a building (or one area of a building or plant) at one time, such as when 20 percent of the lamps have burned out. The percentage will vary with costs and efficiency requirements. When labor costs are considered, the extra cost of the new lamps is less than the cost of replacing lamps one at a time.

TUNGSTEN-HALOGEN LAMPS

Conventional incandescent fixtures are seldom used in commercial or industrial locations today, so coverage of this topic is largely confined to Chapter 14. However, a special type of incandescent lamp, the tungsten-halogen type (the basic operation is covered in Chapter 14), does have industrial applications and deserves additional coverage in this chapter.

In the general-purpose type, tungsten-halogen lamps are available in sizes from 100 to 2000 watts. The bulb is tubular, has a diameter of less than $\frac{1}{2}$ in., and the lengths for general lighting types range from $3\frac{1}{8}$ to $10\frac{1}{18}$ in. Bases vary a great deal with the size and type; two common types are shown in Fig. 29–1. One of these has a single contact at each end. They are also available in the PAR shape flood or spot, or even in conventional bulb shapes, in which case the quartz tube

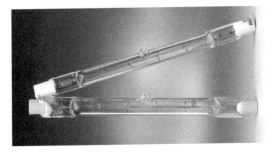

Fig. 29-1 Tungsten-halogen lamps are long and slim but are sometimes enclosed in larger glass bulbs, sometimes of the PAR shape. *(Osram Sylvania)*

is enclosed in a larger glass bulb. Special-purpose lamps are available in other sizes, in different-sized bulbs, and with different kinds of bases, including some types that have merely a wire lead at each end.

FLUORESCENT LIGHTING BECOMES MORE OF A SCIENCE

Chapter 14 discussed fluorescent lighting in some detail, since this lighting system is steadily making inroads into the residential market. However, there are additional styles of fluorescent lighting that primarily address nonresidential locations, and this chapter covers these additional applications, along with some of the more sophisticated control systems for this lighting.

Bases on fluorescent lamps   Figure 29–2 shows the various kinds of bases used on fluorescent lamps, depending on their size and starting method. Later paragraphs define the particular base used on each kind of lamp. The bi-pin base is shown in *A* of Fig. 29–2; there are three diameters (miniature, medium, and mogul) depending on the diameter of the lamp. At *B* is shown the recessed double-contact type (made in two diameters), and at *C* the single-contact type. The four-pin type shown at *D* is used only on circular lamps. In addition, compact fluorescent bases are among several possibilities for the new fluorescent lighting systems that use small profile lamps comparable in size to incandescent lamps.

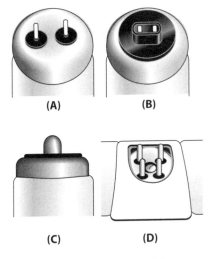

Fluorescent type designations The numbering scheme used for fluorescent lamps can be confusing. Two basic schemes are used, one for lamps with bi-pin bases, and another for lamps with single-pin or recessed double-contact bases.

A typical designation for a bi-pin lamp is F30T8. The "F" means "fluorescent";

Fig. 29-2 Common types of bases on fluorescent lamps.

"30" means "30 watts"; "T" means "tubular"; "8" means tube diameter is ⅜ or 1 in. (In the combination preheat/rapid-start type, which is 12/8 or 1½ in. in diameter, you would expect the designation TI2, but it does not appear, the lamp designation being only F34; note that the old F40 succumbed to government concerns about efficacy.) The type designation often has additional letters indicating color or other special construction.

In lamps with single-pin or recessed double-contact bases, the number in the designation, instead of standing for the watts rating of the lamp, indicates the nominal length in inches; the watt designation does not appear at all. Thus an F48TI2 lamp is a fluorescent 48 in. long, in a tubular bulb with a 12/8 or 1½ in. diameter.

Energy efficiency and fluorescent lighting As concern for scarce (and expensive) energy resources increases, a great deal of attention is being directed at the traditional 4-ft fluorescent lamp. Although this form of lighting is inherently a far more efficient source than incandescent lamps, there are so many of these fixtures in place that even a small improvement in lamp efficiency would translate into major savings. By using thinner tubes (T8, 1-in. diameter instead of T12, 1½-in. diameter) that put the phosphors closer to the arc, and using special phosphors, the same fixture will deliver 8 to 12% less light, but at an energy savings of 12 to 14%.

The slight light reduction can often be completely counteracted, or somewhat improved over the original output through the use of special, highly polished reflectors. Be sure to use reflector conversion kits that are classified by a qualified test lab for this use. For example, if UL did the classification, the marking would say CLASSIFIED BY UNDERWRITERS LABORATORIES INC. FOR USE ONLY WITH FLUORESCENT FIXTURES IDENTIFIED IN MANUFACTURER'S INSTRUCTIONS. Then examine the directions that come with the kit, and make sure it is compatible with the fixture at hand. The most common applications for these retrofits is in commercial troffer-style fixtures that measure 2 ft by 2 ft or 4 ft and drop into suspended ceiling grids.

The other piece of the energy efficiency picture involves the ballast. This chapter focuses on the most energy efficient types of ballasts. Ballasts work using inductive coupling, and the higher the frequency, the greater the effect. One reason airplanes use 400-Hz systems is to reduce the size and therefore the weight of such components. Modern electronic ballasts also increase the frequency of the current supplying the lamps. This allows for greater overall efficiencies, and the electronics also facilitate better lighting controls, such as occupancy sensors. This is particularly true for those electronic ballasts that have a Class 2 control circuit interface built in, allowing for more sophisticated switching options, such as better quality dimming controls including those integrated with daylighting sensors. These systems are covered in greater detail later in this chapter. The federal government has gotten into the act as well, as covered in Chapter 14 under the heading "Ballast designs respond to occupancy requirements." Electronic ballasts are now available that sense the end of life on a fluorescent lamp and shut it down instead of expending energy fighting a hopeless cause.

Compact fluorescent lighting is now being incorporated into fixture profiles that once would have been used only for incandescent fixtures. These fixtures (see Fig. 29–3) use permanent ballasts connected to bases that mate with compact lamps. In this way only the lamps need be replaced as they burn out, and their life is about 10 times

Fig. 29–3 A modern fixture designed for compact fluorescent lamps. *(Enertron)*

greater than comparable incandescent lamps. They are much more efficient, not only in terms of lumens per watt but also for lower heat output, which means less energy required for air conditioning. For example, incandescent lamps run about 20 lumens per watt consumed, in comparison with compact fluorescent lamps that run at or somewhat above 70 lumens per watt.

U-shaped fluorescent lamps   This lamp, shown in Fig. 29–4, can be used in place of the usual type to provide attractive ceiling patterns and lighting effects.

The length of the tube is about 48 in., but it is bent into the U shape so that the space occupied by the lamp is only about 24 in. Two of these can be installed in a single 24 × 24 in. ceiling module. The output in lumens per watt is somewhat less than

Fig. 29–4 The U-shaped fluorescent lamp has advantages in some types of lighting. *(Osram Sylvania)*

that of the 40-watt straight lamp, but higher than would be obtained by using straight 2-ft, 20-watt lamps that would fit into the same space.

Effect of voltage on fluorescent lamps   Ordinary incandescent lamps must be selected for the circuit voltage on which they are to be operated. Fluorescent lamps on the other hand are not rated in volts; the same lamp is used on a 120-volt circuit as on a 277-volt circuit. However, the ballast used must be carefully selected to match the voltage of the circuit on which the ballast and the lamp are to be used. If the system voltage does not match the ballast rating, the ballast will be destroyed or have a shorter life, but the lamps will be largely unaffected. Some electronic ballasts, however, will work on a complete range of voltages from 120 to 277 volts. If so, they will be marked for this range of service.

Color of light from fluorescent lamps   There are many kinds of white fluorescent lamps. The deluxe varieties of white are about 25% less efficient than the other kinds of white. The cool-white lamp represents about 75% of all the fluorescent lamps sold and is quite satisfactory where color discrimination is not critical.

For measuring the color of a light source, two related but independent scales are used: the Correlated Color Temperature (CCT) and the Color Rendering Index (CRI). It is useful to understand what these terms mean, because modern fluorescent lamp catalogs now list both of these terms right along with the length of the lamps and their wattage. The CCT is a measure of the "whiteness" of the light source. It relates to the temperature of a black body being heated to the point that it gives off visible light. As the temperature increases, the body passes through red to yellow to white heat. The color spectrum produced by the heated body at any given temperature is precisely reproducible, and by convention the temperature is stated as the number of degrees above absolute zero using degree intervals equal to degrees Celsius (which equals the degrees Kelvin; as a reference, the freezing point of water is 273°K and the boiling point is 373°K). Conventional incandescent lamps run about 2700°K, tungsten-halogen lamps about 3000°K, metal-halide lamps about 3500°K, traditional cool-white fluorescent around 4000°K, and so on. By varying the phosphors, fluorescent lamps can be made in a wide range of color temperatures, with the lower temperatures being closer to incandescent lamps and having a warmer effect that emphasizes red, orange, and brown colors, but at some cost to energy efficiency. For T8 lamps, the rated color temperature often appears within the tube description. For example, for F32T8741 the last two digits correspond to the color temperature.

When it is important that lighted objects appear the same as under natural light, with all of their colors discernable, it is useful to understand the information provided by the Color Rendering Index (CRI). This is an international numbering system from 0 to 100 that indicates the relative color-rendering ability of a light source in comparison with a standard reference operating at the same CCT. It expresses the degree to which object colors appear natural to an observer. Tungsten-halogen incandescent lamps have CRIs close to 100. Metal-halide lamps have standard CRIs around 65, but special models approach 90. Ordinary fluorescent lamps run about 75, but here again special models approach 90. In marked contrast are low-pressure sodium lamps with CCTs around 2000 (below incandescent) and CRIs in the 20s. Design specifications, especially in commercial settings such as in merchandise display areas, frequently set minimum CCT and CRI values.

For decorative purposes and specialized lighting, some fluorescent lamps are available in colors such as blue, green, gold, pink, and red. Ordinary incandescent lamps in color are very inefficient, being ordinary lamps with color on the glass that absorbs most of the light and allows only the desired color to pass through. Fluorescent lamps on the other hand have the proper phosphor on the inside of the tube to create primarily the color desired, making them many times as efficient as colored incandescent lamps.

QUALITY OF ILLUMINATION

For offices, institutional facilities, and industrial facilities such as manufacturing sites and warehouses, quality illumination is important because work has to be performed with efficiency and in a safe manner. Some quality factors to consider include:

Luminance; luminance ratio—The amount of contrast between what you are working on and the background substantially determines how well you can distinguish details as you work. A lighting engineer refers to this concept in terms of a ratio of luminances, and the Illuminating Engineering Society (IES) recommends that the luminance in the area close to the task surround should have a maximum bright-to-dark ratio of 3:1 with respect to the task. Away from the immediate task surround (but within the field of view), a maximum ratio of 10:1 with respect to the task is recommended. For example, the lowest illuminance (footcandle level) should be in a circulation area or a corridor adjacent to the area of general illumination in an open office. A task-lighting fixture installed at each workstation provides still higher illuminance levels on the work surface.

In addition, the reflectance values and textures of room finishes affect the light level and apparent brightness of the room. Dark finishes and heavy textures absorb light whereas light finishes reflect the light from ceiling and walls, thus making the lighting system more efficient and making the space appear even brighter. The room boundaries also have a strong influence on people's perceptions of a space and the effectiveness of a lighting system. The room cavity ratio (RCR) is the ratio of the surface area of walls to the floor area and is often used in lighting calculations to determine fixture performance.

Glare and visual comfort—Glare, for these purposes, is a direct or indirect source of light that is uncomfortable or that reduces the ability to see and work on objects or otherwise reduces visual performance. Lighting engineers identify disturbing or disabling glare as the luminance within the field of view exceeding the level to which the eyes have adapted.

Annoying flicker—Arc discharge light sources (fluorescent and HID) using magnetic ballasts operating on alternating current can have lamp flicker. This happens because the arc current is extinguished at each current zero, causing a 120 cycle-per-sec flicker. The best way to mitigate this problem requires paired lamps, one with a lagging lamp current and the other a leading lamp current, so the lamps don't pass through their current zeros at the same time. Modern electronic ballasts largely eliminate this problem because they work at such high frequencies that what flicker exists is imperceptible to the eye.

MODERN CONTROL SYSTEMS ADDRESS ENERGY CONCERNS

In addition to providing energy savings, lighting controls also offer convenience and the ability to change light distribution in response to activities in a space. Consider also that the federal and state governments have lighting power budget requirements and control requirements that define the maximum lighting watts/sq ft allowed, and where control systems must be used.

To achieve energy savings, occupancy sensors allow lighting loads to be turned off in unoccupied rooms. Passive infrared (PIR) sensing devices detect heat from the human body, and the latest technology does better at making sure that someone sitting quietly doesn't end up in darkness until he or she waves frantically at the sensor, while still avoiding nuisance activation. These devices can replace a wall switch, or mount in the ceiling. Devices using ultrasound transmit and receive low intensity sound waves in the 25 to 40kHz frequency range, which is inaudible to the human ear. These switches interpret any change in the signal return time as motion, activating the sensor. Dual technology sensing devices combine both PIR and ultrasonic sensing. The ultrasonic sensor can detect someone in a large chair facing away from the switch, for example; this orientation, being out of a direct sight line to the switch, could otherwise leave the person in the chair in darkness after the unit timed out. These enhanced detectors work well because both the PIR and ultrasonic components must activate in order to close the switch, but as long as either component stays activated, the switch will not open. Outdoor motion sensing devices are usually sold as a component of the lighting fixture being controlled, although in some cases the sensor can be mounted separately. Using PIR technology and constructed for exterior use, these units work well for walkways and entrance lighting.

Dimming control options Dimming control equipment is increasingly important because of the potential energy savings to be realized, plus it enhances the ability to customize the lit environment to the needs of a customer. For example, *architectural lighting control systems* provide a host of operational features for automatically creating lighting scenes. These units allow any group of lighting fixtures, including multiple fixtures energized from different phases of the power system, to be activated at a user-programmed brightness level. *Remote/dimming control* allows an individual office occupant to switch lights on or off or reduce the illumination level. These types of controls allow the user to match the light level to a specific work task.

Automatic control systems are based on either occupancy or daylight harvesting. Infrared or ultrasonic motion detectors automatically turn lights on and off depending on activity in the space, thus saving on energy consumption. A daylight harvesting system continuously senses the level of daylight and turns off or dims the lighting system accordingly. Increasingly, dimming fluorescent systems are being used; this capability can be added to an electronic ballast at little additional cost. Conventional fluorescent dimming uses a 0-to-10V DC analog voltage signal sent over a pair of control conductors, instructing the ballast to operate at a certain level. For daylight dimming, fixtures near windows are wired together to operate as directed by a photocell and controller.

However, the latest development in dimming technology—digital lighting control products—offers complete flexibility in programming and grouping fixtures, setting scenes, and even fading from one scene to the next. With digital control, fixtures are wired together in "free topology" (any combination of series and parallel connections). Group assignment is handled by a software-driven polling/flashing routine, allowing an individual fixture to belong to any group or to all groups. After fixtures are placed in groups, scenes can be programmed by setting groups at various dimming levels or by adding a sensor input to the control scheme.

The emerging standard for digital lighting control is a nonproprietary protocol that offers both dimming *and* switching functions. Called a Digital Addressable Lighting Interface (DALI), it is a recent effort by a group of U.S. and European lamp/ballast/fixture manufacturers to provide an economical control method. The cost of a DALI control system is between that of a 0-to-10V analog control system and the more complex bus control used in building automation systems.

Fixture selection   A fixture directs, diffuses, or modifies the light produced by a light source. In addition to a lamp socket, the components of the light fixture may include a reflector, enclosing materials, a ballast (for arc discharge sources), and stems and canopies.

The best source for selecting a fixture is the manufacturer's fixture data sheet. An indoor fixture data sheet, or report, generally deals with the distribution of light in many directions, because the light will be reflected from the surfaces of a room or space and from interior objects. Along with the fixture identification and general information, this sheet usually contains a photometric report. For an indoor model, the data generally includes candlepower distribution, the cut-off angle (or shielding angle), total fixture efficiency, spacing to mounting height ratio, luminance data, and coefficient of utilization. See Fig. 14–4 for an example of the type of information supplied as part of typical cut-sheet information. Note that, in contrast, an outdoor fixture report is concerned with light distribution in only very specific directions—to a flat surface, such as a building wall, or directly downward to a street or pavement.

An exterior fixture is categorized as either floodlight equipment or roadway type equipment. Floodlight beam spreads and their effective projection distances extend from Type 1 to Type 7; in all cases as the distance from the floodlight to the illuminated area increases, the beam spread becomes wider. All classifications in roadway photometrics are based on lighting a strip running perpendicular to the crossarm of a fixture serving a roadway.

HIGH-INTENSITY DISCHARGE (HID) LIGHTING

In ordinary incandescent lamps, the current flows through a tungsten filament, heating it to a high temperature, often above 4000°F. At that temperature light is produced. In an electric-discharge lamp the current flows in an arc inside a glass or quartz tube from which the air has been removed, and a gas or mixture of gases introduced. The most ordinary example is the fluorescent lamp already discussed. Here the current is relatively low, from a small fraction of an ampere to a maximum of 1½ amps. It flows through a relatively long length, up to 96 in. The gas pressure in the tube is relatively low, almost a vacuum.

Advantages and disadvantages of HID lamps HID lamps, in contrast to fluorescent types, have a current flow as high as 10 amps, but it flows through a very short arc tube, just a few inches long. The gas pressure in the tube is very much higher than in a fluorescent lamp. There are three different kinds in common use, described separately in later paragraphs.

In all HID lamps, the arc tube operates at a very high temperature, which is necessary for the proper operation of the lamp. Air currents must not be allowed to affect the temperature, so the small arc tube is enclosed in a much larger glass bulb, which determines the overall dimensions of the lamp. None of these lamps can be connected directly to an electric circuit; they must be provided with a ballast to match the type and size of lamp involved and the voltage of the circuit. When first turned on, the lamp starts at low brightness, gradually increasing; it may require 10 minutes for the lamp to reach full brilliancy. If turned off it must cool off for several minutes, perhaps as much as 15 minutes in a very high-wattage lamp that operates at very high temperatures, before it will relight.

Although HID lamps have a long life, that life depends on the number of times the lamp is turned on. Hours of life shown are based on burning the lamp at least 5 hours every time it is turned on. If burned continuously, the life is much longer. The number of lumens shown for each lamp is the figure after burning 100 hours, as for fluorescent lamps; the lumens per watt is based on the watts consumed by the lamps not including the ballasts. In the larger sizes the ballast watts are very roughly 10% of the lamp watts, but a higher percentage on the smaller sizes.

The principal advantages of HID lamps are high efficacy (lumens per watt), very long life, and high watt output (and high number of lumens) from single fixtures, thus reducing installation and maintenance costs. They are widely used in factories, service stations, gymnasiums, parking lots, street lighting, and generally in locations where large areas must be lighted.

HID lamps are made in many types, three of which are in common use: (1) mercury-vapor, (2) metal-halide, and (3) high-pressure sodium. Low-pressure sodium lamps, with an extreme yellowish-orange light, are even more efficient (running from 100 to 185 lumens per watt, as much as 50% higher than high-pressure sodium), but due to problems of color rendition their principal use is for roadway and parking lot lighting. Available in wattages from 18 to 180W, and an average life of 14 000 to 18 000 hours, the lamp has excellent lumen maintenance, but the longest warm-up time among all HID sources, from 7 to15 minutes. All HID lamps operate on the principles already outlined, but the details of each will be discussed separately.

Mercury-vapor (MV) lamps   Mercury lamps have a short arc tube with some argon gas in it, and some mercury, which is a liquid. As with all HID lamps, the starting procedure (about 10 minutes) involves a small arc that vaporizes the mercury, allowing the arc to fill the tube at full brilliancy. Mercury lamps are available in sizes from 50 to 1000 watts; the 175- and 400-watt sizes are the most common. The bottom lamp in Fig. 29–5 shows the 400-watt size,

which has an overall length of about 11¼ in. Lamp life is long, about 24 000 hours (almost three years of continuous use), but efficacies run in the 25 to 55 lumen-per-watt range, which is not competitive with other HID sources. Very few MV fixtures are being installed as part of original installations today.

Metal-halide (MH) lamps   This type was first introduced in 1964. In appearance they are quite similar to ordinary mercury lamps; Fig. 29–5 at the top shows the 400-watt size. The principle of operation is substantially that of the mercury lamp, except that the arc tube, in addition to argon gas and mercury, contains other ingredients such as sodium iodide, thallium iodide, indium iodide, or scandium iodide. This leads to a very high efficacy of 60 to 115 lumens per watt. Suitable ballasts are required, and the restarting time is about 10 minutes. Some of these lamps must be lit in a certain orientation (base up, base down, or horizontal); watch for any specifications from the manufacturer.

These lamps range in size from 20 to 1500 watts. With the exception of some units designed for sports lighting, lamp life ranges from 6000 to 30 000 hours. Having a warm-up time of 2 to 5 minutes, the MH source offers a white light output, making it suitable for locations where accurate color rendering is important. In addition,

Fig. 29–5 Three types of HID lamps: metal-halide, high-pressure sodium, and mercury. *(Osram Sylvania)*

since the MH lamp produces light from a fairly small arc tube, it can direct and focus its light output in a fairly precise manner using lenses and reflectors within a fixture. For sports lighting—and stadium illumination in general—it is the only practical light source, because its output (or spectral energy distribution) can be matched to the color sensitivity of television cameras.

At the end of about two-thirds of their life, the lumen output drops to about 80% of initial output. Their cost is higher than that of mercury lamps. In general, don't attempt to relamp a mercury-vapor fixture with a metal-halide bulb, although a few such combinations have been listed and could be used.

The MH arc tube contains a starting gas (typically argon or xenon), and some mercury and metallic salts, generally iodides of sodium, scandium, and dysprosium. Except for the salts, the MH arc tube is similar to the MV arc tube, but it operates at much higher temperatures and pressures. While MV arc tubes operate typically at temperatures of 600 to 800°C and under contained pressures of 3 to 5 atmospheres, MH lamps operate at temperatures of 900° to 1100°C with contained pressures of 5 to 30 atmospheres.

For lamps above 150W rating, when the ballast is initially energized, full voltage from the ballast is applied between one of the operating electrodes and a smaller adjacent starting electrode, less than an inch away. This voltage creates an emission of electrons, setting up a "local glow." The mercury then slowly vaporizes, allowing the arc to strike between the two operating electrodes. A bi-metal switch disconnects the starting electrode once the lamp is warmed up. It takes a number of minutes for the arc to reach its full light output, for the lamp current to stabilize, and for the arc tube to reach its operating pressure.

Unique design constraints apply to HID lighting systems   Any interruption in the power supply, or even a voltage dip of a few cycles, causes the MH lamp to lose arc conduction. Then the arc tube has to cool down and the tube's internal vapor pressure has to decrease before the arc can restrike. This restrike time, together with the warm-up period, adds up to an 8 to 15 minute period before full light output returns.

To address this problem, the *NEC* requires a back-up lighting system in the event that power is momentarily interrupted to an HID lighting system, in instances where they will be used for emergency systems (*NEC* Article 700; see discussion beginning on page 446). For example, an industrial MH fixture can be specified with an integral tungsten-halogen lamp, using a separate power source, usually 120VAC, connected to an emergency or standby (see discussion of the differences on page 447) power source. A time-delay following return of normal power maintains the tungsten lamp illumination until the HID lamp returns to full output. In addition, a ballast component, called a hot restrike device, can be specified for some single- and double-ended MH lamps. The hot restrike feature delivers a high voltage pulse to one of the electrodes.

Because of the MH lamp's high internal pressure, the potential always exists for arc tube rupture, especially at the end of lamp life because over time, crystallization can occur within the arc tube body (called devitrification), creating stresses in the material. If a crack occurs after the arc tube reaches full operating wattage, it can rupture with enough force to fracture the outer bulb (for which the industry coined the wonderful euphemism "nonpassive failure").

Therefore, the American National Standards Institute (ANSI) divides MH lamps into three classifications:

- E-type lamps are for use only in suitably rated enclosed fixtures.

- S-type lamps, limited to specific models in the 350- to 1000-watt range, are used in open fixtures, when the lamp is operated in a near vertical position.

- O-type lamps with quartz arc tubes comply with ANSI Standard C78.387 for containment testing and may be used in open fixtures. Currently, O-type lamps have a protective glass sleeve over the arc tube.

The 2005 *NEC* essentially addressed these restrictions for the first time, at 410.73(F)(5).

High-pressure sodium (HPS) lamps   This type of HID lamp was introduced in 1965. The arc tube operates at an exceedingly high temperature, around 1300°C (about 2300°F), so it is made not of glass or quartz, but of a very special translucent ceramic material (aluminum oxide) that can withstand the temperature involved. The arc tube, in addition to xenon gas and mercury, also contains sodium. While lamps containing sodium normally produce a very yellowish light, the high-pressure sodium lamps produce light that is quite rich in orange and red, much like that of warm-white fluorescents or the smaller sizes of ordinary incandescent lamps.

The principle of starting is quite different from that of other HID lamps; special ballasts and ignitors are required because HPS arc tubes must be shaped in a way that does not accommodate an additional starting electrode. The ignitor creates a pulse of several thousand volts, allowing the starting arc to strike directly across the principal electrodes. A lampholder (socket) used in these fixtures must have an ability to withstand those voltages. They are usually porcelain and have been evaluated for open circuit voltages accordingly, generally being rated for use with up to a 5kV momentary voltage pulse. It lights to full brilliancy in less time than other HID lamps (3 to 4 minutes). It is available in a variety of sizes from 35 to 1000 watts. The 400-watt size is shown in the middle photo in Fig. 29–6; it is a little over 9 in. long. Its light output is the highest of the HIDs discussed in this part of the chapter: 100 to 130 lumens per watt, almost double that of fluorescent or mercury lamps, and about five times that of 500-watt incandescent lamps. The life of the lamps is about 24 000 hours in the 400-watt size. Its output in lumens is maintained at a very high level, at about 90% after 8000 hours.

While this is an expensive lamp, its very high output in lumens per watt makes the cost of the electric power for the light very low. The small size of the arc tube permits narrow beams to be projected, making the lamp suitable for either flood-lighting or spotlighting. It is often used in buildings such as factories with high ceilings, and for outdoor locations such as street lighting, parking lots, and athletic fields, and for "washing" (floodlighting) the exterior walls of a building.

High-pressure sodium lighting is showing up in some commercial and educational (such as gymnasiums) applications because the economics are compelling and its color rendition steadily improves as more research goes into the design. One major industrial customer cut his connected lighting load from 59.5 kW to 42.7 kW by changing from high-output fluorescent fixtures to high-pressure sodium, and in the process doubled the light in his plant from 50 fc to 100 fc. His production rose 7% and the rejection rate dropped 40% after the change.

Continuing developments in design   Over the last decade, manufacturers have developed smaller arc tube shapes (formed body is one example) offering better control of the metallic salts in MH lamps. Because the much smaller arc tube does not have room for the starting electrode, a high voltage starting pulse, provided by an ignitor within the ballast, is applied across the arc tube to activate these lamps, which are called pulse-start MH lamps. A pulse-start ballast/metal-halide (PS MH) lamp system can provide qualities comparable to HPS systems with respect to high light output that continues over time (lumen maintenance) and long life, but avoiding the poor color rendering quality of the HPS lamp. The PS MH system also exceeds conventional MH systems by a quarter to a half in terms of light output (lumens per watt), and that advantage continues over the life of the lamp because the higher lumen output is maintained, and the lamp life increases by up to 50%. The PS MH systems warm up faster and restrike more quickly after an outage. Finally, PS MH systems are more consistent from lamp to lamp in terms of color, and the rated color rendering index value for these systems approaches 85, which is excellent. Note that even this is not the last word. There are now special "white" HPS lamps available. These lamps compete with MH lamps for applications in retail marketing.

Height-to-spacing considerations   Industrial high-intensity-discharge (HID) fixtures are generally divided into two categories. The first, a high-bay unit uses a spacing-to-mounting height ratio of 1.0 or less and a mounting height of not less than 25 ft. The second, a low-bay unit, uses a spacing-to-mounting height ratio of more than 1.0 and the mounting height is less than 25 ft.

In addition to HID fixtures, twin-tube, quad-tube, and linear fluorescent fixtures are also widely specified for manufacturing facilities and warehouse applications, and they can be placed under the same high-bay and low-bay categories. In the past, the choice was usually between metal-halide (MH) and high-pressure sodium (HPS) lamps, and HPS was frequently selected. Recently, the advent of pulse-start (high wattage 175 to 1000W) and low wattage (35/39W to 150 W), and ceramic metal-halide lamps have blurred the line between these choices.

Dimming options for HID systems   Dimming can be used in HID lighting systems to save energy, such as when periods of non-occupancy in a warehouse occur. Because of the warm-up time of the HID lamps, as mentioned previously, shut-off and restart based on occupancy is not practical. In addition, lamp manufacturers rate HID lamp life at a minimum of 10 hours per start.

The dimming procedure can be either a step-level method or continuous dimming method. Step-level, or bi-level dimming, which inserts or removes a capacitor within a magnetic ballast circuit, is an economical procedure. When, for example, a warehouse is unoccupied, the lamp output is reduced to about an 80% level.

Continuous reduction of lamp wattage in an HID fixture is handled with either a panel level procedure or with an electronic ballast. The panel level method reduces the power supplied to the branch circuit using one of three procedures: a variable step transformer (typically coupled with a constant wattage autotransformer CWA ballast in the fixture); a variable reactor; or by waveform modification. In the third procedure, an electronic control device trims, or chops, the peak of the AC waveform, thus reducing the RMS voltage to the fixtures on the branch circuit.

Electronic ballasts for the metal halide source offer new benchmarks in performance and energy savings. Lumen maintenance over conventional HID systems is improved by 30 to 50% and maintained lumens are increased by up to 56% (for a 400W system). Some of these models allow dimming down to 50%, using control devices such as relays and occupancy sensors. Improved color consistency and longer lamp life results from the increased control of the arc tube wattage over the lamp's life. Further enhancements of certain electronic HID ballasts include universal and/or multi-voltage input, multi-wattage capability, end of life detection and shutdown, and thermal protection and shutdown.

INCORPORATE FIXTURE CHARACTERISTICS IN THE LIGHTING DESIGN

Chapter 14 discusses the fundamental fact that one lumen of light falling on one square foot of area always produces one footcandle of illumination. From this statement it is easy to jump to some very wrong conclusions. For example, if a room with 100 sq ft of floor area is lighted by a lamp producing 5000 lumens, there are obviously 50 lumens for every sq ft of floor area, but the illumination will be very much less than 50 fc. That is because only a portion of the light that is generated reaches the intended surfaces. If you multiply the theoretical answer for required lumens per foot by 4, you will probably be "in the ball park" and not hopelessly wrong.

NEC 220.12 requires service and feeder capacity for a minimum of anywhere from ¼ to 3½ VA per sq ft for lighting in various occupancies. It would be very useful to be able to translate "volt-amperes per sq ft" directly into a predictable level of illumination, but unfortunately this can't be done. Of the total light produced by a lamp, some is absorbed by the fixture and some by the room surfaces. The part that reaches the intended surfaces depends on many factors. There is no simple, reliable method that will give you the level of illumination in a room if you know only the light sources. Volt-amperes per sq ft has been used as the basis for rule-of-thumb estimates, but accurate predictions using this method depend on years of experience.

Comparative lumen output of various kinds of lamps   The output of each kind of lamp in lumens per watt has already been outlined in detail in other parts of this chapter. Remember that for fluorescent and HID lamps the figures given were for lamps after 100 hours of use and were based on the watts of the lamps only, not including the power consumed by ballasts or other auxiliary equipment. Table 29–1 lists performance aspects of some fluorescent and MH lamp designs to make comparisons easier to see (data based on 40% of rated life elapsed).

Table 29–1 LAMP COMPARISONS*

Lamp type	Rated input watts	Rated life (hours)	Initial lumens	Initial lumens per watt	@40% of rated life	% of initial (design lumens)
MH standard	250	12,000	20,000	82	13,500	66%
MH pulse start	250	18,000	23,800	95	16,600	67%
Fluor T8 standard	32	24,000	2,900	95	2,750	95%
Fluor T5HO	54	22,000	5,000	93	4,740	95%
CFL long	50	14,000	4,300	86	3,870	90%

*Adapted from *Lighting Industrial Facilities*, ANSI/IESNA RP-7-01, Illuminating Engineering Society of North America.

Task lighting   In addition to general lighting, it is customary to provide additional localized lighting where work of an exacting nature is performed. Such locations are common in manufacturing establishments: assembly of very small parts, inspection stations, toolmakers' stations, etc. In work such as that shown in Fig. 29–6, 500 fc is by no means too high; investment in such lighting pays dividends. Special reflectors or reflector-type lamps described earlier in this chapter are often used.

The illumination in the concentrated area should not be excessive compared with the immediately surrounding area. With extreme contrast, glare will become a factor and much of the advantage of the additional lighting will be lost.

Lighting of athletic fields   The lighting of an athletic field is a much bigger project than is usually realized. For example, consider what is required for a

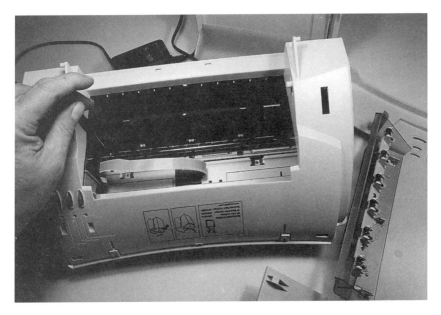

Fig. 29–6 In areas where exacting detail work is performed, provide 300 to 500 fc of light.

baseball field for major league games. Not many years ago, 150 fc for the infield and 100 fc for the outfield were considered acceptable. Today the American League standard calls for 250 fc in the infield, and 200 fc in the outfield, and many major league team managements are specifying even higher levels because they televise better. These levels are impractical with inefficient lighting; special hot-start HIDs have been developed that restart after a power outage without needing to cool down first.

Here again, the more efficient light sources increase lighting while decreasing energy costs. For example, at Cleveland Municipal Stadium the original installation consisted of 1318 incandescent lamps, each consuming 1500 watts, a total of roughly 2000 kilowatts. The installation was revamped, retaining 398 of the 1500-watt lamps, but replacing 920 of them with 1000-watt metal-halide lamps. The total power consumption dropped from about 2000 to about 1500 kilowatts. In spite of the reduction in power, the lighting on the infield increased from about 180 to almost 300 fc, and in the outfield from about 140 to almost 200 fc. Although the lighting has been even further improved since, and that particular stadium is no longer used for baseball, the lesson is still a compelling one.

From this example you can see that by using HID lamps, or a combination of incandescent and HID, it is possible to greatly increase the footcandles of lighting without increasing (or instead actually reducing) the total amount of power used, and thus avoid the extremely expensive procedure of installing larger wires, conduits, transformers, and so on, that would be required if additional incandescent lighting were added. These designs are beyond the scope of this book, but they illustrate the effect of technology over time on the energy requirements of these applications. The actual design of such installations is not for the beginner.

REMOTE SOURCE LIGHTING (FIBER OPTICS)

A fiber optic lighting system delivers illumination from a light source in one location to another completely separate location, with the light traveling through glass or plastic tubing. Typically, a fiber optic (FO) lighting system consists of:

- An illuminator, which generates light and then contains and directs it into the end of a fiber optic light guide. Illuminators typically include a lamp and a reflector. More sophisticated units include an IR or UV filter, and you can find some with dichroic glass color filter wheels so you can provide a continuous or fixed change of color.

- A light guide, which transmits the light to where it is needed. This consist of optical fiber cables, made up of a bundle of plastic or glass fibers or a single plastic fiber. The science of fiber optics uses a phenomenon called total internal reflection, discussed in greater detail below. If light is delivered to the end of the fiber, it is called an end-emitting fiber. If light is emitted along the entire length of the fiber, it is called a side-emitting fiber, which looks similar to neon tube lighting.

- A fixture (or fitting). Various types of reflective or diffusing fittings at the end of the fiber allow for a variety of light distributions.

FO lighting systems are completely unobtrusive, because only the end housing hardware is visible, and the hardware fittings can protrude from the ceiling, walls, or shelves. They are also available in a side-emitting configuration, where the light is designed to spill out the sides of the tubing with much the same effect as the tubing in a neon sign, but without the problems of high-voltage operation required for neon applications. In contrast, the thin optical fiber cabling can be snaked into the ceiling or wall of an existing structure. FO systems are useful for decorative applications, and where the relative high cost per lumen delivered is acceptable. Glass fiber basically lasts forever unless it breaks. Plastic fiber is less susceptible to breakage, but IR and UV energy from the light source can degrade the fiber with age.

Total internal reflection makes FO systems work. When a wave-like phenomenon (such as sound or light) moves at an angle between one medium and another through which it passes at different speeds, it changes direction at a known angle based on the difference of travel speeds through the different media. This is why objects under water in a lake appear to be in a different location than their real position; as the light leaves the lake at an angle it bends toward the water surface because its slower speed through water bends it toward the water as it emerges into the air. This is why spear fishermen aim below the apparent position of a fish in order to hit it. If you put your eye right at the water's edge on a calm day, the water will look like a mirror and you won't see anything underwater at all, because any light trying to get out of the lake and reach your eye is refracted back into the water. Light passing through a fiber optic cable does the same thing; having entered the cable almost parallel to the long axis of the cable, it never leaves unless it hits some imperfection within the cable. The cables also use cladding on the cable surface, which, using the same principle of the effect of different speeds of light on refraction, achieves the same effects as an air/cable interface, but more reliably since the underside of the cladding isn't affected by environmental conditions.

LIGHT-EMITTING DIODES (LEDs)

Light-emitting diodes are an important new technology in light generation, and they already are replacing incandescent lamps in specialized applications, such as displays, gardens, walkways, and some accent illumination.

A typical LED consists of a diode chip mounted into a reflector cup held by a mild steel lead frame that is connected to a pair of electrical wires and then enclosed (encapsulated) in epoxy. The diode chip is generally about 0.25 millimeters square. When current flows across the junction of two different (semiconductor) materials, light is produced from within the solid crystal chip. The shape of the emitted light beam (its narrowness or wideness) is determined by a number of factors: the reflector cup configuration, the size of the chip, and the shape and position of the lens relative to the LED chip. Material composition determines the wavelength and color of light.

Advantages of LEDs include a relatively high efficacy, reasonably long life, small size, rugged construction, fast on (60nsec vs. 10msec for incandescent), ease of dimming and control, and since they operate from a low-voltage power supply, safety. In addition, this solid-state light source does not use mercury.

Presently, white light LEDs cost many times more than an incandescent lamp with the same lumen output and are about twice as efficient, in lumens per watt, as incandescent lamps. However, advances in technology are continuously occurring. Driven by energy concerns, the near future holds dramatic changes in the way we use electricity to light our world.

CHAPTER 30
Industrial and Commercial Motor Applications

CHAPTER 15 COVERS the wiring of ordinary types of motors as used in residential and farm applications. Review that chapter now, because some of the points covered there are not repeated here. This chapter covers the wiring of motors in commercial and industrial applications (at not over 600 volts) so that when installed the motor will have proper operating characteristics, proper control, and proper protection.

This chapter opens with concepts that apply to all motor circuits, required motor ratings being one example. There is a comprehensive review of the terminology used in motor circuit layouts, including all switching devices recognized by the *NEC* for motor control purposes. Then the chapter presents a review of *NEC* requirements for the simplest motor circuits, those with only one motor on them, and follows with how those requirements interact with other motors or even nonmotor loads on the same circuit, which commonly occurs at the feeder level. The chapter closes with a discussion of two special motor applications that vary from the normal rules—motors immersed in refrigerant (hermetic refrigerant motor-compressors) and fire pump motors, which are intended to run to failure and therefore purposely omit or trade off many normal protective requirements for motors.

MOTOR RATINGS AND MOTOR CIRCUIT TERMINOLOGY

In reading this chapter, you will gain greater understanding if you think in terms of one specific motor, then reread the chapter thinking about a much smaller or a much larger motor or a different kind of motor.

The *NEC* requirements for wiring motors are very detailed. The wiring of motors ranging from a fraction of a horsepower to thousands of horsepower can never be reduced to a few simple rules. There is no attempt in this chapter to explore every detail covered by more than 35 full-sized pages in Articles 430 and 440 of the 2005 *NEC*. Rather this chapter pertains to the more ordinary types of motors and installations, which covers a very large percentage of all motors being installed. For the more difficult and intricate types of installations, consider this chapter a preview of the kinds of problems that will be met; you will not be called upon to design such installations until you have had considerable experience in the more ordinary jobs. But it will be useful for you to understand the basic principles behind the rules so you can use the *NEC* itself to better advantage.

Voltage ratings of motors   While circuits used in ordinary wiring are rated at 120, 240, and 480 volts, motors are still rated at 115, 230, 460, and 575 volts, but are designed to operate at voltages 5% above or below their rated voltage. Accordingly this chapter refers to 115-, 230-, 460-, and 575-volt circuits, instead of 120-, 240-, 480-, and 600-volt circuits as in other chapters. To some extent this reflects practical voltage drop considerations, particularly when the motor is under load and drawing its rated current through the circuit wires. This does not mean that the circuits are any different; they are only called 115-, 230-, 460-, and 575-volt circuits. Motors are also available specifically designed for use on 200- or 208-volt systems.

Ampere ratings of motors   In general the *NEC* insists that you design the electrical system generically, and not to any specific brand of motor. At the end of *NEC* Article 430 there is a series of tables that have columns organized by motor voltage, and rows organized by motor horsepower. The bodies of those tables read in amperes. All elements of the motor circuit (disconnect sizes, wire ampacities, etc.) must be designed around those current values, except for running overload protection, which is covered later in this chapter. This principle applies to the overwhelming majority of motors you will install, but there are some exceptions. Multispeed motors, adjustable-voltage motors, and torque motors (which operate at all times under essentially locked-rotor conditions) all involve such significant departures from generic values that the *NEC* requires you to work from the nameplate. In addition, many motor-operated appliances have exaggerated horsepower ratings intended to increase market appeal. Test labs don't police those ratings, but they do police the claims for full-load current. Therefore, in the case of appliances, use the nameplate full-load current and ignore any horsepower claims.

"In sight from" requirements   Review the discussion under this same heading on page 276.

Time rating   *NEC* 430.7(A) requires a considerable assortment of information on the nameplate of a motor. One of the required items is the time rating. Each motor must be identified as continuous rated, or 5-, 15-, 30-, or 60-minute rated. A motor with a "continuous" rating may be used even if not expected to run continuously, but a motor not rated as continuous must never be used if it could under some circumstances run continuously, as explained in the next paragraph. Such use would in many cases lead to a burned-out motor.

Continuous-duty motors   The term "continuous duty" as applied to a motor must not be confused with "continuous load" (discussed on page 476) as applied to loads other than motors. As applied to motors, *NEC* 430.33 states: "Any motor application shall be considered to be for continuous duty unless the nature of the apparatus which it drives shall be such that the motor cannot operate continuously with load under any condition of use." That means if it is possible for the motor to run continuously, it must have a continuous-duty rating—the time rating on the nameplate must state "continuous." This is true even if the motor is not intended to run continuously. For example, automatic controllers can fail in the on position, and manual controllers can be forgotten in the on position. Either of these contingencies would result in the motor running continuously. In practice, just about every motor must be considered "continuous duty" except special types discussed in the next paragraph.

Non-continuous-duty motors   Examples of such motors are those on hoists or elevators; they can't operate continuously under load. Elevator motors, for example, being in buildings of finite height, cannot pull the car against gravity continuously—the *NEC* classifies them as an intermittent-duty application. Such motors are considered protected by the branch-circuit overcurrent protection, provided its rating does not exceed that for "more than 1-horsepower" continuous-duty motors. But it is wise to protect such motors with motor-running overcurrent protection as required for the continuous-duty type, if this can be done without nuisance blowing of fuses or tripping of breakers.

Motor branch-circuit components   The elements that make up a motor branch circuit are lettered here for easy reference to Fig. 30–1, which shows a motor that is on a separate circuit. Frequently in practice one element can be made to serve two or even three different purposes. At other times, two separate elements may be installed in a single cabinet.

A. *Motor branch-circuit wires*—The wires from the circuit's source of supply, such as a panelboard, to the motor.

B. *Disconnecting means*—The switch or circuit breaker or other recognized disconnecting means to disconnect all the ungrounded wires to the motor and its control. Additional information on the recognized types is given later in this chapter.

C. *Motor branch-circuit short-circuit and ground-fault protection*—The circuit breaker or set of fuses that protects the wires, the controls, and the motor itself against overcurrent due to short circuits or ground faults only.

D. *Motor controller*—The device itself (magnetic contactor or manual switch) that is used to start and stop the motor, or reverse it, control its speed, and so on.

E. *Control circuit*—Wires leading from a controller to a device such as a push-button station at a distance. It does not carry the main power current, but only small currents to signal the controller how to operate.

F. *The push-button station*—As mentioned in preceding sentence.

G. Motor overload protection—A device to protect the motor, its controller, and its circuit only against possible damage from higher than normal currents caused by overloading the motor, or in case the motor does not start when turned on. Such devices are not capable of opening short-circuit or ground-fault currents.

H. Secondary control circuit—The wires to a separate controller by which the speed of a wound-rotor motor can be regulated. This type of motor is not used often enough to warrant discussion in this book.

I. Secondary controller—The controller mentioned in the preceding paragraph.

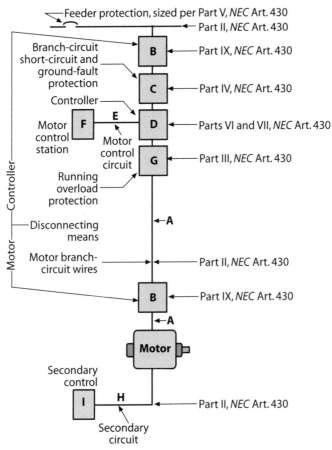

Fig. 30–1 Elements of a motor branch circuit.

Motor current variations   The current consumed by a motor varies with the circumstances. The starting current is very high, but it gradually drops as the motor speeds up. When the motor reaches full speed, delivering its rated horse-power at rated voltage, the motor carries what is called "full-load current." But the current increases further if the motor is overloaded.

Full-load current   This is also called "running current." This is the current consumed by the motor while delivering its rated horsepower at its rated voltage. It is the current in amperes stamped on the nameplate of the motor. When an installation is being planned, usually the motor has not yet been obtained and the nameplate information is not available. Therefore consult *NEC* Table 430.248 for single-phase motors and Table 430.250 for three-phase motors. (*NEC* Table 430.247 covers dc motors, and Table 430.249 covers two-phase motors; neither type is discussed in this book so the tables are not shown in the Appendix.) If you are using motors rated at 200 or 208 volts, the *NEC* now includes additional columns for these voltages instead of a percentage adjustment.

Overload current   As discussed at the beginning of Chapter 15, no motor should be installed with the expectation that it will continuously deliver more than its rated horsepower. The motor can deliver more than its rated horsepower, but it will consume more current and overheat. If the overload is large enough the motor will burn out. The *NEC* requires a motor overload device that will disconnect the motor if the overload is large enough to damage the motor if continued indefinitely. It will also disconnect a motor that fails to start when turned on. Overload devices are described later in this chapter.

Starting current   When a motor is first connected to the line, the current consumed is very much higher than after the motor comes up to speed. The ratio of starting current (also called "locked rotor" or "stalled rotor" current) to running current varies greatly with the size and type of motor and may be as high as 600%. An ordinary single-phase, 1-hp, 230-volt motor consuming only 8 amps while running at full speed and full load will require as much as 48 amps for a second or two when first started. For larger motors the ratio is usually lower, but starting current is nevertheless always higher than running current.

Due to continuing concerns about energy efficiency, motors, just as lighting fixtures, have received intense scrutiny due to their obvious share in the national energy budget. To this end, motor efficiencies are going up, but the higher-efficiency motors require significant changes in the electrical system design, principally due to much higher starting currents. The way to engineer a highly efficient motor is to reduce resistance in its windings, which thereby reduces I^2R losses in the motor. However, at standstill (and start-up) the only impedance in the motor is the winding resistance; there isn't any counter emf to reduce the inrush current. This means that highly efficient motors tend to have extremely high inrush currents, on the order of 800% of full-load current. In addition, their key performance characteristics, such as starting torque, may significantly differ. Unlike changing to a high-efficiency fluorescent fixture ballast, changing a motor to a different design letter frequently involves substantial redesign of many elements of the circuit.

The ratio of starting current to running current for any given motor also depends on the kind of machinery the motor is driving. The starting amperes will be higher for a hard-to-start load than for an easy-to-start load. Moreover, the higher starting amperage will persist for a longer time because the motor will not come up to full speed as quickly as with an easy-starting machine.

Code letters   All ac motors of ½ hp or more manufactured from 1940 until the 1996 *NEC*, and most motors manufactured subsequently, carry on their nameplates a "Code letter" that indicates the approximate kilovolt-amperes consumed per horsepower with the motor in a locked-rotor condition. Locked-rotor means the motor is locked so the rotor cannot turn. Of course a motor is never operated that way, but at the moment a motor is first connected to the power line, it is not turning, so the locked-rotor condition does exist until the motor starts to turn. The current consumed by the motor until it starts to turn is very high. Then the current drops off until the motor reaches full speed, at which time normal current is established. From the Code letter you can determine the maximum current required by a motor while starting, which is useful information in establishing the size of the various components in a motor circuit. The Code letter may no longer be required, but the information it conveys is; the *NEC* now allows motor manufacturers the option of indicating a Code letter or indicating the locked-rotor current in amperes. Most motors still carry Code letters. See Table 30–1 for interpretation of Code letters.

Design letters   Generally speaking, the larger the motor, the lower the locked-rotor current per horsepower. The 1996 *NEC* introduced the concept of design letters, which must now be marked on motors as well. A motor "design" charac-

---

Table 30–1 MOTOR LOCKED-ROTOR CODE LETTERS [from *NEC* Table 430.7(B)]*

Code letter	Kilovolt-amperes per horsepower with locked rotor	Code letter	Kilovolt-amperes per horsepower with locked rotor
A	Under 3.14	L	9.00 to 9.99
B	3.15 to 3.54	M	10.00 to 11.19
C	3.55 to 3.99	N	11.20 to 12.49
D	4.00 to 4.99	P	12.50 to 13.99
E	4.50 to 4.99	R	14.00 to 15.99
F	5.00 to 5.59	S	16.00 to 17.99
G	5.60 to 6.29	T	18.00 to 19.99
H	6.30 to 7.09	U	20.00 to 22.39
J	7.10 to 7.99	V	22.40 and up
K	8.00 to 8.99		

*Reprinted with permission from NFPA 70-2005 *The National Electrical Code,* © NFPA 2004.

teristic takes into account locked-rotor current, torque, and other performance characteristics. At present there are four principal design classifications: A, B, C, and D. Design B also involves, in many instances, a classification of "Design B energy efficient," which covers the energy-efficient motors described just previously, and for which some requirements in Article 430 are more forgiving in order to permit them to start.

To establish the approximate maximum amperes required by a specific motor while starting, first determine the actual Code letter on the nameplate of the motor; as an example assume that it is J and that the motor is 3 hp. Refer to the table and you will find that J shows from 7.10 to 7.99 kVA per hp. Call the average 7.55 kVA; or 7550 VA. For 3 hp the total is 7550×3 or 22,650 VA. If the motor is single-phase, divide by the voltage. This 22 650/230 produces a result of about 98 amps. In other words as the motor is first thrown on the line, there is a momentary inrush of about 98 amps, diminishing gradually to about 17 amps (*NEC* Table 430.248) when the motor is running at full speed and delivering its rated 3 hp. If the motor doesn't use a Code letter, just refer to the locked-rotor current rating which must be marked instead.

If the motor is a three-phase motor, first multiply the volts by 1.73 (see Figs. 3–5 and 3–6 and related text). In the case of a 230-volt motor, the result is 397.9; use the number 400, which is easy to remember. Then divide 22 650 by 400, giving about 57 amps, which is the inrush when the motor is first turned on, diminishing to about 8 amps when the motor reaches full speed.

Types of motor disconnecting devices   The *NEC* recognizes the following devices as suitable disconnecting means for motors:

■ *Motor-circuit switch*—A switch, rated in horsepower, capable of interrupting the maximum operating current of a motor of the same horsepower rating as the switch at the rated voltage. The device must be listed to qualify. For Design E motors (over 2 hp), the switch must be specifically listed for Design E service, or derated to account for the increased locked-rotor current. For Design E motors rated 3 to 100 hp, the switch horsepower rating must be at least 1.4 times the motor horsepower, and 1.3 times the motor horsepower for larger motors.

■ *Circuit breaker*—A device designed to open and close a circuit by nonautomatic means, and to open the circuit automatically on a predetermined overcurrent without injury to itself, when properly applied within its rating. (A circuit breaker is a combination switch and an overcurrent device. It can be manually closed or opened, but trips automatically on overload.)

▶ *Inverse time (as applied to circuit breakers)*—A qualifying term indicating that there is purposely introduced a delay in the tripping action of the circuit breaker, which delay decreases as the magnitude of the current increases.

▶ *Instantaneous trip (as applied to circuit breakers)*—A qualifying term indicating no delay is purposely introduced in the tripping action of the circuit breaker. It must be used as an integral part of a listed combination controller. This is why these devices carry component recognition only (explained in Chapter 1); it would be an *NEC* violation for a test lab to grant a listing to such a device.

▪ *Molded case switch*—A nonautomatic circuit breaker. Molded case switches contain the switching mechanism and manual-operable handle of circuit breakers, but no thermal or magnetic sensing mechanism that would cause an automatic trip.

▪ *Self-protected combination controller*—This combination device controls the motor's performance, and qualifies as a disconnecting means. The device must be listed to qualify.

▪ *Manual motor controller additionally marked* SUITABLE AS MOTOR DISCONNECT— Figure 30–4 shows a manual motor controller. Even though it is controlled manually, and says OFF and ON, it does not qualify as a disconnecting means without meeting additional qualifications. Motor controllers, being designed as the manual equivalents of automatic controllers generally wired on the load side of a conventional disconnect, don't have as robust internal spacings as full-fledged disconnect switches. The *NEC* allows these devices, if so listed, to be used as formal disconnects in two circumstances. The first, covering small motors, allows them to be used as disconnects for motors of 2 hp or less, just as snap switches (described later in this list). The second, covering larger motors, allows them to be used as disconnects provided they are on the load side of the final branch-circuit short-circuit and ground-fault protective device. In either case, their horsepower rating must not be less than the motor.

▪ *General-use switch*—A switch intended for use in general distribution and branch circuits. It is rated in amperes, and is capable of interrupting its rated current at its rated voltage. See Fig. 30–2 for an example. Its ampere rating must be not less than twice the full-load current rating of the motor. It generally cannot be used for a motor larger than 2 hp, unless it additionally qualifies as a motor-circuit switch, as described earlier. It also qualifies if the installation (2 hp up to 100 hp) involves an autotransformer-type controller and the motor is driving a generator

Fig. 30–2 This 30-amp switch may be used as the disconnecting means for small motors. This switch also has a horsepower rating and qualifies as a motor-circuit switch. *(Square D)*

with overload protection, provided the controller has (1) no-voltage release (shuts off when power is discontinued, with manual restart only), (2) running overload protection limited to 125% of motor full-load current, and (3) the capability to interrupt locked-rotor current. In addition, the branch circuit must include separate fuses or an inverse-time circuit breaker not over 150% of the motor full-load current.

▪ *General-use snap switch*—A form of general-use switch constructed so that it can be installed in flush device boxes or on outlet box covers, or otherwise used in conjunction with wiring systems recognized by the *NEC*. (In other words, the ordinary switches used in controlling lights in ordinary house wiring.) They are for ac motors only. To qualify, the switch must be rated ac-only (general-use ac-dc snap switches are not acceptable) and the motor full-load current must not exceed 80% of the ampere rating of the switch.

▪ *Isolating switch*—A switch intended for isolating an electric circuit from the source of power. It has no interrupting rating, and it is intended to be operated only after the circuit has been opened by some other means. This is permitted only for dc motors over 40 hp and for ac motors over 100 hp.

▪ *Plug and receptacle*—A cord- and plug-connected motor need not have an additional disconnecting means if the plug and receptacle have a suitable horsepower rating. Refer to Figs. 20–6 and 20–7 for horsepower ratings of common receptacle configurations. For the complete list, refer to the UL guide card information under the heading "Receptacles for Attachment Plugs and Plugs" (Guide Card Designator RTRT), and manufacturer's literature. The separate horsepower rating is not required for appliances, room air conditioners, or portable motors rated ⅓ hp or less.

▪ *System isolation equipment*—New in the 2005 *NEC*, this concept uses a contactor on the load side of a motor circuit switch, circuit breaker, or molded case switch as a disconnecting means. Because contactors can be closed inadvertently, these devices must be listed and include means to redundantly monitor the contactor position. This equipment is designed for extremely large industrial applications, typically involving multiple personnel entry points to a piece of equipment and therefore making remote lockout provisions in a control circuit desirable. These applications are far beyond the scope of this book.

START WITH THE BASICS: ONE MOTOR AND NO OTHER LOAD ON A CIRCUIT

If the motor is the only load on the circuit there is no feeder, only a branch circuit, whether it starts from the service equipment or from a panelboard that is fed by a feeder from the service equipment. Sometimes a feeder supplies more than one motor; information concerning such cases is furnished later in this chapter. In the unusual case where a single motor is a significant portion of the entire load on the panelboard where the motor's branch circuit originates, and where that panelboard is a power panelboard that does not require individual overcurrent protection (review the page 527 discussion on power panelboards), the procedure that applies to sizing the panelboard supply wires is similar to that for multiple motors supplied by a feeder.

Branch-circuit wire size and voltage drop   This subject is covered by Part II of *NEC* Article 430. Branch-circuit wires for one motor must have a minimum ampacity of 125% of the full-load current of the motor as determined from the nameplate on the motor or from *NEC* Tables 430.248 or 430.250 (see Appendix). Why must the wires have a larger ampacity than the load they are expected to carry? The overload device (which is also installed in a circuit and is explained later in this chapter) usually has a rating 25% higher than full-load current to allow for nominal overloads. Thus the circuit wires must be large enough to carry at least 25% more than the normal full-load current.

Assume you are going to install a 2-hp, 230-volt, single-phase motor; its full-load current per *NEC* Table 430.248 is 12 amps. Multiply that by 1.25 (125%) to reach 15 amps. Use wire with an ampacity of at least 15. If you are going to install a 10-hp, 230-volt, three-phase motor, its full-load current per *NEC* Table 430.250 is 28 amps. Multiply that by 1.25 to reach 35 amps. Use wire with an ampacity of at least 35.

If the motor is not a continuous-duty motor, the minimum size of the motor-circuit wires ranges from 85 to 200% of the full-load current, as shown in *NEC* Table 430.22(E). Consult this table in your copy of the *NEC*. The various types of motors that are not continuous duty are defined in *NEC* Article 100.

The *NEC* does not take voltage drop into consideration in specifying the minimum size of wires to be used. It is not considered good practice to permit voltage drop over 2½% in a motor circuit (with the motor running at full speed at normal power) because voltage drop is wasted power that may cost a considerable sum during the life of the motor. Too great a voltage drop also means the motor will not deliver full power,[1] start heavy loads, or accelerate as rapidly as under full rated voltage. There is additional voltage drop in the feeder, if one is involved.

Don't overlook the starting current of the motor. If you have figured your circuit wires size for a 2½% drop during normal running, but the motor consumes four times normal current while it is starting, the drop during that period will be 10%, not 2½%. This may be of little importance if the motor starts without load and the load is applied after the motor comes up to speed. But if the motor must start against a heavy load, voltage drop can become quite important. There is one compensating factor. Since the wires must have an ampacity of at least 125% of the normal full-load current, the voltage drop will be less than when figured on

---

1. A motor develops power in proportion to the square of the voltage. At 90% of rated voltage, the motor will deliver 0.90 × 0.90 or 0.81 (81%) of the power it would deliver at rated voltage.

the basis of full-load current. So in quite short runs, the minimum size of wire will probably not result in excessive drop; in longer runs you will be wise to calculate the drop. How to do so is explained in the topic on reducing voltage drop, beginning on page 107.

Motor disconnecting means—purpose, minimum hp, and current ratings

This subject is discussed in Part IX of *NEC* Article 430. A motor disconnecting means (circuit breaker or switch) is required to disconnect the motor and its controller from the circuit. This is in the interest of safety so the motor and its controller can be totally disconnected from the electric power when someone is working on the motor or on the machinery that it drives. The disconnecting means must plainly indicate whether it is in the closed (on) or the open (off) position. It must be readily accessible (if there are more than one, at least one must be readily accessible). It must open all ungrounded wires to the motor and its controller. A disconnecting means must always be "in sight from" the controller. If these devices are remote from the motor, you will probably need an additional disconnect in sight from the motor. Review the page 276 discussion of "in sight from" requirements, which covers the major change in the 2002 *NEC* on this point.

Required ratings for motor disconnecting means   The disconnecting means for any motor must have an ampere rating of not less than 115% of the full-load current of the motor. In most cases, it has to have a higher rating than that; and, as previously stated, it must be much higher (200%) where a general-use switch is used. Possible locked-rotor conditions require a disconnecting means to have a high interrupting capacity. In addition, for disconnecting means rated in horsepower, the horsepower rating must at least equal that of the motor that is to be disconnected. In the case of a nonfused motor-circuit switch rated in both horsepower and amperes, if the horsepower rating is adequate the current rating may drop below the 115% value, but only if the switch is listed. Product standards assure sufficient interrupting capability based on horsepower.

Height and access requirements for motor (and other) disconnecting means

Since disconnects may need to be operated in an emergency, it follows that they should not be located where they will be out of reach. This subject is covered by *NEC* 404.8(A). Motor and other equipment disconnects, service disconnects, and circuit breakers "used as switches" (and it is difficult to imagine a time when any conventional circuit breaker is not used in this way) must not be located more than 6 ft 7 in. above the floor or working platform for the equipment. This height translates to 2.0 meters, to correlate with international requirements. The *NEC* specifies that this height must be measured to the center of the operating handle grip.

Two exceptions to this rule address instances where, although the switch is higher, it can still be operated from the floor, namely, hookstick-operable isolating switches and busway switches located on the busway but arranged for operation from the floor. A third exception covers disconnects installed adjacent to the equipment supplied, which can be at the equipment even if you need a ladder to reach them. This protects those working on equipment or motors near a ceiling from being endangered by someone at the floor closing the disconnect. However, 430.107 still applies, which requires that at least one of the disconnecting means in a motor circuit be "readily accessible." (Review the discussion of this critical *NEC* terminology on page 58).

Motor branch-circuit and ground-fault protection   This subject is covered by Part IV of *NEC* Article 430. This motor branch-circuit overcurrent protection is necessary to protect the wires of the circuit against overloads greater than the starting current, in other words, against short circuits and grounds. At the same time, it protects the motor controller, which is designed to handle only the current consumed by the motor and which would be damaged by a short circuit or ground if branch-circuit protection were not separately provided. Grounds are often practically equivalent to short circuits. As in other branch circuits, the protection in a motor branch circuit may take the form of either fuses or circuit breakers. The maximum ampere rating permitted by the *NEC* for the overcurrent device in a motor branch circuit is sometimes higher when fuses are used than when circuit breakers are used.

Maximum rating of motor branch-circuit overcurrent protection   This subject is covered by *NEC* 430.52. The maximum rating depends on the type of motor, the kind of overcurrent protection provided, and the kind of starter used. For both single- and three-phase motors most often used, if they are provided with starters of the kind discussed in this book, the limits expressed as a percentage of the full-load motor current are according to *NEC* Table 430.52 and as shown here in Table 30–2.

To determine the maximum rating for less common types of motors, or starting methods other than those discussed in this book, first refer to *NEC* Tables 430.248

Table 30–2 MAXIMUM SIZE OF SHORT-CIRCUIT AND GROUND-FAULT PROTECTION FOR MOTOR BRANCH CIRCUITS [from *NEC* Table 430.52]*

	Percentage of Full Load Current			
Type of motor	**Non-time-delay fuse**	**Time-delay fuse**	**Inverse time breaker**	**Instantaneous trip breaker**
Single-phase motor	300	175	250	800
Polyphase (not wound rotor or synchronous)—				
Not Design B energy efficient	300	175	250	800
Design B energy efficient	300	175	250	1100

*Reprinted with permission from NFPA 70-2005 *The National Electrical Code*, © NFPA 2004.

or 430.250, then refer to *NEC* Table 430.52 to determine the maximum rating of the overcurrent device.

Where the ratings shown in *NEC* Table 430.52 will not allow the motor to start, such as where a motor may have to start under extremely heavy load conditions, time-delay fuses may be increased up to 225% of full-load current. Circuit breakers may be increased up to 400% of full-load current if that current is not over 100 amps, and up to 300% if it is over 100 amps. If applying these various percentages leads to a nonstandard rating of fuse or breaker, the next larger standard rating may be used.

Since *NEC* 430.22(A) requires the motor branch-circuit wires to have an ampacity of at least 125% of the motor full-load current, and since 430.52 permits overcurrent protection in some cases up to 400% of the motor full-load current, it follows that the branch-circuit overload protection may be as much as $^{400}/125$, or 320% of the ampacity of the wire. This is totally contrary to general practice and it should be well understood that this is permitted only in the case of wires serving motors that have motor-running overload protection rated at not over 125% of the full-load motor current, which protects both the motor and wires from damage caused by overload.

All the discussion under this heading refers to the maximum setting permitted for the overcurrent device. In practice it should be set as low as possible and still carry the maximum current required by the motor while starting or running. However, *NEC* 430.57 requires that when fuses are used, the fuseholder must be capable of holding the largest fuse permitted. Occasionally this will necessitate using an adapter to permit, for example, 60-amp fuses to be used in fuseholders designed for the larger 70- or 100-amp fuses. This requirement is waived if time-delay fuses are used. It makes sense to use only time-delay fuses.

Instantaneous-trip ground-fault and short-circuit protective devices   These devices are used only as part of listed combination motor controllers. They must be set not to nuisance trip, and that may be a very high value compared with motor full-load current. In the case of traditional motors, these settings may need to be as high as 1300% of full-load currents. It gets worse for Design B energy efficient motors, which may approach 1700%. The *NEC* allows these numbers, but only if an engineering evaluation substantiates the application. Conventional motors now get 800% and Energy Efficient Design B now get 1100% by right. The move to 1300% (conventional motors) and 1700% (energy efficient motors) is what requires engineering supervision. The *NEC* now recognizes three of these devices:

- Instantaneous-trip circuit breakers (already covered).

- Motor short-circuit protectors. Often confused with instantaneous-trip circuit breakers, these devices have fusible elements.

- Self-protected combination controllers (already covered).

Overload relay restrictions   Occasionally a motor control manufacturer may decide that the allowable ratios of short-circuit settings to motor full-load current are a little too generous for their product, specifically the overload relays, which take the brunt of a short circuit or ground fault until cleared. If the manufacturer decides their product is not safely compatible with the upper end of the protective ratios, they may mark their relay table accordingly with maximum sizes and styles of overcurrent protective devices, and those instructions (if given) supersede the ratios given in the *NEC*.

Multispeed motor protection   The basic rule is for one overcurrent device to be acceptable for the entire motor, provided you can show it meets the ratio limitations for both windings considered individually. If that is problematic, an exception allows for one device sized to the highest current winding only. To use this exception, each winding must have individual overload protection. In addition, the wires going to the lowest current winding and the controller serving that winding must both be sized in accordance with the highest current winding.

Solid-state controls   Solid-state motor controller systems, such as some variable speed drives, require special fuses for protection, usually very high speed. The replacement fuses required for these controls must be marked adjacent to the fuseholders.

Motor controllers start and stop motors   This subject is covered by Part VII of *NEC* Article 430. A motor controller according to the *NEC* definition in 430.2 is "any switch or device normally used to start and stop a motor by making and breaking the motor circuit current." The controller may be a manually operable device, or it may be an automatic device as found in refrigerators, air conditioners, furnace motors, and similar appliances. Under the category of automatic controllers are the traditional magnetic contactors that are either open or closed, and electronic controllers that can control voltage and frequency, allowing the motor to ramp up to speed gently, among many other features.

Requirements for controllers   In general the controller "shall be capable of starting and stopping the motor it controls, and shall be capable of interrupting the locked-rotor current of the motor." These basic requirements are of chief interest to manufacturers. Users and contractors will find that approved controllers supplied by manufacturers will meet these requirements if the proper selection is made by the user. Nevertheless it will be useful to be familiar with the requirements.

The usual form of controller is commonly called a motor starter; it can be the manual type or the magnetic type. The starter also contains the motor overload protection devices (which are discussed separately, and which must not be confused with branch-circuit overcurrent devices). Figure 15–3 shows small manual starters. A magnetic starter is shown in Fig. 30–3, and Fig. 30–4 shows a separate start-stop station that may be located at a distance from the motor.

◀ **Fig. 30–3** A magnetic starter, including overload devices. *(Square D)*

Fig. 30–4 A start-stop push button station used with magnetic starters. *(Square D)*
▼

Controllers have sturdy contactors that will withstand many openings and closings of the contacts as the motor is started and stopped; they last very much longer than the contacts of ordinary switches would under the arcing that occurs when a motor is stopped.

Reduced voltage and variable speed controllers The types described connect the motor directly to the full voltage of the power supply. There are other kinds that start the motor at a reduced voltage, the full voltage being applied only as the motor approaches full speed. These types are used for large motors to limit the effects of large inrush currents on plant voltage and utility distribution systems. They are beyond the scope of this book. Modern motor controllers often involve speed regulation, which in the case of ac motors involves varying the frequency of the power circuit to the motor. The methods of doing this are also beyond the scope of this book. Always follow any instructions that come with the drives, particularly in terms of relating the size of the motor to the rating of the drive. In general, the drive becomes the motor for the purposes of setting branch-circuit sizing requirements, and therefore 430.122(A) requires the supply conductor ampacity to be 125% of the "rated input to the power conversion equipment." These drives frequently include bypass circuits, so the motor can be operated across the line if the drive fails for any reason, and the bypass circuits must include all the protective elements of conventional motor circuits. The 2005 *NEC* has gathered the principal requirements for these systems in a new Part X of Article 430.

Which wire to open   Unless the same circuit breaker is used as both controller and disconnecting means (which is discussed later), the controller does not need to open all ungrounded wires, but only enough to start and stop the motor. On a single-phase motor therefore it need open only one wire; on a three-phase motor only two wires. But more often the controller disconnects all ungrounded wires. The controller may also open the grounded wire if it is constructed to simultaneously open all the ungrounded wires. In the case of three-phase motors supplied by a corner-grounded, delta-connected supply, it also opens the grounded wire.

Controller "in sight from"   The controller must be in sight from and within 50 ft of one of the qualifying disconnecting means already discussed, and there are no exceptions for work on 600-volt and lower systems. Except for a group of coordinated controllers governing motors comprising parts of a single machine, each controller must have its own disconnect, which must also be in sight from the motor and its driven machinery unless (1) the motor disconnecting means can be locked in the open (off) position or (2) an additional manually operable disconnecting switch is installed in sight from the motor location. This is covered by NEC 430.102. The purpose is safety—to make sure a motor is not started accidentally while somebody is working on the motor or its driven machinery.

Type of controller required   In general a controller or starter must have a horsepower rating not lower than that of the motor. The most important exceptions are:

■ *Stationary motor of ⅛ hp or less*—For a stationary motor rated at ⅛ hp or less that is normally left running and is constructed in a way that it cannot be damaged by overload or failure to start, such as clock motors and the like, the branch-circuit overcurrent device may serve as the controller.

■ *Portable motor of ⅓ hp or less*—For a portable motor rated at ⅓ hp or less, the controller may be an attachment plug and receptacle.

■ *Stationary motor of 2 hp or less*—For a stationary motor rated at 2 hp or less, and 300 volts or less, the controller may be a general-use switch having an ampere rating at least twice the full-load current rating of the motor. A general-use ac-only switch may also be used, provided the full-load current of the motor does not exceed 80% of the rating of the switch; the switch must have a voltage rating at least as high as that of the motor.

■ *Circuit breaker or molded case switch as controller*—A molded case switch or a branch-circuit circuit breaker, rated in amperes only, may be used as a controller. When this circuit breaker is also used for running overload protection (impossible for a molded case switch), it must also meet the general requirements for running overload protection.

How motor starters operate   Some manually operated starters for small motors look like large toggle switches, as shown in Fig. 15–3, and are turned on and off in the same way. Others, as shown to the right in Fig. 15–3, have push buttons in the cover. A mechanical linkage from the buttons closes the contacts when you push the start button, opens them when you push the stop button. Each type contains one or more overload devices that open the contacts if the motor is considerably overloaded for some time, or if it fails to start when turned on. If these devices stop the motor, find out why and correct the problem, then reset by turning the switch back on or by pushing a stop-reset button.

Magnetic starters, as shown in Fig. 30–3, can be stopped by start-stop buttons located at a distance, sometimes also by buttons in the cover. Inside the starter is a magnetic coil or solenoid that closes the starter contacts when the start button is pushed. When the stop button is pushed, the circuit to the coil is interrupted, the coil loses its energy, the starter contacts open, and the motor stops. The motor overload device(s) in the starter stop the motor on continued overload or failure to start. After a trip-out, reset by pushing the reset button.

The principle of operation of a magnetic starter can be better understood by studying the diagram of Fig. 30–5, which is for a three-phase motor. The wires shown as heavy lines carry the power current; those shown as light lines carry only a very small signal current flowing through the start-stop buttons. Note the difference between the two buttons: The start button is the normally open type, and pushing the button closes the circuit; the stop button is the normally closed type, and pushing the button opens the circuit.

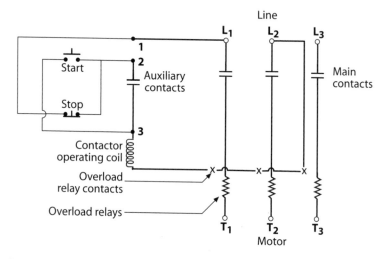

Fig. 30–5 Circuit diagram for the starter and push button shown in Figs. 30–3 and 30–4.

The magnetic starter is an electrically operated switch. When the operating coil is energized by pushing the start button, it closes the three main contacts and starts the motor. It also closes a small auxiliary contact at the same time, often referred to as the "sealing" contact. (In the diagram, the operating coil is not energized and the motor is not running.) Study the diagram and you will see that pushing the start button lets current flow through the operating coils, which energizes the start button, thus closing all four contacts. That starts the motor. Removing your finger from the start button appears to open the circuit, but the coil remains energized because the circuit through the coil, at first energized by the current flowing through the start button, now remains energized by the current flowing through the auxiliary contact. At all times while the motor is running, the current that energizes the coil flows through the stop button, which remains closed until the motor is stopped by pushing the stop button.

For multiple points of control, simply add more start-stop stations in parallel, as shown in Fig. 30–6. Furthermore, no matter how complicated the ladder diagram may be, most elements of motor control work involve items that cause a motor either to start or stop. As such, they involve contacts that simulate pushing either a start button or a stop button, and usually involve some form of sealing circuit so the stimulus that initiates the motor operation need not continue. Of course, some circuits must open when a stimulus ceases, and that may involve omission of the sealing circuit. But learn the elements of this simple control circuit well and you will be able to cope with surprisingly complex control systems.

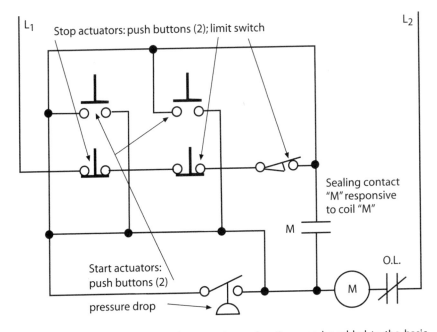

Fig. 30–6 An indefinite number of start and stop functions can be added to the basic control circuit. The start actuators are in parallel, and the stop actuators in series.

Pushing the stop button interrupts the current flowing through the coil, de-energizing it, and all four contacts open, stopping the motor. If the voltage fails or drops to a very low value while the motor is running, the coil will not have enough power to keep the contacts closed. The contactor will "drop out," stopping the motor, just as if you had pushed the stop button. It can be started again only by pushing the start button. This is a safety measure, because a motor running at a greatly reduced voltage would probably burn out. Also, if the contactor did not drop out, the motor would automatically restart when the power is restored, leading to a dangerous situation if motor-driven machinery starts unexpectedly. This feature is called "low-voltage release."

Motor control circuits have special protective provisions   This subject is covered by Part VI of *NEC* Article 430. A motor control circuit is defined as "The circuit of a control apparatus or system that carries the electric signals directing the performance of the controller, but does not carry the main power current." In other words, it consists of the wires between a controller and one or more start-stop stations (in Fig. 30–7 the start-stop station is shown next to the controller, but it could be in the same case with the controller; the station is usually located at a distance from the controller).

Voltage limitations   Motor control circuits are good examples of control and signaling circuits covered in *NEC* Article 725. Usually (but not necessarily) they don't comply with the severe energy limitations imposed on Class 2 or Class 3 circuits (review Chapter 19), which means they are by default Class 1 control circuits. In general, Class 1 control circuits have no energy limitations beyond being limited to 600 volts. Although 120 volts is a common motor control circuit specification, basic start-stop stations are frequently rated for and commonly run at line voltage, even on 600-volt, three-phase motors.

Grounding   Motor control circuits require system grounding if their voltage meets the basic rules for system grounding in *NEC* 250.20. For example, a 120-volt control circuit separately derived from a control transformer with a 600-volt primary must be grounded, because *NEC* 250.20(B)(1) requires such systems to be grounded. As long as the control transformer doesn't exceed 1kVA, however, the grounding electrode connection may be to the equipment ground from the transformer supply, and the size need not exceed 14 AWG. In practice, this means picking one of the transformer secondary terminals and running a 14 AWG bonding jumper to the enclosure. Remember to use white wire from this point out into the control circuit, at all points where the control circuit maintains a solid connection to the grounded terminal.

System grounding for control circuits is more extensive than commonly realized. For example, even a Class 2 circuit requires system grounding if the primary of the supply transformer operates over 150 volts to ground, and with the system grounding rule comes mandatory equipment grounding; refer to *NEC* 250.112(I). Furthermore, with mandatory equipment grounding come certain mandatory colors. *NEC* 250.122(A) allows the size to drop to that of the circuit conductors, so the Class 2 cable need not incorporate an oversize grounded or grounding conductor. However, many control wiring specifications for 277-volt, duct-heater controls originating at a Class 2, 277/24-volt transformer never anticipated the ways in which the colors white and green were spoken for.

When you are wiring a grounded control circuit, analyze the schematic to be sure that an accidental ground that occurs in any portion of the control circuit that is wired remote from the motor controller will not start the motor and will not bypass any manually operated shutdown devices or automatic safety shutdown devices. The provision about "remote from the motor controller" reflects the nearly universal practice of wiring the running overload relay contacts on the other side of the contactor coil from all other control functions. In the case of a grounded control circuit, the white wire would be brought to those contacts. That means, technically, that a fault to ground in the control circuit between the coil and the overload relays would disable those relays; if they operated, the motor would still run because the coil would still have a complete circuit through the enclosure. The Code-making panel decided that given typical spacings the risk is so remote it wasn't worth forcing a redesign of almost all overload relay control circuiting arrangements. See Fig 30–7 for an example of how to apply these rules.

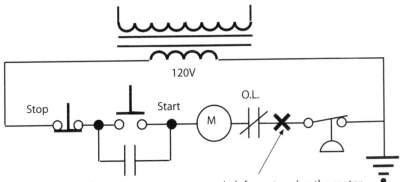

Accidental ground at "X" prevents pressure switch from stopping the motor.

Fig. 30–7 This control circuit would not operate to shut off the motor if a ground occurred at point *X*.

System separation   Class 1 control circuits need not be divorced from power raceways the way Class 2 and 3 circuits are. However, Class 1 circuits must be functionally related to the equipment supplied by power circuits within the common raceway. Be aware that there are two common and diametrically opposite interpretations of this rule [*NEC* 725.26(B)(1)] when the power circuits in the raceway supply unrelated loads. Do the Class 1 conductors need to be related to all the power wires, or only some of them? Since no general *NEC* rule prohibits running power circuits supplying different equipment in a common raceway (there are specific prohibitions such as fire pump wiring, but no general rule), in the context of the entire *NEC* it would appear that a functional relationship to one set of power wires should be enough, but your local inspector may take the opposite view. It is difficult to think of a safety objection that would not apply with even more force to the power wires, and plainly they have no such general limitation.

Overcurrent protection   Motor control circuit wires require overcurrent protection, although not always very much. In fact, if opening the control circuit introduces likely further hazard, such as in the case of a fire pump, the control circuits need have no overload protection. They are assumed to be properly connected by the motor-circuit overcurrent protection however large it may be. The basic rule is protection in accordance with ampacity, which is the normal *NEC* Table 310.16 values plus the flexible cord ampacities for 18 and 16 AWG wires in *NEC* Table 400.5A (7 and 10 amps respectively). If your control circuit originates at a transformer, you may have to provide this protection, usually with a fuse at the source. But if your control transformer is a straight two-wire/two-wire primary/secondary, the *NEC* allows you to base your secondary side exposure on the winding ratio of the transformer. For example, 14 AWG is good for (short-circuit and ground-fault protection only) 45 amps. A 20-amp overcurrent device ahead of a 240-volt, three-phase motor will also protect the 14 AWG control circuit. The 2:1 winding ratio means that a 20-amp overcurrent device will respond to a 40-amp requirement on the secondary side. However, these limits don't apply if you meet other, more forgiving conditions, and you will frequently meet one of the other conditions in those cases where the control circuit is tapped directly from the motor branch circuit.

The other conditions are based on the premise that for these control circuits, short-circuit protection is enough and overload protection isn't necessary. For straight control circuits governed only by *NEC* 725.24(C), which covers control circuits tapped from the power circuits they are intended to control, that limitation is 300%. Motor control circuits tapped from the motor branch circuits are essentially the same, although that similarity is anything but obvious. If the conductors leave the enclosure, the limitations are the same 7 and 10 amps for 18 AWG and 16 AWG, but 14 AWG, 12 AWG, and 10 AWG control wires are acceptable if protected at not over 300% of the *NEC* 240.4(D) overcurrent limitation for the same wires. Thus, as 14 AWG is limited to 15-amp overcurrent protection (a limitation that does not, ironically, apply to motor circuits) it need not have separate overcurrent protection if the motor branch circuit from which it is tapped does not exceed 45 amps.

If the control circuit is tapped from the motor branch circuit but never leaves the enclosure, the allowable protection ratio is inconsistent but runs about 400%. This situation arises in combination controllers where the disconnect, short-circuit and ground-fault protection, controller, and running overload protection are all in a single enclosure, frequently with push buttons on the door. In this case even the 18 AWG and 16 AWG are judged acceptable on 25 and 40 amps respectively. From there the limits go to 400% of the 90°C ampacity column for these wires, so 14 AWG (ampacity 25 amps) is acceptable tapped from a 100-amp motor circuit, etc., but beyond 10 AWG the limits go to 400% of the 60°C column, which means literally that an 8 AWG has the same limitation as a 10 AWG, that being 160 amps. Control circuit wires larger than 14 AWG or perhaps 12 AWG are very unusual.

Disconnection and mechanical protection   The motor control circuit wires should be protected against physical damage; running them through a raceway is one protective method. That advice becomes mandatory if damage to the control circuit would introduce a hazard. Control circuits must be disconnected by the motor circuit disconnecting means. In cases where they are tapped from the motor circuit, or fed from a separately derived system whose primary is tapped from the same location, disconnection is automatic. Further, any control circuit transformer within a controller enclosure must be disconnected by the local disconnecting means. However, some control systems use outside power sources. In such cases either the control circuit disconnect goes next to the motor disconnect, or the motor disconnect itself adds one or more additional contacts for this purpose.

If there are more than 12 control circuit conductors to be disconnected, the *NEC* allows the control circuit disconnects to be located remotely. Access to energized parts must be restricted to all but qualified persons, such as by location in a locked electrical room only they can enter. In addition, the equipment enclosure door must duly warn those working on the equipment that the control circuit disconnecting means are remote, and state where and how to find and recognize each of those disconnects. In addition, if opening a control circuit disconnect could result in unsafe conditions, that disconnect may be remotely located but only under the same conditions of restricted access and warning signs for personnel.

Motor and branch-circuit running overload protection   This subject is covered by Part III of *NEC* Article 430. As discussed earlier in this chapter, the overcurrent protection in the motor branch circuit protects the circuit, the motor, and its control apparatus against short circuits and ground-fault current only. But that protection is usually no protection at all for the motor itself against current higher than normal running current caused by overloading the motor or by failure of the motor to start when turned on.

A motor that requires, for example, 10 amps while delivering its rated horsepower may require 40 amps while starting. Once the motor has come up to speed, if it is not overloaded, the current will drop to 10 amps. But there may be times when the machine that the motor drives will need more than the rated horsepower of the motor; the motor will in almost all cases be capable of delivering more than its rated horsepower for a short time, but will consume a much higher current while doing so. Under overload, the motor normally consuming 10 amps may draw 15 amps or more. If allowed to draw 15 amps continuously or for a considerable period of time, the motor will in all likelihood be damaged, reducing its life, or it might even burn out.

So far, only motor branch-circuit overcurrent protection has been discussed, and that may be as high as 400% of the normal running current of the motor, or 40 amps in this case. Obviously a 40-amp overcurrent device will in no way protect the motor against damage due to an overload that makes the motor consume 15 amps instead of 10 amps. Therefore the *NEC* requires (in addition to the branch-circuit overcurrent protection) another device to protect the motor and its controller against current smaller than short-circuit or ground-fault current, but nevertheless higher than the normal full-load current of the motor, and caused by overloading the motor or failure of the motor to start when turned on. The *NEC* calls this "motor and branch-circuit overload protection," which is much too long a phrase to use repeatedly, so to distinguish it from breakers and fuses in general (and specifically those used as branch-circuit overcurrent protection), in this book it is simply called an overload device. Just remember that when an "overload device" is mentioned, it refers only to the device that protects the motor and its controller against too high a current caused by overloading the motor or the failure of the motor to start. The device can be integral with (built into) the motor, or separate. It is usually installed in the motor starter.

When required   Overload devices are always required except in the case of (1) a non-continuous-duty motor as defined earlier in this chapter; (2) a manually started motor of 1 hp or less if not permanently installed but located in sight from its controller (however, any manually started motor may be used on any 120-volt circuit that has branch-circuit overcurrent protection rated at not over 20 amps); (3) a motor that is part of an approved assembly that does not normally subject the motor to an overload, as long as the motor is protected against failure to start either by a protective device integral with the motor or by the safety controls of the assembly itself (such as the safety combustion controls of a domestic oil burner); or (4) a motor that has sufficient impedance[2] in its windings to prevent overheating due to failure to start, but only if manually started and installed as part of an approved assembly that includes provisions that preclude overheating.

In which wire   If fuses are used as overload devices, one must be inserted in each ungrounded wire. In the case of three-phase motors supplied by a corner-grounded delta system, one must also be inserted in the grounded wire.

---

2. There are many impedance-protected ac motors, usually 1/20 hp or less, such as clock motors and some fan motors. Motors in UL-listed assemblies of this type are now marked on their nameplates as IMPEDANCE PROTECTED or by the letters ZP (the symbol for impedance is "Z").

In the case of three-phase motors, one consideration in favor of not using fuses as overload devices (or in the motor branch circuit) lies in a characteristic of all three-phase motors. Three wires run to every three-phase motor; if only two were connected, the motor would not start. But if the motor is once started, and one of the three wires is disconnected (as in the case of a blown fuse), the motor will continue to run; it is said "to single phase." It will deliver less than its rated horse-power and draw more than normal current. If not quickly stopped by overload devices, it could be seriously damaged in a very short time.

If overload devices other than fuses are used, you must install one in the ungrounded wire to any 115-volt, single-phase motor; one, in either wire, to any 230-volt, single-phase motor; three, one in each wire, to any three-phase motor.

Types of overload devices   Although fuses are permitted they are generally not used except for smaller motors, and usually only if the branch-circuit fuse is rated as required for a motor branch circuit and is not rated higher than permitted for overload protection. Ordinary fuses are totally unsuitable, but time-delay fuses will sometimes serve the purpose.

Circuit breakers may be used. However, the overload limits do not provide for rounding up to any next higher standard sizes. This means that few circuit breakers will fall into any given fairly narrow window of acceptable overload devices for a particular motor.

Another type of overload device is the integral, built-in device in a motor, discussed under a separate heading.

Fig. 30–8 An overload unit (heater coil) used in motor starters to protect against overloads. *(Square D)*

The most usual overload device is shown in Fig. 30–8. It is called by various names, such as thermal units or heater units. In almost all cases they are installed on motor controllers or starters; they could be in separate cabinets. While starters are rated in horsepower, the overload devices are rated in amperes, and must be selected to correspond with the running current of the motor. Suppliers of controllers have information that will permit selection of proper ratings of the overload devices. In general, motors of 1 hp or less if automatically started, and all motors of more than 1 hp, may have overload devices rated at not more than 125% of the motor running current if the motor has a service factor of 1.15 or more, or is marked for a temperature rise of not over 40°C; if not marked with one of these two factors, the limit is 115%.

Note that you should enter a manufacturer's relay table directly with the full-load nameplate current on the particular motor to be installed. The relay tables already have the service factor figured into the selections. If you add

15% or 25% to the full-load nameplate current, and then enter the relay table, you will oversize the relay elements.

If, however, overload devices so rated do not permit the motor to start, the next higher rating may be used, provided it does not exceed 140% where 125% is the normal maximum, and 130% where 115% is the normal maximum. Today many motor controllers use electronic controls that can be set with a dial, but the basic principles don't change.

The overload devices are basically heating elements and, when properly installed in a controller, carry the full current supplied to the motor. But each overload device also has a set of separate contacts, not carrying the current supplied to the motor. If the current in the overload device increases too much as a result of an overload on the motor, or if the motor does not start, the heat actuates the separate contact. That contact is connected in series with the operating coil of the contactor, and if the contact opens, it opens the circuit to the coil and stops the motor. In the case of a three-phase motor, the three separate contacts are all in series, as shown in Fig. 30–5. No matter which of the three overload units operates to open the contact, the control circuit is opened and the motor stops just as if the stop button had been pushed.

The motor can be started again only by first pushing a reset button to close the overload-device contact; this can be done only after the overload device has cooled off. It is very important to make certain that if a motor shuts off due to an overload condition, it must not restart if restarting could result in injury to people. For example, a table saw should not be allowed to restart at random whenever it cools off.

Integral overload devices Some motors have built-in overload protection, making the need for separate overload protection unnecessary. This built-in device is a component of the motor, and operates not only from the heat created by the current flowing through it, but also heat conducted to it from the windings of the motor. If the motor is already hot from a long period of running at full load, the device will disconnect the motor more quickly if an overload develops than if the motor started cold and immediately became overloaded to the same degree. When such devices shut off a motor, they usually must be manually reset after the motor cools off. If such a device is capable of automatic restart, its use is governed by the same principle of preventing injury to people that applies to automatic restarts on conventional overload protective devices installed separately. The *NEC* requirements for such integral protectors are of interest primarily to motor manufacturers.

Combining several components of motor branch circuits As long as the various protective requirements are met, there is no safety reason why several protective devices can't be rolled into one. The following paragraphs provide examples of where this economy has been working well for many years.

Branch-circuit overcurrent protection and overload device combined As previously discussed, the motor branch-circuit overcurrent protection may in most cases have an ampere rating several times as great as the maximum permitted for the overload device. But some motors start so readily that their starting current isn't much higher than their running current. In that case, under *NEC* 430.55, the overload device may be omitted if the rating of the branch-circuit overcurrent protection is not higher than the maximum permitted for the overload device. When this situation exists, the motor branch-circuit overcurrent protection can be a circuit breaker or time-delay fuses. Ordinary fuses will not serve the purpose.

Controller and overload device combined   If ordinary motor starters of the type shown in Figs. 15–3 and 30–3 are installed, they usually include overload devices of the type shown in Fig. 30–8, or similar ones. In that case the controller serves both purposes; it would be more proper to say that the overload devices are combined with the controller. This is covered in *NEC* 430.39.

If the controller is a circuit breaker (or fused switch) and its ampere rating does not exceed the maximum permitted for the overload device, no separate overload device is required; see *NEC* 430.83(A)(2).

Disconnecting means, branch-circuit overload protection and controller combined   According to *NEC* 430.111, if a circuit breaker is installed meeting the conditions required for branch-circuit overcurrent protection, it may also be used as the disconnecting means and as the controller (not overlooking the "in sight" and similar requirements already discussed). The breaker must be operable by using your hand on a lever or handle.

If, however, a fused switch is used (to replace the overcurrent protection ordinarily provided by the breaker), the fuses must be of an ampere rating not exceeding that required for protection of the motor branch circuit.

Disconnecting means, branch-circuit overcurrent protection, controller, and overload devices combined   Often a single cabinet will contain a circuit breaker serving as the disconnecting means and branch-circuit overcurrent protection (or a fused switch serving the same purpose), and the motor controller, which usually contains the overload devices. This is not really a case of one device serving several purposes, but rather a case of several components, each meeting the requirements that would apply to each component if installed separately, being installed in a single enclosure. This arrangement is often referred to as a "combination starter."

NOW ADD COMPLEXITY: TWO OR MORE MOTORS ON ONE CIRCUIT

The *NEC* permits two or more motors on one circuit, but the requirements for the components of the circuit can become quite complicated. Unless it is desirable to have several motors on the same circuit, it is generally best to provide a separate circuit for each motor. Even in cases where an entire coordinated process must shut down at the same time for safety reasons, the best approach is to use appropriate logic in the control circuit design rather than rely on a single circuit. This can be done by running a feeder to the general location of the motors, often to a motor control center which is essentially a switchboard comprised of combination motor controllers connected to a common power bus. However, there are cases where multiple motors can run on a single branch circuit.

"Listed for group installation"   In any discussion concerning two or more motors on a single circuit, you will frequently see the expression "listed for group installation." As previously discussed, controllers and overload devices ordinarily used in a circuit supplying one motor are not capable of opening short-circuit or ground-fault currents; they are protected by the motor branch-circuit overcurrent device. But when several motors are connected on a single circuit, that overcurrent device must have a rating high enough to protect the entire group of motors, and then might exceed 400% of the running current of the smallest motor in the group, and a short circuit or ground-fault current would damage or destroy the controller or overload device for that motor. To offset this possibility, special controllers and overload devices (including thermal cutouts and overload relays) have been developed that are capable of opening a much higher current than would an ordinary controller or overload device. Such devices are marked LISTED FOR GROUP INSTALLATION.

Some of these devices are simply precombined by the manufacturer into a listed factory assembly that either (1) includes the short-circuit and ground-fault protective device, or (2) is marked with the appropriate device to be connected in the field. However, if the various components are combined in the field, they must all be listed for such applications, and the manufacturer's directions must clearly anticipate, through reciprocal instructions, that they can be used together in this way. In addition, each circuit breaker employed must be listed for group installation (look for the marking). Each controller listed in conjunction with components allowed for group installation must be marked with the maximum rating of the breaker or fuses that may be used as the branch-circuit overcurrent protection in the circuit serving the group of motors. Those devices must protect the running overload relays for the smallest motor in the group, and they must also not exceed the size of the overcurrent device that would apply to the largest motor plus the sum of the full-load currents of all other motors in the group. If that calculation turns out to be less than the ampacity of the supply conductors, the next higher standard size overcurrent protective device can be used.

Several motors on general-purpose branch circuit   This subject is covered by *NEC* 430.42, which makes reference to several other sections that must be considered in interpreting it.

If the motors are cord and plug connected, and are 1-hp or smaller, the attachment plug may not be rated at more than 15 amps at 125 volts, or 10 amps at higher voltages; separate overload protection is not required. But if it is larger than 1 hp, overload protection is required but it must be integral with the motor or the motor-driven appliance.

If the motors are not cord and plug connected, and are not larger than 1 hp they may be connected to a circuit with branch-circuit overcurrent protection not over 20 amps at not over 125 volts, or 15 amps at more than 125 but not over 600 volts, but only if (1) the full-load rating of each motor is not over 6 amps; and (2) each motor has individual overload protection if it would be required for the motor if installed on a separate circuit, as discussed earlier in this chapter; and (3) the controllers are approved for group installation; and (4) the branch-circuit overcurrent protection is not larger than the maximum shown on any controller.

If the motors are not cord and plug connected but are larger than 1 hp, each must have (1) individual overload protection that would be required if installed on a separate circuit, but approved for group installation; and (2) a controller approved for group installation; and (3) branch-circuit overcurrent protection not larger than the maximum stated on the controller and the overload device.

Several motors on one motor branch circuit   This subject is covered by *NEC* 430.53. If each motor is not larger than 1 hp, the conditions are as already outlined for several motors on a general-purpose branch circuit.

Two or more motors of any size (or one or more motors plus other load), may be connected to a motor branch circuit if all the following conditions are complied with: (1) Each motor has an overload device and controller listed for group installation and for use with each other, showing the maximum permissible rating of the branch-circuit overcurrent device; and (2) any circuit breaker in the circuit is of the inverse-time type approved for group installation; and (3) the branch-circuit fuses or circuit breakers are not larger than would be required for the largest motor if the motor were connected to a separate circuit, plus the full-load current of each of the other motors, plus the ampere ratings of any other loads on the circuit.

In this connection, note that while ordinary overload devices are not suitable for interrupting short-circuit and ground-fault currents, the types approved for group installation have a much higher interrupting capacity than those not so approved. Thus the branch-circuit overcurrent protection may be larger if the motors are equipped with controllers and overload devices approved for group installation, than would otherwise be permitted. When several motors are connected on the same circuit, you will see references to the largest motor in the group. The largest motor is not necessarily the one with the highest horsepower. It is the one having the highest full-load current per *NEC* Tables 430.248 and 430.250. If two motors are identical, consider one of them as the "largest."

The following paragraphs describe the various components of a motor circuit supplying two or more motors, in the same order as discussed in the case of a single motor on its own circuit.

Feeders for two or more motors   The ampacity of the feeder is computed this way: Start with 125% of the full-load current, according to *NEC* Tables 430.248 and 430.250, of the largest motor supplied; add the full-load current of all the other motors supplied; add the number of amperes required for any other load supplied. The total is the minimum ampacity of the feeder wires, but if the circuitry is interlocked so that all the motors cannot start and operate at the same time, consider only the largest combination that can operate at one time. This subject is covered by *NEC* 430.24 and 430.25.

Special overcurrent rules apply to motor feeders   The overcurrent protection required for such a feeder is discussed in Part V of *NEC* Article 430. The maximum rating of the overcurrent device is the sum of (1) the maximum rating permitted by *NEC* Table 430.52 for the largest motor in the group and (2) the total of full-load currents of all the other motors in the group. If in addition to motors the feeder also supplies other loads, the amperes needed to supply those loads may be added to the sum of (1) and (2) above.

Make the calculation differently if one or more of the motor branch-circuit protective devices consists of instantaneous trip circuit breakers or motor short-circuit protectors. Mentally replace any such devices with the same type of overcurrent protective device as will be used to protect the feeder. Then recalculate the largest branch-circuit overcurrent protective device accordingly. For example, suppose a feeder protected with time-delay fuses supplies two 30-hp, 460-volt motors, full-load current 40 amps. One motor has an instantaneous trip breaker with adjustable settings, set based on the allowable 800% multiplier (320 amps), and the other has an inverse-time breaker set at 250%, or 100 amps. Don't use a 350-amp fuse (320 + 40 = 360 amps) for the feeder protection. Instead, assume the first motor has a 70-amp fuse, based on the 175% allowed for such devices. Now the largest device would be the 100-amp breaker. Adding 100 amps to 40 amps gives 140 amps. Feeder protective device sizing is a not-to-exceed sizing, so instead of using a 350-amp fuse, use a 125-amp fuse.

Motor-feeder circuit taps   Now see Fig. 30–9. This could be interpreted as three motors on one branch circuit or as three motors on three separate circuits supplied by feeder *A-B-C*. Suppose devices *B* and *C* would end up near the ceiling. The *NEC* generally requires overcurrent devices to be readily accessible, and *NEC* 430.28 provides the solution for this example. Omit the overcurrent protection at the points where the circuits begin, and install it at the ends of the taps (at *D*, *E*, or *F*), provided the tap for each motor has the ampacity required for a motor on its own circuit and, additionally, meets one of the following four conditions:

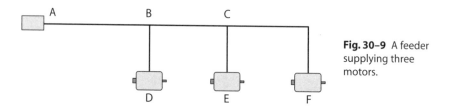

Fig. 30–9 A feeder supplying three motors.

- Not over 10 ft long and enclosed in a raceway or motor controller enclosure, and, if installed in the field, sized such that the overcurrent device next back on the feeder from which the tap is made does not exceed 1000% of the tap ampacity.

- Not over 25 ft long, with an ampacity of at least one-third the ampacity of the wire in the feeder, and protected from physical damage or enclosed in a raceway.

- Occurs in a high-bay manufacturing building (over 35 ft high at the walls; tap not less than 30 ft above the floor) under conditions of qualified supervision and maintenance, and extends not more than 25 ft horizontally and not more than 100 ft in total length, provided the ampacity is at least one-third that of the feeder being tapped; the tap terminates in a single overcurrent device (or set of fuses) complying with all normal requirements for short-circuit and ground-fault protective devices for motor branch circuits or feeders as would be applicable; the tap is protected from physical damage and installed in a raceway; the tap is continuous without splices and is at least 6 AWG (or 4 AWG aluminum); and the tap does not penetrate walls, floors, or ceilings.

- The tap wire has the same ampacity as that of the wire in the feeder. Following this procedure avoids the complications inherent when several motors are supplied by one branch circuit.

Branch circuits for two or more motors   The ampacity of such a branch circuit is determined exactly as just outlined for a feeder. The overcurrent protection for the circuit is also computed as for a feeder.

Combination equipment that has two or more motors (or a combination of motor and other loads) of the type that operates as a unit must be marked to show the maximum rating of the branch-circuit breaker or fuses that may be used. Such equipment must also be marked to show the minimum branch-circuit ampacity.

Disconnecting means   Each motor may have its own disconnecting means, but the *NEC* permits a group to be handled by a single disconnecting means under one of three conditions:

- If several motors drive different parts of a single machine, as for example metalworking or woodworking machines, cranes, hoists.

- If several motors are in a single room within sight of the disconnecting means.

- If several motors are protected by a single branch-circuit overcurrent device.

The rating of such a common disconnecting means must not be smaller than would be required for a single motor of a horsepower equal to the sum of the horsepower of all the individual motors (or a full-load current equal to the sum of the full-load currents of all the individual motors). This calculation is very involved. You have to do the calculations three times, and pick the worst case. The first step is to develop an equivalent horsepower based on the total current. Add up the full-load currents of the loads that can possibly operate at the same time, motor and nonmotor. Take this ampere rating into the motor full-load current table, and come up (after rounding up) with an equivalent horsepower rating for your combined load.

Then go into the locked-rotor current table and add up all the locked rotor currents in your combined load, together with all other loads. Compare this summation with locked-rotor currents in *NEC* Table 430.251A or B as applicable, and then come up with another (after rounding up) equivalent horsepower rating for your combined load. Compare these two results; the largest is the equivalent horsepower to use.

Finally, sum all the full-load currents, and multiply by 1.15. Your disconnect (if a switch; circuit breakers and molded case switches don't have horsepower ratings) must have a horsepower rating at least equal to the worst-case equivalent horsepower, and an ampere rating at least equal to the 115% calculation. There is an exception for listed nonfused motor-circuit switches. If they comply with the equivalent horsepower calculation, they need not comply with the ampere calculation.

HERMETIC REFRIGERANT MOTOR-COMPRESSORS HAVE THEIR OWN *NEC* ARTICLE

In Article 440, the *NEC* deals with electric-motor-driven air-conditioning and refrigerating equipment,[3] the major component of which is the hermetic refrigerant motor-compressor, which is defined in *NEC* 440.1 as "a combination consisting of a compressor and motor, both of which are enclosed in the same housing, with no external shaft or shaft seals, the motor operating in the refrigerant." In this book it is referred to simply as a motor-compressor.

Ordinary motors are cooled by radiating into the surrounding air the heat that develops during operation. But in a motor-compressor, the motor is closely coupled to the compressor, and both are sealed in a common case. See Fig. 30–10. While the motor-compressor is running, the refrigerant is continuously entering the case in a gaseous state at a temperature lower than the ambient, and it cools the motor. Because of that unusually efficient cooling, a motor of any particular physical size can safely consume more amperes (and deliver more horsepower) than an ordinary motor of the same physical size and cooled in the usual way. But when such a motor-compressor is started after a period of idleness, the refrigerant is at the ambient temperature and provides relatively poor cooling; the motor then heats up faster than an ordinary motor of the same size. For that reason, the requirements for motor-compressors differ in some respects from those covering other motors.

---

3. If such equipment is driven by an ordinary motor, not a motor-compressor, install the motor as explained in the previous 22 pages. Room air conditioners of the type discussed in Chapter 20, as well as equipment such as household refrigerators, freezers, drinking water coolers, and beverage dispensers, are subject to the rules discussed there, not to the rules in this chapter.

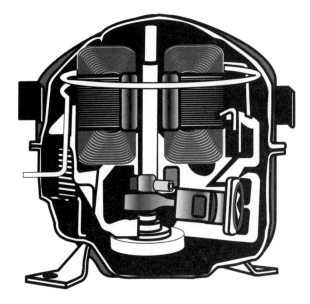

Fig. 30–10 Cross-sectional view of a typical motor-compressor.

Location of disconnecting means   Unlike motors generally, which under some conditions can avoid a local disconnecting means when the controller disconnect can be locked open, motor-compressors must have a disconnect within sight and readily accessible from the air-conditioning or refrigeration equipment. Frequently that disconnecting means is within or on the equipment. Attachment plugs and mating receptacles used as disconnects must be accessible, but need not be readily accessible. There is one other exception that applies only to motor-compressors that provide refrigeration or air conditioning essential to an industrial process: their controller disconnects (with lock-open capability) need not be in sight of the equipment, provided there is qualified maintenance and supervision. These processes are beyond the scope of this book and will not be discussed further.

Information on nameplate   A horsepower rating appears on the nameplate of an ordinary motor, but never on a motor-compressor. Instead, the nameplate on a motor-compressor shows:

- *Rated-load current*—This is the current that results when a motor-compressor is operated at rated load, rated voltage, and rated frequency of the equipment it serves.

- *Branch-circuit selection current*—Sometimes this is shown, sometimes not. If required, the manufacturer of the motor-compressor will show it on the nameplate. If it does appear, you must always use it in place of rated-load current when computing some of the components of the circuit supplying the motor-compressor.

- *Locked-rotor currents*—If the motor-compressor is three-phase, this must always be shown. If it is single-phase, it is required only if the rated-load current is over 9 amps at 115 volts, or over 4½ amps at 230 volts.

Nameplate data governs selection of circuit elements   In installing an ordinary motor, the ratings of various components in the circuit are dependent on the horsepower rating of the motor. But the motor-compressor does not have a horsepower rating. Where the requirements for a component are different from those of an ordinary motor, they will be pointed out in following paragraphs.

As an example, assume you are going to install a 60-Hz, 230-volt, three-phase motor-compressor with a rated-load current of 28 amps (and in some cases a branch-circuit selection current of, for example, 35 amps). Remember, if it has both, always use the higher of the two. Assume it has a locked-rotor current of 125 amps.

Motor branch-circuit wires   The minimum ampacity of motor branch-circuit wires must be 125% of the rated-load current of 28 amps, or 35 amps. But since the motor-compressor in this sample installation also has a branch-circuit selection current of 35 amps, the minimum ampacity becomes 125% of 35 or 44 amps.

Motor branch-circuit overcurrent protection   The branch-circuit fuses or circuit breaker must be able to carry the starting current of the motor-compressor. It must normally have a rating not over 175% of the rated-load current, or the branch-circuit selection current, whichever is greater. In the case of the motor-compressor under discussion, that would be 175% of 35 amps, or 61 amps. If that will not carry the starting current, it may be increased to 225%, or 79 amps. These parameters are "not-to-exceed" parameters, which means you can't automatically go to the next higher standard size unless the calculation comes out below 15 amps. In this case, that would mean trying a 60-amp overcurrent device, and possibly increasing it to a 70-amp device.

Disconnecting means   Ordinarily the disconnecting means depends on the horsepower of the motor involved, but the motor-compressor under discussion does not have a horsepower rating, so it is necessary to work backward. For the moment, imagine it does not have a branch-circuit selection current, but only a rated-load current of 28 amps. Turn to *NEC* Table 430.150 in the Appendix and look in the "230 volt" column. As it turns out, there is an exact match at 10 hp, so temporarily call it a 10-hp motor. But the motor-compressor under discussion does have a branch-circuit selection current of 35 amps, and *NEC* Table 430.150 does not show a motor corresponding to 35 amps, so use the next higher figure of 42 amps which corresponds to a 15-hp motor. So temporarily call it a 15-hp motor.

But the motor-compressor under discussion has a locked-rotor current of 125 amps. Go to *NEC* Table 430.151B (use the Design B, C, D column), look in the three-phase, 230-volt column for 125 amps. You won't find it, but the next higher number is 127 amps which corresponds to a 7½-hp motor.

If *NEC* Tables 430.150 and 430.151 lead to two different answers, you must use the larger of the two. If the motor-compressor under discussion did not have a branch-circuit selection current, you would provide a disconnecting means suitable for a 10-hp motor; because it does have a branch-circuit selection current you must provide disconnecting means suitable for a 15-hp motor.

Controller   When you have determined the answer for the disconnecting means, you also have the answer for the controller. For this motor-compressor, provide the same controller that you would provide for a 10-hp (or 15-hp) motor. In most cases, as for example in a household air-conditioning system, the controller will be part of the air-conditioning unit.

Motor-running overload protection   Proceed as in the case of an ordinary motor but base your conclusions on the branch-circuit selection current of the motor-compressor, if it has one; if it does not have one, base it on the rated-load current. The maximum rating is 140% if a thermal protector is used, or 125% if a circuit breaker is used. Usually the overload device is part of the equipment, frequently integral with (built into) the motor. Equipment that is thermally protected must be marked THERMALLY PROTECTED or ZP as explained in this chapter.

FIRE PUMP MOTOR "PROTECTION" ISN'T LIKE ANYTHING ELSE IN THE *NEC*

Fire pumps differ from all other motors because they must be allowed to run to failure—putting out the fire is the greater good.

Electrical supply must be highly reliable   Fire pump motors must not have running overload protection, and the overcurrent protection that is on their supply side must provide short-circuit and ground-fault protection only, although the circuit conductors can be sized at the usual 125% of full-load current. Fire pump controllers and the pump motors must both be listed for this service, a rare example of listed motors. In an extremely rare example of a required voltage drop calculation, the *NEC* prohibits voltage drop at the controller that:

■ exceeds 15% under starting conditions, which prevents the contactor from chattering when the starter is attempting to engage it, and

■ exceeds 5% under operating conditions with the motor running at 115% of its full-load current.

The power supply must be assured of reliability by being directly connected to a service, or through an intervening transformer dedicated to this function. If the inspector doubts the reliability of the utility supply, he may ask for an on-site generator to back it up, in which case you would also need to add a listed fire pump transfer switch (available separately or as combined equipment with the controller).

The supply conductors, where in the form of unprotected service conductors, must not pose a threat to the building. Either route them outside the building, or keep them under the floor slab or encased in 2 in. of concrete. The usual permission for service conductors to run through a transformer vault does not apply because a transformer fire could destroy them. Supply conductors on the load side of a service disconnect and having short-circuit and ground-fault protection pose no threat to the building, but they must be protected from the building. They must be independent of other wiring, and they must run in 1-hour fire rated construction and be protected from likely damage from fire, structural failure, or operational accident.

Install electrical circuit protective systems correctly   The supply wires may also utilize listed electrical circuit protective systems that assure circuit survivability for at least 1 hour under fire conditions. These systems are in the *UL Fire Resistance Directory,* and notably include Type MI cable with enhanced support provisions and a requirement for all terminations at least 1 ft into a protected space. These systems absolutely do not include beam sprays designed to prevent steel girders from reaching the point of softening and deflecting (above 1000°F) in a fire. Beam sprays are completely worthless in assuring the survivability of electrical circuits; thermoplastic insulation lasts about 1 minute under such conditions. Nevertheless there are miles of conduits coated with these beam sprays in complete ignorance of what this requirement really involves.

The allowable wiring methods consist of rigid and intermediate metal conduit, Type MI cable, or for flexibility, liquidtight flexible metal or nonmetallic conduit (Type LFNC-B only). The control circuits must be wired to fail in the on position, so the fire pump would keep running and never fail to start if so directed by another circuit even if a given set of control wires breaks or shorts. There are many additional rules in the fire pump article (*NEC* Article 695), notably a large number aimed at assuring equivalent reliability on large industrial or institutional campus-type wiring systems where the nearest service connection might be thousands of feet away from the building, and at a very high voltage. These distributions are beyond the scope of this book.

CHAPTER 31

Wiring Specific Locations and Occupancies

IN A SINGLE BOOK OF THIS KIND it is not possible to provide complete information about the wiring of every type of location or occupancy. The design of wiring installations of the kinds covered in this chapter should be undertaken only by those who have had much experience on more ordinary installations, or who have worked under thoroughly experienced people on installations like those discussed here. The basic outlines of the wiring problems are covered to acquaint you with at least the fundamental nature of the problems involved.

Use the methods already explained in other chapters to determine the requirements for services, feeders, panelboards, or circuits.

See *NEC* Tables 220.12 for the minimum volt-amperes per unit area that must be allowed for lighting. Using the minimum figures will rarely provide the degree of illumination desirable. The levels of illumination mentioned in this chapter for various occupancies are those recommended by authorities in the field. Be sure to review Chapters 14 and 29, both on lighting.

OFFICE SPACES MUST PROVIDE FLOOR-PLAN FLEXIBILITY

Only individual offices and smaller buildings with relatively small numbers of individual offices are considered in this discussion. A small, modern, well-appointed office building will of course have adequate lighting and an assortment of electric equipment such as water coolers and copying machines, in addition to computers and peripheral equipment such as printers and facsimile machines.

Lighting   The *NEC* used to require an allowance of 5 VA per sq ft except for hallways and storage areas, where the requirements are ½ and ¼ VA, respectively. That basic requirement now stands at 3½ VA per sq ft, with an additional allowance of 1 VA for general purpose receptacles when the actual number is not known. This may not provide the footcandles considered acceptable in a modern office, unless very efficient lighting sources are installed. For example, with old fluorescent lighting, 5 watts per sq ft might provide 50 to 80 fc, depending on the many factors outlined in Chapter 29. The recommended minimum level is 75 fc;

100 fc is better. For accounting and drafting work, the minimum recommendations are 150 and 200 fc respectively.

Nevertheless, modern offices are being wired today that achieve those levels of illumination where they are needed. In fact, it is not uncommon for engineers to admit, seldom for public attribution, that the NEC requires oversized lighting feeders under today's building codes. It is very important to track the requirements in NEC 210.11(B) that directly bear on this point. When a per-square-foot load calculation governs the amount of load in a given area, as is indeed the case with office lighting, that load must be included in the feeder sizing to the area. However, branch circuits need only be installed to serve the actual lighting installed. Despite building code requirements forcing substantial reductions in energy use per unit area, the lighting industry seems to be more than keeping up through technology aimed at both light source efficiency and control of use. After all, the most energy efficient light is one turned off when not needed. But it would be unrealistic to assume further technological leaps forward at this point until they are proven. In addition, there is always the human factor; witness the reluctance to abandon incandescent lighting in many quarters as an example.

Receptacles   The NEC contains no specific requirements for the number of receptacles in offices. Most have entirely too few. Every individual office should have no fewer than two; one for every 10 lineal ft of wall space is not too many considering the wide use of desktop computers and other electronic equipment. Outside the individual offices, don't overlook receptacles for equipment such as copying machines, network printers, water coolers, and maintenance equipment. Remember that even though the NEC has no specific placement rules for receptacles, NEC 210.50(B) requires receptacles in any occupancy to be installed where flexible cords with attachment plugs will be used.

Relocatable partitions   If a sizable floor area is involved, which will accommodate a considerable number of people, it is not likely that the original arrangement of individual office spaces will long remain unchanged. Large office space is often subdivided into many individual offices by relocatable prewired partitions, covered in NEC Article 605. These partitions are shifted as the need arises, and they typically include task illumination along with receptacles. Although they usually don't go all the way to the ceiling, they may do so (but not penetrate the ceiling) with the inspector's permission. These partitions can be connected to each other using extra-hard-usage cord, not longer than necessary (2 ft maximum) and equipped with strain relief. The partitions can be cord- and plug-connected in mechanically contiguous sections not over 30 ft long (or hard wired in any length). The supply receptacle must be on a circuit that supplies only such partitions. It must be located not more than 1 ft from the partition it supplies. The interconnected partitions must not total more than thirteen 15-amp, 125-volt receptacles, and no cord-connected partitions can contain multiwire branch circuits.

Suspended ceilings   In addition to strategies for achieving flexibility around physical partitions, modern building design makes extensive use of suspended ceiling cavities for all utilities as well as, in some designs, functioning as return-air paths. With that much action going on unseen overhead, the *NEC* includes numerous requirements to make sure the electrical system functions properly. There are three principal issues to be considered in this context.

Fire separation between floors, ceiling penetrations   The *NEC* requires that in cases where the electrical system penetrates a fire-rated partition, the fire rating of the partition must be preserved. A good example of this principle involves so-called "poke through" fittings (Fig. 31–1). In this case, the suspended ceiling contributes to wiring flexibility in the floor above, allowing a receptacle to be added in the middle of the floor as required. After a cylindrical hole is cut through the concrete, these listed fittings are installed with a tight fit to the concrete. In addition, they usually incorporate intumescent material that expands

when subjected to heat. In the process, they maintain a complete seal even if the insulation burns off the wires. Some even expand to the point of filling the void created when a nonmetallic conduit burns away, and are listed accordingly. They are listed in up to 4-hour fire ratings.

Fire separation between floors, ceiling as integral element   Some ceiling designs actually incorporate the suspended ceiling into the fire separation between floors. There is a great deal of confusion surrounding exactly what this means. When a building code requires a certain fire separation between points, it is actually saying that if a prescribed fire (there are nationally accepted standards specifying exactly how hot for how long, etc.) takes place on one side of a specified barrier, the other side of that barrier won't exceed specified maximum values during the allotted time.

Fig. 31–1 A fire-rated poke-through assembly, designed for a 3-in. hole bored in a concrete floor. The wiring connections are all in the floor below, usually in a suspended ceiling cavity. *(Hubbell Electrical Products)*

These ratings are not determined by theoretical engineering calculations. UL, and some other test labs, have giant furnaces in which they can construct sample walls or ceilings, start a prescribed fire, and measure the outcome. Only constructions that pass the actual required fire tests are recorded and published in the *UL Fire Resistance Directory*, with a design number assigned.

Many years ago, various construction interests decided to find out if a suspended ceiling could be used to assist in establishing the fire rating of a ceiling. In other words, suppose you need a 2-hour ceiling, and you would like to avoid pouring the 4 in. of concrete normally required. If you are going to have a suspended ceiling anyway because of the flexibility it provides, could you pour less concrete?

The answer to that question is a qualified yes. UL has actually tested these constructions. After some period the suspended ceiling does give way and allow the fire to impinge directly on the concrete. However, the suspended ceilings stay intact long enough to allow a much thinner concrete slab to pass the test. For example, for a 2-hour separation using Design A012, you can reduce the concrete slab from 4 in. to 2½ in. So the concept does work. But UL has also found that to assure success nearly every variable has to be controlled. Accordingly, the design information involves some of the most extensive fine print you will find anywhere, right down to the style and thickness of the T-bars, the thickness and mounting pattern of the support wires, the exact type of ceiling tile, etc. For example, if you want to include lay-in fluorescent lighting, the lighting must be of a certain type with a particular pattern of direct support to the structural ceiling. If you want to use ductwork for conditioned air, you must use a certain type of ductwork supported in a specific way. If speakers for background music are included, they must be a certain type of speaker supported in a certain way.

Now, in this context, try installing wiring in such a suspended ceiling cavity. If you stay on the structural ceiling, or on supports hung from that ceiling, that won't affect the fire performance at all. But as soon as you attach to any element of the suspended ceiling itself, you compromise the design, because UL could not possibly anticipate how that loading would affect the ceiling during a fire scenario. The *NEC* requires that when a suspended ceiling is part of a fire-rated floor-ceiling (or roof-ceiling) design, the wiring stays off the ceiling support wires that are part of the required design. If you need to use support wires, add your own that are in addition to the ones required as part of the original design, and attach to those. The *NEC* allows this, but only if the additional wires are tagged or color coded, etc., so the inspector can be sure whose wires you landed on.

There is an exception for wiring that is included as part of the rated design, but that would mean going back to UL for retesting, which would be completely impractical. The nature of the trade is that no matter how you anticipate running a wire in the field, that wire goes somewhere else. Furthermore, having UL build another system (including curing the specified concrete mixture for 30 days) in their furnace for retesting would be prohibitive.

The *NEC* is more forgiving of suspended ceilings that function simply for cosmetic purposes. For these ceilings you also have to add your own support wires, but you need not identify them; mixing up one or two isn't nearly as critical in these cases. The *NEC* also includes an exception to use the support wires installed by the ceiling installer, if the ceiling manufacturer included directions to this effect when supplying the ceiling. It's rare that a suspended ceiling manufacturer actually does this, so plan on making your own supports.

Air-handling ceilings   Many suspended ceilings are used as a conditioned air return. Warm (or cool) air is ducted into the room from a central supply, and ceiling louvers allow the return air to move above the ceiling and back to a central collection point. Some suspended ceiling fluorescent fixtures are specially designed and listed with louvers along their perimeter to allow room air to enter the ceiling without introducing additional grillwork.

Any fire in such a space will generate enormous quantities of smoke subject to dispersal throughout the building. To reduce the fuel and smoke load, the *NEC* in 300.22(C) requires that only wiring methods with metallic construction be used for these locations. This would prohibit Type NM cable, for example, but allow Type MC (without any nonmetallic covering) and Type AC cables in these spaces, which the *NEC* terms "other space used for environmental air." Such spaces also include panned stud and joist cavities used as cold air returns, and the same restrictions apply, with an important exception. The restrictions do not apply to such cavities in dwelling units, but only where the wiring passes directly across the spaces perpendicular to their long dimensions. Do not attempt to place a switch or receptacle in these locations unless the wiring methods fully comply with *NEC* 300.22(C).

Elements of a manufactured wiring system are shown in Fig. 31–2. These systems typically use prefabricated wiring subassemblies for power, signaling, and communications purposes, frequently consisting of segments of Type MC cable to feed equipment with mating fittings. The systems are widely used in suspended ceilings because they are perfectly designed to match the layout flexibility of equipment within a suspended ceiling. The wiring methods in the manufactured wiring system are compatible with other spaces for environmental air. This equipment may also be extended from a suspended ceiling into hollow walls to allow for switch and receptacle connections.

Fig. 31–2 A lighting fixture with a factory-installed receptacle for compatibility with a manufactured wiring system component. The system is energized through a plug designed to be connected to a specially configured receptacle mounted in a $4^{11}/_{16}$-in.-sq box cover.

RETAIL STORES—LIGHTING AS A MERCHANDISING TOOL

Retail stores come in all sizes, from small specialty shops or corner groceries to multistory department stores or shopping centers. The principles explained here can be more readily understood if applied to the smaller stores. Don't overlook the *NEC* restriction on circuit ratings if the loads are continuous, as is normally the case in any store.

In-store lighting   *NEC* Table 220.12 requires 3 VA per sq ft in areas open to shoppers. But if you plan your installation based on the *NEC* minimum, you probably will not achieve the level of illumination needed for good merchandising. The minimum lighting in the average store must be at least 80 fc in areas open to the public to provide attractive displays. In larger stores and in better locations, the minimum is 100 fc. For self-service counters 150 to 200 fc may be needed. If these levels of illumination seem high, remember that good lighting draws attention to the merchandise on display, encouraging customers to buy what they want rather than just the things they need. Good lighting can be considered an investment. General lighting is provided by recessed fluorescent fixtures. Showcases and wall cases are downlighted from recessed ceiling fixtures using PAR incandescent lamps. Some of the recessed fixtures are HIDs to boost overall lighting levels efficiently.

Today the cost of energy may make those lighting levels difficult, but the levels are achievable if the high-intensity lighting is focused where actually needed. It's wise to consult with a lighting engineer who specializes in this work.

In many cases, one or more branch circuits may be needed to supply receptacles for plugging in showcases lighted by fixtures inside the case.

Show-window lighting   The *NEC* in 220.14(G) and 220.43(A) requires an allowance of at least 200 watts for every linear foot of show window, measured horizontally at the base. For better stores, this is not enough. Usually numerous Type R or PAR lamps are very effective. For an effective merchandise display, the installation must be made so the light source can't be seen from the street. The show-window lights can be controlled manually, but it is better to install a time switch of the general type shown in Fig. 31–3, which automatically turns the lights on and off at preset times. The particular switch shown has a "Sunday and holiday cutoff" device so that any particular day of the week can automatically be skipped. It also has an "astronomical dial" allowing the switch to be set to turn the lights on and off at a predetermined interval before and after sundown. Once set, it automatically compensates for longer or shorter days as the seasons change.

Types of lighting equipment   For a modern store located in perhaps the most competitive area of a large city, there is little choice except to design the lighting system specifically to suit the size, structure, and layout of the building, the type of merchandise sold, the effects desired, and many other factors. This can be

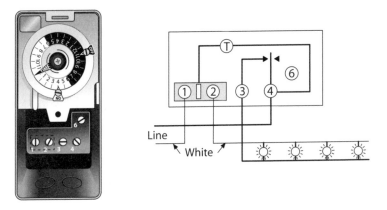

Fig. 31–3 Automatic time switches are convenient for controlling show-window lighting.

accomplished only by a competent lighting specialist, who will select the type, number, and arrangement of fixtures after analyzing all factors involved.

In general, lighting fixtures will be of the fluorescent type, with some supplementary incandescent lighting. The fixtures might be suspended from the ceiling, or mounted on or recessed in the ceiling. For fixtures to be installed on a low ceiling, select styles that direct most of the light downward. For fixtures suspended below higher ceilings, it is better to select fixtures that direct part of their light upward to avoid a cavernous appearance resulting from unlighted ceilings. Often it is desirable to direct extra light on some particular item on display. Recessed ceiling fixtures with incandescent lamps are useful for the purpose. Types R and PAR lamps are particularly effective.

When installing fluorescent fixtures, be sure to select the particular kind of "white" that goes best with the type of merchandise on display. Discussions of the color of light produced by fluorescent lamps begin on pages 261 and 564.

PLACES OF ASSEMBLY ARE ALLOWED COMPARATIVELY FEW WIRING METHODS

This subject is covered in *NEC* Article 518. A place of assembly includes buildings or portions of buildings or structures designed or intended for use by 100 or more persons for assembly purposes, such as dining, meetings, entertainment, lectures, bowling, worship, dancing, or exhibitions, and includes museums, gymnasiums, armories, group rooms, mortuaries, skating rinks, pool rooms, places for awaiting transportation, places of deliberation (court rooms), places for sporting events, and similar purposes. A key concept of the assembly nature of a use is that the use is vulnerable to self-reinforcing panic that is so dangerous in large gatherings. This risk is so low in, for example, a large supermarket, even with an occupancy load well over 100 persons, that it is not classified as a place of assembly. The 100-person threshold applies even where the applicable building code uses a lower threshold (typically 50). The *NEC* addresses different concerns than those addressed by building codes, which primarily focus on egress requirements.

The chief restriction is that in places of assembly the wiring must be in metal raceways, MI cable, or Types MC, or AC cable (the latter restricted to the style having a separate green wire in it). Nonmetallic raceways are allowed, but only if encased in 2 in. of concrete. But if the buildings or parts thereof are not required to be of fire-rated construction by the applicable building code, most other wiring methods may be used. If an occupancy includes a stage for theatrical or musical productions, that area, including the associated audience seating area, must be wired in accordance with the rules for theaters in *NEC* Article 520, which are beyond the scope of this book.

Some limited places of assembly, including restaurants, hotel conference rooms, dining facilities, and church chapels, now have the option of using nonmetallic raceways without concrete encasement. The raceways (either rigid nonmetallic conduit or ENT) must be concealed behind the same 15-minute fire finish as for the use of ENT in high-rise construction, and the new exemption for fully sprinklered facilities does not apply to these locations. (Refer to the ENT discussion on page 184.)

If the building or part thereof is used as an exhibition hall (as for trade shows), the wiring to the display booths may be approved cables or cords used as temporary wiring. The cords must be suitable for hard or extra-hard usage if laid on the floor, and must be protected from general public contact.

GOOD SCHOOLS SHOWCASE GOOD LIGHTING DESIGN

Experience indicates that good lighting contributes to raising the grade averages of students. Often those who have lagged behind respond when proper illumination is provided. Good lighting is therefore a tremendous asset for the student, and has been found to reduce the cost of providing education on a "per student per year" basis.

Footcandles required   The number of footcandles required varies greatly with the nature of the work done in a particular area. Classrooms need 50 fc for reading reasonably large print, 100 fc for smaller print or handwritten material. Domestic science rooms need 50 fc except for sewing, which needs 150 fc. Shops and drafting rooms need 100 fc. For corridors 20 fc is usually considered sufficient. Figure 31–4 shows a well-lighted classroom.

For cafeterias, 30 fc is adequate; auditoriums (except for the stage) require 15 fc, and gymnasiums 30 to 50 fc. But it would be wise to provide those areas with circuits permitting much higher levels of illumination even if the suggested footcandle levels are not normally exceeded. Such areas often are pressed into temporary use as study halls, lecture rooms, for giving examinations, and similar purposes; often such areas must be permanently converted if the number of students increases beyond original estimates.

In auditoriums, it is good practice to install dimmers so the light can be controlled as the area is used for various purposes. Dimmers are available for fluorescent lighting, the type most widely used.

NEC requirements for schools

The *NEC* requires a lighting load allowance of 3 VA per sq ft of area, except only 1 VA is required for assembly halls and auditoriums, ½ VA in corridors, and ¼ VA in storage areas. The demand factor is always 100%, and the lighting should be considered continuous (see the definition and wiring rules for protecting devices under continuous loads beginning on page 476). But if you provide circuits only for the *NEC* required

Fig. 31–4 A well-lighted school using fluorescent lighting. The shadow line indicates efficiency because most of the light is on the work surfaces.

minimums, you probably will not achieve the levels of illumination discussed. To provide adequate lighting, be sure to install enough circuits and feeders.

In addition to the lighting circuits, you must provide circuits for other loads, and in modern schools that constitutes a very significant total. Take into consideration not only usual loads such as heating, ventilation, possibly air conditioning, but other loads: video equipment; computers and peripherals in classrooms, offices, and in computer labs; appliances in domestic science departments; special loads in shops and laboratories, etc. Don't overlook plenty of receptacle outlets, including those in corridors where they are essential for maintenance crews.

CHURCH LIGHTING PRESENTS UNIQUE CHALLENGES

The *NEC* requires an allowance of 1 VA per sq ft for lighting, with 100% demand factor. That will not provide the footcandles of illumination now considered minimum. Illumination below 10 to 15 fc in the main worship area would be considered inadequate. To provide this level may require 2 to 4 VA per sq ft, depending on many factors such as ceiling height, reflectance of ceilings and walls, and especially the type of lighting fixtures selected.

In older churches, one frequently sees lighting fixtures that seem to have been designed primarily to conform to the architectural scheme of the church, with less thought given to good lighting. Sometimes such fixtures are designed to use a number of small lamps, providing much less light than a single lamp consuming the same total number of watts.

Today in many newer churches, the major part of the lighting is usually provided by simple, unobtrusive fixtures at the ceiling. They are installed pointing forward and not visible to people seated in the pews unless they happen to look toward the rear of the church. By using lamps of the PAR type, good light distribution is obtained without using large, clumsy reflectors. But because the lamps are not easy to replace, it may be well to operate them at less than their rated voltage, thus prolonging their life, and certainly lamps with longer than usual rated life should be used (for example the metal-halide Quartzline type). Additional general lighting is often provided by suspended fixtures or lanterns that fit into the architectural scheme of the particular church.

The sanctuary and the altar constitute the focal point and must receive attention. This area should be lighted at 50 to 100 fc or even higher; lighting is usually provided by floodlights, or Types R and PAR lamps, concealed from the congregation. This may require a minimum of 1500 watts depending on the area, the arrangement, and the desired effect. Figure 31–5 shows a well-lighted church; it is a tasteful blend of unobtrusive, efficient downlighting mixed with cove lighting and period chandeliers. But each church represents an individual problem, and the design of the installation should be left to one well versed in the art of church lighting.

Fig. 31-5 An unusually well-lighted church. Note the combination of downlighting and fixtures for general illumination.

If the church has a choir loft, at least 50 fc or more should be provided. An organ motor will probably require its own circuit. Conveniently located receptacles must be installed to operate pulpit lights, public address systems, and similar equipment, not overlooking those required for maintenance work. Outdoor receptacles are desirable for lighted displays during festival seasons.

WIRING IN WET, CORROSIVE, AND OUTDOOR LOCATIONS

Wiring methods have to stand up to their environment, and these locations can deteriorate even non-metallic wiring methods if they aren't suitable for the location.

Wiring in wet locations When steel boxes or cabinets are installed outdoors, *NEC* 300.6(A) requires that they be protected from corrosion both inside and out. If the protection involves organic coatings (including paint) they must be marked RAINTIGHT or RAINPROOF or OUTDOOR TYPE. For wiring installed in locations that are more or less permanently wet, such as laundries and dairies (creameries), or in locations where walls are frequently washed or have wet or damp absorbent surfaces such as damp wood, *NEC* 300.6(D) requires that all boxes, raceways, cabinets, fittings, etc., be mounted so there will be an air space of at least ¼ in. between the equipment and the supporting surface. This rule also applies to such locations that involve nonabsorbent surfaces, such as tile or concrete. On such surfaces, nonmetallic wiring methods do not require the ¼-in. air space.

Wiring in corrosive conditions   *NEC* 300.6 specifies that boxes, cabinets, raceways, cable armors, etc., must be suitable for the environment in which they are installed. Ordinary materials used elsewhere are not suitable or acceptable in locations where corrosive conditions exist. Such locations are described in the extensive fine print note following *NEC* 300.6, and include (but are not limited to) areas where acids and alkali chemicals are handled or stored, meat-packing plants, some barns (see the page 430 discussion of poultry, livestock, and fish confinement systems), and areas immediately adjacent to a seashore.

The circuit components mentioned must have special finishes suitable for the purpose, or be made of materials such as brass or stainless steel that require no further protection. For example, Type MC cable is available with nonmetallic jacketing, and rigid conduit can be made of nonferrous materials including brass. It is also available with a bonded nonmetallic coating. Be aware that the *NEC* generally requires circular raceways (such as conduit and tubing) to be listed, and no manufacturers are presently producing listed brass conduit. Essentially the same product (heavy-wall brass) is available as water pipe, without a listing, but you'll have to obtain a local approval in such a case.

Outdoor wiring involves wet locations generally exposed to sunlight if above grade   This subject is covered by *NEC* Article 225. Underground wiring is discussed in Chapter 16. Some overhead wiring is discussed in Chapter 16 in connection with services and in Chapter 22 concerning farm wiring (see Figs. 22–3 and 22–4 and related text).

Also review in those chapters discussion regarding the minimum size of wires between supports, and their clearances above ground, above roofs, and from windows, porches, and the like. Other aspects of outdoor wiring are discussed here.

The ampacity of overhead spans is as shown in *NEC* Tables 310.17 and 310.19 unless the wires are run together on a messenger. If they are on a messenger, another ampacity table (*NEC* Table 310.20) applies to that condition. Those ampacities are lower than the free-air versions, but still higher than the customary values as given in Table 310.16. But a few words of caution: When you use the higher-ampacity tables, watch for excessive voltage drop if the wires are loaded to their full ampacity, especially if the distance is substantial. Review voltage-drop discussions in Chapter 7. In addition, don't forget that the higher ampacities can't be used to judge the allowable sizes to be used at terminations; review Chapter 26.

Festoon lighting   This is defined in *NEC* Article 100 as "a string of outdoor lights that is suspended between two points." Such a string of lights consists of two or three wires (depending on whether a two- or three-wire circuit is used, but usually it is a two-wire circuit) with weatherproof sockets or pin-type sockets attached to the wires at close intervals. Figure 31–6 shows a weatherproof socket (whose leads are 6 to 8 in. long, though shown shortened here), and Fig. 31–7 shows a pin-type socket.

The minimum size wire permitted for festoon lighting is 12 AWG, unless the string is supported by a messenger wire, in which case 14 AWG is the minimum size permitted. For spans over 40 ft, a messenger wire must be used. (A messenger wire is a nonelectric supporting wire from which the wires are suspended.) The wires are individually supported by single insulators or by a rack with an insulator for each wire. The means of support must not be a fire escape, downspout, or plumbing equipment.

Fig. 31–6 A weatherproof socket for outdoor use.

For a permanent festoon-lighting installation, use weatherproof sockets as pendants suspended from the festoon wires. The socket leads are connected to the wires in staggered fashion by means of small mechanical splicing devices of the type shown in Fig. 8–15. The joint is then insulated with tape. For a temporary festoon installation, the pin-type sockets may be used only if the festoon wires are stranded. The socket is opened up, the wires are laid into the grooves in the lower section, and the top section is then screwed tight, causing the pins in the socket to puncture the insulation and penetrate into the strands of the stranded conductors. For installations with a messenger wire, the upper portion of the pin-type socket has a wire rack or hook to hang over the messenger wire. These hooks support the sockets and the festoon wires. Such installations are frequently used for temporary outdoor display areas. The permanent type is often used above outdoor used-car sales lots.

Fig. 31–7 Using sockets of this type, simply lay the wires into the grooves and screw on the cover. Use only with stranded wires.

Be aware that for temporary lighting on construction sites, branch-circuit wiring is not permitted in the form of individual conductors, and this would preclude the sort of festoon wiring depicted in Fig. 31–7 from being used for that purpose. A construction worker in Pennsylvania was electrocuted after an ungrounded conductor that was part of festoon lighting on the site became damaged and energized a section of metal framing members. There are cable assemblies available with lampholders molded to the cord, and with bulb protectors, all designed to meet *NEC* requirements for these more demanding environments.

WIRING FOR SIGNS AND OUTLINE LIGHTING—UNIQUE GROUNDING PROVISIONS

This topic is covered by *NEC* Article 600. Such equipment is made and installed by competent sign shops. The circuit to a sign may be rated at not over 20 amps, unless it supplies only transformers for electric-discharge lighting (neon signs for example), in which case the maximum is 30 amps.

Each commercial building or occupancy must be provided with at least one 20-amp branch circuit, with an accessible outlet for use with the sign located at the entrance to each tenant space. This rule formerly referred to outdoor outlets; the present rule recognizes the prevalence of shopping malls. When doing a load calculation for a commercial occupancy, don't forget to include a minimum load of 1200 VA for a sign, unless you know the one to be installed is more.

Each sign must be provided with a disconnecting means to disconnect all ungrounded wires from the sign. It can be either a switch or circuit breaker; if not in sight from any part of the sign, as when the sign continues around a corner, it must be capable of being locked in the off position. If the sign is controlled by a remote electronic or electromechanical controller, the disconnect can be placed there. In addition to the disconnecting means, the sign may have other control devices, such as flashers, which must be of a type rated for inductive loads, or have an ampere rating at least double the ampere rating of the transformer it controls.

Signs must be grounded, like most other electrical equipment, but the sign article has provisions unique in the *NEC*, particularly on the secondary (high-voltage) side of the sign transformer for neon signs. In order to reduce the voltage stress on the insulation, when the secondary is run in rigid nonmetallic conduit the grounding conductor must be run outside the conduit, spaced at least 1½ in. away for conventional transformer supplies and 1¾ in. for the new electronic power supplies operating the circuit over 100 Hz. The nonmetallic raceway is limited to 50 ft from the power supply to the first sign connection; in the case of grounded metal raceways the length limitation stays at the 20-ft limit that had been in the *NEC* for many years. Transformers and electronic power supplies, in most cases, now must be listed and incorporate secondary ground-fault protection. This is a major safety initiative, essentially mandating a low-level GFPE. Review the discussion that begins on page 544 relating to ground-fault protection of equipment (GFPE) on 480-volt services (and feeders) for these high-voltage secondaries. The same principles are at work, except that these devices work at small fractions of an ampere.

HAZARDOUS (CLASSIFIED) LOCATIONS REQUIRE VERY HIGH WORKMANSHIP STANDARDS

Locations are classified as hazardous (dangerous) because of the presence or handling of explosive gases, liquids, flammable dusts, or easily ignitable fibers. The basic rules for wiring in such locations are covered by *NEC* Articles 500 through 506, and more specific information is covered in Articles 510 through 517—altogether more than 85 pages. This book can discuss only the barest details of this intricate subject.

Before you attempt to install any wiring or equipment in a hazardous location, you will need to have considerable experience under the supervision of experts, and you will need to become completely familiar with the *NEC* requirements involved. The intent of the information set forth here is to provide some of the fundamental facts of the various classifications and the *NEC* rules applying to them.

The *NEC* uses three basic classes to indicate the type of hazard involved, and Classes I and II are further subdivided into groups to indicate the exact type of hazard. Each of the three classes is divided into two divisions to indicate the degree of hazard. Area classifications are seldom performed by an electrical inspector; they are done by qualified engineers subject to review. The reviewer may be the chief of the fire department or some other official. The *NEC* now requires that hazardous (classified) locations be completely documented, and that those documents be made available to the inspector.

Class I locations   Class I locations are those "in which flammable gases or vapors are or may be present in the air in quantities sufficient to produce explosive or ignitable mixtures."

Class I, Division 1 locations are those in which hazardous concentrations of gas (1) exist continuously or intermittently under normal operations, (2) exist frequently because of maintenance or leakage, or (3) might exist because of breakdowns or faulty operation that might also result in simultaneous failure of electric equipment in such a way as to cause the electrical equipment to become an ignition source. Examples are paint-spraying areas, systems that process or transfer hazardous gases or liquids, portions of some cleaning and dyeing plants, and hospital operating rooms in countries that still use ethers and related flammable anesthetic agents. These anesthetizing materials are no longer used in the United States, but they are still used in some countries.

Class I, Division 2 locations are locations (1) in which flammable gases or volatile flammable liquids are normally confined within containers or closed systems from which they can escape only in case of breakdown, rupture, or abnormal operation; (2) in which the hazardous concentrations are prevented by mechanical ventilation from entering but which might become hazardous upon failure of the ventilation; or (3) that are adjacent to a Class I, Division 1 location and might occasionally become hazardous unless prevented by positive-pressure ventilation from a source of clean air by a ventilation system that has effective safeguards against failure. Examples are storage in sealed containers and piping without valves.

Class II locations   Class II locations are those having combustible dust.

Class II, Division 1 locations are those in which combustible dust is or may be in suspension in the air in sufficient quantity to produce an explosive or ignitable mixture (1) continuously or intermittently during normal operation, (2) as a result of failure or abnormal operation that might also provide a source of ignition by simultaneous failure of electric equipment, or (3) in which combustible dust that is electrically conductive may be present. Examples are grain elevators, grain-processing plants, powdered-milk plants, coal-pulverizing plants, and magnesium-dust-producing areas.

Class II, Division 2 locations are those in which combustible dust is not normally in suspension in the air and is not likely to be under normal operations in sufficient quantity to produce an explosive or ignitable mixture but in which accumulations of dust (1) might prevent safe dissipations of heat from electric equipment or (2) might be ignited by arcs, sparks, or burning material escaping from electric equipment. Examples are closed bins and systems using closed conveyors and spouts.

Class III locations   Class III locations are those in which there are easily ignitable fibers or flyings, such as lint, but in which the fibers or flyings are not in suspension in the air in sufficient quantity to produce an ignitable mixture.

Class III, Division 1 locations are those in which easily ignitable fibers or materials that produce combustible flyings are manufactured, used, or otherwise handled, such as textile mills, cotton gins, and woodworking plants.

Class III, Division 2 locations are those in which such fibers are stored, such as a warehouse for baled cotton, yarn, and so forth.

Groups   Equipment for use in Class I locations is divided into Groups A, B, C, and D; for Class II into Groups E, F, and G. Each group is for a specific hazardous material, which you can determine by consulting your copy of the *NEC*. All equipment for hazardous locations must be marked to show the class and the group for which it is approved; some equipment is suitable for more than one group and is so marked. Note that for Group E, which encompasses the electrically conductive metallic dusts, there are no Division 2 locations. These dusts are so hazardous that if they are present in any quantities, no matter how infrequently, the location must be wired to Class II, Division 1 standards, using equipment evaluated for Group E exposure.

Basic principles applied to equipment   Enclosures for equipment in Class I, Division 1 locations and enclosures for devices with ignition-capable arcing contacts in Class I, Division 2 locations must be explosionproof. This does not mean that explosive gases or vapors cannot enter into them. They are designed to allow such gases or vapors to enter and potentially ignite and explode, but the enclosure is made to withstand and contain the force of the explosion. Moreover, the hot exploded gas does escape, but not until it has passed through a tight joint that is either threaded or has a wide ground-finish flange. In either case, before it finally escapes to the outside of the enclosure it has cooled to a temperature below the ignition temperature of the gas in the surrounding atmosphere. The cooling takes place while the gas passes through the long circuitous path of a threaded joint, or across the wide, tight-fitting, ground-finish flange.

For Class II locations, enclosures must be dust-ignition-proof. Dust can be prevented from entering enclosures by means of gaskets, and enclosures can be made with large exposed surfaces for more rapid heat dissipation.

If a Class II dust-ignition-proof enclosure is used in a Class I location, gas can get in, explode, and blow the enclosure to pieces, possibly setting off a larger explosion in the general area, leading to fire or injury. Likewise, if a Class I explosionproof enclosure is installed in a Class II location, it can overheat when blanketed with dust and start a fire.

For a Class III location, equipment only has to be totally enclosed to prevent the entry of fibers and flyings, and to prevent the escape of arcs, sparks, or hot particles.

Other protective approaches   The best approach is to use ingenuity and keep electrical equipment out of the hazardous environment. Failing that, another approach is to keep the power circuits somewhere else, and just maintain sensors near the combustible material. If the energy level can be assured to be sufficiently low, any failure in the wiring system won't be ignition capable. These energy levels are very low. For example, an energy level of just ¼ milliwatt-second will ignite methane, and levels one-tenth of that will ignite unsaturated hydrocarbons like ethylene. Nevertheless, many sensors can and do function on what are called "intrinsically safe circuits" (covered in *NEC* Article 504). These circuits use zener diodes in their supplies, arranged so that even a direct short circuit with its associated potential arcing and heating would not be ignition capable. These circuits can be used in hazardous (classified) locations for which they are rated Division 1 or Division 2 in ordinary wiring.

Nonincendive wiring is closely related to intrinsically safe wiring. Under all normal operating conditions it can't produce ignition-capable energy, but under very unusual (but foreseeable) operating conditions, might produce such an effect. These circuits are allowed in ordinary wiring in Division 2 locations, but not Division 1. The principle here is that given two low-likelihood probabilities, the probability of them occurring at the same moment is infinitesimal—the two low probabilities in this case being a Division 2 vapor release and a nonincendive circuit failing in a way that is ignition capable.

Another strategy is to exclude all gases and vapors through positive pressure, called "purged and pressurized" in the *NEC*. This equipment must be connected to a source of clean air under constant pressure. It must be arranged with automatic means to disconnect power if the air supply fails, and it must be equipped with a time-delay function so that its interior can purge prior to the restoration of power. There are three levels of purged and pressurized enclosures. Type X purging reduces a Division 1 environment to unclassified, Type Y purging reduces from Division 1 to Division 2, and Type Z purging goes from Division 2 to unclassified.

Other strategies involve hermetic seals to exclude the hazardous atmosphere, and oil-immersed contacts, which also exclude the hazardous gas or vapor. These two methods are allowable in Division 2 locations only.

Zone classification system (*NEC* Article 505)   A number of international standards take the Class and Division system a step further, dividing Division 1 into areas where the specified environment exists routinely and areas where it would exist only for brief periods, probably not long enough to be taken up into the interior of a wiring system and exploding, conditions for which Division 1 equipment must be designed. The result is the Zone system, which is now in the *NEC* as a parallel classification system. Zone 0 is the worst case, and all power wiring is excluded from it. Zone 1 gets the balance of the traditional Division 1, and since the routine or continual exposure doesn't apply, most explosionproof requirements don't apply to Zone 1 either. Zone 2 is essentially the same as Division 2.

You can design to the Zone system, or to the Class and Division system, but there must be no overlap or intersection between them, except that Zone 2 and Division 2 locations can adjoin but not overlap. You are permitted to reclassify an existing Class- and Division-classified area under the Zone system, provided the entire space classified on the basis of a single flammable gas or vapor is reclassified under Zone system requirements. The Zone system classifications and equipment selection must be under the control of a Registered Professional Engineer.

The 2005 *NEC* (Article 506) has extended the zone concept to environments with combustible dust, comparable to Class II under the traditional system, with supervisory requirements comparable to Article 505 applications. These requirements will likely remain largely theoretical for a very long time. By way of comparison, Article 505 first appeared in the 1996 *NEC*, and very few such applications are in actual use even yet. The traditional Class and Division system has proven very durable, in large part because Zone 2 and Division 2 are functionally identical. The principal economic benefit of the Zone system derives from the more lenient treatment of Zone 1 wiring in comparison to Division 1. However, the overwhelming majority of hazardous (classified) locations today are Division 2, because environmental concerns have made the sort of chemical releases that create Division 1 environments largely nonexistent, or confined to extremely small areas from which it is quite practical to exclude most wiring other than instrumentation. Most facilities are finding no net benefit in changing classification systems.

Proper maintenance   In all cases, enclosures such as lighting fixtures and panelboard enclosures must be kept reasonably cleaned of accumulations of residue, fibers, dust, or whatever is contained in the atmosphere. And they must be properly installed. For example, if one of four screws in the cover of an explosionproof switch enclosure is left untightened, or if the flanged joint of a cover or box is scratched, exploded gas will escape before it has sufficiently cooled, and an explosion or fire could result.

A threaded joint must have at least five full threads fully engaged (four and a half threads if factory threaded). Figure 31–8 shows an explosionproof box with threaded hubs for threaded metal rigid conduit, and a threaded cover. Figure 31–9 shows an explosionproof receptacle and plug. The box containing the receptacle has a ground-finish flange joint with four screws, and a threaded hub for conduit. The plug and receptacle are designed so the plug can be withdrawn only part of the way, which breaks down the circuit; an arc, if it forms, will explode the small amount of vapor in the interior. This takes place during the brief time it takes to twist the plug before it can be withdrawn.

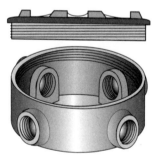

◀ **Fig. 31–8** A typical explosionproof box.

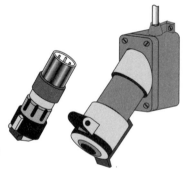

Fig. 31–9 An explosionproof receptacle and plug to fit. ▶

Some plugs and receptacles are designed so the plug cannot be inserted or removed unless a switch is in the off position, and the switch cannot be turned on unless the plug is fully inserted.

An explosionproof incandescent light fixture is shown in Fig. 31–10. Explosionproof fluorescent lighting fixtures are also available, as are explosionproof panelboards, disconnecting means, circuit breakers, motors, and various other equipment—even telephones.

PEW HAZLOC INC.
Metal Halide
100 Watt
Class I Div. 1&2
Groups B* C, D
Class II Div. 1&2
Groups E, F, G
Class III
120V-277V
*restricted

Fig. 31–10 An explosionproof lighting fixture.

Sealing fittings   Explosive gases or vapors can pass from one enclosure to another through conduit. The conduit itself can contain a substantial amount of such material. To minimize the quantity of explosive vapors that can accumulate in one place, sealing fittings of the general type shown in Fig. 31–11 must be installed. The sealing fitting is installed in the conduit adjacent to the enclosure, and thereby completes the enclosure. Then after the wires are pulled into place, the fitting is filled with a sealing compound that effectively prevents explosive or exploded gases from passing from one part of the electrical installation to another, or from passing from a hazardous location to a nonhazardous location where unexploded gases might reach a source of ignition. Note the qualification of the word "prevents." The actual *NEC* text uses the term "minimizes" because seals will pass some gases over time, particularly if the gas is pressurized. The principal function of these seals is to isolate sections of conduit or enclosures in the event of an internal explosion; seepage over long periods must be addressed through other approaches. A worker was killed and a catastrophic fire resulted from a disconnect being opened in an enclosure that had accumulated a hazardous quantity of gas resulting from this type of seepage.

The fitting shown in Fig. 31–11 is for a vertical run of conduit. A different type consists of a T-shaped explosionproof fitting or box with a screw-on cover having a spout, with a plug to close the end of the spout. This kind (Fig. 31–12) can be used in either vertical or horizontal runs of conduit. Seals must always be accessible after installation. Seals are also required at hazardous (classified) location boundaries, but those at the boundary between a Division 2 and an unclassified location need not be explosionproof.

The product standards for seal fittings were reorganized under the 1968 *NEC*, which at the time set a limit of 25% wire fill for new work, and 40% for old work. Since seals cannot be opened and reused, the standards presumed a 25% fill. When the *NEC* changed to allow 40% fill for even new installations, these standards were overlooked until comparatively recently. The *NEC* now requires seals to be figured on the basis of 25% fill unless listed otherwise. A number of manufacturers now have seals with larger bodies that will accommodate a 40% fill. In addition, the seal in Fig. 31–12 clearly allows for easy spreading of the conductors, solving the major problem of undersized seals. A seal with two wires

Fig. 31–11 By pouring sealing compound into a fitting of this kind, one run of conduit is sealed off from another, minimizing danger of explosion.

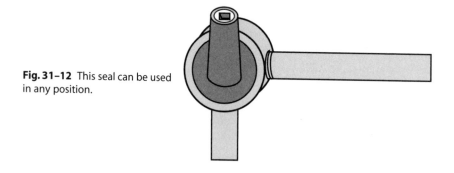

Fig. 31–12 This seal can be used in any position.

touching, and the seal compound incomplete at that point, is no seal at all. A major manufacturer has developed a new sealing compound, however, that can be squirted into its seal fitting with no damming or conductor separation, and it will harden into an acceptable seal. This should be very beneficial.

Wiring methods for hazardous locations   Threaded rigid metal conduit, intermediate metal conduit, and Type MI cable are acceptable in all locations. Rigid nonmetallic conduit is permitted for underground use below 24 in. of cover, with steel conduit ends over the final 2 ft of the run in both directions, allowing a conventional threaded steel end to mate a seal to, because in many cases the hazardous environment is assumed to extend to the point of conduit emergence. In industrial locations with restricted public access, a special type of Type MC cable with a gas/vapor-tight corrugated aluminum armor under a nonmetallic jacket can be used in hazardous locations. Where flexibility is required, there are explosionproof flexible fittings. If additional flexibility is required, extra-hard-usage cord is permitted under conditions of qualified maintenance and supervision, guarding or equivalent protection, and suitable seals where terminated.

In Division 2 locations, most metal-clad or metallic wiring methods are permitted. Where flexibility is required, the usual metallic flexible wiring methods are allowable, along with extra-hard-usage flexible cord.

Commercial garages   This subject is covered by *NEC* Article 511. A commercial garage is a location used for servicing and repairing self-propelled vehicles that use flammable liquid fuels, including cars, trucks, buses, or tractors. The 2005 *NEC* has greatly revised many key portions of this material in order to correlate it with the NFPA 30A standard covering these facilities. Under NFPA 30A, repair garages fall into one of two principal types, described as "major" and "minor." A minor facility does tune-ups, oil and other fluid changes, brake repairs, tire rotations, etc. This classification includes the popular "quick lube" type of facility. In contrast, a major facility does (in addition) engine overhauls, body and painting work, and (most critically in terms of area classification) repairs that require draining of a fuel tank. Area classification for these facilities focuses on three areas (floor, ceiling, and subfloor pits), and the requirements in each depend on whether specified ventilation is provided.

Major repair garages   The general floor areas in major repair garages are considered as Class I, Division 2 to a height of 18 in. above the floor unless the space is ventilated to provide at least four air changes per hour, or one cubic foot per minute of fresh air per square foot of floor area. The ventilation must be taken from no higher than 1 ft above the floor, and arranged to exchange air across the entire floor area. Ceiling areas are classified to the same extent (that is, unclassified if there is 1 cfm/ft^2 ventilation provided, otherwise, Division 2), except the 18 in. measurement for Division 2 extends down from the ceiling. The classification does not apply if the facility will not be working on vehicles using lighter-than-air fuels such as hydrogen or natural gas. If classification does apply, the exhaust port must be within 18 in. from the ceiling high point. For pit areas, they are unclassified if they are separately exhausted at the rate of 1 cfm/ft² (taken from no higher than 1 ft above the floor of the pit). Pits not meeting this robust ventilation threshold become Class I, Division 1 locations.

Minor repair garages   The general floor areas in minor repair garages are unclassified as long as there are no subfloor work areas or pits. If the work area includes these subfloor areas, the floor areas are classified exactly the same as in the case of major repair garages. Ceiling areas are unclassified unless lighter-than-air fuels are transferred. For pit areas, they are unclassified if they are separately exhausted in a similar manner as major repair garages. Pits not so ventilated in a minor repair garage become Class I, Division 2 locations.

Office areas, stock rooms, toilets, etc., are all classified as nonhazardous locations if they have at least four air changes an hour, or if they are effectively cut off from the repair area by partitions. Areas that are adjacent to the repair area but not cut off by partitions and do not have four air changes an hour may be classified as nonhazardous locations if they have either sufficient ventilation for ordinary needs, an air pressure differential, or spacing that presents no hazard in the judgment of the authority having jurisdiction. (A pressure differential means that the air pressure is slightly higher in the adjacent area than in the hazardous [repair] area so that no vapors will flow from the hazardous area to the adjacent area.)

Wiring methods   The wiring in nonhazardous areas of commercial garages may be any type discussed in earlier chapters. Exceptions: In an area above a hazardous location that is not cut off by a ceiling, the wiring must be in a metal raceway, rigid nonmetallic conduit, ENT, Type AC or MC or MI cable, or manufactured wiring systems, or Type TC cable, or for signaling purposes, Type PLTC. EMT is very commonly used in such areas.

Lighting fixtures in the area above vehicle lanes must be not less than 12 ft above the floor, or be constructed in a way that prevents sparks or hot particles from falling to the floor. Where fluorescent lighting is used and the fixtures are neither 12 ft above the floor nor equipped with a glass or plastic bottom, there are plastic sleeves and similar devices available to hold a broken fluorescent tube (lamp) and prevent the hot cathode ends of the lamp from falling to the floor. Other equipment above the hazardous location, such as motors or switches, must be totally enclosed if less than 12 ft above the floor. Drop lights, unless restrained against coming within 18 in. of the floor, must be suitable for Class I, Division 1 locations.

Wiring in the hazardous location must be of a type described in this chapter for Class I, Division 1 or Division 2 (whichever it is). A sealing fitting is required to be installed where the wiring enters or leaves the hazardous location so that there is no coupling or other fitting between the sealing fitting and the boundary between the hazardous and the nonhazardous locations. A sealing fitting is also required at an equipment enclosure, such as a receptacle or switch box, that is within the hazardous location, but it is common practice to keep all such wiring and equipment above the 18-in. level where practical. Equipment can almost invariably be kept above the 18-in. level, but sometimes the wiring is brought in from underground or installed in the floor slab. In all such cases, install threaded rigid metal conduit, use gasoline- and oil-resistant wire, and seal the conduit so there is no coupling or other fitting between the sealing fitting and the 18-in. level. Install a sealing fitting at the panelboard end as the last fitting above grade.

Battery chargers and their control equipment must not be installed in the hazardous location. If batteries are charged in a separate compressor room, tire-storage room, or similar area, the room must be well ventilated to allow dissipation of the hydrogen gas. (There is not enough hydrogen gas involved for a battery room to be classified as a hazardous location, but the small amount that is developed must be allowed to dissipate.)

Service stations *NEC* Article 514 covers gasoline-dispensing and service stations. The service area (such as a "lubritorium"), office, storage room, toilets, etc., are treated the same as similar areas in a commercial garage. The gasoline-dispensing areas are classified as follows:

The area within the gasoline dispenser (pump) is no longer classified by language in the *NEC*; due to differences in modern dispenser designs, this area classification is now left to the product standards as part of the listing process. However, any pit or below-grade space that extends upward into a classified location within the dispenser is Class I, Division 1. In addition, the entire area within 18 in. horizontally of the dispenser enclosure, or of any liquid-handling components within the enclosure, is Class I, Division 2. Moving outward from the dispenser, the entire area extending 20 ft horizontally in all directions from the dispenser enclosure and extending up to 18 in. above grade is a Class I, Division 2 location. This includes any indoor area not suitably cut off by partitions. The below-grade area beneath the Class I, Division 2 location is a Class I, Division 1 location.

This underground classified location extends horizontally (for *NEC* enforcement purposes) to the point of conduit emergence above grade. This allows the seal, which must always be accessible and within a conduit length of the classification boundary, to be at a convenient location. Otherwise installers would be installing handholes in areas of unbroken pavement to accommodate sealing requirements.

The area within 10 ft horizontally of a storage-tank fill pipe (for a loose-fill connection) or 5 ft horizontally of the tight-fill connections more common today because of environmental concerns, and extending to 18 in. above grade is a Class I, Division 2 location; the below-grade area beneath this is a Class I, Division 1 location. For storage tank vent pipes, the spherical volume within a 3-ft radius of the discharge point of a storage-tank vent pipe is a Class I, Division 1 location; the spherical volume between the 3-ft and 5-ft radius is a Class I, Division 2 location.

Any wiring to pole lights, signs, etc., must be kept out of the hazardous locations or be made to comply with the rules for the location, which include a sealing fitting at each end of a conduit that passes through a hazardous area. It also means that any below-grade wiring within any hazardous location must be in rigid metal conduit and be nylon-covered gasoline- and oil-resistant wire, since gasoline can and often does get into below-grade conduit in hazardous locations. (Under some conditions, as noted in the general discussion, rigid nonmetallic conduit and Type MI cable may also be installed underground.) For wiring to a gasoline dispenser, a sealing fitting must be the first fitting in the conduit where the conduit emerges from below grade at the dispenser and also where the conduit emerges from below grade at the panelboard. An explosionproof flexible fitting is frequently required between the sealing fitting and the equipment inside the dispenser, since the conduit may not quite line up with the internal fittings, and there is not much room for bending rigid metal conduit underneath the dispenser.

Any branch circuit supplying a dispenser or passing through a dispenser (such as to supply a lighting standard near the dispenser) must have a circuit breaker or switch that disconnects all conductors of the circuit, including the grounded (white) conductor. There are special circuit breakers made for this specific purpose; they have a switching pole for the grounded conductor as well as for the ungrounded conductor, but there is no overcurrent device in the grounded conductor pole of the breaker. If such a circuit breaker is not used, a switch without a fuse in the grounded conductor, but that opens all conductors (including the grounded conductor) simultaneously, must be used.

Paint-spraying booths   *NEC* Article 516 covers finishing processes, which include paint-spraying booths and areas. Some repair garages have a body shop with a paint-spraying booth. No electric equipment, and no wiring except threaded rigid metal (or intermediate metal) conduit, or Type MI cable, or metal boxes (or fittings) provided they contain no splices or terminal connections, are permitted in a paint-spraying booth unless the equipment and the splice and terminal enclosures are approved for both Class I, Division 1, Group D locations, *and* for accumulations of readily ignitable residues. Lighting fixtures are the only equipment available (as of this writing) meeting both these requirements. To wire such a booth, proceed as follows:

Lighting fixtures (in conjunction with transparent baffles, exhaust-air movement, and careful placement) can often be installed where they will not be covered with paint residue, but it is sometimes difficult to do and not always acceptable to some inspectors. So use lighting fixtures that are approved for both Group D and for paint residue. Then wire them with threaded rigid (or intermediate) metal conduit (or Type MI cable) without splice boxes or switches inside the booth. Explosionproof boxes, as shown in Fig. 31–8, may be used as long as there are no splices in the box; and switches must be installed outside the booth.

If fixtures other than the type that are approved for residue are used, install them outside the booth so that light will enter the booth interior through extra-strength glass panels in the roof or sides of the booth (or both). Wire these fixtures as required for the garage location in which they are located.

The exhaust fan (which must have nonferrous blades, that is, not iron or steel but rather aluminum or brass) must have the fan motor installed outside the exhaust stack and connected to the fan by a metal shaft or by a staticproof belt. Wire the motor as required for the location. That is, if it is inside the garage repair area, wire it accordingly; if it is outdoors, wire it for an outdoor location. (In some cases one wall of a booth may be the exterior building wall; so in some cases, the motor could be outdoors.)

Caution: Remember, all wiring in hazardous areas is subject to many rules and restrictions, many of which involve other standards outside the *NEC*. Pay close attention to all instructions supplied by the manufacturer. Make sure you follow the very highest standards of workmanship. Good workmanship is more than just a good idea; the *NEC* requires it in most electrical work. It matters more in these locations than anywhere else.

Appendix

In some cases the format has been slightly modified to accommodate the size of this book.

Table A–1 [*NEC* Table 310.16]† ALLOWABLE AMPACITIES OF INSULATED CONDUCTORS RATED 0 THROUGH 2000 VOLTS, 60°C THROUGH 90°C (140°F THROUGH 194°F) NOT MORE THAN THREE CURRENT-CARRYING CONDUCTORS IN RACEWAY, CABLE, OR EARTH (DIRECTLY BURIED), BASED ON AMBIENT TEMPERATURE OF 30°C (86°F)

Size	Temperature Rating of Conductor (See Table 310.13)						Size
	60°C (140°F)	75°C (167°F)	90°C (194°F)	60°C (140°F)	75°C (167°F)	90°C (194°F)	
AWG or kcmil	Types TW, UF	Types RHW, THHW, THW, THWN, XHHW, USE, ZW	Types TBS, SA, SIS, FEP, FEPB, MI, RHH, RHW-2,THHN, THHW, THW-2, THWN-2, USE-2, XHH, XHHW, XHHW-2, ZW-2	Types TW, UF	Types RHW, THHW, THW, THWN, XHHW, USE	Types TBS, SA, SIS, THHN, THHW, THW-2, THWN-2, RHH, RHW-2, USE-2, XHH, XHHW, XHHW-2, ZW-2	AWG or kcmil
	COPPER			ALUMINUM OR COPPER-CLAD ALUMINUM			
18	—	—	14	—	—	—	—
16	—	—	18	—	—	—	—
14*	20	20	25	—	—	—	—
12*	25	25	30	20	20	25	12*
10*	30	35	40	25	30	35	10*
8	40	50	55	30	40	45	8
6	55	65	75	40	50	60	6
4	70	85	95	55	65	75	4
3	85	100	110	65	75	85	3
2	95	115	130	75	90	100	2
1	110	130	150	85	100	115	1
1/0	125	150	170	100	120	135	1/0
2/0	145	175	195	115	135	150	2/0
3/0	165	200	225	130	155	175	3/0
4/0	195	230	260	150	180	205	4/0
250	215	255	290	170	205	230	250
300	240	285	320	190	230	255	300
350	260	310	350	210	250	280	350
400	280	335	380	225	270	305	400
500	320	380	430	260	310	350	500
600	355	420	475	285	340	385	600
700	385	460	520	310	375	420	700
750	400	475	535	320	385	435	750
800	410	490	555	330	395	450	800
900	435	520	585	355	425	480	900
1000	455	545	615	375	445	500	1000
1250	495	590	665	405	485	545	1250
1500	520	625	705	435	520	585	1500
1750	545	650	735	455	545	615	1750
2000	560	665	750	470	560	630	2000
	CORRECTION FACTORS						
Ambient Temp. (°C)	For ambient temperatures other than 30°C (86°F), multiply the allowable ampacities shown above by the appropriate factor shown below.						Ambient Temp. (°F)
21–25	1.08	1.05	1.04	1.08	1.05	1.04	70–77
26–30	1.00	1.00	1.00	1.00	1.00	1.00	78–86
31–35	0.91	0.94	0.96	0.91	0.94	0.96	87–95
36–40	0.82	0.88	0.91	0.82	0.88	0.91	96–104
41–45	0.71	0.82	0.87	0.71	0.82	0.87	105–113
46–50	0.58	0.75	0.82	0.58	0.75	0.82	114–122
51–55	0.41	0.67	0.76	0.41	0.67	0.76	123–131
56–60	—	0.58	0.71	—	0.58	0.71	132–140
61–70	—	0.33	0.58	—	0.33	0.58	141–158
71–80	—	—	0.41	—	—	0.41	159–176

*See 240.4(D).
†Reprinted with permission from NFPA 70-2005 *The National Electrical Code,* © NFPA 2004.

Table A–2 [*NEC* Table 310.15(B)(2)(a)]† ADUSTMENT FACTORS FOR MORE THAN THREE CURRENT-CARRYING CONDUCTORS IN A RACEWAY OR CABLE	
Number of Current-Carrying Conductors	**Percent of Values in Tables 310.16 through 310.19 as Adjusted for Ambient Temperature if Necessary**
4–6	80
7–9	70
10–20	50
21–30	45
31–40	40
41 and above	35

†Reprinted with permission from NFPA 70-2005 *The National Electrical Code* © 2004, National Fire Protection Association, Quincy, MA 02269. This reprinted material is not the complete and official position of the National Fire Protection Association on the referenced subject which is represented only by the standard in its entirety.

Table A–3 [*NEC* Table 430.248]† FULL-LOAD CURRENTS IN AMPERES, SINGLE-PHASE ALTERNATING-CURRENT MOTORS

The following values of full-load currents are for motors running at usual speeds and motors with normal torque characteristics. Motors built for especially low speeds or high torques may have higher full-load currents, and multispeed motors will have full-load current varying with speed, in which case the namplate current ratings shall be used. The voltages listed are rated motor voltages.

The currents listed shall be permitted for system voltage ranges of 110 to 120 and 220 to 240 volts.

Horsepower	115 Volts	200 Volts	208 Volts	230 Volts
1/6	4.4	2.5	2.4	2.2
1/4	5.8	3.3	3.2	2.9
1/3	7.2	4.1	4.0	3.6
1/2	9.8	5.6	5.4	4.9
3/4	13.8	7.9	7.6	6.9
1	16	9.2	8.8	8.0
1 1/2	20	11.5	11.0	10
2	24	13.8	13.2	12
3	34	19.6	18.7	17
5	56	32.2	30.8	28
7 1/2	80	46.0	44.0	40
10	100	57.5	55.0	50

†Reprinted with permission from NFPA 70-2005 *The National Electrical Code* © 2004, National Fire Protection Association, Quincy, MA 02269. This reprinted material is not the complete and official position of the National Fire Protection Association on the referenced subject which is represented only by the standard in its entirety.

Table A–4 [*NEC* Table 430.250]† FULL-LOAD CURRENT THREE-PHASE ALTERNATING-CURRENT MOTORS

The following values of full-load currents are typical for motors running at speeds usual for belted motors and motors with normal torque characteristics.

Motors built for low speeds (1200 rpm or less) or high torques may require more running current, and multispeed motors will have full-load current varying with speed. In these cases, the nameplate current rating shall be used.

The voltages listed are rated motor voltages. The currents listed shall be permitted for system voltage ranges of 110 to 120, 220 to 240, 440 to 480, and 550 to 600 volts.

Horsepower	Induction Type Squirrel Cage and Wound Rotor (Amperes)							Synchronous-Type Unity Power Factor* (Amperes)			
	115 Volts	200 Volts	208 Volts	230 Volts	460 Volts	575 Volts	2300 Volts	230 Volts	460 Volts	575 Volts	2300 Volts
½	4.4	2.5	2.4	2.2	1.1	0.9	—	—	—	—	—
¾	6.4	3.7	3.5	3.2	1.6	1.3	—	—	—	—	—
1	8.4	4.8	4.6	4.2	2.1	1.7	—	—	—	—	—
1½	12.0	6.9	6.6	6.0	3.0	2.4	—	—	—	—	—
2	13.6	7.8	7.5	6.8	3.4	2.7	—	—	—	—	—
3	—	11.0	10.6	9.6	4.8	3.9	—	—	—	—	—
5	—	17.5	16.7	15.2	7.6	6.1	—	—	—	—	—
7½	—	25.3	24.2	22	11	9	—	—	—	—	—
10	—	32.2	30.8	28	14	11	—	—	—	—	—
15	—	48.3	46.2	42	21	17	—	—	—	—	—
20	—	62.1	59.4	54	27	22	—	—	—	—	—
25	—	78.2	74.8	68	34	27	—	53	26	21	—
30	—	92	88	80	40	32	—	63	32	26	—
40	—	120	114	104	52	41	—	83	41	33	—
50	—	150	143	130	65	52	—	104	52	42	—
60	—	177	169	154	77	62	16	123	61	49	12
75	—	221	211	192	96	77	20	155	78	62	15
100	—	285	273	248	124	99	26	202	101	81	20
125	—	359	343	312	156	125	31	253	126	101	25
150	—	414	396	360	180	144	37	302	151	121	30
200	—	552	528	480	240	192	49	400	201	161	40
250	—	—	—	—	302	242	60	—	—	—	—
300	—	—	—	—	361	289	72	—	—	—	—
350	—	—	—	—	414	336	83	—	—	—	—
400	—	—	—	—	477	382	95	—	—	—	—
450	—	—	—	—	515	412	103	—	—	—	—
500	—	—	—	—	590	472	118	—	—	—	—

*For 90 and 80 percent power factor, the figures shall be multiplied by 1.1 and 1.25, respectively.
†Reprinted with permission from NFPA 70-2005 *The National Electrical Code*, © NFPA 2004.

Table A–5 [NEC Chapter 9, Table 8]† CONDUCTOR PROPERTIES

Size (AWG or kcmil)	Area mm²	Area (Circular Mils)	Stranding Quantity	Stranding Diameter mm	Stranding Diameter (in.)	Overall Diameter mm	Overall Diameter (in.)	Overall Area mm²	Overall Area (in.²)	Copper Uncoated (ohm/km)	Copper Uncoated (ohm/kFT)	Copper Coated (ohm/km)	Copper Coated (ohm/kFT)	Aluminum (ohm/km)	Aluminum (ohm/kFT)
18	0.823	1620	1	—	—	1.02	0.040	.0823	0.001	25.5	7.77	26.5	8.08	42.0	12.8
18	0.823	1620	7	0.39	0.015	1.16	0.046	1.06	0.002	26.1	7.95	27.7	8.45	42.8	13.1
16	1.31	2580	1	—	—	1.29	0.051	1.31	0.002	16.0	4.89	16.7	5.08	26.4	8.05
16	1.31	2580	7	0.49	0.019	1.46	0.058	1.68	0.003	16.4	4.99	17.3	5.29	26.9	8.21
14	2.08	4110	1	—	—	1.63	0.064	2.08	0.003	10.1	3.07	10.4	3.19	16.6	5.06
14	2.08	4110	7	0.62	0.024	1.85	0.073	2.68	0.004	10.3	3.14	10.7	3.26	16.9	5.17
12	3.31	6530	1	—	—	2.05	0.081	3.31	0.005	6.34	1.93	6.57	2.01	10.45	3.18
12	3.31	6530	7	0.78	0.030	2.32	0.092	4.25	0.006	6.50	1.98	6.73	2.05	10.69	3.25
10	5.261	10380	1	—	—	2.588	0.102	5.26	0.008	3.984	1.21	4.148	1.26	6.561	2.00
10	5.261	10380	7	0.98	0.038	2.95	0.116	6.76	0.011	4.070	1.24	4.226	1.29	6.679	2.04
8	8.367	16510	1	—	—	3.264	0.128	8.37	0.013	2.506	0.764	2.579	0.786	4.125	1.26
8	8.367	16510	7	1.23	0.049	3.71	0.146	10.76	0.017	2.551	0.778	2.653	0.809	4.204	1.28
6	13.30	26240	7	1.56	0.061	4.67	0.184	17.09	0.027	1.608	0.491	1.671	0.510	2.652	0.808
4	21.15	41740	7	1.96	0.077	5.89	0.232	27.19	0.042	1.010	0.308	1.053	0.321	1.666	0.508
3	26.67	52620	7	2.20	0.087	6.60	0.260	34.28	0.053	0.802	0.245	0.833	0.254	1.320	0.403
2	33.62	66360	7	2.47	0.097	7.42	0.292	43.23	0.067	0.634	0.194	0.661	0.201	1.045	0.319
1	42.41	83690	19	1.69	0.066	8.43	0.332	55.80	0.087	0.505	0.154	0.524	0.160	0.829	0.253
1/0	53.49	105600	19	1.89	0.074	9.45	0.372	70.41	0.109	0.399	0.122	0.415	0.127	0.660	0.201
2/0	67.43	133100	19	2.13	0.084	10.62	0.418	88.74	0.137	0.3170	0.0967	0.329	0.101	0.523	0.159
3/0	85.01	167800	19	2.39	0.094	11.94	0.470	111.9	0.173	0.2512	0.0766	0.2610	0.0797	0.413	0.126
4/0	107.2	211600	19	2.68	0.106	13.41	0.528	141.1	0.219	0.1996	0.0608	0.2050	0.0626	0.328	0.100

†Reprinted with permission from NFPA 70-2005 The National Electrical Code © 2004, National Fire Protection Association, Quincy, MA 02269. This reprinted material is not the complete and official position of the National Fire Protection Association on the referenced subject which is represented only by the standard in its entirety.

Table A–5 [NEC Chapter 9, Table 8]† CONDUCTOR PROPERTIES (Continued)

Size (AWG or kcmil)	Area		Conductors Stranding			Overall				Direct-Current Resistance at 75°C (167°F)					
						Diameter		Area		Copper				Aluminum	
				Diameter						Uncoated		Coated			
	mm²	(Circular Mils)	Quantity	mm	(in.)	mm	(in.)	mm²	(in.²)	(ohm/km)	(ohm/kFT)	(ohm/km)	(ohm/kFT)	(ohm/km)	(ohm/kFT)
250	—	—	37	2.09	0.082	14.61	0.575	168	0.260	0.1687	0.0515	0.1753	0.0535	0.2778	0.0847
300	—	—	37	2.29	0.090	16.00	0.630	201	0.312	0.1409	0.0429	0.1463	0.0446	0.2318	0.0707
350	—	—	37	2.47	0.097	17.30	0.681	235	0.364	0.1205	0.0367	0.1252	0.0382	0.1984	0.0605
400	—	—	37	2.64	0.104	18.49	0.728	268	0.416	0.1053	0.0321	0.1084	0.0331	0.1737	0.0529
500	—	—	37	2.95	0.116	20.65	0.813	336	0.519	0.0845	0.0258	0.0869	0.0265	0.1391	0.0424
600	—	—	61	2.52	0.099	22.68	0.893	404	0.626	0.0704	0.0214	0.0732	0.0223	0.1159	0.0353
700	—	—	61	2.72	0.107	24.49	0.964	471	0.730	0.0603	0.0184	0.0622	0.0189	0.0994	0.0303
750	—	—	61	2.82	0.111	25.35	0.998	505	0.782	0.0563	0.0171	0.0579	0.0176	0.0927	0.0282
800	—	—	61	2.91	0.114	26.16	1.030	538	0.834	0.0528	0.0161	0.0544	0.0166	0.0868	0.0265
900	—	—	61	3.09	0.122	27.79	1.094	606	0.940	0.0470	0.0143	0.0481	0.0147	0.0770	0.0235
1000	—	—	61	3.25	0.128	29.26	1.152	673	1.042	0.0423	0.0129	0.0434	0.0132	0.0695	0.0212
1250	—	—	91	2.98	0.117	32.74	1.289	842	1.305	0.0338	0.0103	0.0347	0.0106	0.0554	0.0169
1500	—	—	91	3.26	0.128	35.86	1.412	1011	1.566	0.02814	0.00858	0.02814	0.00883	0.0464	0.0141
1750	—	—	127	2.98	0.117	38.76	1.526	1180	1.829	0.02410	0.00735	0.02410	0.00756	0.0397	0.0121
2000	—	—	127	3.19	0.126	41.45	1.632	1349	2.092	0.02109	0.00643	0.02109	0.00662	0.0348	0.0106

Notes:
1. These resistance values are valid only for the parameters as given. Using conductors having coated strands, different stranding type, and, especially, other temperatures changes the resistance.
2. Formula for temperature change: $R_2 = R_1[1 + \alpha(T_2 - 75)]$ where: $\alpha_{cu} = 0.00323$, $\alpha_{Al} = 0.00330$ at 75°C
3. Conductors with compact and compressed stranding have about 9 percent and 3 percent, respectively, smaller bare conductor diameters than those shown. See Table 5A for actual compact cable dimensions.
4. The IACS conductivities used: bare copper = 100%, aluminum = 61%.
5. Class B stranding is listed as well as solid for some sizes. Its overall diameter and area is that of its circumscribing circle.

    FPN: The construction information is per NEMA WC8-1992
    or ANSI/UL 1581-1998. The resistance is calculated per
    National Bureau of Standards Handbook 100, dated 1966,
    and Handbook 109, dated 1972.

Table A–6 ABBREVIATIONS

A, amp.	amperes
AWG	American Wire Gauge
B&S	Brown & Sharpe (wire gauge)
c, cyc	cycles (see Hertz)
c.m., cmil	circular mils
C	centigrade, Celsius (temperature scale)
cb, CB	circuit breaker
cps	cycles per second (see Hertz)
EMT	electrical metallic tubing (thin-wall conduit)
F	Fahrenheit (temperature scale)
fc	footcandles
ft	foot, feet
GFI, GFCI	ground-fault circuit-interrupter
h	hours
HID	high-intensity-discharge (lamps)
hp	horsepower
Hz	Hertz (= cycles per second)
in	inches
k	kilo- (a thousand , watts, etc.
kcmil	thousand circular mils
kHz	thousands of Hertz
kVA	kilovolt-amperes
kW	kilowatts
kWh	kilowatthours
lm	lumens
ma	milliamperes
MCM, mcm	thousand circular mils
mega-	million (watts, volts, etc.)
micro-	one-millionth (volt, ampere, etc.)
mil	mils (thousandths of an inch)
milli-	one-thousandth (volt, ampere, etc.)
NEC	*National Electrical Code*
p	pole (1-p = single-pole, etc.)
ph, ϕ	phase
RI	repulsion-induction (motor)
r/min	revolutions per minute
UL	Underwriters Laboratories
V	volts
VA	volt-amperes
W	watts
wp, wpf	waterproof
μ (Greek mu)	micro- (one millionth)
ϕ (Greek phi)	phase
Ω (Greek omega)	ohms
1-p	single-pole
2-p	two-pole, double-pole
3-p	three-pole

About the authors

The late Herbert P. Richter, long regarded as "the electrician's electrician," first wrote *Practical Electrical Wiring* in 1939.

Frederic P. Hartwell is nationally recognized for his editorial and technical skills on *NEC* topics. As the *NEC* specialist for EC Online, the online electrical trade magazine division of VerticalNet, he has had primary editorial responsibility for *National Electrical Code* coverage. He was formerly the senior editor and *National Electrical Code* expert with *Electrical Construction and Maintenance Magazine*. He has been responsible for nearly one thousand successful proposals and public comments regarding changes in the *NEC* over the past seven Code-making cycles, including complete rewrites of Article 680 (covering swimming pools, fountains, spas and hot tubs) in the 2002 *NEC* and Article 695 (covering fire pumps) in the 1999 *NEC*. He is responsible for the new Example 3(a) in the 2005 *NEC*, which is the first example to cover conductor selection, and the first to use industrial topics and voltages. He continues as a principal member of the *National Electrical Code* Committee, serving on Code Making Panel 9. He is a licensed master electrician and former head electrician at a college in Massachusetts. He is the secretary of the Massachusetts Electrical Code Advisory Committee.

Hartwell is the author of *Illustrated Changes in the 1999 National Electrical Code®* (Engineer's and Electrician's editions), published by EC&M Books, PRIMEDIA Intertec Group, and edited *Understanding NE Code Rules on Grounding and Bonding* and *Understanding NE Code Rules on Emergency and Standby Power Systems*.

Further reading

General
American Electrician's Handbook, 14th edition, Croft, Terrell, and Wilford I. Summers (eds), 2002, McGraw-Hill, New York.

NEC 2005, National Electrical Code®, National Fire Protection Association, 2004, Quincy, Massachusetts.

NEC 2005 National Electrical Code® Handbook, National Fire Protection Association, 2005, Quincy, Massachusetts.

Specialized topics
Agricultural Wiring Handbook, 13th edition, 2003, National Food and Energy Council, Columbia, Missouri.

EC&M's Benfield Conduit Bending Manual, 2nd ed., Benfield, Jack, 1996, Intertec Publishing Corp., Overland Park, Kansas.

EC&M's Fundamentals of Telecom/Datacom, Rosenberg, Paul, 2000, Intertec Publishing Corp., Overland Park, Kansas.

EC&M's Practical Guide to Emergency, Standby, and Other Auxiliary Power Systems, 2nd edition, 2000, Intertec Publishing Corp., Overland Park, Kansas.

Electrical Installations in Hazardous Locations, 2nd edition, Schram, Peter J. and Earley, Mark W., 1998, National Fire Protection Association, Quincy, Massachusetts.

Electrical Motor Controls, Rockis, Gary and Mazur, Glen A., 2nd edition, 2001, American Technical Publishers, Inc., Homewood, Illinois.

Farm Buildings Wiring Handbook, 2nd edition, 1992, Midwest Plan Service, Iowa State University, Ames, Iowa. MWPS-28. [Based on an obsolete edition of the *NEC*, but contains useful design information.]

IAEI Soares Book on Grounding, 8th edition, 2002, International Association of Electrical Inspectors, Richardson, Texas.

Stallcup's Electrical Design Book, Stallcup, James G., 2005, National Fire Protection Association, Inc., Quincy, Massachusetts.

Stallcup's Generator, Transformer, Motor and Compressor Book, Stallcup, James G., 2005, National Fire Protection Association, Inc. Quincy, Massachusetts.

Index

A

OVERDUE FINE .10 PER DAY

Bellmore Memorial Library

2288 Bedford Avenue

Bellmore, New York

BE

2/05